Nanotechnology in the Life Sciences

Series Editor

Ram Prasad
School of Environmental Science and Engineering,
Sun Yat-Sen University, Guangzhou, China

Amity Institute of Microbial Technology,
Amity University Uttar Pradesh, Noida, UP, India

Nano and biotechnology are two of the 21st century's most promising technologies. Nanotechnology is demarcated as the design, development, and application of materials and devices whose least functional make up is on a nanometer scale (1 to 100 nm). Meanwhile, biotechnology deals with metabolic and other physiological developments of biological subjects including microorganisms. These microbial processes have opened up new opportunities to explore novel applications, for example, the biosynthesis of metal nanomaterials, with the implication that these two technologies (i.e., thus nanobiotechnology) can play a vital role in developing and executing many valuable tools in the study of life. Nanotechnology is very diverse, ranging from extensions of conventional device physics to completely new approaches based upon molecular self-assembly, from developing new materials with dimensions on the nanoscale, to investigating whether we can directly control matters on/in the atomic scale level. This idea entails its application to diverse fields of science such as plant biology, organic chemistry, agriculture, the food industry, and more.

Nanobiotechnology offers a wide range of uses in medicine, agriculture, and the environment. Many diseases that do not have cures today may be cured by nanotechnology in the future. Use of nanotechnology in medical therapeutics needs adequate evaluation of its risk and safety factors. Scientists who are against the use of nanotechnology also agree that advancement in nanotechnology should continue because this field promises great benefits, but testing should be carried out to ensure its safety in people. It is possible that nanomedicine in the future will play a crucial role in the treatment of human and plant diseases, and also in the enhancement of normal human physiology and plant systems, respectively. If everything proceeds as expected, nanobiotechnology will, one day, become an inevitable part of our everyday life and will help save many lives.

More information about this series at http://www.springer.com/series/15921

Kamel A. Abd-Elsalam • Ram Prasad
Editors

Nanobiotechnology Applications in Plant Protection

Springer

Editors
Kamel A. Abd-Elsalam
Plant Pathology Research Institute
Agricultural Research Center (ARC)
Giza, Egypt

Ram Prasad
School of Environmental Science
and Engineering
Sun Yat-Sen University
Guangzhou, China

Amity Institute of Microbial Technology
Amity University
Noida, UP, India

ISSN 2523-8027 ISSN 2523-8035 (electronic)
Nanotechnology in the Life Sciences
ISBN 978-3-030-08189-8 ISBN 978-3-319-91161-8 (eBook)
https://doi.org/10.1007/978-3-319-91161-8

Printed on acid-free paper

This Springer imprint is published by the registered company Springer Nature Switzerland AG.
The registered company address is: Gewerbestrasse 11, 6330 Cham, Switzerland

Preface

Plant diseases are caused by bacteria, fungi, insects, nematodes, phytoplasmas, and viruses; the diseases provoked by these pests cause financial losses by reducing attainable yields, product quality, and/or shelf life; only in the United States, over $600 million is expended annually on fungicides in challenge to control plant pathogens. Traditional plant protection strategies often prove insufficient, and application of chemical-based pesticides has negative effects on animals and human beings apart from causing decline in soil fertility. Recent industrial advancements have led to the fabrication of nanomaterials of diverse sizes and shapes. These innovations are the base for further engineering to create unique properties targeted toward specific applications. Nanotechnology would deliver green and efficient alternatives for the management of plant diseases without harming the nature, while the most favorable strategies, in recent scenario, are the use of micro- and nanotechnology to promote a more efficient assembly and then release of specific and environmental sustainable active principles. The wide range of nanotechnology applications in agriculture also includes nanopesticides for the control of plant pathogen interactions and provides new techniques for crop disease control. However, use in agriculture, especially for plant protection and production, is an under-explored area in the research community. Nanotechnology has many applications in all stages of production, processing, storing, packaging, and transport of agricultural products. Nanotechnology will revolutionize agriculture and food industry by innovation of new techniques such as precision farming techniques, enhancing the ability of plants to absorb nutrients, improving seed germination and growth, more efficient and targeted use of inputs, plant protection, pathogen detection, control diseases, pesticide/herbicide residue detection, and withstand environmental pressures and effective systems for processing, storage, and packaging.

This book deals with the application of nanotechnology for quicker, more cost-effective, and precise diagnostic procedures of plant diseases. Additionally, the combination of nanotechnology with microfluidic systems has been effectively applied in molecular plant pathology and can be adapted to detect specific pathogens and toxins. Moreover, the application of nanotechnology in plant disease control, antimicrobial mechanisms, and nanotoxicity on plant ecosystem have been discussed in detail.

The first chapter by Sabry and Ragaei reviews nanotechnology and its applications in insect's pest control. Chapter 2 highlights the nanoparticles-based plant disease management tools for agricultural sustainability presented by Yadav and Yadav. In Chap. 3, Gabal et al. describe copper nanostructures and their applications in plant protection. Nanoantimicrobials for phytopathogens control by mechanistic approaches and potential applications are described by Mohamed and Abd–Elsalam in Chap. 4. In Chap. 5, Kaushal highlights on the role of microbes in plant protection using intersection of nanobiotechnology. Chapter 6 highlights on the role of nanoemulsions as antimicrobial agents in plant protection by Hashim et al. In Chap. 7, Mohamed et al. describe the application of nano-carbon in plant growth promotion and protection.

In Chap. 8, Jampílek and Kráľová highlight on benefits and risks factor of nanotechnology applications in crop protection. In Chap. 9, Gupta et al. highlight on applications of silver nanoparticles in plant protection. Tahsin Shoala details on positive impacts of nanoparticles in plant resistance against different stimuli in Chap. 10. In Chap. 11, Mostafa et al. give an overview of nanoantimicrobials mechanism of action. Sustainable nanotechnology approaches for mycotoxin detection and protection are discussed in Chap. 12 by Thipe et al. Finally, application of chitosan-based nanostructures in plant protection has been discussed by Al-Dhabaan et al. in Chap. 13.

We wish to thank Springer officials, particularly William F. Curtis, Eric Schmitt, Eric Stannard, and Sanjana MeenakshiSundaram, and Anup Kumar, for their generous support and efforts in accomplishing this volume. We are highly delighted and thankful to all our contributing authors for their vigorous support and outstanding cooperation to write altruistically these authoritative and valuable chapters. We specially thank our families for consistent support and encouragement.

With a bouquet of information on different aspects of plant protections from nanomaterials, editors hope this book is a valuable resource for the students of different divisions; researchers and academicians, working in the field of nanoscience, nanotechnology, plant sciences, agriculture microbiology, and fungal biology; and the scholars interested in strengthening their knowledge in the area of nanobiotechnology.

Giza, Egypt
Guangzhou, China
Noida, UP, India

Kamel A. Abd-Elsalam
Ram Prasad

Contents

Contributors

Kamel A. Abd-Elsalam Plant Pathology Research Institute, Agricultural Research Center (ARC), Giza, Egypt

Fahad A. Al-Dhabaan Department of Biology, Science and Humanities College, Shaqra University, Alquwayiyah, Saudi Arabia

Mousa A. Alghuthaymi Department of Biology, Science and Humanities College, Shaqra University, Alquwayiyah, Saudi Arabia

Hassan Almoammar National Centre for Biotechnology, King Abdulaziz City for Science and Technology (KACST), Riyadh, Saudi Arabia

Amal-Asran Plant Pathology Research Institute, Agricultural Research Center (ARC), Giza, Egypt

Esraa Gabal Agricultural Sciences and Resource Management in the Tropics and Subtropics (ARTS), Faculty of Agriculture, University of Bonn, Bonn, Germany

Nomita Gupta Amity Institute of Microbial Technology, Amity University, Noida, India

Ayat F. Hashim Food industries and Nutrition Division, National Research Center, Giza, Egypt

Josef Jampílek Department of Pharmaceutical Chemistry, Faculty of Pharmacy, Comenius University, Bratislava, Slovakia

Kattesh V. Katti Department of Physics and Astronomy, University of Missouri, Columbia, MO, USA

Institute of Green Nanotechnology, University of Missouri, Columbia, MO, USA

Manoj Kaushal International Institute of Tropical Agriculture, Ibadan, Oyo State, Nigeria

Marshall Keyster Department of Biotechnology, University of the Western Cape, Bellville, South Africa

Katarína Kráľová Institute of Chemistry, Faculty of Natural Sciences, Comenius University, Bratislava, Slovakia

Mohamed A. Mohamed Plant Pathology Research Institute, Agricultural Research Center (ARC), Giza, Egypt

Manal Mostafa CIHEAM IAMB - Mediterranean Agronomic Institute of Bari, Valenzano, BA, Italy

Plant Pathology Research Institute, Agricultural Research Center (ARC), Giza, Egypt

Microbiology Department, Faculty of Agriculture, Cairo University, Giza, Egypt

Ram Prasad School of Environmental Science and Engineering, Sun Yat-Sen University, Guangzhou, China

Amity Institute of Microbial Technology, Amity University, Noida, India

Mohamed Ragaei Pests and Plant Protection Department, National Research Centre, Cairo, Egypt

Mohamed M. Ramadan Plant Pathology Research Institute, Agricultural Research Center (ARC), Giza, Egypt

Al-kazafy Hassan Sabry Pests and Plant Protection Department, National Research Centre, Cairo, Egypt

Tahsin Shoala College of Biotechnology, Misr University for Science and Technology, Giza, Egypt

Amar Singh Lal Bahadur Shastri Memorial College, Jamshedpur (Kolhan University, Chaibasa), Jamshedpur, India

Velaphi C. Thipe Department of Chemistry, University of Missouri, Columbia, MO, USA

Institute of Green Nanotechnology, University of Missouri, Columbia, MO, USA

Chandrama Prakash Upadhyaya Department of Biotechnology, DR Harisingh Gour Central University, Sagar, Madhya Pradesh, India

Anurag Yadav Department of Microbiology, College of Basic Science & Humanities, S.D. Agricultural University, Banaskantha, India

Kusum Yadav Department of Biochemistry, University of Lucknow, Lucknow, India

Khamis Youssef Plant Pathology Research Institute, Agricultural Research Center (ARC), Giza, Egypt

About the Editors

Kamel A. Abd-Elsalam is a head researcher at Plant Pathology Research Institute, Agricultural Research Center, Giza, Egypt. Dr. Kamel's research interests include developing, improving, and deploying plant biosecurity diagnostic tools, understanding and exploiting fungal pathogen genomes, and developing eco-friendly hybrid nanomaterials for controlling toxicogenic fungi and plant diseases. He has published 8 book chapters, 6 review articles, 1 translated book, and more than 120 research articles in international peer-reviewed journals including *Fungal Diversity*, *Fungal Biology*, *FEMS Review Microbiology*, *PLOS One*, and *PLOS Genetics*. He is associate editor for *Mycosphere* and review editor for *Frontiers in Genomic Assay Technology* and referees for journals, including the *Plant Pathology*, *Journal of Phytopathology*, *Crop Protection*, *IET Nanotechnology*, *Fungal Diversity*, *BMC Genomics*, and *Foodborne Pathogens and Diseases.* Dr. Kamel has also served as molecular mycologist for 5 years in the Department of Botany and Microbiology Department, College of Science, King Saud University, Saudi Arabia. He received the Federation of Arab Scientific Research Councils' prize for distinguished scientific research in biotechnology (fungal genomics) during 2014 (first ranking). Dr. Kamel has pursued his Ph.D. in Molecular Plant Pathology from Christian Albrechts University of Kiel (Germany) and Suez Canal University (Egypt) and then been awarded postdoctoral fellowship from Christian Albrechts University of Kiel in 2008. Dr. Kamel served as visiting associate professor in Institute of Excellence in Fungal Research, Mae Fah Luang University, Thailand; Institute of Microbiology, Technical University of Munich (TUM), Germany; Laboratory of Phytopathology, Wageningen University, the Netherlands; and Plant Protection Department, University of Sassari, Italy.

Ram Prasad, Ph.D is associate with Amity Institute of Microbial Technology, Amity University, Uttar Pradesh, India, since 2005. His research interest includes plant-microbe interactions, sustainable agriculture, environmental microbiology, and microbial nanobiotechnology. He has more than hundred publications to his credit, including research papers, review articles, book chapters; five patents issued or pending; and edited or authored several books. Dr. Prasad has 12 years of teaching experience and has been awarded the Young Scientist Award (2007) and Prof. J.S. Datta Munshi Gold Medal (2009) by the International Society for Ecological Communications; FSAB fellowship (2010) by the Society for Applied Biotechnology; the American Cancer Society UICC International Fellowship for Beginning Investigators, USA (2014); Outstanding Scientist Award (2015) in the field of Microbiology by Venus International Foundation; BRICPL Science Investigator Award (ICAABT-2017); and Research Excellence Award (2018). Previously, Dr. Prasad served as Visiting Assistant Professor in the Department of Mechanical Engineering, Whiting School of Engineering, Johns Hopkins University, USA, and presently, working as Research Associate Professor at School of Environmental Science and Engineering, Sun Yat-Sen University, Guangzhou, China.

Chapter 1
Nanotechnology and Their Applications in Insect's Pest Control

Al-kazafy Hassan Sabry and Mohamed Ragaei

1.1 History of Nanotechnology in Insect's Control

It was estimated that about 2.5 million tons of conventional pesticides are used in agriculture against insects each year. This quantity is expected to increase and cause worldwide hazards due to high toxicity, long persistence of pesticides, lack of scientific formulations, leaching, and loss during application going into soil, water bodies, and atmosphere causing pollution or remnants on the crop surfaces affecting ecology and public health. On the other hand, the use of integrated pest management (IPM) is not enough to reduce the insect population under the threshold economy. So, it is badly needed to develop new and modern strategies for the management of insect pest. This new strategy is a nanotechnology in agriculture. By this approach it can use nanomaterials against insect pest infestation.

The potential uses and benefits of nanotechnology are enormous. These include management of insect pests through the new insecticide formulations based on nanomaterials (Ragaei and Sabry 2014). It was known that one nanometer is one billionth of a meter or one millionth of a millimeter or one thousandth of a micrometer. That is about 1/80,000 of the diameter of a human hair or ten times the diameter of a hydrogen atom.

The American physicist Richard Feynman lectured, "There's Plenty of Room at the Bottom," at an American Physical Society meeting at Caltech on December 29, 1959, which is often held to provide attention for the field of nanotechnology. Feynman had discovered a process by which the ability to manipulate individual atoms and molecules might be developed, using one set of precise tools to build and operate another proportionally smaller set, down to the needed scale, and so on. In the course of this, Feynman stated that scaling issues would arise from the changing magnitude of various physical phenomena: gravity would become less important,

---

A.-k. H. Sabry (✉) · M. Ragaei
Pests and Plant Protection Department, National Research Centre, Cairo, Egypt

K. A. Abd-Elsalam, R. Prasad (eds.), *Nanobiotechnology Applications in Plant Protection*, Nanotechnology in the Life Sciences,
https://doi.org/10.1007/978-3-319-91161-8_1

and surface tension and van der Waals attraction would become more important (Gribbin and Gribbin 1997). So, it can be said that Richard Feynman is the father of nanotechnology.

Nanoinsecticides are plant protection products used against insect pests. Nanotechnology is employed to enhance the efficacy or reduce the environmental contamination. Nanotechnology started in the sixth revolutionary technology in the current era after the Industrial Revolution of the mid-1700s (Rosen and Abribat 2005), Nuclear Energy Revolution of the 1940s, the Green Revolution of the 1960s, Information Technology Revolution of the 1980s, and Biotechnology Revolution of the 1990s.

Nanotechnology has developed in the past decade and was able to make many new nanomaterials used in all fields of life. These fields include nanoinsecticides. Some of those nanoinsecticides are used against harmful insects. After the Second World War, the organic insecticides such as chlorinated hydrocarbon, organophosphorus, carbamates, etc. were used. Due to the extensive use of all conventional insecticides against insect pests, the insects acquired resistance to all of these insecticides. The scientists researched for alternative methods in insect control such as new and unconventional insecticides to face the insect outbreak. Unfortunately, these new methods cannot suppress the insect outbreak. Scientists have been extremely working over the past decade to develop new insecticide products based on nanotechnology. Nanopesticide research was used at high speed at the agrochemical labs; however, these topics have not reached public awareness or state authorities so far nor are any products available at the market. Since those nanoinsecticides have new or enhanced properties, this will change in the near future and will inevitably result in both new risks and new benefits to human and environmental health (Kah et al. 2013; Prasad et al. 2014, 2017a). At this scale, particles have a disproportionately large surface area relative to their overall size. The high surface area makes the difference, because the greater surface area ratio means more of the total volume of pesticide comes into contact with the pests and that in turn means being able to reduce the amount of pesticide needed.

Field application of nanoinsecticides would be spread in large quantities of nanoparticles into the environment. Innovation always results in both drawbacks and benefits for human and environmental health. Nanoinsecticides may reduce the environmental pollution through the reduction of insecticide application rates and reduced losses. On the other hand, nanoinsecticides may also make new types of soil contamination and waterways due to enhanced transport, longer persistence, and higher toxicity. Many studies carried out on the nanoparticles and reported that these nanomaterials were effective against plant pathogens, insects, and other pests. So, these nanomaterials are not only used in the preparation of new formulations like pesticides and insecticides but also as insect repellants (Barik et al. 2008; Gajbhiye et al. 2009; Goswami et al. 2010; Owolade et al. 2008; Bhattacharyya et al. 2016; Prasad et al. 2017b; Sangeetha et al. 2017). Bhattacharyya et al. (2010) stated that nanotechnology will revolutionize agriculture including pest management in the near future.

1.2 Formulations of Nanoinsecticides

Nanoformulations are miraculous surfactants that can simply multiply the effects of formulation sprayed on the plants with its unique properties of surface tension reduction. These insecticides formulations can help in reducing the number of field applications of insecticides due to enhancement of efficacy, making maximum output of money, and time spent on these insecticides.

There are at least four different nanoformulation types of nanoinsecticides: nanoemulsions, nanosuspension, nanocapsules, and nanoparticles.

The aims of all of them are to improve the efficacy of the active ingredient in the insecticides, improve the safety of the products in the environment, and control active ingredient release.

1.2.1 Nanoemulsions

Nanoemulsion has been identified as a promising delivery system for various chemical compounds including insecticides. Nanoemulsions are defined also as isotropic and kinetically stable emulsion systems in which the oil droplets containing the hydrophobic component are stabilized by a thin layer of emulsifier. They appear to be either transparent (droplet diameter < 200 nm) or milky (droplet diameter ≈ 500 nm) with the mean droplet diameters ranging from 50 to 1000 nm.

Generally, emulsion is usually a homogenous mixture consisting of various oils and/or fats intimately dispersed throughout the aqueous continuous phase in the presence of an emulsifying agent (also called emulsifier) (Fig. 1.1). So, it can be said that emulsion is the agent that makes the oil and water mix. Or an emulsifier is defined as a molecule with two different ends, one that loves water (hydrophilic) and a second one that loves lipids and hates water (hydrophobic).

It was known that most insecticides used in agriculture or used against medical insects (mosquitoes and housefly) are oily. By using of nanoinsecticides, the using of nanoformulations are suitable with nanoactive ingredient and nanoemulsifier is needed.

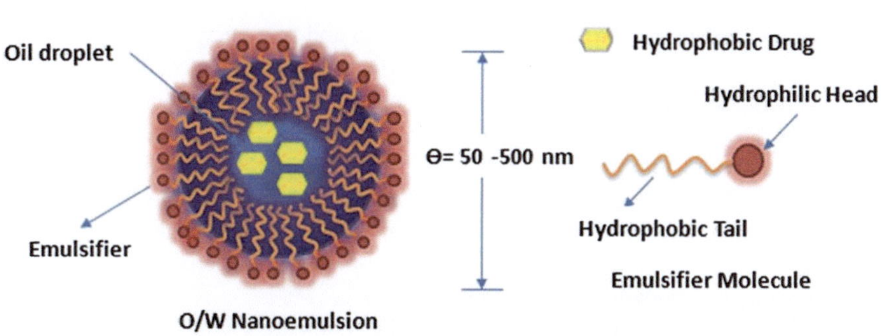

Fig. 1.1 Emulsion contents. (Cited from Praveen and Divya 2015)

The types of emulsion formed, normally oil-in-water (o/w) or water-in-oil (w/o), were commonly determined by the volume ratio of the two liquids and also by the phase addition sequence and by the nature of emulsifier. The average droplet sizes of nanoemulsion, say oil-in-water type, fall typically in the range of 100–500 nm. The former type finds a wide area of applications in oral and topical formulations of oil-soluble drugs as it is not only pleasant to take/use but also provides a remarkable masking effect with the inclusion of a suitable flavor.

The Differences Between O/W and W/O

O/W means water is the dispersion medium and oil is the dispersed phase, nongreasy and easily removable, used externally to provide cooling effect, and preferred for internal use. The oil component influences curvature by its ability to penetrate. Stability in high temperature can be used as carriers for wide range of organic compounds.

W/O means oil is the dispersion medium and water is the dispersed phase, greasy and not water washable, used externally to prevent evaporation of moisture from the surface and preferred for external use. So, it can be used on the wax layer of some plant leaves.

As shown in Fig. (1.1), nanoemulsion consists of oils, emulsifiers, and co-emulsifiers as recipients in their preparation or formulation strategy. It should be noted that the combination of these three components are essential to impart and control physical stability and consistency of nanoemulsion in the formulations. The oil phase of nanoemulsion plays a vital role in nanoemulsion formulation as it can solubilize highly potent lipophilic compound to facilitate their transport via the intestinal lymphatic system.

The Differences Between Emulsion and Nanoemulsion

There are many differences between the normal or microemulsion and nanoemulsion. All these differences showed that the nanoemulsion is better than the normal emulsion. These differences include:

1. The nanoemulsion is more stable than the normal emulsion. This means that the nanoinsecticides are more effective and stable than the conventional insecticides which have nanoemulsion.
2. The color of nanoemulsion is transparent, while normal emulsion is milky in color.
3. The nanoemulsions have a wide surface area and free energy than the normal emulsion; this means that little quantity of nanoemulsion can cover a wide area.
4. The nanoemulsions don't produce the inherent creaming, flocculation, coalescence, and sedimentation. So, nanoinsecticides will be homogenous.
5. The nanoemulsions are in the sub 100 nm size range not more than 100 nm.
6. Nanoemulsions are attractive for the aforementioned applications because they are relatively the least sensitive to physical and chemical changes.
7. Increase the rate of absorption. This means that the nanoactive ingredient was more conjugated with the nanoemulsion.
8. Dispersibility of nanoemulsion is very high compared with normal emulsion because the small droplet prevents the flocculation of droplets, and this process makes the system dispersed without separation and homogenous.

9. Nanoemulsion formulation provides a rapid penetration of active ingredients through the plant leaves. So, this formulation is very effective against sucking insects.
10. These formulations may be used to increase the bioavailability of poor water-soluble insecticides (Wagner et al. 1966; Kim et al. 2001).
11. Nanoemulsion can be formulated into different dosage forms with ease of manufacture and scale-up.
12. Nanoemulsion can be used as a template for producing nanoparticles and nanocapsules.

1.2.1.1 Preparation of Nanoemulsion

Emulsion is a heterogeneous mixture of lipid (hydrophobic) and aqueous phase (hydrophilic). The stability of this mixture is created by using a suitable material known as emulsifying agents. So, the nanoemulsion is a translucent system compared with the ordinary emulsion or some time microemulsion. Nanoemulsion has been demonstrated that with the help of nanoemulsion as a delivery system, retention time of nonchemical in the environment can be increased, so low amount of chemical is required for the treatment.

Two factors must be made when preparing nanoemulsion:

1. The size of droplets must be ranged between 60 and 100 nm.
2. The stability of nanoemulsion formulation.

The size of particles in nanoemulsion is a very important trait because it influences its optical, rheological, stability, and released characteristics. The particle size distribution of nanoemulsion can be controlled by varying emulsion preparation conditions and/or system composition. For example, in high-energy methods, the droplet size depends on the intensity and duration of the energy input, the type and concentration of emulsifier used, the interfacial tension, and the relative viscosities of the disperse and continuous phases (McClements 2004; Jafari et al. 2006). Smaller droplets can be produced by increasing the intensity or duration of homogenization, by increasing the concentration of emulsifier used, or by controlling the viscosity ratio (Wooster et al. 2008). The size of the droplets in an emulsion can be reduced by using solvent displacement and/or solvent evaporation methods (Horn and Rieger 2001; Chu et al. 2007).

In general, droplets stabilized by nonionic surfactants such as Tweens and Spans should have no droplet charge, but in practice it often has a significant negative charge which may be due to the presence of free fatty acids or other ionic impurities within them. Droplets stabilized by anionic surfactants have a negative charge such as lecithin, DATEM, CITREM, and fatty acids, whereas those stabilized by cationic surfactants have a positive charge (e.g., lauric arginate) (Mun et al. 2005).

The droplets of nanoemulsions often have an electrical charge because of adsorption of ionized emulsifiers, mineral ions, or biopolymers to their surfaces (McClements 2005). The sign and magnitude of the electrical charge on the droplets of nanoemulsion play an important role in the functional performance and stability, e.g., aggregation stability, interaction with other food components, and ability to adhere to biological surfaces such as the waxed plant leaves. The electrical characteristics of nanoemulsion droplets can be controlled by careful selection of particular emulsifier types.

1.2.1.2 Components of Nanoemulsion

Nanoemulsion consists of three main components:

1. Oils such as oleate, sesame oil, castor oil, arachis oil, corn oil, lanolin, jojoba oil, etc.

   The oil component was used to increase the ability of penetration and swill that tail group region of the surfactant monolayer.
2. Emulsifier as natural lecithins from plant or animal source
3. Some additive as antioxidants (ascorbic acid), tonicity modifier (glycerol or sorbitol), pH adjustment (NaOH or KOH), and preservative

1.2.1.3 Types of Nanoemulsion

1. **Oil-in-water (O/W):** means water is the dispersion medium and oil is the dispersed phase.
2. **Water-in-oil (W/O):** means oil is the dispersion medium and water is the dispersed phase.
3. **Bi-Contentious nanoemulsion:** Microdomains of oil and water are dispersed within the system.

1.2.1.4 Methods of Nanoemulsion Preparation

Phase Inversion Temperature (PIT)

This method was created by changing in phase by using a higher temperature to a microemulsion (El-Aasser et al. 1986; Pouton 1997). The high temperature converted the microemulsion to nanoemulsion.

This method uses an oil-in-water emulsion (O/W) initially formed by homogenizing an organic phase (lipid and organic solvent) with an aqueous phase (water and hydrophilic emulsifier) (Fig. 1.2). The organic solvent is selected according to its water miscibility, boiling point, safety, and legal status.

Izquierdo et al. (2004) studied the effect of phase behavior on nanoemulsion droplet size and stability. The obtained results showed that although the breakdown of droplet process of nanoemulsions was due to the oil transference from the smaller to the bigger droplets, the increase in stability is found with the increase in surfactant concentration. This result may be attributed to the higher surfactant excess, favoring of the oil micellar transport between the emulsion droplets.

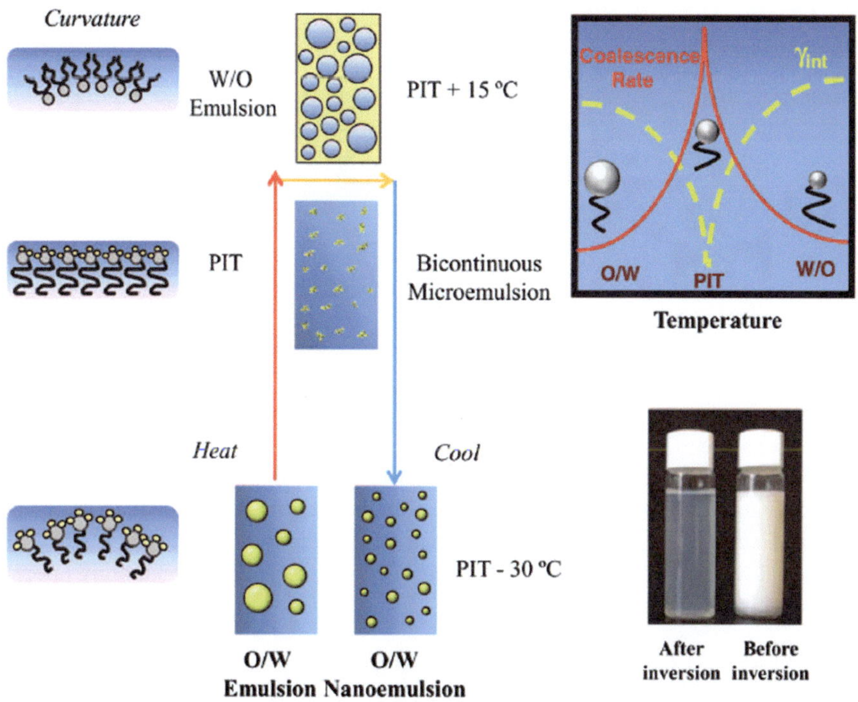

Fig. 1.2 Schematic diagram of the temperature dependence of the spontaneous curvature of surfactant monolayers and their influence on emulsion properties. (Cited from McClements and Rao 2011)

This method depends on changes in solubility of polyoxyethylene-type surfactant with temperature. The surfactant becomes lipophilic (soluble in lipids) by the increase of temperature. These changes are attributed to dehydration of polymer chain. On the other hand, at low temperature, the surfactant monolayer has a large positive spontaneous curvature forming oil-swollen micellar solution phase (Shinoda and Saito 1968).

By this method nanoemulsions can be formed continuously by varying the temperature-time profile of certain mixtures of oil, water, and nonionic surfactant. This type of phase inversion is usually used to the controlled transformation of an emulsion from one type to another (e.g., W/O to O/W or vice versa) through an intermediate liquid crystalline or bicontinuous microemulsion phase (McClements and Rao 2011).

Sonication Method

In this method the size of droplet in normal emulsion is reduced in size by the sonication mechanism. Only a small quantity of nanoemulsion can be prepared by this method. This method is not suitable for large quantity (Walstra 1996). The energy is provided through sonotrodes called as sonicator probe. As the tip of sonicator

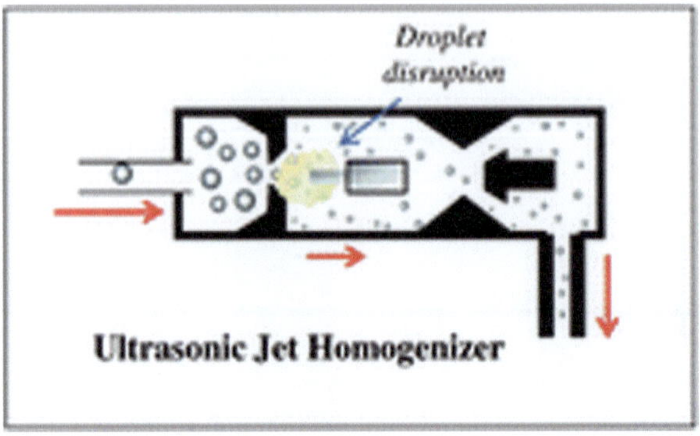

Fig. 1.3 Using of ultrasonic in nanoemulsion formation. (Cited from McClements 2011)

contacts the liquid, it produces a mechanical vibration and cavitations occur. Cavitations mean that the formation and collapse of vapor cavities in liquid are used. Thus, ultrasound can be directly used to produce nanoemulsion. This method is mainly used in laboratories where emulsion droplet size as low as 0.2 μm can be obtained (Jaiswal et al. 2015). In this method, the sample was homogenized and made to flow through a channel containing an element capable of generating intense ultrasonic waves (Fig. 1.3).

High-Pressure Homogenizer

This method is conducted by using a high pressure over the system having oil phase, aqueous phase, and surfactant or co-surfactant. The pressure is obtained by the help of homogenizer (Fig. 1.4). In this method some problems associated with homogenizer are poor productivity, component deterioration attributed to generation of much heat. According to this method, only oil-in-water (O/W) liquid nanoemulsion of less than 20% oil phase can be prepared, and cream nanoemulsion of high viscosity or hardness with a mean droplet diameter lower than 200 nm cannot be prepared.

Microfluidization

In this method a high-pressure positive displacement pump (500–200 PSI) which forces the product through the interaction chamber, consisting of small channels called micro channels, is used (Fig. 1.5). The microemulsion flows through the micro channels toward to an impingement area resulting in very small particles. Both aqueous phase and oily phase are combined together and processed in an inline homogenizer to produce a course emulsion. The course emulsion is converted to a micro fluidizer where it is further processed to obtain a stable nanoemulsion.

Microfluidization typically involves forming a coarse emulsion and then passing it into an interaction chamber, where the emulsion is split into two streams that are made to impinge to one another at high velocity (Jafari et al. 2007).

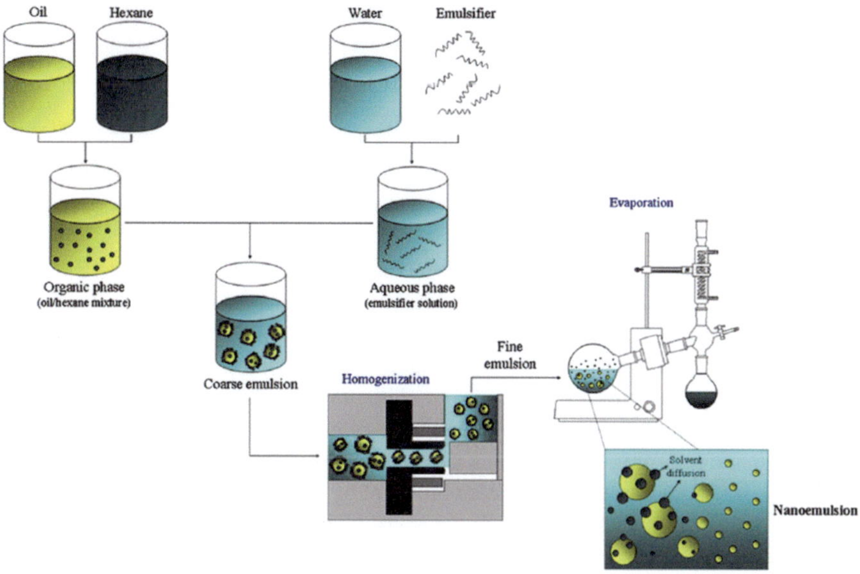

Fig. 1.4 Schematic representation of the nanoemulsion preparation using high-pressure homogenization and solvent evaporation stages. (Cited from Troncoso et al. 2012)

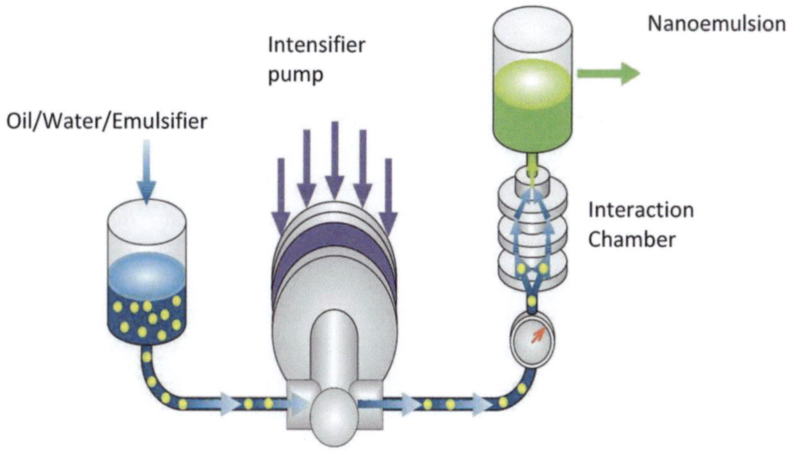

Fig. 1.5 Microfluidizer for nanoemulsion formation. (Cited from Panagiotou and Fisher 2012)

1.2.1.5 Nanoinsecticides with Nanoemulsion Formulation

Hexane-soluble fraction from ethanolic crude extract from fruits of *Manilkara subsericea* and its triterpenes was used as an effective extract against a cotton pest (*Dysdercus peruvianus*). Unfortunately, most of the plant extracts which have an insecticidal activity have poor water solubility, including triterpenes. So, nanotechnology has solved this problem by a good alternative nanoemulsion formulation (Fernandes et al. 2014). The authors made a new extract formulation extracted from *Manilkara subsericea* fruits. This formulation consists of 5% of a polar fraction from *M. subsericea*, 5% of oil (octyldodecyl myristate as a fatty acid), 5% of surfactants (sorbitan monooleate/polysorbate 80%), and 85% of water. The droplets size of this new formulation was 155.2 ± 3.8 nm. This formulation increased the mortality of *D. peruvianus* pest. On the other hand, this formulation has no effects on mammalian (nontarget organisms).

The plant extract of Copaiba (*Copaifera duckei* Dwyer, Fabaceae) oleoresin is an important plant extract. This extract is poor water solubility. By using nanotechnology, this problem was solved, especially oil-in-water nanoemulsions. The nanoemulsion of these extracts was effective against *Aedes aegypti* larvae. The size of formulation droplets was 145.2 ± 0.9 nm. This nanoemulsion consists of 90% of distilled water, 5% of oil, and 5% of surfactants. The main advantage of this product is low cost and ecofriendly green natural-based nanoformulations and considered a promising insecticidal products (Veerakumar et al. 2014). *Eucalyptus* is a diverse genus of flowering trees and shrubs in the Myrtle family, Myrtaceae. The oil extraction of eucalyptus leave has an allelopathic property and prevents insects from attacking it, thereby, acting as a natural insecticide (Brooker and Kleinig 1990; Batish et al. 2008).

Sugumar et al. (2014) prepared a nanoinsecticide extracted from plant extracts. The oil-in-water nanoemulsion was prepared by using eucalyptus oil, non-ionic surfactant (Tween 80), and water. The concentration of eucalyptus oil (6%, v/v) was fixed for all formulations. Firstly, coarse emulsion was prepared by adding water to organic phase containing oil and surfactant in different ratios, 1:1, 1:2, and 1:3 (v/v) using a magnetic stirrer, which was then subjected to ultrasonic emulsification using a 20 kHz Sonicator (Ultrasonics, USA) with a power output of 750 W. Energy input was given through sonotrode containing a piezoelectric crystal with a probe diameter of 13 mm. Sonicator probe generates disruptive forces that reduce the droplet diameter converting coarse emulsion to nanoemulsion.

Nanoemulsion formulation of pyrethroids was prepared such as γ-cyhalothrin, β-cypermethrin, nanopermethrin, etc. Nanopermethrin was prepared by oil-in-water emulsion (O/W) which proved to have larvicidal property. β-Cypermethrin nanoemulsion covers the leaves uniformly due to its small-sized droplets. Wetting, spreading, and penetration of pesticide are also enhanced.

1.2.2 Nanosuspensions

The main usage of nanosuspensions is for the compounds that are water insoluble (hydrophobic) and soluble in oil. The nanosuspension insecticides have an easy dispersibility, high stability, and high bioavailability. There are many disadvantages for the powder insecticides including dust drift, environmental pollution resulting

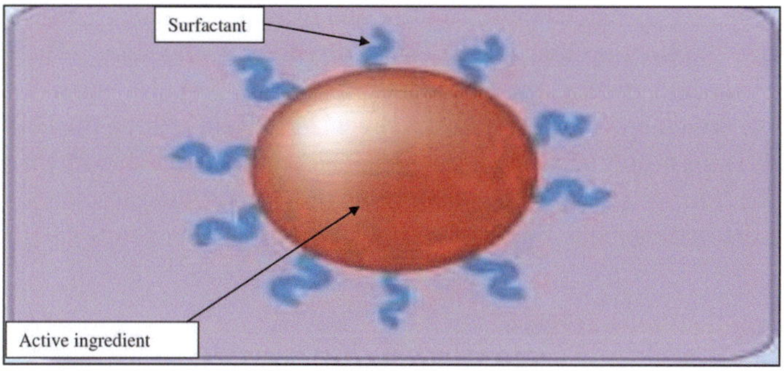

Fig. 1.6 Nanosuspension content. (Cited from Talekar et al. 2013)

from the use of organic solvent, and poor dispersion, further decreasing their pest control efficacy (Qian and Li 2005).

Recently, nanosuspensions, consisting of nanoparticles less than 1 μm in diameter, are suspended in aqueous phase using surfactants as stabilizers (Fig. 1.6). The nanosuspensions have attracted a large amount of attention as an effective formulation (Dubey 2006).

Advantages of Nanosuspension Compared with Normal Suspension

1. Increase the bioavailability of materials used
2. Easy and simple methods for preparation
3. Solving the problem of poor aqueous solubility
4. Suitable for hydrophilic insecticides (Liversidge and Cundy 1995)
5. Increase the stability of insecticides used
6. Reduction in insecticide concentrations used

1.2.2.1 Preparation of Nanosuspensions

High-Pressure Homogenization (HPH)

According to Kalvakuntla et al. (2016), the nanosuspension can be prepared by using aqueous solutions of stabilizers such as Tween 80, Poloxamer 188, PVA, and SLS. The nanosuspension was prepared by different concentrations using purified water. Aprepitant (125 mg) was diluted and suspended in 10 mL of the Tween 80. The dispersion was homogenized using high-speed homogenizer (Polytron PT 3100, Kinematica) at 10,000 rpm for 10 min to form homogeneous microsuspension. This microsuspension was subjected to probe sonication (Vibracell VCX130; Sonis, USA) at amplitude of 80% and pulse 4 s for 15 min to form presuspension. During this sonication, the temperature does not exceed 0 °C by using an ice bath. This presuspension was added dropwise to the remaining stabilizer solution and homogenized. Firstly, the first step was conducted at 5000 psi for five cycles using high-pressure homogenizer (Emulsiflex-C3, Avestin, USA). Then an HPH step was applied at 15,000 psi for ten cycles.

Grau et al. (2000) stated that there were two methods for nanosuspension preparation called "bottom-up technology" and "top-down technology." The top-down technology method involves the disintegration of larger particles into nanoparticles, examples of which are high-pressure homogenization and milling methods.

The bottom-up approach is more advantageous than the top-down approach because the former has a better chance of producing nanostructures with less defects, more homogenous chemical composition, and better short- and long-range ordering.

H96 (Lyophilization + HPH)
The H96 process includes both of lyophilization and HPH techniques. The amount of active ingredient and organic solvent was the most important factor to be considered for lyophilization in H96 process (Salazar et al. 2012). In the last step of material (aprepitant) synthesis, no crystallization of the synthesis material was formed, but the material solution was made by using methanol as a solvent. Methanol was the solvent of aprepitant and observed to be freely soluble in it. The prepared solution was then lyophilized. In the next step, the lyophilized product was dispersed in different stabilizer solutions in many concentrations by using purified water, which was immediately released through a homogenizer 15,000 psi for five cycles. Four different stabilizers (Tween 80, Poloxamer 188, PVA, and SLS) were conducted in different concentrations, alone and in combinations. After freeze-drying, the dry powder was investigated for possible aggregation by visual inspection. Shape and surface morphology of the freeze-dried nanocrystals was studied using SEM (JEOL, JSM 50A, Tokyo, Japan). An appropriate amount of freeze-dried nanocrystals was mounted on metal (aluminum) stubs; the samples were mounted onto aluminum specimen stubs using double-sided adhesive tape and fractured with a razor blade. The samples were sputter-coated with gold/palladium for 120 s at 14 mA under argon atmosphere for secondary electron emissive SEM and observed for morphology, at acceleration voltage of 20 kV.

1.2.2.2 Preparation of the Lambda-Cyhalothrin Nanosuspension

Lambda-cyhalothrin is the most important insecticide used against some crop pests. This insecticide belongs to the pyrethroid insecticide group. This insecticide is poorly water-soluble (5 μg/L at 21 °C); lambda-cyhalothrin nanosuspension was produced by the melt emulsification-HPH method (Pan et al. 2015). The powder of lambda-cyhalothrin was dispersed in deionized water-containing surfactants at 80 °C (this melting point is above the lambda-cyhalothrin melting point) and emulsified with C25 shearing machine (ATS, Germany) at 10,000 rpm for 2 min. After that the emulsified mixture was homogenized using AH100D high-pressure homogenizer (ATS Engineer Inc., Canada) under different pressures in the range of 200–1000 bar for 15 cycles. Finally, the nanosuspension was obtained by cooling down the homogenized mixture to room temperature in order to solidify the melted droplets into nanoparticles (Fig. 1.7).

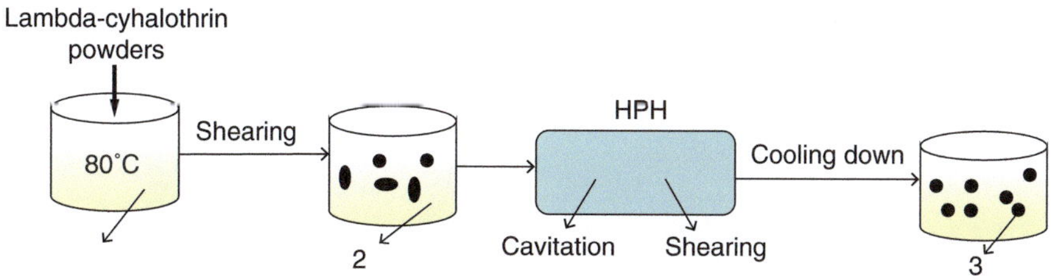

Fig. 1.7 Preparation of nanosuspension of lambda-cyhalothrin. (Cited from Pan et al. 2015)

Pyridalyl is a new insecticide intended for the control of lepidopterous larvae and thrips (Thysanoptera) as part of insect resistance management (IRM) and integrated pest management (IPM) programs. This insecticide inhibits cellular protein synthesis in insect without any effects on mammalian.

Pyridalyl nanosuspension is prepared by using sodium alginate two- and sixfold more effective as stomach poison against *Helicoverpa armigera* than the technical product and the commercial formulation, respectively (Saini et al. 2014).

1.2.3 Nanocapsules

A nanocapsule is a nanoscale cover made from a nontoxic substance called polymer (Fig. 1.8). It was made of a polymeric membrane which encapsulates an inner liquid or powder core at the nanoscale to protect the active ingredient from the environment factors, control of material release, and precision targeting (Ezhilarasi et al. 2012). The size of nanocapsule ranged from 10 to 1000 nm. This size is according to use of nanocapsule. Nanocapsules (<1 μm in diameter) containing a rubber-like consistency solid core are loaded with any insecticides. Active ingredient is found in the solid core at a controlled and determined rate. Consequently, the active ingredient can be delivered gradually over months.

The nanocapsules are defined as nano-vesicular systems that exhibit a typical core-shell structure in which the compound is confined to a reservoir or within a cavity surrounded by a polymer membrane or coating (Letchford and Burt 2007; Anton et al. 2008).

Nanocapsules are vesicular systems in which specific compounds, solubilized in an aqueous or oil core, are covered by a single polymeric membrane (wall material) (Couvreur et al. 2002). After synthesis, the evaluation of the nanocapsule stability is crucial mainly in relation to important parameters, such as size, polydispersity index (PDI), zeta potential, morphology, pH, and release profile.

Advantages of nanocapsules

- Longer period of effective action in a single use of any insecticides effectively reduces the level of insect infestation.

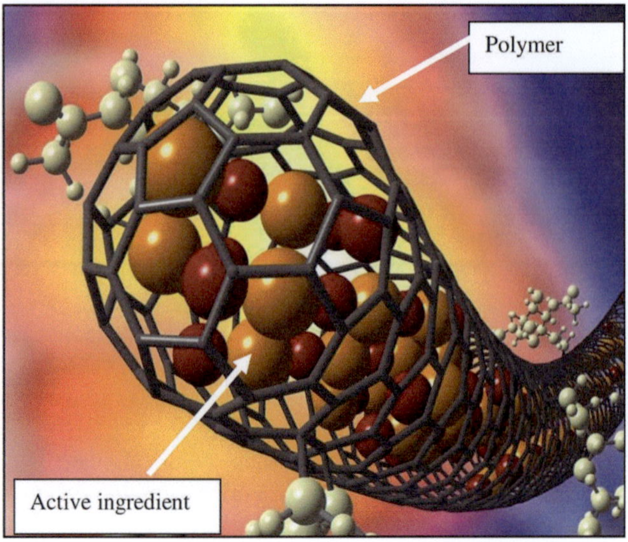

Fig. 1.8 Nanocapsules. (Cited from Wilton 2010)

- Operational safety: The active ingredients are gradually and successively released by the core and polymer walls of the capsules. This makes it possible to maintain a low concentration of those ingredients in sprayed places.
- Easier removal.
- Nonstain effect of the sprayed surface.

1.2.3.1 Preparation of Insecticide Nanocapsules

The nanocapsules are colloidal-sized, vesicular systems in which the active ingredient is confined to a reservoir or within a cavity surrounded by a polymer membrane or coating (Soppimath et al. 2001). Any nanoparticle needs a suitable protectant to prevent them from undergoing aggregation during the reaction that is to be catalyzed. Polymer–nanoparticle contents are suitable materials used for the purpose of exploiting or enhancing the unique properties of the nanoparticles, while the polymer matrix can control host–guest interactions to ensure the well-defined spatial distribution of nanoparticles (Gao et al. 2007). The most important component of insecticide nanocapsules is the polymer. There are many sources of polymers such as polysaccharides (e.g., chitosan, alginates, starch) and polyesters (e.g., poly-ε-caprolactone, polyethylene glycol) that have been used for the synthesis of nanoinsecticides. The first formulation containing polymer for controlled release of bioinsecticides started in the early 1970s. With the growing awareness for environmental pollution, application of biodegradable and biocompatible polymers of natural origin is preferred over the synthetic ones. The metabolites produced from the degradation of such polymers are of little concern. It was known that most of nanoinsecticides are prepared as nanoemulsion. Due to biodegradation and environmental factors affecting these formulations, the nanocapsulated insecticides may be better than this formulation.

1.2.3.2 Preparation of Polymer

Polymer is a Greek word divided into parts: poly means many, and meres means parts so this word means many parts.

Polymers are defined as large molecules made with hundreds of monomers conjugated to form long chains.

The ideal polymers must be:

1. Nontoxic
2. Compatible with environment
3. Not expensive
4. Good mechanical strength

Lactic acid (2-hydroxypropanoic acid) is the simplest 2-hydroxycarboxylic acid with a chiral carbon atom and exists in two optically active stereo isomers, namely, L and D enantiomers (S and R in absolute configuration, respectively). These L- and D-lactic acids are generally synthesized by fermentation using suitable microorganisms. Racemic DL-lactic acid (RS configuration) consisting of the equimolar mixture of D- and L-lactic acids shows characteristics different from those of the optically active ones. DL-lactic acid is conveniently synthesized by chemical method rather than fermentation.

For example, polymerization of citric acid onto the surface of oxidized multiwalled carbon nanotubes is leading to MWCNT-graft-poly (citric acid) (MWCNT-g-PCA) hybrid materials. Due to the presence of conjugated citric acid branches, synthesized MWCNT-g-PCA hybrid materials were not only soluble in water but also able to trap water-soluble chemical species and metal ions.

Polylactic acid (PLA) is synthesized from lactic acid and naturally occurring organic acid that can be produced by fermentation. Polylactic acid is different than most polymers in that it is derived from renewable resources like cornstarch or sugarcane.

PLA production is a popular idea as it represents the fulfillment of the dream of cost-efficient, nonpetroleum plastic production. The huge benefits of PLA as a bioplastic are versatility and the fact that it naturally degrades when exposed to the environment. PLA and its copolymers are being used in many biomedical trends as a biomaterial substance in the form of implants or devices due to its excellent biocompatibility and biodegradability. PLA production is a simple idea as it represents the fulfillment of the dream of cost-efficient, nonpetroleum plastic production.

There are two important methods for PLA synthesis:

1. Ring-opening polymerization of lactic acid cyclic dimmer, known as lactide
2. Direct polycondensation of lactic acid

The most common polymerization technique is known as ring-opening polymerization. This is a process that utilizes metal catalysts in combination with lactide to create the larger PLA molecules. The condensation process is similar with the principal difference being the temperature during the procedure and the by-products (condensates) that are released as a consequence of the reaction.

Polycondensation of polymer formation is processed by linking small molecules (monomers) together, accompanied by elimination of by-products (e.g., water and alcohols). In the case of PLA, polycondensation of lactic acid by connecting carboxyl and hydroxyl groups produces water by-product simultaneously. Due to the difficulty in removing by-products completely from the highly viscous reaction mixture, polymer produced through direct polycondensation is usually of low molecular weight (<50,000 g/mol) and low quality. In order to overcome this main disadvantage, numerous newly developed polycondensation methods have been proposed. In recent years, azeotropic polycondensation (AP) and solid-state polymerization (SSP) are two main directions.

1.2.3.3 Insecticides with Nanocapsule Formulation

Methomyl Nanocapsule

Methomyl is a c conventional insecticide which belongs to the carbamate group. This insecticide was used in control of many insect pests and has good efficacy against these pests. Although they have advantages, these insecticides have some disadvantages such as a short residual activity and an easy decomposition into non-toxic compounds during 3 days after application (Sun et al. 2014). Methomyl nanocapsules were prepared by adding 10.0 mg Az CMC Sand to 5.0 mg methomyl and were dissolved into 20.0 mL distilled water to obtain a solution. pH of all mixture must be adjusted to 4.0. After that the solution was sonicated for 3 min using a probe-type sonifier (JY92-2D, made by Ningbo Xinzhi Bio-tech Co. LTD.) at 100 W using a pulse function (pulse on 2.0 s, pulse off 2.0 s) to produce a methomyl-loaded nanocapsule dispersion solution. After that methomyl insecticide-loaded nanocapsule dispersion solution was exposed to UV light for 5 min using an ultraviolet lamp (20 W, 253.7 nm) to get a methomyl-loaded shell cross-linked nanocapsule solution. The nanocapsule methomyl had a hollow spherical configuration and a relatively uniform size, indicating that incorporation of methomyl had little effect on the topology and distribution of the nanocapsules.

Temephos

Temephos is a synthetic insecticide which belongs to organophosphate group. This insecticide was used against mosquito larvae and adults. The stock solutions of temephos were obtained by diluting 0.01 mL of insecticide in 1000 mL water to obtain a concentration of 10 mL/L. Different parts of polyethylene glycol (PEG)

(6000), viz., 49.5, 49.0, 48.0, 46.0, and 42.0 g, were heated separately at 65 °C. To these molten PEG, different parts 1%, 2%, 4%, 8%, and 16% (w/v) of temephos were mixed and stirred gently with the help of glass rod to ensure even distribution of the mixture. The mixtures were allowed to cool at room temperature, and the solidified mass was grounded completely in a mortar and sieved using a 200 mesh sieve. The resultant powders were finally placed in airtight, self-sealable polyethylene containers and stored at 25 °C in desiccators containing calcium chloride to prevent moisture absorption prior to further experiments (Bhan et al. 2015a,b).

Imidacloprid

Imidacloprid is a very important new insecticide which belongs to neonicotinoid insecticide group. It acts as a systemic insecticide, and the main target in insect is a nicotinic acetylcholine receptor stimulator. This insecticide is less toxic to humans and highly effective against pests. It was known worldwide for these qualities (Guan et al. 2008; Zhou et al. 2006).

PEG (Mn¼1000) and citric acid (molar ratio of CA/PEG¼1/10) were used for synthesis of PCA–PEG–PCA copolymers using the method described by Naeini et al. (2010). Three steps were involved in the polymerization process. The first step, the temperature of polymerization was increased to 110 °C for 20 min so that a transparent viscose compound was formed. In the second step, the temperature was increased to 130 °C to the melting state for 15 min. In the third step, the temperature was maintained at 150 °C for 30 min, and mixture was stirred vigorously. During these processes, water was removed from the reaction medium by a vacuum pump. After these steps, these contents were dissolved in tetrahydrofuran and then precipitated in diethyl ether several times. The PCA–PEG–PCA copolymers formed a viscous yellow compound with a yield of 8%.

Imidacloprid dissolved in acetone (1 g/100 mL) and PCA–PEG–PCA copolymers dissolved in ethanol as a basic solvent (1 g/20 mL) were mixed at room temperature and stirred for 8 h. Applying a dialysis bag (Mn cutoff 2000), free imidacloprid was separated, and then resultant solution containing nano-imidacloprid was maintained at 4 °C.

In general it can be said that preparation of nanocapsules involving the organic phase which consists of solvent, polymer, oil, and active ingredient is penetrated into the pores of an ultrafiltration membrane by the filtrate side, and then it is pressed. The aqueous phase consists of water, and surfactant circulates inside the membrane module and removes the nanocapsules forming at the pore outlets (Fig. 1.9).

Neem Oil

Neem oil was extracted from the seed of neem plants and used as a botanical extract against some insect pests. The efficacy of neem oils against insects is due to the presence of some active ingredients such as limonoids and azadirachtin which is

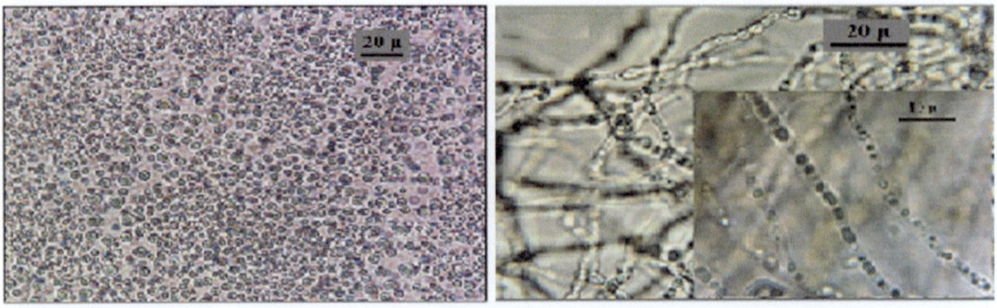

Fig. 1.9 Optical microscopy images of globular segments resulting from ethanol solution of PCA–PEG–PCA copolymers (12.5 g/L) and imidacloprid (2500 mg/L), after 20 days. (D) Optical microscopy images of fibers resulting from water solution of PCA–PEG–PCA copolymers (250 mg/L) and imidacloprid (50 mg/L), after 10 days. (Cited from Memarizadeh et al. 2014)

considered the most complex and potent agent (Mordue and Nisbet 2000). Azadirachtin acts as antifeedant activity due to its efficacy on insect chemoreceptors, ecdysteroid, and juvenile hormone titers by blocking morphogenetic peptide hormone flow, resulting in growth and molting disturbance, and has histopathological effects on insect muscles, fat body, and gut epithelial cells (Mordue and Blackwell 1993). Giongo et al. (2016) used nanocapsule of neem oil against fall armyworm, *Spodoptera frugiperda*.

Cypermethrin Nanocapsule

Cypermethrin is an important insecticide which belongs to pyrethroid insecticide group. The sodium and potassium gate in nervous cells is a main target for this insecticide. Cypermethrin nanocapsules were prepared by mixing microemulsion polymerization with the emulsifiers (DNS-86/OP-10). The effects of emulsifier contents, mixture ratio of emulsifiers, shell-core ratios, and temperature of emulsification on the particle sizes and size distribution of the nanocapsules were studied by Xiaomiao et al. (2009). The results showed that the microemulsion is very stable. This stability was due to the better emulsification of the mixed emulsifiers. Excellent nanocapsules can be obtained when the emulsifier content is 9.29%, the ratio of emulsifiers DNS-86: OP-10 is 4:3, the ratio of pesticide and monomer is 3:2.5, and the temperature of polymerization is 60.

1.2.4 Nanoparticles

Nanoparticles is defined as a small material that behaves as one unit with its transport and properties. The size of nanoparticles ranged between 1 and 100 nm. The main objective of nanoparticle insecticides is to improve insecticide distribution on the plant leaf surface. Silica particles (Fig. 1.10) were transformed to nanoparticles to use in insect pest control.

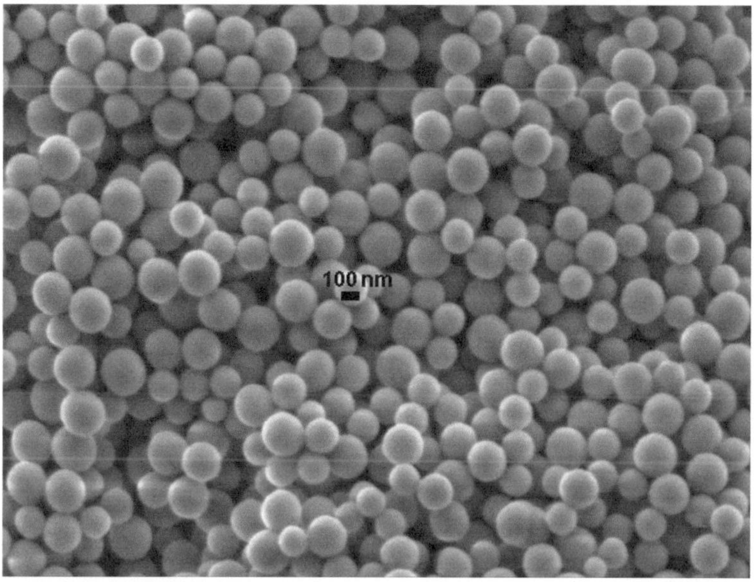

Fig. 1.10 Silica nanoparticles. (Cited from Hernandez-Leon et al. 2017)

1.2.4.1 Types of Nanoparticles

There are two types of nanoparticles:

1. Organic nanoparticles
   These include micelles, dendrimers, liposomes, and hybrid and compact polymeric nanoparticles.
2. Inorganic nanoparticles

These include fullerenes, quantum dots, silica, and gold nanoparticles.

Araj et al. (2015) used five sources of silver nanoparticles (Ag NPs) and sulfur nanoparticles (SNPs) on larval, pupal, and adults of the fruit fly *Drosophila melanogaster* under laboratory conditions. The authors used nanoparticles of silver and sulfur in different concentrations (10, 50, 100, 200 ppm) against *D. melanogaster*. The results showed that silver nanoparticles (Ag NPs) were most effective against larvae, pupae, and adults' mortality and egg suppression. The results also showed that silver nanoparticles can be used as an effective method in integrated pest management of *D. melanogaster*.

1.2.4.2 Synthesis of Silver Nanoparticles

Bae et al. (2010) synthesized the silver nanoparticles as in Fig. (1.11) by adding silver nitrate solution to sodium citrate and sodium borohydride stirred at room temperature. The color of sodium nitrate changed from yellow to red after stirring.

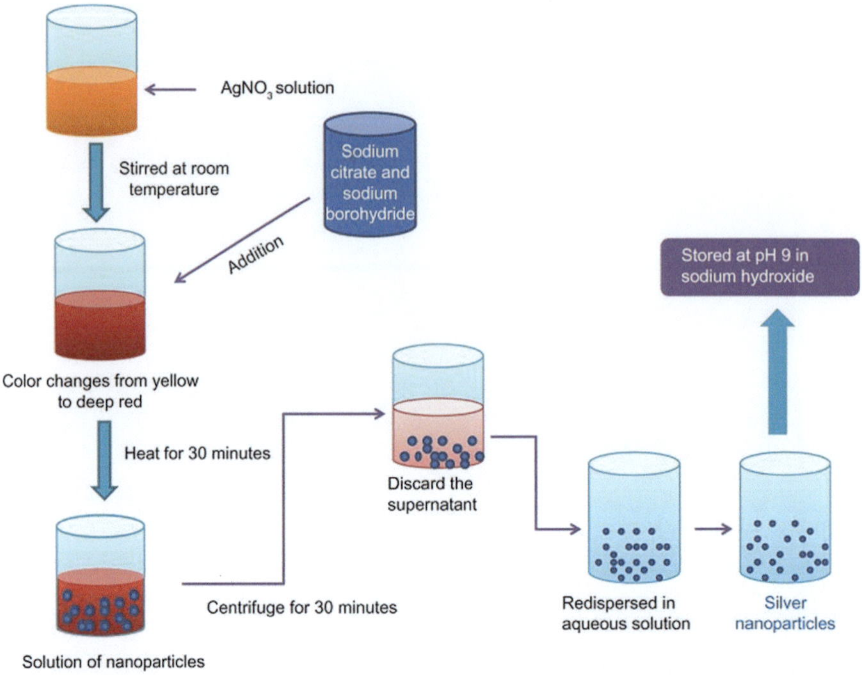

Fig. 1.11 Preparation of silver nanoparticles using a citrate synthesis method (Bae et al. 2010)

The mixture solution was heated for 30 min to produce solution nanoparticles. Then the solution is put in centrifuge for 30 min. After that the red color decreased, and the solution changed into transparent solution, and the silver nanoparticles was produced.

On the other hand, Araj et al. (2015) prepared silver nanoparticles by adding 5 mL of plant leaf extract (olive, fig, loquat, citrus, pistachio, and mulberry) to 100 mL of 1 mM of aqueous solution of silver nitrate ($AgNO_3$) with stirring magnetically with the room temperature or heating at 20–80 °C. The produced solution became brown and then changed to gray-black in color. The changing in color indicated that the silver nanoparticles were formed (Ag NPs). The concentrations of silver nitrate solutions, the quantity of plant extract, and the temperature were also varied at 1–4 mM, 5–10% by volume, and at temperature 20–80 °C. The silver nanoparticles were centrifuged at 15,000 rpm for 5 min and subsequently dispersed in sterile distilled water to get rid of any uncoordinated biological materials.

Rouhani et al. (2012b) used the silver nanoparticles against the oleander aphid, *Aphis nerii*. The authors compared the effectiveness of silver nanoparticles with the traditional insecticide imidacloprid. The lethal concentration of 50% for insect population (LC_{50}) values was 0.13 μL/mL and 424.67 mg/mL for imidacloprid silver nanoparticles, respectively. The authors recommended that the silver nanoparticles can be used in integrated pest managements. The use of imidacloprid and silver nanoparticles in combination increases the potency of imidacloprid. So, the silver nanoparticles have a synergist action.

Babu et al. (2014) synthesized the silver nanoparticles by using marine bacteria *Shewanella algae* bangaramma in the laboratory.

The silver nanoparticles are characterized by using UV–vis spectrum, TEM, FTIR, EDAX, XRD, and AFM analysis. The synthesized silver nanoparticles are spherical, crystalline, and 5–30 nm in diameter. The maximum LC_{50} and LC_{90} values were 4.529 and 9.580 mg/mL against the third instar larvae of *Lepidiota mansueta* (Burmeister). Rouhani et al. (2012a) synthesized the silica and silver nanoparticles through a solvothermal method and used different concentrations against *Callosobruchus maculatus*. In this experiment, the LC_{50} value for $SiO_2$and Ag nanoparticles was 0.68 and 2.06 g/kg cowpeas on adults and 1.03 and 1.00 g/kg on larvae, respectively.

Sadowski et al. (2008) synthesized the silver nanoparticles from microorganisms (*Penicillium* strain). The inoculated fungi were put in Petri dish and male extracts added to 0.5% yeast extract in a room temperature. The fungi were grown aerobically in liquid medium containing (g/L): KH_2PO_4 7.0, K_2HPO_4 2.0, $MgSO_4 \times 7H_2O$ 0.1, $(NH_4)_2SO_4$ 1.0, yeast extract 0.6, and glucose 10.0. All of them were inoculated in Erlenmeyer flasks with spores and incubated at 25 °C with shaking (150 rpm) for 72 h. After the cleaning steps, $AgNO_3$ (1 mM of final concentration) was mixed with cell-free filtrate in an Erlenmeyer flask and agitated at 25 °C in dark.

Yasur and Rani (2015) tested the effect of silver nanoparticles (AgNPs) on growth and feeding responses of two lepidopteron insects, namely, Asian armyworm, *Spodoptera litura*, and castor semilooper, *Achaea janata* L. (Lepidoptera: Noctuidae). The larvae were fed on PVP-coated AgNP-treated castor leaf at different concentrations. The efficacy of silver nanoparticles was compared to that of silver nitrate ($AgNO_3$)-treated leaf diets. Larval and pupal body weights decreased along with the decrease in the concentrations of AgNPs and $AgNO_3$ in both the test insects. On the other hand, Debnath et al. (2012) found that silica nanoparticles (SNPs) could effectively kill second-stadium larvae of *S. litura*.

1.2.4.3 Synthesis of Sulfur Nanoparticles

In agriculture fields sulfur is used as a fertilizer, insecticides, antimicrobial agents and/or fumigants (Ober 2003). Also, nanosulfur was used in pharmaceuticals and synthesis of nanocomposites for lithium batteries (Kobayashi et al. 2008). Chaudhuri and Paria (2010) prepared sulfur nanoparticles by acid-catalyzed precipitation of thiosulfate in aqueous surfactant solutions: the nanoparticles size is generated in the presence of CTAB.

On the other hand, Salem et al. (2016) synthesized the nanosulfur by using the pomegranate extract. By this method sulfur nanoparticles were synthesized as follows: 24.8 g of sodium thiosulfate pentahydrate ($Na_2S_2O_3{\cdot}5H_2O$) was added to 500 mL of pomegranate extract under mild stirring for 10 min at room temperature and then diluted with 500 mL deionized water. After that 10% HCl was added drop by drop under stirring for allowing the sulfur precipitations uniformly. The obtained suspended sulfur particles were centrifuged at 5000 rpm for 10 min at ambient temperature. The supernatant was discarded, and the precipitate was repeatedly washed with distilled water and absolute ethanol to get rid of any biological materials. The product was finally dried in a vacuum at 60 °C for 4 h. In pomegranate extract and acidic solution, sodium thiosulfate ($Na_2S_2O_3$) undergoes through a disproportion reaction to sulfur and sulfonic acid according to:

$$Na_2S_2O_3 + 5H_2O + 2HCl \rightarrow 2NaCl + S\downarrow + SO_2 + 6H_2O$$

Araj et al. (2015) synthesized sulfur nanoparticles (SNPs) also, by mixing amount of sodium thiosulfate ($Na_2S_2O_3$ + $5H_2O$) dissolved in 100 mL of sterile deionized water in a beaker 250 mL under mild stirring with magnetic stirrer at room temperature and atmospheric pressure. Ten milliliter of aqueous solution of citrus leaf extract acidified with dilute hydrochloric acid (HCl) was added to the aqueous solution of sodium thiosulfate with rate 1 mL/min with mild stirring for allowing the sulfur precipitations uniformly. The suspended sulfur particles produced were separated by centrifugation at 1000 rpm/min for 5 min and then repeatedly washed with sterile distilled water to remove any biological materials. The sulfur nanoparticles are divided into two parts. In the first part, the sulfur nanoparticles remained in the sterile distilled water without any additives added. In the second part, the sulfur nanoparticles after purification were dried in a vacuum at 80 °C for 2 h.

Rao and Paria (2013) used sulfur nanoparticles (SNPs) against two phytopathogens, *Fusarium solani* and *Venturia inaequalis*. The authors found that the small-sized particles are very effective in preventing the fungal growth.

1.2.4.4 Silica Nanoparticles

Silica is the name related to a group of minerals consisting of silicon and oxygen. These two elements are the most abundant elements in the earth's crust. Silica is found commonly in the crystalline state and rarely in an amorphous state. Silicon dioxide is mostly obtained by mining and purification of quartz. Quartz comprises more than 10% by mass of the earth's crust (Flörke et al. 2008). It consists of one atom of silicon and two atoms of oxygen resulting in the chemical formula SiO_2. Silica nanocomposites have received much attention because of its thermal degradation behavior and applications in chromatography, medicine, optics, etc. Silica, a new and safety material, was used as a safety and effective material against the second-instar larvae of tomato leaf miner, *Tuta absoluta* (Ragaei et al. 2013). Hunt et al. (2008) found that silica acts on insect pests by reducing its digestibility, not just palatability. How? One possibility is that the silica acts chemically, preventing digestion or absorption. Another mode of action is by reducing the amount they chew when eating high-silica grass, to avoid excessive abrasion. Subramanyam and Roesli (2000) hypothesized that silica nanoparticle (SNP)-based insecticide is physically active against some insects and causes damage to the cuticular water barrier of the insects mostly by abrasion and to some extent due to adsorption. Insect death occurs due to desiccation. Silica is suitable for many purposes, while for others chemical processing is needed to make a modification or otherwise more suitability. Silicon (Si) was used in agriculture that has been started since the 1970s (Laing et al. 2006). It can be said that silicon application can significantly act on insect pest and disease resistance in plants and cause yield increases in many crops.

Miller et al. (1960) found that when the stems of wheat (*Triticum aestivum* L.) contained high levels of silicon, these stems are not infested by Hessian fly (*Phytophaga destructor* Say) larvae. Moreover, in a greenhouse experiment, the authors showed that the most resistant wheat varieties had dark shapes of silicon depositions ranging from round to oblong and a relatively dense and grainy covering of silicon over the entire surface of the leaf sheath.

Preparation of Silica Nanoparticles

According to Park et al. (2002), silica nanoparticles were prepared by dissolving tetraethyl orthosilicate (TEOS) in ethanol to control particle size (Fig. 1.12).

Lovingood et al. (2012) prepared silica nanoparticles by dissolving tetraethyl orthosilicate (TEOS) in 50 mL acetone in conical tube. This amount is added to 850 μL of the 1 mM HCl in conical tube. By micropipette add 150 μL of TMOS to the plastic conical tube containing the 850 μL of 1 mM HCl. The mixture resulted in colorless layers. By flipping the mixture, this mixture will turn into clear colorless solution. Add acetone to the plastic conical tube containing the silicic acid solution and dilute to a total volumc of 40 mL.

On the other hand, Sugimoto (2000) found that silica nanoparticles can be prepared by hydrolysis of tetraethyl orthosilicate (TEOS) in water and added ethanol. This reaction will give silicon tetra hydroxide $Si(OH)_4$. Silicon tetra hydroxide suspension in ethanol will give nano SiO_2.

$$\underset{\textbf{Tetra ethyl orthosilicate (TEOS)}}{Si(OC_2H_5)_4 + 4H_2O} \xrightarrow[pH\ 11-12\ (NH_3)]{Suspension\ in\ ethanol}$$

$$\underset{\textbf{Silicon tetra hydroxide}}{Si(OH)_4} \xrightarrow[pH\ 11-12\ (NH_3)]{Suspension\ in\ ethanol} \underset{\textbf{Nano silica}}{\text{Nano } SiO_2 + 2HO}$$

Role of Silica Nanoparticles in Insect Control

Rouhani et al. (2012a) tested the efficacy of silica nanoparticles against the larvae and adults of *Callosobruchus maculatus*. The results showed that the silica nanoparticles were very effective against both larvae and adults. Vani and Brindhaa (2013)

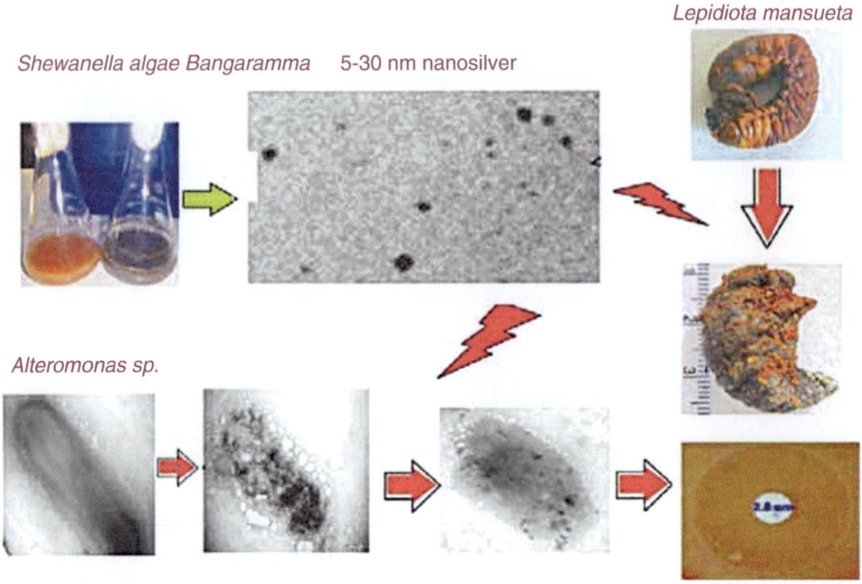

Fig. 1.12 Synthesis of silver nanoparticles by *Shewanella algae* (Babu et al. 2014)

evaluated the efficacy of silica nanoparticles against the stored grain pest *Corcyra cephalonica*. The percentages of mortality reached at 100%.

References

Anton N, Benoit JP, Saulnier P (2008) Design and production of nanoparticles formulated from nano-emulsion templates—a review. J Control Release 128:185–199

Araj SA, Salem NM, Ghabeish IH, Awwad AM (2015) Toxicity of nanoparticles against *Drosophila melanogaster* (Diptera: Drosophilidae). J Nanomater 5:1–9

Babu MY, Devi VJ, Ramakritinan CM, Umarani R, Taredahalli N, Kumaraguru AK (2014) Application of biosynthesized silver nanoparticles in agricultural and marine pest control. Curr Nanosci 10(3):374–381

Bae DR, Han WS, Lim JM (2010) Lysine-functionalized silver nanoparticles for visual detection and separation of histidine and histidine-tagged proteins. Langmuir 26(3):2181–2185

Barik TK, Sahu B, Swain V (2008) Nano-silica—from medicine to pest control. Parasitol Res 103:253–258

Batish DR, Singh HP, Kohli RK, Kaur S (2008) Eucalyptus essential oil as a natural pesticide. For Ecol Manage 256:2166–2174

Bhan S, Mohan L, Srivastava CN (2015a) Photosensitization of nanoencapsulated temephos and *Cuscuta reflexa* combination on mosquito larvae. Int J Pharm Res Biosci 4(1):94–110

Bhan S, Mohan L, Srivastava CN (2015b) Combinatorial studies on thermosensitization of nano-encapsulated temephos and *Cuscuta reflexa*. Int J Pharm Res Biosci 4(1):20–35

Bhattacharyya P, Samal AC, Majumdar J, Santra SC (2010) Accumulation of arsenic and its distribution in rice plant (*Oryza sativa* L.) in Gangetic West Bengal, India. Paddy Water Environ 8(1):63–70

Bhattacharyya A, Duraisamy P, Govindarajan M, Buhroo AA, Prasad R (2016) Nano-biofungicides: Emerging trend in insect pest control. In: Prasad R (ed) Advances and applications through fungal nanobiotechnology. Springer International Publishing, Cham, pp 307–319
Brooker M, Kleinig D (1990) Field guide to the Eucalyptus, vol 1, Revised edn. Inkata Press, Melbourne and Sydney
Chaudhuri RG, Paria S (2010) Synthesis of sulfur nanoparticles in aqueous surfactant solutions. J Colloid Interface Sci 343:439–446
Chu BS, Ichikawa S, Kanafusa S, Nakajima M (2007) Preparation of protein-stabilized β-carotene nanodispersions by emulsification–evaporation method. J Am Oil Chem Soc 84(11):1053–1062
Couvreur P, Barratt G, Fattal E, Vauthier C (2002) Nanocapsule technology: a review. Crit Rev Ther Drug Carrier Syst 19:99–134
Debnath N, Das S, Patra P, Mitra S, Goswami A (2012) Toxicological evaluation of entomotoxic silica nanoparticle. Toxicol Environ Chem 94(5):944–951
Dubey R (2006) Pure drug nanosuspensions – impact of nanosuspension technology on drug discovery and development. Drug Del Tech 6:65–71
El-Aasser MS, Lack CD, Vanderhoff JW, Fowkes FM (1986) Miniemulsification process-different form of spontaneous emulsification. Colloids Surf 29:103–118
Ezhilarasi PN, Karthik P, Chhanwal N, Anandharamakrishnan C (2012) Nanoencapsulation techniques for food bioactive components: a review. Food Bioproc Tech 6(3):628–647
Fernandes CP, de Almeida FB, Silveira AN, Gonzalez MS, Mello CB, Feder D, Apolinário R, Santos MG, Carvalho JCT, Tietbohl LAC, Rocha L, Falcão DQ (2014) Development of an insecticidal nanoemulsion with *Manilkara subsericea* (Sapotaceae) extract. J Nanobiotechnology 12(1):22–31
Flörke OW, Graetsch HA, Brunk F, Benda L, Paschen S, Bergna HE, Roberts WO, Welsh WA, Libanati C, Ettlinger M, Kerner D, Maier M, Meon W, Schmoll R, Gies H, Schiffmann D (2008) Silica in Ullmann's Encyclopedia of Industrial Chemistry. Wiley-VCH, Weinheim
Gajbhiye M, Kesharwani J, Ingle A, Gade A, Rai M (2009) Fungus mediated synthesis of silver nanoparticles and its activity against pathogenic fungi in combination of fluconazole. Nanomedicine 5(4):282–286
Gao Y, Ding X, Zheng Z, Cheng X, Peng Y (2007) Template-free method to prepare polymer nanocapsules embedded with noble metal nanoparticles. Chem Commun:3720–3722
Giongo AMM, Vendramim JD, Forim MR (2016) Evaluation of neem-based nanoformulations as alternative to control fall armyworm. Ciência Agrotec 40(1):26–36
Goswami A, Roy I, Sengupta S, Debnath N (2010) Novel applications of solid and liquid formulations of nanoparticles against insect pests and pathogens. Thin Solid Films 519:1252–1257
Grau MJ, Kayser O, Muller RH (2000) Nanosuspensions of poorly soluble drugs – reproducibility of small scale production. Int J Pharm 196:155–157
Gribbin J, Gribbin M (1997) Richard Feynman: a life in science. Dutton, London, p 170 ISBN 0-452-27631-4
Guan H, Chi D, Yu J, Li X (2008) A novel photodegradable insecticide: preparation, characterization and properties evaluation of nano-Imidacloprid. Pestic Biochem Physiol 92:83–91
Hernandez-Leon SG, Sarabia-Sainz JA, Montfort GRC, Guzman-Partida AM, Robles-Burgueño MR, Vazquez-Moreno L (2017) Novel synthesis of core-shell silica nanoparticles for the capture of low molecular weight proteins and peptides. Molecules 22(10):1–12
Horn D, Rieger J (2001) Organic nanoparticles in the aqueous phase–theory, experiment, and use. Angew Chem Int Ed Engl 40(23):4330–4361
Hunt JW, Dean AP, Webster RE, Johnson GN, Ennos AR (2008) A novel mechanism by which silica defends grasses against herbivory. Ann Bot 102:653–656
Izquierdo P, Esquena J, Tadros TF, Dederen JC, Feng J, Garcia-Celma MJ, Azemar N, Solans C (2004) Phase behavior and nanoemulsion formation by the phase inversion temperature method. Langmuir 20(16):6594–6598
Jafari SM, He YH, Bhandari B (2006) Nano-emulsion production by sonication and microfluidization—a comparison. Int J Food Prop 9(3):475–485

Jafari M, He Y, Bhandari B (2007) Optimization of nano-emulsions production by microfluidization. Eur Food Res Technol 225(5–6):733–741

Jaiswal M, Dudhe R, Sharma PK (2015) Nanoemulsion: an advanced mode of drug delivery system. 3 Biotech 5(2):123–127

Kah M, Beulke S, Tiede K, Hofmann T (2013) Nanopesticides: state of knowledge, environmental fate, and exposure modeling. Crit Rev Environ Sci Technol 43(16):1823–1867

Kalvakuntla S, Deshpande M, Attari Z, Kunnatur K (2016) Preparation and characterization of nanosuspension of aprepitant by H96 process. Adv Pharm Bull 6(1):83–90

Kim CK, Cho YJ, Gao ZG (2001) Preparation and evaluation of biphenyl dimethyl dicarboxylate microemulsions for oral delivery. J Control Release 70:149–155

Kobayashi T, Imade Y, Shishihara D, Homma K, Nagao M, Watanabe R, Yokoi T, Yamada A, Kanno R, Tatsumi TJ (2008) All solid-state battery with sulfur electrode and thio-LISICON electrolyte. Power Sources 182:621

Laing MD, Gatarayiha MC, Adandonon A (2006) Silicon use for pest control in agriculture: a review. Proc S Afr Sug Technol Ass 80:278–286

Letchford K, Burt H (2007) A review of the formation and classification of amphiphilic block copolymer nanoparticulate structures: micelles, nanospheres, nanocapsules and polymersomes. Eur J Pharm Biopharm 65:259–269

Liversidge GG, Cundy KC (1995) Particle size reduction for improvement of oral bioavailability of hydrophobic drugs. Part 1: absolute oral bioavailability of nanocrystalline danazol in beagle dogs. Int J Pharm 125:91–97

Lovingood DD, Owens JR, Seeber M, Kornev KG, Luzinov I (2012) Controlled microwave-assisted growth of silica nanoparticles under acid catalysis. ACS Appl Mater Interfaces 4:6875–6883

McClements DJ (2004) Food emulsions: principles, practice and techniques, 2nd edn. CRC Press, Boca Raton, FL

McClements DJ (2005) Food emulsions: principles, practice, and techniques. CRC Press, Boca Raton, FL

McClements DJ (2011) Edible nanoemulsions: fabrication, properties, and functional performance. Soft Matter 7:2297–2316

McClements DJ, Rao J (2011) Food-grade nanoemulsions: formulation, fabrication, properties, performance, biological fate, and potential toxicity. Crit Rev Food Sci Nutr 51(4):285–330

Memarizadeh N, Ghadamyari M, Adeli M, Talebi K (2014) Preparation, characterization and efficiency of nanoencapsulated imidacloprid under laboratory conditions. Ecotoxicol Environ Saf 107:77–83

Miller BS, Robinson RJ, Johnson JA, Jones ET, Ponnaiya BWX (1960) Studies on the relation between silica in wheat plants and resistance to Hessian fly attack. J Econ Ent 53:995–999

Mordue AJ, Blackwell A (1993) Azadirachtin: an update. J Insect Physiol 39(11):903–924

Mordue AJ, Nisbet AJ (2000) Azadirachtin from the neem tree *Azadirachta indica*: its action against insects. Anais Soc Entomol Bras 29(4):615–632

Mun S, Decker EA, McClements DJ (2005) Influence of droplet characteristics on the formation of oil in-water emulsions stabilized by surfactant chitosan layers. Langmuir 21:6228–6234

Naeini AT, Adeli M, Vossoughi M (2010) Poly(citric acid)-block-poly(ethylene glycol) copolymers-new biocompatible hybrid materials for nanomedicine. Nanomedicine 6:556–562

Ober JA (2003) Materials flow of sulfur: US Geological Survey Open File Report 02-298

Owolade OF, Ogunleti DO, Adenekan MO (2008) Titanium dioxide affects disease development and yield of edible cowpea. Electron J Environ Agri Food Chem 7(50):2942–2947

Pan Z, Cui B, Zeng Z, Feng L, Liu G, Cui H, Pan H (2015) Lambda-cyhalothrin nanosuspension prepared by the melt emulsification-high pressure homogenization method. J Nanomater:296–302

Panagiotou T, Fisher R (2012) Improving product quality with entrapped stable emulsions: from theory to industrial application. Challenges 3:84–113

Park SK, Kim KD, Kim HT (2002) Preparation of silica nanoparticles: determination of the optimal synthesis conditions for small and uniform particles. Colloids Surf A 197(1–3):7–17

Pouton CW (1997) Formulation of self emulsifying drug delivery systems. Adv Drug Deliv Rev 25:47–58
Prasad R, Kumar V, Prasad KS (2014) Nanotechnology in sustainable agriculture: present concerns and future aspects. Afr J Biotechnol 13(6):705–713
Prasad R, Bhattacharyya A, Nguyen QD (2017a) Nanotechnology in sustainable agriculture: recent developments, challenges, and perspectives. Front Microbiol 8:1014. https://doi.org/10.3389/fmicb.2017.01014
Prasad R, Gupta N, Kumar M, Kumar V, Wang S, Abd-Elsalam KA (2017b) Nanomaterials act as plant defense mechanism. In: Prasad R, Kumar V, Kumar M (eds) Nanotechnology. Springer, Singapore, pp 253–269
Praveen KG, Divya A (2015) Nanoemulsion based targeting in cancer therapeutics. Med Chem 5:272–284
Qian L, Li T (2005) Review of reproductive toxicity of environmental chemical pollutants. J Environ Occup Med 22:167–171
Ragaei M, Sabry KH (2014) Nanotechnology for insect pest control. Int J Sci Environ Technol 3(2):528–545
Ragaei M, Sabry KH, El-Rafei Amira M (2013) Towards using of new and safety material against tomato leafminer, *Tuta absoluta* (Meyrick). Arch Phytopathol Plant Prot 46(20):2450–2458
Rao KJ, Paria S (2013) Use of sulfur nanoparticles as a green pesticide on *Fusarium solani* and *Venturia inaequalis* phytopath. RSC Adv 3:10471–10478
Rosen H, Abribat T (2005) The rise and rise of drug delivery. Nat Rev Drug Discov 4:381–385
Rouhani M, Samih MA, Kalantari S (2012a) Insecticidal effect of silica and silver nanoparticles on the cowpea seed beetle, *Callosobruchus maculatus* F. (Col.: Bruchidae). J Entomol Res 4(4):297–305
Rouhani M, Samih MA, Kalantari S (2012b) Insecticide effect of silver and zinc nanoparticles against *Aphis nerii* Boyer De Fonscolombe (Hemiptera: Aphididae). Chil J Agri Res 72(4):590–594
Sadowski Z, Maliszewska IH, Grochowalska B, Polowczyk I, Koźlecki T (2008) Synthesis of silver nanoparticles using microorganisms. Mater Sci Poland 26(2):219–224
Saini P, Gopal M, Kumar R, Srivastava C (2014) Development of pyridalyl nanocapsule suspension for efficient management of tomato fruit and shoot borer (*Helicoverpa armigera*). J Environ Sci Health B 49:344–351
Salazar J, Ghanem A, Muller RH, Moschwitzer JP (2012) Nanocrystals: comparison of the size reduction effectiveness of a novel combinative method with conventional top-down approaches. Eur J Pharm Biopharm 81(1):82–90
Salem NM, Albanna LS, Abdeen AO, Ibrahim QI, Awwad AM (2016) Sulfur nanoparticles improves root and shoot growth of tomato. J Agric Sci 8(4):179–185
Sangeetha J, Thangadurai D, Hospet R, Harish ER, Purushotham P, Mujeeb MA, Shrinivas J, David M, Mundaragi AC, Thimmappa AC, Arakera SB, Prasad R (2017) Nanoagrotechnology for soil quality, crop performance and environmental management. In: Prasad R, Kumar M, Kumar V (eds) Nanotechnology. Springer Nature Singapore Pte Ltd, Singapore, pp 73–97
Shinoda K, Saito H (1968) The effect of temperature on the phase equilibria and the type of dispersion of the ternary system composed of water, cyclohexane and nonionic surfactant. J Colloid Interface Sci 26:70–74
Soppimath KS, Aminabhavi TM, Kulkarni AR, Rudzinski WE (2001) Biodegradable polymeric nanoparticles as drug delivery devices. J Control Release 70:1–20
Subramanyam BH, Roesli R (2000) Inert dusts. In: Alternatives to pesticides in stored product IPM. Kluwer Academic Publishers, Dordrecht, pp 321–380
Sugimoto T (2000) Fine particles-synthesis, characterization, and mechanism of growth, Surf Sci Ser 92. Marcel Dekker, New York
Sugumar S, Clarke SK, Nirmala MJ, Tyagi BK, Mukherjee A, Chnadrasekaran N (2014) Nanoemulsion of eucalyptus oil and its larvicidal activity against *Culex quinquefasciatus*. Bull Entomol Res 104:393–402
Sun C, Shu K, Wang W, Ye Z, Liu T, Gao Y, Zheng H, He G, Yin Y (2014) Encapsulation and controlled release of hydrophilic pesticide in shell cross-linked nanocapsules containing aqueous core. Int J Pharm 463:108–114

Talekar M, Ganta S, Amiji M, Jamieson S, Kendall J, Denny WA, Gar S (2013) Development of PIK-75 nanosuspension formulation with enhanced delivery efficiency and cytotoxicity for targeted anti-cancer therapy. Int J Pharm 450(1–2):278–289

Troncoso E, Aguilera JM, McClements DJ (2012) Fabrication, characterization and lipase digestibility of food-grade nanoemulsions. Food Hydrocoll 27:355–363

Vani C, Brindhaa U (2013) Silica nanoparticles as nanocides against *Corcyra cephalonica* (s.), the stored grain pest. Int J Pharm Bio Sci 4(3):1108–1118

Veerakumar K, Govindarajan M, Hoti SL (2014) Evaluation of plant-mediated synthesized silver nanoparticles against vector mosquitoes. Parasitol Res 113:4567–4577

Wagner JG, Gerard ES, Kaiser DG (1966) The effect of the dosage form on serum levels of indoxole. Clin Pharmacol Ther 7:610–619

Walstra P (1996) Emulsion stability. In: Becher P (ed) Encyclopedia of emulsion technology. Marcel Dekker, New York, pp 1–62

Wilton P (2010) Nanocapsule delivers radiotherapy. University of Oxford, Oxford

Wooster TJ, Golding M, Sanguansri P (2008) Impact of oil type on nanoemulsion formulation and ostwald ripening stability. Langmuir 24:12758–12765

Xiaomiao C, Jin Y, Yifeng Z, Wangyan N (2009) Application of mixed emulsifiers for preparation of cypermethrin nanocapsules. Spec Petrochem 2009-05

Yasur J, Rani PU (2015) Lepidopteran insect susceptibility to silver nanoparticles and measurement of changes in their growth, development and physiology. Chemosphere 124:92–102

Zhou Q, Ding Y, Xiao J (2006) Sensitive determination of thiamethoxam, imidacloprid and acetamiprid in environmental water samples with solid phase extraction packed with multiwalled carbon nanotubes prior to high performance liquid chromatography. Anal Bioanal Chem 385:1520–1525

Chapter 2
Nanoparticle-Based Plant Disease Management: Tools for Sustainable Agriculture

Anurag Yadav and Kusum Yadav

2.1 Interactions Between NPs, Pathogens, and Plants

2.1.1 NP Interaction with Plants

Plant parts remain protected from external components by organ-specific coatings. However, NPs being smaller can easily penetrate cell wall to show useful or harmful effects, as plant cell wall is nearly porous to 3.5–20 nm range macromolecules. NPs penetrate the plant cells via cytoplasmic membrane proteins and ionic channels or through endocytosis by forming vesicles around the transport component. Plant roots permeate solutes through tips via root hairs; the rest of the root surface is waterproof due to suberin. NPs can easily be absorbed and transported through these openings. As a general assumption, NPs may be transported inside leaves through stomatal openings. However, it should be noted that stomata are present on the dorsal side of leaves, making difficult for airborne particles to penetrate inside (Chichiriccò and Poma 2015). NPs of size $\leq$43 nm can easily penetrate stomata (Eichert et al. 2008; Schreiber 2005). Table 2.1 shows the positive and negative effect of NPs on the plant.

A. Yadav (✉)
Department of Microbiology, College of Basic Science & Humanities, S.D. Agricultural University, S.K. Nagar, Banaskantha, India

K. Yadav
Department of Biochemistry, University of Lucknow, Lucknow, India

K. A. Abd-Elsalam, R. Prasad (eds.), *Nanobiotechnology Applications in Plant Protection*, Nanotechnology in the Life Sciences,
https://doi.org/10.1007/978-3-319-91161-8_2

Table 2.1 Effect of NPs on plant

S. no.	Plant	Nanoparticle	Concentration	Growth medium	Effect	References
1.	*Allium cepa*	ZnO	0.2, 0.4, and 0.8 g L^{-1}	Hydroponics	Loss of membrane integrity, increased chromosome aberrations, micronucleus formation, DNA strand breaks, and cell cycle arrest at the G2/M checkpoint in root meristem	Ghosh et al. (2016)
2.	*Allium cepa*	ZnO	25, 50, 75, and 100 μg mL^{-1}	Hydroponic	Increased mitotic index, lipid peroxidation, pyknotic cells, and micronucleated cells	Kumari et al. (2011)
3.	*Arabidopsis*	Ag and Ag+	0.01 to 100 mg L^{-1}	Hydroponics	Enhanced root elongation and fresh weight at moderate concentrations	Wang et al. (2013)
4.	*Arabidopsis*	CdSe/ZnS–QDs	5 μg mL^{-1} Cd^{2+}, 5 μg mL^{-1} SeO_3^{2-}, and 5.8 nm QD (5 μg mL^{-1} in Cd^{2+})	Hydroponics	Oxidative stress	Navarro et al. (2012)
5.	*Arabidopsis*	TiO_2	100 μm	Glass chamber	Disorganization of the microtubules and isotropic growth of epidermal cell roots	Wang et al. (2011b)
6.	*Arabidopsis*	MWNTs		Leaf cell cultures	Decrease in superoxide dismutase (SOD) activity	Lin et al. (2009)
7.	*Arabidopsis* and poplar	Ag	0.1 mg L^{-1}	Liquid growth medium	Root elongation, fresh weight, and evapotranspiration at sublethal concentrations	Wang et al. (2013)
8.	*Arabidopsis thaliana*	CeO_2	250 ppm		Biomass increased	Ma et al. (2013)
9.	*Arabidopsis thaliana*	Au	10 μg mL^{1}	Murashige and Skoog (MS) medium under glasshouse	Improved seed germination rate, vegetative growth, and free radical scavenging activity	Kumar et al. (2013)

10.	*Arabidopsis thaliana*	CeO_2 and In_2O_3	0–2000 ppm	MS medium and under glasshouse	Decrease in growth rate and chlorophyll content	Ma et al. (2013)
11.	Barley	Ag	1, 2.5, 5, and 10 mg L^{-1}	Hydroponics	Increased root length	Gruyer et al. (2013)
12.	Barley (*Hordeum vulgare*)	CeO_2 and TiO_2	0, 500, 1000, and 2000 mg L −1	Petri plates containing filter paper	NPs influenced reactive oxygen species (ROS) generation and ATP content	Mattiello et al. (2015)
13.	*Boswellia ovalifoliolata*	Ag	10 to 30 μg ml^{-1}	MS medium	Enhanced seed germination and plant height	Savithramma et al. (2012)
14.	Corn, alfalfa, soybean	CeO_2	0.5, 1, 2 and 4 g L^{-1}	Petri plates containing germination paper	Increased root and stem growth	López-Moreno et al. (2010)
15.	Cucumber (*Cucumis sativus*)	CeO_2	10 ± 1 nm and 8 ± 1 nm, respectively	Hydroponics	Reduced yield by 31.6%	Zhao et al. (2013)
16.	Duckweed (*Lemna minor*)	Al	30 mg L^{-1}	Magenta vessels containing Hutner's medium	Enhanced biomass	Juhel et al. (2011)
17.	Fennel (*Foeniculum vulgare* Mill.)	TiO_2	0, 5, 20, 40, 60, and 80 mg L^{-1}	Petri plates containing filter paper	Enhanced germination value, vigor index, and mean daily germination	Feizi et al. (2013)
18.	Flax (*Linum usitatissimum* L., cv. Electra), ryegrass (*Lolium perenne* L., cv. Tove), and two-rowed barley (*Hordeum vulgare* L., cv. Annabell)	Fe and Ag	100, 250, 500, 1000, 2000, and 5000 mg nZVI kg^{-1} soil	Aqueous suspension and soil	Reduction in shoot growth and lower seed germination	El-Temsah and Joner (2012)

(continued)

Table 2.1 (continued)

S. no.	Plant	Nanoparticle	Concentration	Growth medium	Effect	References
19.	Flax, ryegrass, and barley	Ag	20–100 mg kg^{-1}	Petri plates containing filter paper and soil	Lower seed germination, reduced shoot growth	El-Temsah and Joner (2012)
20.	Garlic (*Allium sativum* L.)	ZnO	10–50 mg L^{-1}	Hydroponics	Decreased mitosis index and root growth	Shaymurat et al. (2012)
21.	Lettuce (*Lactuca sativa*)	Ag	1, 2.5, 5, and 10 mg L^{-1}	Soil	Decreased root length	Gruyer et al. (2013)
22.	Maize (*Zea mays*)	MWCNTs	20 mg L^{-1}	Growth chamber	Improved water absorption, plant biomass, and the concentrations of the essential Ca and Fe nutrients	Tiwari et al. (2014)
23.	Maize (*Zea mays* L.) and *cabbage* (*Brassica oleracea* var. *capitata* L.)	Citrate-nAg, nZnO, $AgNO_3$, and ZnSO	0.01–1000 μg mL^{-1}	Petri plates containing filter paper	Cell erosion in maize due to $AgNO_3$	Pokhrel and Dubey (2013)
24.	Mung bean (*Vigna radiata*) and sorghum (*Sorghum bicolor*)	Ag	0–40 mg L^{-1}	Agar and soil medium	Seedling growth inhibition	Lee et al. (2012)
25.	Mustard (*Brassica juncea*)	Au	0, 10, 25, 50, and 100 ppm	Field	Enhanced growth and seed yield	Arora et al. (2012)
26.	Parsley (*Petroselinum crispum*)	TiO_2	10, 20, 30, and 40 mg mL^{-1}	MS medium	Enhanced root-shoot length and chlorophyll content	Dehkourdi and Mosavi (2013)
27.	Pumpkin (*Cucurbita mixta*)	Fe_3O_4	30, 100, and 500 mg L^{-1}	Hydroponics	Root elongation	Wang et al. (2011a)
28.	Radish (*Raphanus sativus*)	CO_3O_4	5 g L^{-1}	Petri plates containing filter paper	Increased root growth	Wu et al. (2012)

29.	Rice (*Oryza sativa*)	MWNTs	0, 10, 20, 40, and 80 mg L^{-1}	Leaf cell culture in MS medium	Decrease in superoxide dismutase (SOD) activity	Tan et al. (2009)
30.	Rice (*Oryza sativa*)	CuO	0.5 mM, 1.0 mM, and 1.5 mM	Cotton pad placed on plastic tray	Loss of root cell viability, declined carotenoid level, and stress-induced oxidative damages	Shaw and Hossain (2013)
31.	Ryegrass (*Lolium perenne* L.) and pumpkin (*Cucurbita mixta*)	Fe_3O_4	30, 100, and 500 mg L^{-1}	Hydroponics	Root-shoot oxidative stress	Wang et al. (2011a)
32.	Sorghum (*Sorghum bicolor*)	Ag	200 mg kg^{-1}	Agar medium and soil	Lower seedling growth	Lee et al. (2012)
33.	Soybean (*Glycine max*)	ZnO and CeO_2	500, 1000, 2000, and 4000 mg L^{-1}	Petri plates containing germination paper	Root elongation	López-Moreno et al. (2010)
34.	Spinach (*Spinacia oleracea*)	TiO_2	0.25–4%	Glasshouse	Increased plant dry weight, chlorophyll formation, the ribulose bisphosphate carboxylase/oxygenase activity, and photosynthetic rate	Zheng et al. (2005)
35.	Spinach (*Spinacia oleracea*)	TiO_2	0.25%	Glasshouse	Enhanced photosynthesis	Hong et al. (2005)
36.	Tomato (*Solanum lycopersicum*)	MWNTs	10, 20, and 40 μg mL^{-1}	MS medium	Higher germination rate, increased biomass, and longer stems	Srinivasan and Saraswathi (2010)
37.	Tomato (*Solanum lycopersicum*)	MWNTs	40 μg mL^{-1}	MS medium	Enhanced seed germination	Villagarcia et al. (2012)
38.	Broccoli (*Brassica oleracea* var. *botrytis*)	ZnO nanorods		MS medium and greenhouse	Enhanced seed germination and biomass	Singhal et al. (2017)

2.1.1.1 NP-Induced Plant Growth

Several studies demonstrated the effect of metal NPs on plant growth. For example, AuNPs significantly enhanced seed germination of *Arabidopsis thaliana* (Kumar et al. 2013) and *Boswellia ovalifoliolata* (Savithramma et al. 2012), and zinc oxide nanorods enhance broccoli plant (Singhal et al. 2017). Similarly AgNPs increase root length of maize, cabbage (Pokhrel and Dubey 2013), and barley plants (Gruyer et al. 2013) and enhance a number of leaves, leaf area, and plant height in borage (Sah et al. 2011) and chlorophyll content of asparagus (An et al. 2008). Also, few other NP types interact with plants to alter their growth. Metal oxide NPs help tomato plant growth by controlling *Verticillium*, a wilt fungus (Elmer and White 2016). Carbon NPs are unique due to their thermal, electrical, and chemical properties. In an experiment on tomato plants exposed to multiwalled carbon nanotubes (MWCNTs), a significant improvement in seed germination was recorded due to NP-mediated water uptake from the cell wall of seed coat; also enhancement in vegetation biomass was recorded by the author (Khodakovskaya et al. 2009). The same has been reconfirmed on tomato (Villagarcia et al. 2012) and maize plants (Tiwari et al. 2014).

2.1.1.2 NP-Induced Phytotoxicity

The NP interaction with the plant is a poorly studied domain. Much of the interaction aspects are still unclear, especially with plant biomolecules; but it can be presumed that NPs produce a mixed effect on the plant. NPs interact with plant through dissolution, direct contact, and co-transport with other materials. Direct contact of NPs with plant influences their growth, germination, and reactive oxygen species (ROS) production (Prasad et al. 2017b). Most of the time, such interactions show a negative effect on the plant by lowering seed germination rate (El-Temsah and Joner 2012), root elongation (Gruyer et al. 2013), and germination index (Zafar et al. 2016). NPs show phytotoxicity inside the plant body by easily penetrating inside. For example, carbon-based NPs can easily enter leaves through stomata and penetrate chloroplast to be accumulated in thylakoids and stroma (Giraldo et al. 2014). NPs can induce cytotoxic effects on roots by chemically changing their vicinity or by blocking absorption. They can reduce the photosynthesis rate in plants. In an experiment on *Zea mays* magnetic Fe_3O_4NPs, the production of chlorophylls a and b and carotenoids was stimulated at 10–50 μL/L but inhibited at higher concentrations with overall low photosynthesis rate (Racuciu and Creanga 2007). Toxic heavy metals when associated with NPs alter their toxicological profile. Also, they alter pesticide uptake in plants (Mukherjee et al. 2016).

2.1.1.3 NP-Induced Genotoxicity

Genotoxicity refers to the effect on genetic material or alteration in gene expression of the cell that affects its integrity (Chichiriccò and Poma 2015). The toxic effects of NPs on plants are not perfectly measurable because the impacts depend on NP type, physicochemical properties, and the plant type. The initial report state that CuONPs damage DNA of some grassland plants (*Raphanus sativus*, *Lolium perenne*, and *Lolium rigidum*) (Atha et al. 2012) perhaps by accumulating inside the cells, causing DNA mutagenic lesions to inhibit the plant growth. In tomato plants, CNT-dependent activation of stress-related genes or genes regulating plasma membrane-based water channels has been shown (Cañas et al. 2008). In an experiment on *Allium cepa*, the ZnONP treatment caused increased mitotic index, lipid peroxidation, pyknotic cells, and micronucleated cells (Kumari et al. 2011). In a similar experiment on the same plant, ZnONPs induced loss of membrane integrity, increased chromosome aberrations, micronucleus formation, DNA strand breaks, and cell cycle arrest at the G2/M checkpoint (Ghosh et al. 2016). In another experiment, treatment of *Hordeum vulgare* L. with CeO_2NPs influenced ROS generation and ATP content (Mattiello et al. 2015).

2.1.2 NP Interaction with Microbes

It is proposed that bacteria may be exposed to natural NPs without adverse effect. Most of the studies focused on medical applications of NPs, especially antibacterial properties. Few other reports show that microbial communities are easily influenced by CuO or ZnO (Rousk et al. 2012); CeO_2, Fe_3O_4, or SnO_2 (Antisari et al. 2013; Pawlett et al. 2013); or Ag-based NPs (Gitipour et al. 2013; Mirzajani et al. 2013). The CuONPs inhibit and ZnONPs enhance pyoverdine levels in *Pseudomonas chlororaphis* O6 cells (Dimkpa et al. 2012). The Ag, CuO, and ZnONPs variably alter the major metabolic pathways of bacteria (Dimkpa et al. 2012). The antimicrobial role of AgNPs is well established in the scientific community (Kim et al. 2007; Swamy and Prasad 2012; Prasad and Swamy 2013; Aziz et al. 2015, 2016). However, their antimicrobial properties differ in field experiments because components present in soil negate their antimicrobial property which is proportional to the number of free Ag^+ as Ag ions get precipitated as AgCl to lose toxicity. Very less is known about carbon nanotube (CNT) interaction with microbial communities. Carbon nanomaterials prove toxic to soil microbes, alter nutrient availability to microbes, or may reduce the toxicity of metabolites (Dinesh et al. 2012). Although their exact mechanism is unknown, however, it is proposed that CNTs destroy microbial cell by interacting with cell membranes. Such interactions disrupt the membrane integrity by oxidative stress or physical damage (Jackson et al. 2013; Prasad et al. 2017b). The antimicrobial effect of NPs has opened a new avenue for the development of nanotechnology-based formulations for targeting plant pathogens (Dimkpa et al. 2012). They are attractive alternatives against agrochemicals, and their use to control plant pathogens indirectly helps plant growth. Most of these studies were generally done in liquid or solid defined bacteriological media under optimum growth conditions. However, it can be assumed that effect of NPs in environmental conditions will be more intense due to limited nutrient and sub-optimal growth conditions.

The effect of NPs on plants is dynamic and depends on soil and water components. The soluble metal NPs interact with plants to show particle-based as well as ion-related toxicities. Since plants rely on microbial activity for their growth and metabolism, the introduced NPs indirectly affect the root-associated soil microbial community (Dimkpa 2014; Sangeetha et al. 2017a). Also, NPs directly affect soil microorganisms by altering their metabolism through modification of synthesized metabolic products. Microorganisms mitigate such effects through the factors secreted on their surface, components of root exudates, and particles present in soil pores. Such metabolites help bacteria to interact with other soil microbes and plants by altering their stress responses. Most of the time, such alterations prove lethal to bacteria. NPs at sub-toxic level form complexes with microorganisms, which are detectable through sensors. Many of these microorganism-specific metabolites have been harnessed commercially. But, often such experiment shows variable results when reproduced due to the soil factor (Dimkpa et al. 2012).

2.2 Plant Disease Diagnosis Using NPs

Conventionally, plant diseases are identified visually by human raters followed by microscopic evaluation. Microbial diagnostic involves pathogen morphology (spore shape, color and arrangement, mycelium, and fruiting bodies). Pathogens later can be isolated and cultivated on specific culture medium for further diagnosis and studies. Such practices, which are routinely done in research labs and industries, are time-consuming, need experienced workers, are expensive, and require a laboratory. These bottlenecks have restricted their widespread use in disease diagnosis, especially in developing countries. In addition, plant diseases are identified by molecular, serological, and microbial diagnostics. A variety of direct methods are available for plant disease monitoring like nucleic acid detection based on polymerase chain reaction (PCR), fluorescence in situ hybridization, enzyme-linked immunosorbent assay (ELISA), immunofluorescence, flow cytometry, and indirect methods like thermography, fluorescence imaging, hyperspectral techniques, and gas chromatography (Fang and Ramasamy 2015). For example, molecular methods are used to detect phytopathogens and to study fungicide resistance in wheat. The preliminary information generated from such studies has helped in developing better fungicides and resistant cultivars. Molecular methods have also been applied to study phytopathogen populations and their way of interaction in plants (McCartney et al. 2003). Lateral flow ELISA was used for detecting *Phytophthora infestans* (late blight), *Ralstonia solanacearum* (brown rot), *Erwinia amylovora* (fire blight), *Pepino mosaic virus*, *Tomato mosaic virus*, *Potato virus Y*, and *Potato virus X* (Danks and Barker 2000). Future development in phytopathological disease diagnosis requires techniques that are automated, cheap, portable, precise, reliable, and sensitive. Most of the phytopathological research needs to be focused on developing plant disease diagnostic biosensors. The developed sensors can measure reflectance, temperature, and fluorescence of canopy (Sankaran et al. 2010) and are sensitive to detect changes in plant tissue color, leaf shape, transpiration rate, canopy morphology, plant density, and variation in wavelength of reflected light from the leaves (West et al. 2010).

New NP-based sensors are in the process of development that could diagnose plant pathogen cheaply, efficiently, and rapidly. They are equipping farmers to detect volatiles, chemical residues in crops, pathogens, and environmental changes

(Omanović-Mikličanina and Maksimović 2016). In agriculture, the commonly used NPs are carbon NPs, metal, and metal oxide NPs. Adoption of nanosensing devices along with new state-of-the-art strategies is sure to revolutionize the agriculture. Some of the nanosensor types are described below.

2.2.1 Diagnosis Through Metal NPs

Metal NPs are distinctively unique due to their high melting points, catalysis, toughness, and coloration (Singh et al. 2017). Metal NPs have a higher surface area to volume ratio and are sensitive in detecting pathogens with lower detection limits. In addition, detecting metal NPs in the form of electrochemical signals is simpler and cost-effective compared to enzyme assays. Due to these advantages, they are now replacing enzyme labeling system in phytopathogen diagnostics. The AuNPs and AgNPs along with ZnS, PbS, and CdS are in wide use for sample detection. The AgNPs are the most studied and used NPs for biosystem. AgNPs have the greater surface area, a higher fraction of surface atoms, and more antimicrobial properties compared to bulk silver. Also, AgNPs quickly disperse in water and are highly stable and effective in reducing microorganisms from planter soils and hydroponic systems. They are usable as foliar spray to check fungal and bacterial plant diseases. Also, they help plant growth by inducing disease and stress resistance. Based on the optical properties of AgNPs, some sensor has been developed. Such sensors detect color change of measured. For example, AgNPs show color change from yellow to brown between dispersed and aggregated forms. The intensity of color change can be correlated with the concentration of the analyte (Schofield et al. 2006). These sensors are used for detecting metal ions (Yoosaf et al. 2007), proteins (Schofield et al. 2006), and pesticides (Saini et al. 2017).

2.2.2 Diagnosis Through Metal Oxide NPs

Metal oxides possess a high density of edge surface sites and are used in several fields of plant science. Due to this advantage, metal oxide NPs are used as solid-state gas detecting devices which are applicable in domestic, commercial, and industrial sectors (Sun et al. 2012). With some modifications, such devices can detect pathogen-specific volatile metabolites. Due to additional advantages of lower cost, suitability for electron conduction, and easiness to be shaped at desired dimensions, they are used as sensors for volatile organic compound detection (Fang and Ramasamy 2015). For example, the ZnONPs are used for building gas sensors and biosensors (Sabir et al. 2014). The TiO_2NP and SnO_2NPs attached on carbon electrodes were used in detecting volatile p-ethylguaiacol secreted due to infection of pathogenic fungus *Phytophthora cactorum* in plants (Fang et al. 2014). Due to the extreme sensitivity of NP-based sensors, such devices are successful in detecting the volatile in ultralow concentrations.

2.2.3 Diagnosis Through Magnetic NPs

Magnetic NPs are distinctive in their features as their size is comparable to the dimensions of the magnetic domain. They display dual behaviors as single-domain ferromagnetism and superparamagnetism. The use of magnetic NPs in biomedical science is not a newer concept, but it is only partially explored in plant pathology. Carbon-coated magnetic NPs have been used to visualize the path, deposition, and transport of NPs inside the plant cells (González-Melendi et al. 2008). A novel NP immunoassay was developed to detect real-time mycotoxin level in plants, which was done using magnetic nanotags onto spin valve sensor surface immobilized with capture antibodies (Mak et al. 2010). A reliable and rapid ELISA method was developed using superparamagnetic NPs that reduced coating, enzyme blocking, and competition time of routine ELISA (Radoi et al. 2008). In a unique approach, the intracellular spaces of plant cells can be filled with magnetic NPs loaded with plant protection chemicals. With the help of strong magnets, such particles can be targeted to disease-specific sites in plants. They can be used to build a system to track the movement of internalized magnetic NPs, and to develop advanced, target-specific chemicals for release. For example, magnetic NPs coated with carbon were developed that can visualize the transport channel and deposition of NPs inside the plant (González-Melendi et al. 2008).

2.2.4 Diagnosis Through Polymeric NPs

The "polymer nanoparticle" term specifically refers to nanospheres and nanocapsules made from synthetic polymers like polyacrylamide, polyacrylate, polyanhydrides, and polycaprolactone or natural polymers like albumin, alginates, chitosan, DNA, gelatin, gliadin, and poly(L-lactides) (Chakravarthi et al. 2007; Prasad et al. 2017c). Polymeric nanomaterials are specifically used to prepare biosensors, whereas conductive polymers like polyacetylene, polyaniline, or polypyrrole convert biological signals to electrical signals (Sadanandom and Napier 2010). It has been reported that conjugated polymer nanoparticles are efficient in the transfecting vehicle for small interfering RNA (siRNA) delivery in plant protoplast. Other available transfection methods cause up to 40% loss of viable protoplast. However, CPN-assisted delivery causes only 5–25% protoplast loss during the first 24 h of delivery (Silva et al. 2010). Reports establish application of chitosan nanofibrous membrane in enzyme immobilization due to biocompatibility, high surface area, and large porosity (Bhatia 2016). Henceforth, biosensors can be developed from the chitosan-based system (Huang et al. 2007). Tyrosinase biosensor was used to detect phenolic compounds by employing carbon electrode modified with tyrosinase-Fe_3O_4 magnetic nanoparticles-chitosan nanobiocomposite film.

The role of glutathione as a signaling molecule in plant defense response is well established (Kovacs et al. 2015). Amperometric nanobiosensors have been developed to estimate glutathione concentration by covalently immobilizing glutathione oxidase onto the surface of gold-coated magnetic nanoparticle-modified Pt electrode (Chauhan et al. 2012). The combination of glutathione oxidase/chitosan/gold-coated magnetic nanoparticles was very sensitive in glutathione detection.

2.2.5 Diagnosis Through Quantum Dots

Quantum dots (QDs) are NPs with semiconductor properties that fluoresce when excited by light. They are nanocrystals made of semiconductor material with 2–10 nm range and are used in cell labeling, cell tracking, and DNA detection. QDs exhibit distinctive electronic properties of bulk semiconductors and are unique molecules due to the exceptionally high surface to volume ratio. Fluorescence is the most prominent property of QDs that produces unique colors depending on particle size. Such fluorophores have edge over organic fluorophores in accurately detecting nucleic acid or proteins concentration. QDs (1) have narrow emission peak, (2) have longer fluorescence lifespan, (3) are resistant to photobleaching, and (4) possess 10–100 times greater molar extinction coefficient. Due to these advantages, multicolor QDs can be excited from single fluorescent dye without emission peak overlapping, causing vivid color and brighter probes (Kashyap et al. 2016). QDs are widely used in medical science to detect specific biological markers, but their use in plant pathology is limited (Sharon et al. 2010). In an experiment on *Fusarium*, QDs were shown to be readily absorbed by the fungal hyphae, reflecting their potential in the development of novel control agents in minimum dosage (Rispail et al. 2014). QD-based highly sensitive nanosensor has been developed to detect *Phytoplasma* (*Candidatus Phytoplasma aurantifolia*) from infected lime trees (Rad et al. 2012). QDs were used in disease diagnosis by fluorescence resonance energy transfer (FRET) mechanism through energy transfer between two light reactive molecules (Grahl and Märkl 1996). The method was later adopted for detection and identification of *Aspergillus amstelodami* (Kattke et al. 2011).

2.2.6 Diagnosis Through Carbon NPs

Carbon is one of the most abundant elements on Earth and is the backbone of most biochemical activities. Carbon has some unique advantages over other elements. As a result carbon NPs are the most used NPs among all the synthesized nanomaterials. CNTs are cylindrical carbon allotropes that help plant growth by increasing water uptake. Due to the unique electrochemical properties, CNTs are used as components in biosensors. Their properties include larger length to diameter ratio, mediation of fast electron transfer kinetics (Yu et al. 2003), and ability to attach with any chemical species (Balasubramanian and Burghard 2006).

The CNTs are being employed for plant disease diagnosis which can sense plant metabolites, as all phytopathological diseases can be associated with a change in metabolism of aromatic compounds (Farkas and Kiraaly 1962). Among volatile metabolites, phenolic compounds play a major role in plant disease suppression. Due to the ease of oxidizability, phenols are detected using amperometric and potentiometric devices. However electrochemical sensors harbor a major drawback of electrode corrosion due to phenol exposure which forms dimeric or polymeric oxidation products (Balasubramanian and Burghard 2006). The use of electrodes made of CNT reduce electrode surface fouling effect by accumulating biomolecules like nucleic acids (Wang et al. 2003a, b). As a result, the CNTs either solely or with enzymes are used in phenolic compound detection.

Indole-3-acetic acid (IAA) is the major plant growth-regulating hormone that is associated with countless developmental processes. Sensors for IAA detection have been developed using coating DHP (dihydropyran)-stabilized MWCNT dispersion onto a carbon electrode (Wu et al. 2003). A wide range of electrochemical nanosensors like amperometric enzyme electrodes and DNA hybridization biosensors have been prepared using CNTs.

2.3 Nanodiagnostic Sensors and Equipment

Early detection of crop pathogens is crucial in agriculture for managing polycyclic diseases, which requires effective identification methods and sensitive devices. Plant diseases are generally detected by measuring plant stress response metabolites (jasmonic acid, methyl jasmonate, and salicylic acid), volatiles, molecular or serological methods to detect the pathogen, or analysis of host-induced biomarkers like transcripts, proteins, and volatiles (Martinelli et al. 2015).

Such nanosensors can be used for pathogen detection or through the compounds secreted by pathogens. For example, nanosensors with gold electrode and CuNPs have been developed to detect plant pathogenic fungus *Sclerotinia sclerotioriim*. The nanosensor measures the rate of production of salicylic acid through which the intensity of infection can be estimated (Wang et al. 2010).

2.3.1 Portable Diagnostic Equipment

The requirement of early diagnosis always inspired us to develop effective diagnostic equipment that may detect the minutest of disease traces. These devices should (1) contain competitive sensors, (2) accommodate entire on-chip processing steps for sample preparation, and (3) contain least manipulation steps (Weigl et al. 2008). Due to technological advances made during the last decade, portable plant disease diagnostic devices are becoming a reality in agriculture. Portable devices in the form of immunoprinting kits, lateral flow devices (LFDs), portable PCR, portable genome sequencers, nanodiagnostic kits, and loop-mediated isothermal amplification (LAMP-PCR) are available. These tools are simpler to use, robust, and quicker. But often such devices are costly and not easily available and are mostly under development stage (Khiyami et al. 2014).

Grain spoilage due to pathogen attack is the leading cause of crop loss faced by farmers. The CO_2-based sensors are already in use to check grain spoilage (Neethirajan et al. 2009). A nanosensor has been developed for detecting grain spoilage at very early stage due to sensitivity toward CO_2 emission and odor-causing chemicals from grains at the level of parts per billion (Joyner and Kumar 2015; Neethirajan et al. 2010). The sensor contains a chip that detects insect or fungus causing the spoilage. Another type of sensor, a fabric sensor, is also under development that warns farmers over their mobile phones at the time of pathogen attack and can even start pathogen-specific remediation.

2.3.2 Nanofabrication Imaging

Nanodiagnostic imaging tools can visualize plant tissues and cells to diagnose phytopathogens. The new device has broadened the diagnostic capabilities available to a phytopathologist. This has helped in early detection of diseases (Rosen et al. 2011) by modulating the physical and chemical properties of NPs, which enhanced the contrast, imaging time, tissue specificity, and signal strength. These imaging techniques are important in studying the nature of pathogen, how it interacts with host and starts infection process. A better understanding of the disease mechanism will help to develop effective chemicals to circumvent the effect of pathogens (Meng et al. 2005).

2.4 Types of NPs Used in Nanodiagnostics

2.4.1 Nanobarcodes

Barcodes are tags used for fast, accurate, and easy identification of goods. Likewise, nanoparticle-based barcodes can be used for nanoscalar disease detection by tagging pathogens. Nanobarcodes (NBs) are in use for multiplexed disease detection in agriculture and ecology sectors. They are battery-powered portable devices usually integrated with computing devices, multiplexed with several pathogen detectors. Such devices are still in the developmental phase and have the drawback of limited pathogen detection. Once their development completes, such an interface will help researchers and end farmers to track several species on the real-time basis. These systems if applied with Global Positioning System (GPS) would provide a complete picture of the specific area and will be applicable in getting a holistic view through remote sensing. NBs are already in use for tagging agricultural products. Such systems have equipped us with auto-ID technologies which were practically not achievable with conventional methods (Branton et al. 2008). Microscopic probes or NBs have been developed that can tag multiple phytopathogens and which are easily detectable through fluorimetry. A group of scientists has developed quantum dot (Qd) nanobarcodes useful in quantifying gene expression in multiplexed format, which has wide applications in phytopathology (Eastman et al. 2006).

2.4.2 *Nanosensors*

The demand for onsite and real-time and sensor-based pathogen detection is expanding due to dynamic changes in plant-pathogen types. The biosensor is a device used to detect a sample through a sensitive component, transducer, along with a detector which converts the recognition into an electrical signal (Mody 2011). Although sensors for detecting antibiotic and complementary DNA sequence have been developed for detecting pathogen-infected cells, however, the incorporation of nanomaterials in building biosensors has enhanced its sensitivity, which has helped in devising novel signal detection procedures. Methods have been developed to detect DNA or protein-functionalized gold NPs to be used as target-specific probes. Such methods are based on direct-charge transfer conductometric biosensor (Pal et al. 2008), CNTs (Serag et al. 2013), and silver and gold NPs (Sadowski 2010). Gold NP biosensors can detect pathogens through several optical or electrochemical procedures. Silica-based NPs have been developed that fluoresce under UV radiation when attached to microbial surface antigen-specific antibodies. Some of these techniques are sensitive enough to detect a single microbe (Zhao et al. 2004). A CO_2 sensor was developed using polyaniline boronic acid conducting polymer for detecting real-time spoilage of stored grain. It works on the concept that during grain spoilage, CO_2, heat, and moisture are produced due to the metabolism of grain, pests, and microorganisms. Such sensors are ultrasensitive to CO_2 detection in the range of 380–2400 ppm (Neethirajan et al. 2010). Advances in nanotechnology have influenced enzyme immobilization technology by improving biosensing abilities. The conjunction of NPs with enzymes has enhanced the sensitivity and stability of biosensors. Several NP-based enzymatic biosensors like nanofibers, nanocomposite, graphene, and nanotubes have been developed. Enzyme biosensors are available for detecting organophosphorus and non-organophosphorus pesticides (Zhang et al. 2015). In an experiment, acetolactate synthase-inhibitor herbicides metsulfuron-methyl and imazaquin were detected by atomic force microscope tip functionalized with acetolactate synthase (da Silva et al. 2014). Another nanobiosensor type was made by entrapping acetylcholinesterase in the liposome setup which successfully detected organophosphorus pesticides paraoxon and dichlorvos (Vamvakaki and Chaniotakis 2007). The aforesaid systems provide protection for enzymes inside the nanoenvironment. Nanosensors based on CNT-immobilized enzymes are also in use. For example, nanobiosensor-based MWCNT-immobilized acetylcholinesterase was introduced for pesticide detection (Du et al. 2007). The same setup was used to detect aflatoxin through MWCNT-immobilized aflatoxin oxidase (Chuan Li et al. 2011). Such sensors have a higher sensitivity limit of pesticide residue detection. Due to the higher surface area and catalytic mode of nanomaterial-immobilized enzymes, the stability and reproducibility of the process have improved. Such ultrasensitive enzymatic nanobiosensors have a pesticide detection limit of 50 pg per liter (Verma 2017).

2.5 Nanotechnology in Microbial Plant Pathogen Management

Among plant pathogens fungi bear prime importance as they encompass more than 70% of all major crop disease (Agrios 2012). Farmers continually battle to protect their crops from microbial diseases. For the purpose, they rely on agrochemicals which are costly, toxic to health, and environment unfriendly. Also, such practices are augmenting rather than impeding the situation. For example, pest resistance is now a serious worldwide problem that has increased with pesticide use. This is the high time to reevaluate conventional agricultural practices and replace them with novel alternatives (Bhattacharyya et al. 2016; Ismail et al. 2017).

Agronanotechnology is one such alternative that can mitigate problems in agronomy (Sangeetha et al. 2017a). Plant disease can be controlled and pathogens targeted using NPs or nanotechnology-based tools. Studies have shown the use of NPs in controlling fungal pathogens like *Alternaria alternata* (Kim et al. 2012), *Aspergillus flavus* (Rajiv et al. 2013), *A. niger* (Rajiv et al. 2013), *Bipolaris sorokiniana* (Jo et al. 2009), *Botrytis cinerea* (He et al. 2011; Park et al. 2006), *Colletotrichum* sp. (Lamsal et al. 2011), *Fusarium oxysporum* (Kim et al. 2012), *Magnaporthe grisea* (Jo et al. 2009; Park et al. 2006), *Penicillium expansum* (He et al. 2011), and *Pythium spinosum* (Kim et al. 2012) and bacterial diseases including *Azotobacter chroococcum*, *Bacillus subtilis*, *Pseudomonas syringae*, *Rhizobium tropici*, *Xanthomonas campestris* (Park et al. 2006), and *X. perforans* (Ocsoy et al. 2013). The AgNPs have received great attention due to their multiple ways of inhibitory activities against phytopathogens (Bhaskar et al. 2016; Park et al. 2006; Rabab and EL-Shafey 2013). Silver finds tremendous potential in the antifungal industry due to its safeness over other available fungicides. Silver ions are used in many formulations as Ag and Ag-based composites and are highly toxic to microorganisms. The AgNPs with ampicillin hyperbranched macromolecules are effective in the antimicrobial surface coating. The smaller nanosilver (1–5 nm) are more effective in fungal growth suppression as they easily pass through the fungal cell membrane (Wainwright et al. 1986). The effect of nanosilver increases multifold as it enters the fungal cell, providing dynamic resistance against disease by forming a physical barrier against pathogenic fungi (Kim et al. 2002). Nanosilver application prevents fungal recurrence for a long time. These particles can suppress disease at concentrations as low as 3 ppm (Huang et al. 2015). Besides, NPs of copper, zinc, titanium, magnesium, gold, and alginate have been tested. Use of copper-based NPs dissolved in water is in practice since the early 1930s against grapes and fruit diseases (Hatschek 1931). About 15 nanoforms of micronutrients along with $CuSO_4$ and $NA_2B_4O_7$ were effective against rust disease of peas (Singh et al. 2013). In a study, SiO_2NPs were reported successful in reducing the severity of *F. oxysporum* and *A. niger* on maize plants (Suriyaprabha et al. 2014). Titanium (TiO_2) is a nontoxic compound widely used in the paint industry, cosmetics, ceramics, etc. There is the tremendous potential of TiO_2 in agriculture due to its nontoxicity against plants. The antifungal property of ZnONPs has been reported against fruit mold *P.*

expansum and *Botrytis cinerea* (He et al. 2011) and *A. flavus* and *A. niger* (Jayaseelan et al. 2012). ZnONPs show less toxicity in comparison with AgNPs to the plant and can be applied as nanopesticide. Moreover, the antibacterial activity of ZnONPs has also been reported (Padmavathy and Vijayaraghavan 2008; Bhuyan et al. 2015). The CuNPs with soda-lime glass powder is antimicrobial toward some Gram-positive and Gram-negative bacteria and some fungi (Esteban-Tejeda et al. 2009). Polymer-based nanocomposites of copper NPs are effective against several plant pathogenic fungi (Cioffi et al. 2004). Table 2.2 shows the list of NPs effective against fungal plant pathogens.

The combination of NPs with biocontrol agents, essential oils, and biopolymer-based cupric and sulfur NPs could show a synergistic effect to enhance antimicrobial activity and reduce their net quantity of usage. In some studies, synergism was obtained when antimicrobial NPs were combined with bioorganic pesticides (Dar and Soytong 2014; Xue et al. 2014). Such combinational studies may help in reducing the overall dosage of pesticides and could delay pest resistance. Accommodation of active ingredients that inactivate themselves with solar radiation encourages the development of green "nanocides" with synergistic pesticide activity with pesticide residue photocatalytic degradation ability. Development of such smart delivery system holds the key for improving fungicide efficiency in agriculture for better and targeted crop protection. For example, nano-dispersed formulations as fungicide can be prepared very cheaply, which will be suitable for developing fungicide on a mass scale. Moreover, such formulations will be less toxic than conventional agrochemicals used in plant protection.

Development of resistant plant varieties through breeding or genetic engineering can help in combating fungal pathogens. Use of nanodiagnostic methods (nanofluidics, nanomaterials, bioanalytical nanosensors) in plant breeding (Abd-Elsalam 2015) or nanoparticle-mediated gene transfer to improve crop varieties for disease resistance (Rai et al. 2012) can minimize the expenditure on agrochemicals used in disease management. Also, NP-based plant genetic transformation can be applied (Rai et al. 2012). The use of NPs in plant pathology opens new avenues for plant protection, pathogen detection, and studying plant-pathogen interaction for effective plant disease management (Ismail et al. 2017).

2.6 Nanotechnology in Insect Pest Management

Due to population explosion, the present agricultural system is facing hurdles in meeting the food supply and is limited by several abiotic and biotic factors. Environmental impacts on agriculture like storms, drought, and floods add to the abiotic cause. Among biotic factors, the insect pest is the important limiting component that severely damages crops leading to a loss in the global economy. Insects

Table 2.2 List of NPs effective against fungal plant pathogens

S#	NP	Size (nm)	Quantity used	Application	Plant pathogenic fungi	References
1.	Ag	~20	0.12 10 ppm	In vitro	*Fusarium culmorum*	Kasprowicz et al. (2010)
2.	Ag	10–50	5, 10, and 15 mg L^{-1}	In vitro	*Alternaria alternata*, *Sclerotinia sclerotiorum*, *Macrophomina phaseolina*, *Rhizoctonia solani*, *B. cinerea*, and *Curvularia lunata*	Krishnaraj et al. (2012)
3.	Ag	10 ± 5	50, 100, and 150 ppm	In vitro	*Alternaria alternata*, *A. citri*, *Penicillium digitatum*	Abdelmalek and Salaheldin (2016)
4.	Ag	4–8	10, 30, 50, 100 ppm	In vivo (pepper plant) and in vitro	*Colletotrichum* sp.	Lamsal et al. (2011)
5.	Ag	10–50	5, 10, 15 mg/10 µl	In vitro	*Alternaria alternata*, *Botrytis cinerea*, *Curvularia lunata*, *Macrophomina phaseolina*, *Rhizoctonia solani*, *Sclerotinia sclerotiorum*	Krishnaraj et al. (2012)
6.	Ag	20	0.12–10 ppm	In vitro	*Fusarium culmorum*	Kasprowicz et al. (2010)
7.	$AgNO_3$ and AgCl	20–30	25–200 ppm	In vitro and growth chamber	*Bipolaris sorokiniana* and *Magnaporthe grisea*	Jo et al. (2009)
8.	AgNPs	5–27	6 µg per well in agar containing petri plates	In vitro	*Amylomyces rouxii* strain KSU-09	Musarrat et al. (2010)
9.	Cu	30 ± 5	2.6 and 4 wt%	In vitro	*Escherichia coli*, *Micrococcus luteus*, *Issatchenkia orientalis*	Esteban-Tejeda et al. (2009)

(continued)

Table 2.2 (continued)

S#	NP	Size (nm)	Quantity used	Application	Plant pathogenic fungi	References
10.	Cu	7–25	10,000–50,000 μl mL^{-1}	In vitro	*Alternaria alternata, A. brassicicola, A. solani, Botrytis cinerea, Cladosporium cucumerinum, Corynespora cassiicola, Cylindrocarpon destructans, Didymella bryoniae, Fusarium oxysporum* f.sp. *cucumerinum, F. oxysporum* f. sp. *lycopersici, F. oxysporum, F. solani, Fusarium* sp., *Glomerella cingulata, Monosporascus cannonballus, Pythium aphanidermatum, P. spinosum, Stemphylium lycopersici*	Kim et al. (2012)
11.	CuO	20	250, 500, and 750 mg kg^{-1}	In vitro	*Botrytis cinerea, Colletotrichum graminicola, Colletotrichum musae, Magnaporthe oryzae, Penicillium digitatum, Rhizoctonia solani, Sclerotium rolfsii*	Huang et al. (2015)
12.	CuO and Cu_2O	11–55	35, 54, 110, and 224 g hl^{-1}	In vivo (foliar)	*Phytophthora infestans*	Giannousi et al. (2013)
13.	SiO_2	----		In vitro	*Blumeria graminis*	Huang et al. (2015)
14.	ZnO	70 ± 15	0, 3, 6, and 12 mmol l −1	In vitro	*Botrytis cinerea* and *Penicillium expansum*	He et al. (2011)
15.	ZnO	27 ± 5 (spherical) and 84 ± 2 (hexagonal)	25, 50, 100 μg mL^{-1}	In vitro	*Aspergillus flavus, A. niger, A. fumigatus, Fusarium culmorum,* and *F. oxysporum*	Rajiv et al. (2013)
16.	ZnO	57.72	1.2–2.9 μg mL^{-1}	In vitro	*A. flavus, A niger*, and *C. albicans*	Jayaseelan et al. (2012)

(continued)

Table 2.2 (continued)

S#	NP	Size (nm)	Quantity used	Application	Plant pathogenic fungi	References
17.	ZnO	57.72	1.2–2.9 μg mL^{-1}	In vitro	*Aeromonas hydrophila*, *Enterococcus faecalis*, *Pseudomonas aeruginosa*, *Staphylococcus aureus*, *Streptococcus pyogenes*	Jayaseelan et al. (2012)

cause more damage among all plant-pathogen types leading to 14% loss which equals $2000 billion per year (Pimentel 2009). To counter crop damage, farmers heavily rely on traditional practices like crop rotation, improved crop varieties, changing crop sowing dates, and integrated pest management (IPM). But the conventional methods for insect pest management are now losing efficiency. The associated problems with chemical pesticide use like environmental degradation, pest resistance, and low efficiency of traditional methods are pushing researchers to discover novel ways to counteract pests. The implementation of such methods can be more efficacious in crop protection and could reduce the pesticide input through host-specific targeted delivery of the agents (Prasad et al. 2014, 2017a, b).

2.6.1 Development of Nanoformulations

The requirement for early diagnosis of plant diseases is compelling us to explore "nano-solutions" for safeguarding food and agriculture from pests. There is a requisite to develop techniques that are simpler, portable, accurate, and less time-consuming and do not need complicated procedures. In a step further, integration of GPS for real-time monitoring of soil and field condition of the crop can help in getting a holistic view of crop pattern and disease outspread. Such technology could be made possible by infusing knowledge of biotechnology, electronics, bioinformatics, etc.

These techniques can help in the targeted delivery of pesticides and agrochemicals (Gogos et al. 2012). Although, nanomaterial-based pesticide formulations are available past few decades, however, the focus has now shifted in developing controlled release formulations (CRFs) containing active compounds with some inert material. The inert compound is required for controlled release of the desired product for the required length of time. The release rate of such formulations entirely depends on the chemical nature of agents, i.e., chemical properties of NPs and inert material, their particle size, and the strength of chemical bonds between them. They are released by breaking chemical interaction between insecticide and polymers, which is achieved by hydrolytic reactions that affect polymer-insecticide bounds in a chain reaction. Further study is required for finding the mechanism through which such inert materials control the chemical release. The "slow release" concept envisions surety of pesticide availability after delivery in the field. Such smart delivery system will not let buildup of pesticide surge during field application, which could be toxic to other non-harmful insects or crops. These field-applied pesticides easily end up in the human food chain at toxic levels. By using nanoencapsulation, pesticides can be trapped inside them for requirement-based controlled delivery. Such nanomaterials are available in the form of nanospheres, nanocapsules, nanogels, and micelles (Ragaei and Sabry 2014). There is a need to develop formulations containing insoluble compounds that can be readily dispersed in aqueous solutions. Such formulations due to their solid nature may have reduced drift or leaching in agricultural fields. Such smart delivery agents can afford to be comparatively more selective to insect pests, and microencapsulation ensured the protection of their bioactive compounds (Peteu et al. 2010).

Nanosilica holds several potential benefits as a pesticide. It easily gets absorbed into cuticular lipids of insects causing their death by physical means because lipid-based insect cuticle retains water balance inside and is impermeable to water. However, no side effects on photosynthesis and leaf and stem surface are observed by their application on plants (Barik et al. 2008). In addition, they do not cause gene expression inside insect gut. Nanosilica can be used to control a variety of insect diseases. In a remarkable study, it has been found that insects use Earth's magnetic field to find directions for food and shelter (Riveros and Srygley 2010). These signals are received by insects with the help of magnetic NPs (Abraçado et al. 2005; Wajnberg et al. 2010) present in the abdomen, thorax, and head (Wakeil et al. 2017). The specific insets possess specific kind of nanoparticle in their body parts. In a unique approach, insect pests can be targeted with magnetic nanomaterials to alter their biological magnetic sensors.

A smart agriculture is an emerging concept in modern farming. By using smart nanosensors, the precise amount of plant-specific fertilizers and insecticides can be delivered. Nanosensor integration with the smart delivery system can help to efficiently manage agricultural resources. Designing diagnostic nanosensors that can be integrated with pesticide delivery system for automatically combating insect attack, fungal infection, or drought could revolutionize the agriculture field. Porous hollow silica NPs (PHSNs) complemented with a pesticide have been tested with success in agriculture for prolonged controlled delivery of a chemical agent (Liu et al. 2006).

2.6.2 Nanopesticides

Nanopesticides are crystalline structures which are thermally stable, water-soluble, and biodegradable. Nanopesticides deal with tiny or altered particles with lethal activity against pests. There is an array of nano-chemicals or agents known that are proven pesticides. The synthesis of nanoporous zeolites that can slowly and efficiently release active components on target has opened new horizons in agricultural productivity. In addition, they seem to have the capability to reduce the carbon footprint on the environment. However, our limited understanding of the mode of action is limiting their commercial release.

The great volume of active ingredients of pesticides have limited water solubility and therefore need a delivery system (Kah and Hofmann 2014). In addition, they may be unstable in the environment and need protection from degradation. The size of NPs in nanoformulation also matters for efficient pesticide delivery, as we know that NPs' surface area to volume ratio increases with a decrease in size. Reports demonstrate rapid release of formulation containing very small NPs (Ao et al. 2012), which is not desirable when the slow release of the formulation is required. The designing of better NPs requires thorough understating of their interaction with active agents. It should be kept in mind that the release of active ingredients is influenced by the length of polymer chains (Sarkar et al. 2012), gum to chitosan ratio (Abreu et al. 2012), and cellulose content of nanocrystals (Xiang et al. 2013). Due to the associated problems of nanopesticide delivery and stability, efforts are underway to develop hybrid formulations like encapsulation of nanoemulsions (Jerobin et al. 2012) or liposome coatings (Kang et al. 2012). Novel formulations are available in the form of nanoemulsions, polymer-based nanopesticides comprising of nanospheres, nanogels, and electron spin nanofibers (Kah and Hofmann 2014). To enhance nanopesticide efficacy, either inorganic nanoparticles are associated with organic active ingredients or inorganic nanoparticles as such are active ingredients. The organic active ingredients accommodate mesoporous silica or titanium dioxide associated with polymer matrix. The inorganic NPs include silica, titanium dioxide, silver, copper, and aluminum (Kah and Hofmann 2014). The silica-based NPs are considered biologically active and environmentally safer for nanopesticide development. The silica NPs easily enter inside insects through their hairs; also insects feed on pesticide-filled nanotubes. Such formulations are commercially available from leading pesticide companies. Moreover, the silica-based nanoparticle has been reported to deliver DNA in a plant system, thus creating a cheaper yet powerful tool for targeted plant delivery.

2.7 Nanotechnology in Nematode Management

Nematodes are the most abundant soil organisms that attack several agronomy plants, causing global loss of 125 billion dollars per year in tropics (Chitwood 2003). They feed on plant roots to cause nutrient loss, poor crop yield, and sometimes death of plant (Bhau et al. 2016). Nematode-infected plants are prone to secondary infections from pathogenic bacteria and fungi. Very few nematode-resistant varieties are available as plant breeder needs several years to develop nematode-resistant variety.

Traditionally crop nematodes are controlled through crop rotation or by sowing trap crops, which are later burnt to kill trapped nematodes. Although chemical treatment using nematicides is an effective way to combat nematodes, however, the practice is highly toxic to plants. Moreover, chemical methods do not provide long-term protection against nematodes (Bhau et al. 2016). Very few workers have reported NPs to control nematodes, but the trials were mostly encouraging (Cao et al. 2015; Huang et al. 2008; Kuźniar et al. 2011; Li et al. 2012; Ma et al. 2009; Roh et al. 2010; Wang et al. 2009). In an experiment on nematicide, abamectin, encapsulated with NPs produced from the red clover necrotic mosaic virus on tomato plant, caused a significant reduction in nematode disease incidence (Cao et al. 2015). Abamectin is a combination of two nematicides, avermectin B1a and avermectin B1b, in 4:1 ratio. Abamectin toxicity is based on its action on γ-aminobutyric acid that blocks the neural signal transmission of insects at the neuromuscular junction, causing paralysis and death (Khalil 2013). Nematicidal effects of silver NPs on root-knot nematode (*Meloidogyne* spp.) in Bermuda grass has been established (Cromwell et al. 2014). Similar effects were reported with NPs of ZnO (Ma et al. 2009), CeO_2 and TiO_2 (Roh et al. 2010), and Al_2O_3 (Li et al. 2012). NPs of TiO_2, ZnO, Al_2O_3, and silver are effective against *Caenorhabditis elegans* (Wang et al. 2009). In a unique but interesting study, magnetic Fe_3O_4NPs reduced the mobility and caused differential gene expression and apoptosis in nematode *C. elegans* (Huang et al. 2008).

2.8 Nanotechnology and the Crop Protection Industry

Agriculture is the backbone of most developing nations. The development of such countries could be envisioned if they adopt novel agricultural technologies for smart and future-ready agriculture. Take phytopathology, for instance. It is estimated that plant pathogens damage 10–20% crop yield, causing considerable loss in the global economy. The mounting pressure for increased food production is compelling us to discover novel ways of disease management. Use of nanotechnology tools for targeted delivery of pesticides and fertilizers is paving the path for low waste generation through "precision farming," a concept which has appeared recently.

Nanotechnology can help in achieving the future goal of "precision farming" for maximizing crop yield with minimal input. Precision farming involves monitoring environmental factors in real time and inferring the targeted action, which is done by integrating wireless networking, nanosensors for regulating agricultural practices (Prasad et al. 2014, 2017a). This requires the use of computers, GPS, and remote sensing devices to sense localized environmental conditions, drought, and pests. This site-specific crop management uses "smart dust" comprised of sensors, robots, and transponders that are regulated through a wireless computer network. On detection of pests or drought, the automated pesticides can be applied at irrigation levels. Precision farming involving nanosensors can reduce environmental pollution due to agricultural practices. Such nanosensors will have a strong impact on future farming by helping farmers to make better decisions. LOFAR_Agro has adopted the technique to measure microclimate of potato crop to regulate *Phytophthora* infection (Dwivedi et al. 2016).

For field application, the crop targeted delivery of NP-based agrochemicals is important, which can be achieved with the aid of nanocapsules that are highly stable and biodegradable structures (Jha et al. 2009). The growing use of NPs in agriculture is enhancing the effectiveness of chemical pesticides by targeted delivery, safety, disease, and adherence of pesticide which is eventually reducing the cost involved in plant protection (Anwunobi and Emeje 2011; Prasad et al. 2014, 2017a). Chemical agents mixed with polymers or encapsulated on metal surfaces provide sound delivery systems. Several companies have formulated NPs of 100–250 nm range that can effectively dissolve in water. NPs are being used as a smart pesticide or fungicide delivery agents in agriculture. Several oil-based or water-based nanoemulsions of the range 200–400 nm are also available that have multi-applications in plant pathology (Goswami et al. 2010). Such agents improve pesticide stability and are less toxic and less likely to leach from the soil. NPs added with a constituent for slow release can reduce the frequent application of chemicals. Nanotechnology as a modern science is continually evolving to accommodate enzymes, metabolites, and the entire cell along with nanostructures to develop hybrid systems that are applicable in agriculture and pest control (Bailey et al. 2010).

Agronanotechnology can turn agriculture into more sustainable business. The application of such technologies for plant disease management, gene delivery, development of nanopesticides, and encapsulation will revolutionize the agriculture industry. These technologies are the promising future of agriculture and have potential to reduce the carbon footprint generated from agricultural practices (Sangeetha et al. 2017a).

2.9 Agronanotechnology Limitations and Issues

While dealing with IPM we should look for even better alternatives because in the near future the national regulatory bodies may limit and control the availability of already available agrochemicals (Srinivasan and Tung 2015). As an alternate, nanotechnology has shown several potentials but has not yet been satisfactorily commercialized in agriculture. The farming community seems disinterested in nanotechnology-based biochemicals due to lack of sufficient economic returns. Other factors responsible for sluggish growth of nanotechnology include (1) disinterest from government (Mukhopadhyay 2014), (2) more attention on conventional farming practices, (3) lack of competition in agriculture sector, (4) less fascination of youth, (5) higher input cost, (6) lack of proper farming knowledge, and (7) disinterest in public (Mukhopadhyay 2014).

Agronanotechnology has tremendous potential to alleviate farmer lives by facilitating agriculture. It is important to make people aware of this science to enhance its acceptability that could accelerate the discovery of newer applications in this domain. For nanotechnology to penetrate agriculture, there is a requirement of well-trained human resources who could translate the available techniques into tools of nanotechnology. The nanotechnology as a subject is taught in engineering stream which is not agriculture-specific. There is an urgent need for curricula that may amalgamate nanotechnology with agriculture to cater the agronanotechnology research. Further developments in agronanotechnology require a high degree of multidisciplinary and cross-sector collaboration within and between academic researchers, industry, and government. The application of such technologies in plant disease management, gene delivery, development of nanopesticides, and encapsulation can revolutionize the agriculture industry (Sangeetha et al. 2017a, b). Once it develops as full science, it will need newer tools for visualization, characterization, and fabrication, which is not possible without multidisciplinary collaboration.

In addition, some issues associated with NPs are yet to be resolved, like NPs floating in the air can deposit on leaves and flowers causing blockage of stomata, creating a toxic barrier on stigma to prevent pollen germination and tube penetration in stigma. NPs can also alter plant vascular transport of water, minerals, and photosynthesis (Khan and Rizvi 2014).

2.10 Conclusion

Human population is exploding at an unprecedented rate from the past few decades, causing expansion of industries, shrinkage of arable land, and land urbanization. To feed billions of people, the current agricultural practices like plant breeding and IPM are not sufficient and need smart alternatives that could match our current and future food demands. It is worth investing in novel agronanotechnology science which is only a decade or two old. By employing NPs we can reduce input on plant protective chemicals, minimize nutrient loss, and enhance crop yield. The technology is sufficient in alleviating problems of higher chemical input cost, poor pesticide efficiently, and pesticide contamination in land and groundwater. For example, zerovalent iron NPs could be employed for remediation of pesticide-infested soil as they possess high absorption affinity toward heavy metals and organic compounds. Moreover, FeNPs harbor wonderful soil-binding qualities like $CaCo_3$. In addition, to reduce the environmental footprint, more emphasis should be provided on the use of agricultural waste products as raw material for NM production.

References

Abd-Elsalam KA (2015) Nanodiagnostic tools in plant breeding. Ommega Publishers 2(2):1–8

Abdelmalek GAM, Salaheldin TA (2016) Silver nanoparticles as a potent fungicide for citrus phytopathogenic fungi. J Nanomed Res 3(5):1–8

Abraçado L, Esquivel D, Alves O, Wajnberg E (2005) Magnetic material in head, thorax, and abdomen of *Solenopsis substituta* ants: a ferromagnetic resonance study. J Magn Reson 175(2):309–316

Abreu FO, Oliveira EF, Paula HC, de Paula RC (2012) Chitosan/cashew gum nanogels for essential oil encapsulation. Carbohydr Polym 89(4):1277–1282

Agrios G (2012) Plant Pathology. Academic Press, London

An J, Zhang M, Wang S, Tang J (2008) Physical, chemical and microbiological changes in stored green asparagus spears as affected by coating of silver nanoparticles-PVP. LWT- Food Sci Technol 41(6):1100–1107

Antisari LV, Carbone S, Gatti A, Vianello G, Nannipieri P (2013) Toxicity of metal oxide (CeO_2, Fe_3O_4, SnO_2) engineered nanoparticles on soil microbial biomass and their distribution in soil. Soil Biol Biochem 60:87–94

Anwunobi A, Emeje M (2011) Recent applications of natural polymers in nanodrug delivery. J Nanomed Nanotechnol 4:2–7

Ao M, Zhu Y, He S, Li D, Li P, Li J, Cao Y (2012) Preparation and characterization of 1-naphthylacetic acid–silica conjugated nanospheres for enhancement of controlled-release performance. Nanotechnology 24(3):035601

Arora S, Sharma P, Kumar S, Nayan R, Khanna P, Zaidi M (2012) Gold-nanoparticle induced enhancement in growth and seed yield of *Brassica juncea*. Plant Growth Regul 66(3):303–310

Atha DH, Wang H, Petersen EJ, Cleveland D, Holbrook RD, Jaruga P, Dizdaroglu M, Xing B, Nelson BC (2012) Copper oxide nanoparticle mediated DNA damage in terrestrial plant models. Environ Sci Technol 46(3):1819–1827

Aziz N, Faraz M, Pandey R, Sakir M, Fatma T, Varma A, Barman I, Prasad R (2015) Facile algae-derived route to biogenic silver nanoparticles: synthesis, antibacterial and photocatalytic properties. Langmuir 31:11605–11612. https://doi.org/10.1021/acs.langmuir.5b03081

Aziz N, Pandey R, Barman I, Prasad R (2016) Leveraging the attributes of Mucor hiemalis-derived silver nanoparticles for a synergistic broad-spectrum antimicrobial platform. Front Microbiol 7:1984. https://doi.org/10.3389/fmicb.2016.01984

Bailey K, Boyetchko S, Längle T (2010) Social and economic drivers shaping the future of biological control: a Canadian perspective on the factors affecting the development and use of microbial biopesticides. Biol Control 52(3):221–229

Balasubramanian K, Burghard M (2006) Biosensors based on carbon nanotubes. Anal Bioanal Chem 385(3):452–468

Barik TK, Sahu B, Swain V (2008) Nanosilica—from medicine to pest control. Parasitol Res 103(2):253–258

Bhaskar B, Ahammed SK, Chaitanya BH, Rasheed VA, Prasad TNVKV (2016) Silver nanoparticles: mycogenesis, characterization and its anti plant pathogenic applications. Int J Res Appl Nat Soc Sci 4(10):105–114

Bhattacharyya A, Duraisamy P, Govindarajan M, Buhroo AA, Prasad R (2016) Nano-biofungicides: emerging trend in insect pest control. In: Prasad R (ed) Advances and applications through fungal nanobiotechnology. Springer International Publishing, Cham, pp 307–319

Bhatia S (2016) Marine polysaccharides based nano-materials and its applications. In: Natural polymer drug delivery systems. Springer International Publishing, Switzerland, pp 185–225

Bhau B, Phukon P, Ahmed R, Gogoi B, Borah B, Baruah J, Sharma D, Wann S (2016) A novel tool of nanotechnology: nanoparticle mediated control of nematode infection in plants. In: Singh D., Singh H., Prabha R. (eds) Microbial Inoculants in Sustainable Agricultural Productivity. Springer, New Delhi, pp 253–269

Bhuyan T, Mishra K, Khanuja M, Prasad R, Varma A (2015) Biosynthesis of zinc oxide nanoparticles from *Azadirachta indica* for antibacterial and photocatalytic applications. Mater Sci Semicond Process 32:55–61

Branton D, Deamer DW, Marziali A, Bayley H, Benner SA, Butler T, Di Ventra M, Garaj S, Hibbs A, Huang X (2008) The potential and challenges of nanopore sequencing. Nat Biotechnol 26(10):1146–1153

Cañas JE, Long M, Nations S, Vadan R, Dai L, Luo M, Ambikapathi R, Lee EH, Olszyk D (2008) Effects of functionalized and nonfunctionalized single-walled carbon nanotubes on root elongation of select crop species. Environ Toxicol Chem 27(9):1922–1931

Cao J, Guenther RH, Sit TL, Lommel SA, Opperman CH, Willoughby JA (2015) Development of abamectin loaded plant virus nanoparticles for efficacious plant parasitic nematode control. ACS Appl Mater Interfaces 7(18):9546–9553

Chakravarthi S, Robinson D, De S (2007) Nanoparticles prepared using natural and synthetic polymers. In: Thassu D., Deleers M., Pathak Y.V. (eds) Nanoparticulate drug delivery systems. CRC Press, Boca Raton, pp 51–60

Chauhan N, Narang J, Pundir C (2012) An amperometric glutathione biosensor based on chitosan–iron coated gold nanoparticles modified Pt electrode. Int J Biol Macromol 51(5):879–886
Chichiriccò G, Poma A (2015) Penetration and toxicity of nanomaterials in higher plants. Nanomaterials 5(2):851–873
Chitwood DJ (2003) Research on plant-parasitic nematode biology conducted by the United States Department of Agriculture–Agricultural Research Service. Pest Manag Sci 59(6-7):748–753
Chuan Li S, Hua Chen J, Cao H, Sheng Yao D, Ling Liu D (2011) Amperometric biosensor for aflatoxin B 1 based on aflatoxin-oxidase immobilized on multiwalled carbon nanotubes. Food Control 22(1):43–49
Cioffi N, Torsi L, Ditaranto N, Sabbatini L, Zambonin PG, Tantillo G, Ghibelli L, D'Alessio M, Bleve-Zacheo T, Traversa E (2004) Antifungal activity of polymer-based copper nanocomposite coatings. Appl Phys Lett 85(12):2417–2419
Cromwell W, Yang J, Starr J, Jo Y-K (2014) Nematicidal effects of silver nanoparticles on root-knot nematode in bermudagrass. J Nematol 46(3):261
da Silva AC, Deda DK, Bueno CC, Moraes AS, Da Roz AL, Yamaji FM, Prado RA, Viviani V, Oliveira ON, Leite FL (2014) Nanobiosensors exploiting specific interactions between an enzyme and herbicides in atomic force spectroscopy. J Nanosci Nanotechnol 14(9):6678–6684
Danks C, Barker I (2000) On-site detection of plant pathogens using lateral-flow devices. EPPO Bull 30(3-4):421–426
Dar J, Soytong K (2014) Construction and characterization of copolymer nanomaterials loaded with bioactive compounds from *Chaetomium* species. J Agri Sc Technol 10:823–831
Dehkourdi EH, Mosavi M (2013) Effect of anatase nanoparticles (TiO_2) on parsley seed germination (*Petroselinum crispum*) in vitro. Biol Trace Elem Res 155(2):283–286
Dimkpa CO (2014) Can nanotechnology deliver the promised benefits without negatively impacting soil microbial life? J Basic Microbiol 54(9):889–904
Dimkpa CO, McLean JE, Britt DW, Anderson AJ (2012) Bioactivity and biomodification of Ag, ZnO, and CuO nanoparticles with relevance to plant performance in agriculture. Ind Biotechnol 8(6):344–357
Dinesh R, Anandaraj M, Srinivasan V, Hamza S (2012) Engineered nanoparticles in the soil and their potential implications to microbial activity. Geoderma 173:19–27
Du D, Huang X, Cai J, Zhang A (2007) Comparison of pesticide sensitivity by electrochemical test based on acetylcholinesterase biosensor. Biosens Bioelectron 23(2):285–289
Dwivedi S, Saquib Q, Al-Khedhairy AA, Musarrat J (2016) Understanding the role of nanomaterials in agriculture. In: Singh D., Singh H., Prabha R. (eds) Microbial Inoculants in Sustainable Agricultural Productivity. Springer, New Delhi, pp 271–288
Eastman PS, Ruan W, Doctolero M, Nuttall R, de Feo G, Park JS, Chu JSF, Cooke P, Gray JW, Li S, Chen FF (2006) Qdot nanobarcodes for multiplexed gene expression analysis. Nano Lett 6(5):1059–1064
Eichert T, Kurtz A, Steiner U, Goldbach HE (2008) Size exclusion limits and lateral heterogeneity of the stomatal foliar uptake pathway for aqueous solutes and water-suspended nanoparticles. Physiol Plant 134(1):151–160
El-Temsah YS, Joner EJ (2012) Impact of Fe and Ag nanoparticles on seed germination and differences in bioavailability during exposure in aqueous suspension and soil. Environ Toxicol 27(1):42–49
Elmer WH, White JC (2016) The use of metallic oxide nanoparticles to enhance growth of tomatoes and eggplants in disease infested soil or soilless medium. Environ Sci Nano 3(5):1072–1079
Esteban-Tejeda L, Malpartida F, Esteban-Cubillo A, Pecharromán C, Moya J (2009) Antibacterial and antifungal activity of a soda-lime glass containing copper nanoparticles. Nanotechnology 20(50):505701
Fang Y, Ramasamy RP (2015) Current and prospective methods for plant disease detection. Biosensors 5(3):537–561
Fang Y, Umasankar Y, Ramasamy RP (2014) Electrochemical detection of p-ethylguaiacol, a fungi infected fruit volatile using metal oxide nanoparticles. Analyst 139(15):3804–3810

Farkas GL, Kiraaly Z (1962) Role of phenolic compounds in the physiology of plant diseases and disease resistance. J Phytopathol 44(2):105–150

Feizi H, Kamali M, Jafari L, Moghaddam PR (2013) Phytotoxicity and stimulatory impacts of nanosized and bulk titanium dioxide on fennel (*Foeniculum vulgare* Mill). Chemosphere 91(4):506–511

Ghosh M, Jana A, Sinha S, Jothiramajayam M, Nag A, Chakraborty A, Mukherjee A, Mukherjee A (2016) Effects of ZnO nanoparticles in plants: cytotoxicity, genotoxicity, deregulation of antioxidant defenses, and cell-cycle arrest. Mutat Res Genet Toxicol Environ Mutagen 807:25–32

Giannousi K, Avramidis I, Dendrinou-Samara C (2013) Synthesis, characterization and evaluation of copper based nanoparticles as agrochemicals against *Phytophthora infestans*. RSC Adv 3(44):21743–21752

Giraldo JP, Landry MP, Faltermeier SM, McNicholas TP, Iverson NM, Boghossian AA, Reuel NF, Hilmer AJ, Sen F, Brew JA (2014) Plant nanobionics approach to augment photosynthesis and biochemical sensing. Nat Mater 13(4):400–408

Gitipour A, El Badawy A, Arambewela M, Miller B, Scheckel K, Elk M, Ryu H, Gomez-Alvarez V, Santo Domingo J, Thiel S (2013) The impact of silver nanoparticles on the composting of municipal solid waste. Environ Sci Technol 47(24):14385–14393

Gogos A, Knauer K, Bucheli TD (2012) Nanomaterials in plant protection and fertilization: current state, foreseen applications, and research priorities. J Agric Food Chem 60(39):9781–9792

González-Melendi P, Fernández-Pacheco R, Coronado MJ, Corredor E, Testillano P, Risueño MC, Marquina C, Ibarra MR, Rubiales D, Pérez-de-Luque A (2008) Nanoparticles as smart treatment-delivery systems in plants: assessment of different techniques of microscopy for their visualization in plant tissues. Ann Bot 101(1):187–195

Goswami A, Roy I, Sengupta S, Debnath N (2010) Novel applications of solid and liquid formulations of nanoparticles against insect pests and pathogens. Thin Solid Films 519(3):1252–1257

Grahl T, Märkl H (1996) Killing of microorganisms by pulsed electric fields. Appl Microbiol Biotechnol 45(1):148–157

Gruyer N, Dorais M, Bastien C, Dassylva N, Triffault-Bouchet G (2013) Interaction between silver nanoparticles and plant growth, International symposium on new technologies for environment control, energy-saving and crop production in greenhouse and plant 1037, pp 795–800

Hatschek E (1931) Electro Chem Processes Ltd assignee Brouisol., Britian

He L, Liu Y, Mustapha A, Lin M (2011) Antifungal activity of zinc oxide nanoparticles against *Botrytis cinerea* and *Penicillium expansum*. Microbiol Res 166(3):207–215

Hong F, Zhou J, Liu C, Yang F, Wu C, Zheng L, Yang P (2005) Effect of nano-TiO_2 on photochemical reaction of chloroplasts of spinach. Biol Trace Elem Res 105(1-3):269–279

Huang G, Yeh L, Chen Y (2008) Nanoparticle-enhanced magnetic field induces apoptosis in nematode. NSTI Nanotech 2008, Vol 2, technical proceedings. pp 497–500

Huang S, Wang L, Liu L, Hou Y, Li L (2015) Nanotechnology in agriculture, livestock, and aquaculture in China. A review. Agron Sustain Dev 35(2):369–400

Huang X-J, Ge D, Xu Z-K (2007) Preparation and characterization of stable chitosan nanofibrous membrane for lipase immobilization. Eur Polym J 43(9):3710–3718

Ismail M, Prasad R, Ibrahim AIM, Ahmed ISA (2017) Modern prospects of nanotechnology in plant pathology. In: Prasad R., Kumar M., Kumar V. (eds) Nanotechnology. Springer, Singapore, pp 305–317

Jackson P, Jacobsen NR, Baun A, Birkedal R, Kühnel D, Jensen KA, Vogel U, Wallin H (2013) Bioaccumulation and ecotoxicity of carbon nanotubes. Chem Cent J 7(1):154

Jayaseelan C, Rahuman AA, Kirthi AV, Marimuthu S, Santhoshkumar T, Bagavan A, Gaurav K, Karthik L, Rao KB (2012) Novel microbial route to synthesize ZnO nanoparticles using *Aeromonas hydrophila* and their activity against pathogenic bacteria and fungi. Spectrochim Acta A 90:78–84

Jerobin J, Sureshkumar R, Anjali C, Mukherjee A, Chandrasekaran N (2012) Biodegradable polymer based encapsulation of neem oil nanoemulsion for controlled release of Aza-A. Carbohydr Polym 90(4):1750–1756

Jha AK, Prasad K, Prasad K (2009) A green low-cost biosynthesis of Sb_2O_3 nanoparticles. Biochem Eng J 43(3):303–306

Jo Y-K, Kim BH, Jung G (2009) Antifungal activity of silver ions and nanoparticles on phytopathogenic fungi. Plant Dis 93(10):1037–1043

Joyner JJ, Kumar DV (2015) Nanosensors and their applications in food analysis: a review. Int J Sci Technol 3(4):80

Juhel G, Batisse E, Hugues Q, Daly D, van Pelt FN, O'Halloran J, Jansen MA (2011) Alumina nanoparticles enhance growth of *Lemna minor*. Aquat Toxicol 105(3):328–336

Kah M, Hofmann T (2014) Nanopesticide research: current trends and future priorities. Environ Int 63:224–235

Kang MA, Seo MJ, Hwang IC, Jang C, Park HJ, Yu YM, Youn YN (2012) Insecticidal activity and feeding behavior of the green peach aphid, *Myzus persicae*, after treatment with nano types of pyrifluquinazon. J Asia Pac Entomol 15(4):533–541

Kashyap PL, Rai P, Sharma S, Chakdar H, Kumar S, Pandiyan K, Srivastava AK (2016) Nanotechnology for the detection and diagnosis of plant pathogens. In: Ranjan S., Dasgupta N., Lichtfouse E. (eds) Nanoscience in Food and Agriculture 2. Sustainable Agriculture Reviews, vol 21. Springer, Cham, pp 253–276

Kasprowicz MJ, Kozioł M, Gorczyca A (2010) The effect of silver nanoparticles on phytopathogenic spores of *Fusarium culmorum*. Can J Microbiol 56(3):247–253

Kattke MD, Gao EJ, Sapsford KE, Stephenson LD, Kumar A (2011) FRET-based quantum dot immunoassay for rapid and sensitive detection of *Aspergillus amstelodami*. Sensors 11(6):6396–6410

Khalil M (2013) Alternative approaches to manage plant parasitic nematodes. J Plant Pathol Microbiol 4:e105

Khan MR, Rizvi TF (2014) Nanotechnology: scope and application in plant disease management. Plant Pathol J 13(3):214–231

Khiyami MA, Almoammar H, Awad YM, Alghuthaymi MA, Abd-Elsalam KA (2014) Plant pathogen nanodiagnostic techniques: forthcoming changes? Biotechnol Biotechnol Equip 28(5):775–785

Khodakovskaya M, Dervishi E, Mahmood M, Xu Y, Li Z, Watanabe F, Biris AS (2009) Carbon nanotubes are able to penetrate plant seed coat and dramatically affect seed germination and plant growth. ACS Nano 3(10):3221–3227

Kim JS, Kuk E, Yu KN, Kim J-H, Park SJ, Lee HJ, Kim SH, Park YK, Park YH, Hwang C-Y, Kim Y-K, Lee Y-S, Jeong DH, Cho M-H (2007) Antimicrobial effects of silver nanoparticles. Nanomed Nanotechnol Biol Med 3(1):95–101

Kim SG, Kim KW, Park EW, Choi D (2002) Silicon-induced cell wall fortification of rice leaves: a possible cellular mechanism of enhanced host resistance to blast. Phytopathology 92(10):1095–1103

Kim SW, Jung JH, Lamsal K, Kim YS, Min JS, Lee YS (2012) Antifungal effects of silver nanoparticles (AgNPs) against various plant pathogenic fungi. Mycobiology 40(1):53–58

Kovacs I, Durner J, Lindermayr C (2015) Crosstalk between nitric oxide and glutathione is required for nonexpressor of pathogenesis-related genes 1 (NPR1)-dependent defense signaling in *Arabidopsis thaliana*. New Phytol 208(3):860–872

Krishnaraj C, Ramachandran R, Mohan K, Kalaichelvan P (2012) Optimization for rapid synthesis of silver nanoparticles and its effect on phytopathogenic fungi. Spectrochim Acta A 93:95–99

Kumar V, Guleria P, Kumar V, Yadav SK (2013) Gold nanoparticle exposure induces growth and yield enhancement in *Arabidopsis thaliana*. Sci Total Environ 461:462–468

Kumari M, Khan SS, Pakrashi S, Mukherjee A, Chandrasekaran N (2011) Cytogenetic and genotoxic effects of zinc oxide nanoparticles on root cells of *Allium cepa*. J Hazard Mater 190(1):613–621

Kuźniar T, Ropek D, Lemek T (2011) Impact of multi--walled carbon nanotubes on viability and pathogenicity of enthomopathogenic nematodes. Ecol Chem Eng 18(5-6):757–762

Lamsal K, Kim SW, Jung JH, Kim YS, Kim KS, Lee YS (2011) Application of silver nanoparticles for the control of *Colletotrichum* species in vitro and pepper anthracnose disease in field. Mycobiology 39(3):194–199

Lee W-M, Kwak JI, An Y-J (2012) Effect of silver nanoparticles in crop plants *Phaseolus radiatus* and *Sorghum bicolor*: media effect on phytotoxicity. Chemosphere 86(5):491–499

Li Y, Yu S, Wu Q, Tang M, Pu Y, Wang D (2012) Chronic Al_2O_3-nanoparticle exposure causes neurotoxic effects on locomotion behaviors by inducing severe ROS production and disruption of ROS defense mechanisms in nematode *Caenorhabditis elegans*. J Hazard Mater 219–220:221–230

Lin C, Fugetsu B, Su Y, Watari F (2009) Studies on toxicity of multi-walled carbon nanotubes on Arabidopsis T87 suspension cells. J Hazard Mater 170(2):578–583

Liu F, Wen L-X, Li Z-Z, Yu W, Sun H-Y, Chen J-F (2006) Porous hollow silica nanoparticles as controlled delivery system for water-soluble pesticide. Mater Res Bull 41(12):2268–2275

López-Moreno ML, de la Rosa G, Hernández-Viezcas JÁ, Castillo-Michel H, Botez CE, Peralta-Videa JR, Gardea-Torresdey JL (2010) Evidence of the differential biotransformation and genotoxicity of ZnO and CeO_2 nanoparticles on soybean (*Glycine max*) plants. Environ Sci Technol 44(19):7315–7320

Ma C, Chhikara S, Xing B, Musante C, White JC, Dhankher OP (2013) Physiological and molecular response of *Arabidopsis thaliana* (L.) to nanoparticle cerium and indium oxide exposure. ACS Sustain Chem Eng 1(7):768–778

Ma H, Bertsch PM, Glenn TC, Kabengi NJ, Williams PL (2009) Toxicity of manufactured zinc oxide nanoparticles in the nematode *Caenorhabditis elegans*. Environ Toxicol Chem 28(6):1324–1330

Mak AC, Osterfeld SJ, Yu H, Wang SX, Davis RW, Jejelowo OA, Pourmand N (2010) Sensitive giant magnetoresistive-based immunoassay for multiplex mycotoxin detection. Biosens Bioelectron 25(7):1635–1639

Martinelli F, Scalenghe R, Davino S, Panno S, Scuderi G, Ruisi P, Villa P, Stroppiana D, Boschetti M, Goulart LR (2015) Advanced methods of plant disease detection. A review. Agron Sustain Dev 35(1):1–25

Mattiello A, Filippi A, Pošćić F, Musetti R, Salvatici MC, Giordano C, Vischi M, Bertolini A, Marchiol L (2015) Evidence of phytotoxicity and genotoxicity in *Hordeum vulgare* L. exposed to CeO_2 and TiO_2 nanoparticles. Front Plant Sci 6:1043

McCartney HA, Foster SJ, Fraaije BA, Ward E (2003) Molecular diagnostics for fungal plant pathogens. Pest Manag Sci 59(2):129–142

Meng Y, Li Y, Galvani CD, Hao G, Turner JN, Burr TJ, Hoch H (2005) Upstream migration of *Xylella fastidiosa* via pilus-driven twitching motility. J Bacteriol 187(16):5560–5567

Mirzajani F, Askari H, Hamzelou S, Farzaneh M, Ghassempour A (2013) Effect of silver nanoparticles on *Oryza sativa* L. and its rhizosphere bacteria. Ecotoxicol Environ Saf 88:48–54

Mody HR (2011) Cancer nanotechnology: recent trends and developments. Int J Med Updt 6(1):3

Mukherjee A, Majumdar S, Servin AD, Pagano L, Dhankher OP, White JC (2016) Carbon nanomaterials in agriculture: a critical review. Front Plant Sci 7:172

Mukhopadhyay SS (2014) Nanotechnology in agriculture: prospects and constraints. Nanotechnol Sci Appl 7:63–71

Musarrat J, Dwivedi S, Singh BR, Al-Khedhairy AA, Azam A, Naqvi A (2010) Production of antimicrobial silver nanoparticles in water extracts of the fungus Amylomyces rouxii strain KSU-09. Bioresour Technol 101(22):8772–8776

Navarro DA, Bisson MA, Aga DS (2012) Investigating uptake of water-dispersible CdSe/ZnS quantum dot nanoparticles by *Arabidopsis thaliana* plants. J Hazard Mater 211:427–435

Neethirajan S, Freund M, Jayas D, Shafai C, Thomson D, White N (2010) Development of carbon dioxide (CO_2) sensor for grain quality monitoring. Biosyst Eng 106(4):395–404

Neethirajan S, Jayas DS, Sadistap S (2009) Carbon dioxide (CO_2) sensors for the agri-food industry—A review. Food Bioproc Tech 2(2):115–121

Ocsoy I, Paret ML, Ocsoy MA, Kunwar S, Chen T, You M, Tan W (2013) Nanotechnology in plant disease management: DNA-directed silver nanoparticles on graphene oxide as an antibacterial against *Xanthomonas perforans*. ACS Nano 7(10):8972–8980

Omanović-Mikličanina E, Maksimović M (2016) Nanosensors applications in agriculture and food industry. Bull Chem Technol Bosnia Herzegovina 47:59–70

Padmavathy N, Vijayaraghavan R (2008) Enhanced bioactivity of ZnO nanoparticles—an antimicrobial study. Sci Technol Adv Mat 9(3):035004

Pal S, Ying W, Alocilja EC, Downes FP (2008) Sensitivity and specificity performance of a direct-charge transfer biosensor for detecting *Bacillus cereus* in selected food matrices. Biosyst Eng 99(4):461–468

Park H-J, Kim S-H, Kim H-J, Choi S-H (2006) A new composition of nanosized silica-silver for control of various plant diseases. Plant Pathol J 22(3):295–302

Pawlett M, Ritz K, Dorey RA, Rocks S, Ramsden J, Harris JA (2013) The impact of zero-valent iron nanoparticles upon soil microbial communities is context dependent. Environ Sci Pollut Res 20(2):1041–1049

Peteu SF, Oancea F, Sicuia OA, Constantinescu F, Dinu S (2010) Responsive polymers for crop protection. Polymer 2(3):229–251

Pimentel D (2009) Pesticides and pest control. In: Peshin R., Dhawan A.K. (eds) Integrated Pest Management: Innovation-Development Process. Springer, Dordrecht, pp 83–87

Pokhrel LR, Dubey B (2013) Evaluation of developmental responses of two crop plants exposed to silver and zinc oxide nanoparticles. Sci Total Environ 452:321–332

Prasad R, Bhattacharyya A, Nguyen QD (2017a) Nanotechnology in sustainable agriculture: recent developments, challenges, and perspectives. Front Microbiol 8:1014. https://doi.org/10.3389/fmicb.2017.01014

Prasad R, Gupta N, Kumar M, Kumar V, Wang S, Abd-Elsalam KA (2017b) Nanomaterials act as plant defense mechanism. In: Prasad R, Kumar V, Kumar M (eds) Nanotechnology. Springer, Singapore, pp 253–269

Prasad R, Pandey R, Varma A, Barman I (2017c) Polymer based nanoparticles for drug delivery systems and cancer therapeutics. In: Kharkwal H, Janaswamy S (eds) Natural polymers for drug delivery. CAB International, CABI, pp 53–70

Prasad R, Kumar V, Prasad KS (2014) Nanotechnology in sustainable agriculture: present concerns and future aspects. Afr J Biotechnol 13(6):705–713

Prasad R, Swamy VS (2013) Antibacterial activity of silver nanoparticles synthesized by bark extract of Syzygium cumini. J Nanopart. https://doi.org/10.1155/2013/431218

Rabab EMA, EL-Shafey RAS (2013) Inhibition effects of silver nanoparticles against rice blast disease caused by *Magnaporthe grisea*. Egypt J Agric Res 91(4):1271–1283

Racuciu M, Creanga D-E (2007) TMA-OH coated magnetic nanoparticles internalized in vegetal tissue. Rom J Phys 52(3/4):395

Rad F, Mohsenifar A, Tabatabaei M, Safarnejad M, Shahryari F, Safarpour H, Foroutan A, Mardi M, Davoudi D, Fotokian M (2012) Detection of *Candidatus* Phytoplasma aurantifolia with a quantum dots FRET-based biosensor. J Plant Pathol 94(3):525–534

Radoi A, Targa M, Prieto-Simon B, Marty J-L (2008) Enzyme-Linked Immunosorbent Assay (ELISA) based on superparamagnetic nanoparticles for aflatoxin M 1 detection. Talanta 77(1):138–143

Ragaei M, Sabry A-kH (2014) Nanotechnology for insect pest control. Int J Sci Environ Technol 3(2):528–545

Rai M, Deshmukh S, Gade A (2012) Strategic Nanoparticle-mediated gene transfer in plants and animals-a novel approach. Curr Nanosci 8(1):170–179

Rajiv P, Rajeshwari S, Venckatesh R (2013) Bio-Fabrication of zinc oxide nanoparticles using leaf extract of *Parthenium hysterophorus* L. and its size-dependent antifungal activity against plant fungal pathogens. Spectrochim Acta A 112:384–387

Rispail N, De Matteis L, Santos R, Miguel AS, Custardoy L, Testillano PS, Risueño MC, Pérez-de-Luque A, Maycock C, Fevereiro P (2014) Quantum dot and superparamagnetic nanoparticle interaction with pathogenic fungi: internalization and toxicity profile. ACS Appl Mater Interfaces 6(12):9100–9110

Riveros A, Srygley R (2010) Migration, orientation and navigation: magnetic compasses in insects. In: Encyclopedia of animal behavior, vol 2. Elsevier/Academic Press, Oxford, pp 305–313

Roh J-Y, Park Y-K, Park K, Choi J (2010) Ecotoxicological investigation of CeO_2 and TiO_2 nanoparticles on the soil nematode *Caenorhabditis elegans* using gene expression, growth, fertility, and survival as endpoints. Environ Toxicol Pharmacol 29(2):167–172

Rosen J, Yoffe S, Meerasa A, Verma M, Gu F (2011) Nanotechnology and diagnostic imaging: new advances in contrast agent technology. J Nanomed Nanotechnol 2:115

Rousk J, Ackermann K, Curling SF, Jones DL (2012) Comparative toxicity of nanoparticulate CuO and ZnO to soil bacterial communities. PLoS One 7(3):e34197

Sabir S, Arshad M, Chaudhari SK (2014) Zinc oxide nanoparticles for revolutionizing agriculture: synthesis and applications. Sci World J 2014:8

Sadanandom A, Napier RM (2010) Biosensors in plants. Curr Opin Plant Biol 13(6):736–743

Sadowski Z (2010) Biosynthesis and application of silver and gold nanoparticles. In: Perez DP (ed) Silver nanoparticles

Sah S, Sorooshzadeh A, Rezazadeh H, Naghdibadi H (2011) Effect of nano silver and silver nitrate on seed yield of borage. J Med Plant Res 5(5):706–710

Saini RK, Bagri LP, Bajpai AK (2017) Smart nanosensors for pesticide detection. In: Grumezescu A.M. (ed) New pesticides and soil sensors. Academic Press, United Kingdom, pp 519–559

Sankaran S, Mishra A, Ehsani R, Davis C (2010) A review of advanced techniques for detecting plant diseases. Comput Electron Agric 72(1):1–13

Sangeetha J, Thangadurai D, Hospet R, Harish ER, Purushotham P, Mujeeb MA, Shrinivas J, David M, Mundaragi AC, Thimmappa AC, Arakera SB, Prasad R (2017a) Nanoagrotechnology for soil quality, crop performance and environmental management. In: Prasad R, Kumar M, Kumar V (eds) Nanotechnology. Springer, Singapore, pp 73–97

Sangeetha J, Thangadurai D, Hospet R, Purushotham P, Manowade KR, Mujeeb MA, Mundaragi AC, Jogaiah S, David M, Thimmappa SC, Prasad R, Harish ER (2017b) Production of bionanomaterials from agricultural wastes. In: Prasad R, Kumar M, Kumar V (eds) Nanotechnology, Springer, Singapore, pp 33–58

Sarkar DJ, Kumar J, Shakil N, Walia S (2012) Release kinetics of controlled release formulations of thiamethoxam employing nano-ranged amphiphilic PEG and diacid based block polymers in soil. J Environ Sci Health A 47(11):1701–1712

Savithramma N, Ankanna S, Bhumi G (2012) Effect of nanoparticles on seed germination and seedling growth of *Boswellia ovalifoliolata* an endemic and endangered medicinal tree taxon. Nano Vision 2(1):2

Schofield CL, Haines AH, Field RA, Russell DA (2006) Silver and gold glyconanoparticles for colorimetric bioassays. Langmuir 22(15):6707–6711

Schreiber L (2005) Polar paths of diffusion across plant cuticles: new evidence for an old hypothesis. Ann Bot 95(7):1069–1073

Serag MF, Kaji N, Habuchi S, Bianco A, Baba Y (2013) Nanobiotechnology meets plant cell biology: carbon nanotubes as organelle targeting nanocarriers. RSC Adv 3(15):4856–4862

Sharon M, Choudhary AK, Kumar R (2010) Nanotechnology in agricultural diseases and food safety. J Phytol 2(4):83

Shaw AK, Hossain Z (2013) Impact of nano-CuO stress on rice (*Oryza sativa* L.) seedlings. Chemosphere 93(6):906–915

Shaymurat T, Gu J, Xu C, Yang Z, Zhao Q, Liu Y, Liu Y (2012) Phytotoxic and genotoxic effects of ZnO nanoparticles on garlic (*Allium sativum* L.): a morphological study. Nanotoxicology 6(3):241–248

Silva AT, Nguyen A, Ye C, Verchot J, Moon JH (2010) Conjugated polymer nanoparticles for effective siRNA delivery to tobacco BY-2 protoplasts. BMC Plant Biol 10:291

Singh D, Kumar A, Singh AK, Tripathi H (2013) Induction of resistance in field pea against rust disease through various chemicals/micronutrients and their impact on growth and yield. Plant Pathol J 12(2):36

Singh P, Kumari K, Vishvakrma VK, Mehrotra GK, Chandra R, Kumar D, Patel R, Shahare VV (2017) Metal NPs (Au, Ag, and Cu): Synthesis, stabilization, and their role in green chemistry and drug delivery. In: Singh R., Kumar S. (eds) Green Technologies and Environmental Sustainability. Springer, Cham, pp 309–337

Srinivasan B, Tung S (2015) Development and applications of portable biosensors. J Lab Autom 20(4):365–389

Srinivasan C, Saraswathi R (2010) Nano-agriculture–carbon nanotubes enhance tomato seed germination and plant growth. Curr Sci 99(3):274–275

Sun Y-F, Liu S-B, Meng F-L, Liu J-Y, Jin Z, Kong L-T, Liu J-H (2012) Metal oxide nanostructures and their gas sensing properties: a review. Sensors 12(3):2610–2631

Suriyaprabha R, Karunakaran G, Kavitha K, Yuvakkumar R, Rajendran V, Kannan N (2014) Application of silica nanoparticles in maize to enhance fungal resistance. IET Nanobiotechnol 8(3):133–137

Swamy VS, Prasad R (2012) Green synthesis of silver nanoparticles from the leaf extract of Santalum album and its antimicrobial activity. J Optoelectron Biomed Mater 4(3):53–59

Tan X-m, Lin C, Fugetsu B (2009) Studies on toxicity of multi-walled carbon nanotubes on suspension rice cells. Carbon 47(15):3479–3487

Tiwari D, Dasgupta-Schubert N, Cendejas LV, Villegas J, Montoya LC, Garcia SB (2014) Interfacing carbon nanotubes (CNT) with plants: enhancement of growth, water and ionic nutrient uptake in maize (*Zea mays*) and implications for nanoagriculture. Appl Nanosci 4(5):577–591

Singhal U, Khanuja M, Prasad R, Varma A (2017) Impact of synergistic association of ZnO-nanorods and symbiotic fungus Piriformospora indica DSM 11827 on Brassica oleracea var. botrytis (Broccoli). Front Microbiol 8:1909. https://doi.org/10.3389/fmicb.2017.01909

Vamvakaki V, Chaniotakis NA (2007) Pesticide detection with a liposome-based nano-biosensor. Biosens Bioelectron 22(12):2848–2853

Verma ML (2017) Enzymatic nanobiosensors in the agricultural and food industry. In: Ranjan S., Dasgupta N., Lichtfouse E. (eds) Nanoscience in Food and Agriculture 4. Sustainable Agriculture Reviews, vol 24. Springer, Cham, pp 229–245

Villagarcia H, Dervishi E, de Silva K, Biris AS, Khodakovskaya MV (2012) Surface chemistry of carbon nanotubes impacts the growth and expression of water channel protein in tomato plants. Small 8(15):2328–2334

Wainwright M, Grayston SJ, De Jong P (1986) Adsorption of insoluble compounds by mycelium of the fungus Mucor flavus. Enzyme Microb Technol 8(10):597–600

Wajnberg E, Acosta-Avalos D, Alves OC, de Oliveira JF, Srygley RB, Esquivel DM (2010) Magnetoreception in eusocial insects: an update. J R Soc Interface 7(Suppl 2):S207–S225

Wakeil NE, Alkahtani S, Gaafar N (2017) 7 - Is nanotechnology a promising field for insect pest control in IPM programs? A2 - Grumezescu. In: Mihai A (ed) New pesticides and soil sensors. Academic Press, United Kingdom, pp 273–309

Wang H, Kou X, Pei Z, Xiao JQ, Shan X, Xing B (2011a) Physiological effects of magnetite (Fe_3O_4) nanoparticles on perennial ryegrass (*Lolium perenne* L.) and pumpkin (*Cucurbita mixta*) plants. Nanotoxicology 5(1):30–42

Wang H, Wick RL, Xing B (2009) Toxicity of nanoparticulate and bulk ZnO, Al_2O_3 and TiO_2 to the nematode *Caenorhabditis elegans*. Environ Pollut 157(4):1171–1177

Wang J, Deo RP, Musameh M (2003a) Stable and sensitive electrochemical detection of phenolic compounds at carbon nanotube modified glassy carbon electrodes. Electroanalysis 15(23-24):1830–1834

Wang J, Koo Y, Alexander A, Yang Y, Westerhof S, Zhang Q, Schnoor JL, Colvin VL, Braam J, Alvarez PJ (2013) Phytostimulation of poplars and *Arabidopsis* exposed to silver nanoparticles and Ag^+ at sublethal concentrations. Environ Sci Technol 47(10):5442–5449

Wang J, Liu G, Jan MR, Zhu Q (2003b) Electrochemical detection of DNA hybridization based on carbon-nanotubes loaded with CdS tags. Electrochem Commun 5(12):1000–1004

Wang S, Kurepa J, Smalle JA (2011b) Ultra-small TiO_2 nanoparticles disrupt microtubular networks in *Arabidopsis thaliana*. Plant Cell Environ 34(5):811–820

Wang Z, Wei F, Liu S-Y, Xu Q, Huang J-Y, Dong X-Y, Yu J-H, Yang Q, Zhao Y-D, Chen H (2010) Electrocatalytic oxidation of phytohormone salicylic acid at copper nanoparticles-modified gold electrode and its detection in oilseed rape infected with fungal pathogen *Sclerotinia sclerotiorum*. Talanta 80(3):1277–1281

Weigl B, Domingo G, LaBarre P, Gerlach J (2008) Towards non-and minimally instrumented, microfluidics-based diagnostic devices. Lab Chip 8(12):1999–2014

West JS, Bravo C, Oberti R, Moshou D, Ramon H, McCartney HA (2010) Detection of fungal diseases optically and pathogen inoculum by air sampling. In: Oerke EC., Gerhards R., Menz G., Sikora R. (eds) Precision Crop Protection - the Challenge and Use of Heterogeneity. Springer, Dordrecht, pp 135–149

Wu K, Sun Y, Hu S (2003) Development of an amperometric indole-3-acetic acid sensor based on carbon nanotubes film coated glassy carbon electrode. Sens Actuators B 96(3):658–662

Wu S, Huang L, Head J, Chen D, Kong I, Tang Y (2012) Phytotoxicity of metal oxide nanoparticles is related to both dissolved metals ions and adsorption of particles on seed surfaces. J Pet Environ Biotechnol 3(4):126

Xiang C, Taylor AG, Hinestroza JP, Frey MW (2013) Controlled release of nonionic compounds from poly (lactic acid)/cellulose nanocrystal nanocomposite fibers. J Appl Polym Sci 127(1):79–86

Xue J, Luo Z, Li P, Ding Y, Cui Y, Wu Q (2014) A residue-free green synergistic antifungal nanotechnology for pesticide thiram by ZnO nanoparticles. Sci Rep 4:5408

Yoosaf K, Ipe BI, Suresh CH, Thomas KG (2007) In situ synthesis of metal nanoparticles and selective naked-eye detection of lead ions from aqueous media. J Phys Chem C 111(34):12839–12847

Yu X, Chattopadhyay D, Galeska I, Papadimitrakopoulos F, Rusling JF (2003) Peroxidase activity of enzymes bound to the ends of single-wall carbon nanotube forest electrodes. Electrochem Commun 5(5):408–411

Zafar H, Ali A, Ali JS, Haq IU, Zia M (2016) Effect of ZnO nanoparticles on *Brassica nigra* seedlings and stem explants: growth dynamics and antioxidative response. Front Plant Sci 7:535

Zhang Y, Arugula MA, Wales M, Wild J, Simonian AL (2015) A novel layer-by-layer assembled multi-enzyme/CNT biosensor for discriminative detection between organophosphorus and non-organophosphrus pesticides. Biosens Bioelectron 67:287–295

Zhao L, Sun Y, Hernandez-Viezcas JA, Servin AD, Hong J, Niu G, Peralta-Videa JR, Duarte-Gardea M, Gardea-Torresdey JL (2013) Influence of CeO_2 and ZnO nanoparticles on cucumber physiological markers and bioaccumulation of Ce and Zn: a life cycle study. J Agric Food Chem 61(49):11945–11951

Zhao X, Hilliard LR, Mechery SJ, Wang Y, Bagwe RP, Jin S, Tan W (2004) A rapid bioassay for single bacterial cell quantitation using bioconjugated nanoparticles. Pro Natl Acad Sci USA 101(42):15027–15032

Zheng L, Hong F, Lu S, Liu C (2005) Effect of nano-TiO2 on strength of naturally aged seeds and growth of spinach. Biol Trace Elem Res 104(1):83–91

Chapter 3
Copper Nanostructures Applications in Plant Protection

Esraa Gabal, Mohamed M. Ramadan, Amal-Asran, Mousa A. Alghuthaymi, and Kamel A. Abd-Elsalam

3.1 Introduction

According to global economic reports, the global agribusiness market estimation in 2010 was found to lie in the range of US$ 20.7 billion to US$ 0.98 trillion. However, it is expected that by 2020 such estimation will rise to exceed US$ 3.4 trillion (Hooley et al. 2014). The USA has recently started with a 4-year initiative concerning nanotechnology studies (National Nanotechnology Initiative), with estimated 3.7 USD investments (Hirsh et al. 2014). Moreover, the path adopted by the USA has been followed by other countries including Japan and the European Union with annual funding of 750 million USD and 1.2 billion USD, respectively, adding to contributions possessed by the individual country (Sodano and Verneau 2014). Today, more than 400 companies in the world are active in nanotechnology research and development, and this number is expected to increase more than 1000 in the next 10 years (www.bhartbook.com). With the use of nanotechnology, it became possible to rely on well-known antimicrobial materials (e.g., silver, copper, zinc oxide) for developing and creating engineered nanostructures/nanocomposites (NCs). Such engineered materials showed significant activities on living organisms, encompassing controlled toxicity to humans, and have been generally known as nano-antimicrobials (Cioffi and Rai 2012). Nanotechnology represents a promising technique that can be used in agricultural sectors and food production systems. It is considered

E. Gabal
Agricultural Sciences and Resource Management in the Tropics and Subtropics (ARTS), Faculty of Agriculture, University of Bonn, Bonn, Germany

M. M. Ramadan · Amal-Asran · K. A. Abd-Elsalam (✉)
Plant Pathology Research Institute, Agricultural Research Center (ARC), Giza, Egypt

M. A. Alghuthaymi
Department of Biology, Science and Humanities College, Shaqra University, Alquwayiyah, Saudi Arabia

K. A. Abd-Elsalam, R. Prasad (eds.), *Nanobiotechnology Applications in Plant Protection*, Nanotechnology in the Life Sciences,
https://doi.org/10.1007/978-3-319-91161-8_3

as developed means to improve the quality of produced crop yields through enhancing the management and conservation of field inputs (Garcia et al. 2010; Sekhon 2014). Moreover, nanotechnology can play a crucial role in the sectors of food security and safety using nanosensors to detect pathogens and food packaging systems through the capsulation of food products using nano-compounds (Sekhon 2014; Sonkaria et al. 2012; Pérez-de-Luque and Hermosín 2013; Prasad 2014; Prasad et al. 2014, 2017a). Inorganic nanoparticles (NPs), because of their stability and long shelf life, offer many distinctive advantages in reducing acute toxicity, overcoming resistance, and lowering cost compared with the conventional organic antimicrobial agents and thus have been proposed as a convincing alternative (Whitesides 2003; Weir et al. 2008). Metallic copper (Cu), cupric oxide (CuO), cuprous oxide (Cu_2O) NPs, and composites of Cu/Cu_2O, Cu_2O/CuO, or core–shell NPs, commonly called copper-based nanoparticles (Cu-based NPs), have attracted attention from diverse disciplines because of their widespread use in electronics, optics, sensors, catalysts, and medical applications (Patolsky et al. 2006; Kumar 2009; Giannousi et al. 2017). It has been investigated that nano-formulations can be used in limiting the hazardous effects of pesticides and fertilizers providing an eco-friendly means due to the possessing of antimicrobial activities against a wide range of known plant and animal pathogens. Developing such nanostructures can represent novel applications to be introduced in agricultural systems and disease management (Scrinis and Lyons 2007). It is widely known that plant pathogenic fungi are among the most destructive factors threatening environmental adaptation and hindering plant growth. Interestingly, applying strategies relied on the introduction of nanotechnology, and metal nanoparticles could help in developing estimated data against plant fungal diseases (Park et al. 2006; Wani and Ahmad 2013). Nanotechnology provides a wide range of applications enhancing the agricultural researches through converting agricultural and food wastes to useful energy and other by-products with the help of enzymatic nano-bioprocessing, enhancing agricultural management through controling the development of many diseases. (Carmen et al. 2003; Sangeetha et al. 2017a,b). The high surface-to-volume ratio and unique optical properties of nanomaterials have enhanced their introduction in plant protection and agricultural system management (Ghormade et al. 2011). Hence, copper nanoparticles could gain the recent interest due to their optical, catalytic, mechanical, and electrical properties enabling them to be a good alternative from other metal nanoparticles being introduced in different applications (Kim et al. 2008; Salavati-Niasari et al. 2008; Eastman et al. 2001). A brief overview of the main applications of Cu NPs as a bioactive material is also involved in this chapter (Fig. 3.1).

3.2 Why Copper is the Best Nanoagrochemical?

According to Boehm et al. (2003), it is crucial to have the knowledge required for designing nano-encapsulated agrochemicals guaranteeing their constant effectivity and stability under diverse environmental conditions with the necessity of taking

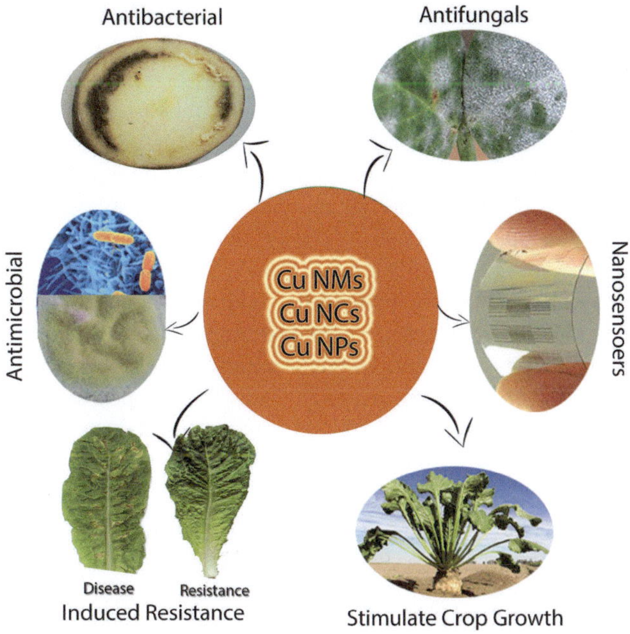

Fig. 3.1 Applications of copper nanomaterials in plant protection

considerations regarding controlled release in response to a specific promoter, enhancing their desired activity and the choice of successful delivery system avoiding the repeat of their application. The prepared nano-encapsulated agrochemicals will avoid the negative effects accompanying applied agrochemicals of polluted water and soil caused by the leaching of chemicals and their degradation through hydrolysis and microbial activity due to the repeated application of these chemicals for reaching an effective concentration since their chemical concentration is usually below the minimum effective concentration sufficient for approaching the crops' target sites.

Copper has a long history for being used as antimicrobial agent since it was used in ancient Egypt (2000 BCE) in sanitation and sterilizing water. Copper cooking utensils had been used for controlling diseases in the Roman Empire as well as by early Phoenicians for increasing the speed of their ships through introducing copper stripes that hindered fouling, while Japanese soldiers used it during World War II for sanitizing their water bottles (Borkow and Gabbay 2009). Moreover, copper was first introduced to agriculture in 1761 when it was investigated that seed-borne fungi could be inhibited by soaking seed grains in weak solution of copper sulfate. Since then, copper was found to adopt fungicidal characteristics, which helped in its introduction in the agricultural management systems (García et al. 2003). Since then, copper has been introduced in agricultural practices such as developing of fungicides with mixed combinations of copper sulfate, lime, and water or copper sulfate and sodium carbonate used in the USA and France by the 1880s (Ingle et al. 2014).

It was in 1882 when Bordeaux mixture was developed in France to be used as fungicide, encompassing a mixture of copper sulfate and calcium found to have a strong inhibitory effect on the fungal growth of *Plasmopara viticola*, the pathogenic fungi of grapes. In addition, Bordeaux mixture is still being used in agriculture for limiting crop damages caused by different pathogenic fungi (Somers 1959). Hence, copper represented a good antimicrobial agent used in generating many agrochemicals to be introduced and applied in agricultural sectors as copper oxychloride [$Cu(OH)_2.CuCl_2$], copper hydroxide [$Cu(OH)_2$], and cuprous oxide [Cu_2O] (Montag et al. 2006).

Copper nanoparticles (Cu NPs) have gained the recent concern as potential antimicrobial agents with applications in plant pathology and disease management due to their biological, physical, and chemical characteristics (Honary et al. 2012; Prasad et al. 2017b). Interestingly, Cu-NPs were found to possess a wide range of applications as bactericidal, fungicidal, and antiviral agents, gas sensors, and catalysts, allowing them to be introduced in diverse fields. Studies have revealed the antibacterial activities provided by Cu NPs, while less is known about their inhibitory effects on fungi and fungi-like organisms (Navarro et al. 2008). Moreover, Kah and Hofmann (2014) have indicated the raised interest regarding the development of copper-based nanopesticides in agricultural markets, principally that of organic farming. According to Chatterjee et al. (2012), the different attitudes and specificity possessed by different forms of Cu-NPs including metallic copper (Cu), cuprous oxide (Cu_2O), and cupric oxide (CuO) shown under in vitro studies have raised the matter of determining the form of copper NPs that fits the target of its application.

However, synthesis of copper nanoparticles is still a sensitive subject, requiring care and enhanced knowledge. The type of synthesized method can affect on the efficiency of copper nanoparticles in their antimicrobial characteristics. Moreover, it was markedly illustrated that efficiency of Cu NPs antimicrobial activity can be reduced by the influence of rust layer formation (oxide layer) on the surface of copper nanoparticles (Jeong et al. 2008; Chen and Sommers 2001; Salzemann et al. 2004). This comes from the fact that copper nanoparticles are characterized with a high surface-to-volume ratio, which enables them to undergo reactions with other particles and microbial membranes with high ratio (El-Sayed 2003).

3.3 Copper Nanostructure Applications in Plant Protection

3.3.1 Nanosensors

Nanoparticles can be used as an identification tool for detecting bacterial, viral, and fungal plant pathogens in agriculture (Boonham et al. 2008; Khiyami et al. 2014; Prasad 2014). As well, they can be exploited in conductivity-based sensors where they can induce a change in the signal upon the attachment of the antibody-tagged nanoparticles with the antigen captured on the sensor surface (Servin et al. 2015).

Wang et al. (2016c) have exploited an indirect stimulus to develop a sensitive electrochemical sensor, using modified gold electrode with copper nanoparticles, to monitor the levels of salicylic acid in oil seeds to detect the pathogenic fungus, *Sclerotinia sclerotiorum*. They successfully and accurately measured salicylic acid using this sensor. Research on similar sensors and sensing techniques needs to be expanded for detecting pathogens and their by-products or monitoring physiological changes in plants due to infections. Copper oxide (CuO) nanoparticles and nanolayers were prepared by solgel and spray pyrolysis techniques, respectively. Both CuO nanoparticles and nanostructural layer biosensors were used for detecting *Aspergillus niger* fungi (Etefagh et al. 2013).

3.3.2 Antimicrobial

Scientists have focused their attention on applying copper NPs in different applications because of their unique properties and antimicrobial activities as well as the low cost of their preparation. It was investigated that cuprous oxide nanoparticles (CuO NPs) have the possibility to be used in the development of formulations required for generating pesticides, thanks to their antimicrobial activities (Ahamad et al. 2014). According to Rai and Kratosova (2015), metal nanoparticles such as that of copper, silver, chitosan, and titanium were found to be beneficial in limiting the development of pathogenic microorganisms and hence hindering their spread among agricultural crops due to the antimicrobial activities possessed. Moreover, copper NPs have shown their significant ability to inhibit the conidial germination and hyphal development of fungi under different experiments. These activities have proved the potentiality of copper nanoparticles as promising tools for managing and detecting diseases with the possibility of involving them in developing nanopesticides and nanoherbicides. Hence, the introduction of nanoparticles in agricultural management and pathogen detection will open up new avenues of sustainable agricultural development with increase in crop productivity (Rai and Kratosova 2015). This section deals with the antimicrobial activity studies carried out with the copper nanoparticles and nanocomposites.

3.3.3 Antifungal

3.3.3.1 Copper Nanoparticles as Antifungal

Copper agrochemicals applied in agriculture have the ability to inhibit fungal growth such as that of *Phytophthora infestans*, the causal agent of late blight disease in tomato (Nelson 2008). However, the relatively high increase of fungicide resistance adopted by a wide range of plant pathogens has raised the necessity for developing other alternatives to be used as fungicidal agents. This target could be fulfilled through the application of nanotechnology, which enabled the transformation of copper metal into nanoparticles showing high efficiency of fungicidal activity (Kanhed et al. 2014).

Nanoparticles have showed antimicrobial activity against various known plant pathogens as *Phytophthora* and *Corticium salmonicolor* (Dang et al. 2010). However, it is a challenging subject to rely on copper NPs for developing agrochemicals as it is a new field with some assumptions regarding the reduction of formulated product concentration once being applied in the field (Young and Santra 2014). Hyperbranched polyamine/Cu NPs, low-polydispersion Cu NPs, functionalized Cu NPs, and Cu-grafted carbon nanotubes are well known as new nano-antimicrobial-coated materials that are applicable in agricultural sectors with advanced functions for adding improvements (Mohan et al. 2011; Wei et al. 2010). The increase in the current concern regarding the use of copper NPs in agricultural management than other metal NPs was a consequence of their wide antimicrobial activities and availability and the cheap facilities required for their generation (Van Acker et al. 2014).

Ramyadevi et al. (2012) have observed the strong antifungal activity performed by Cu-based NPs against fungal strains of *Aspergillus flavus* and *A. niger*, while the antimicrobial activity of copper NPs against four filamentous fungal species of *Alternaria alternata*, *Aspergillus flavus*, *Fusarium solani*, and *Penicillium chrysogenum* was investigated by Ghasemian et al. (2012). Moreover, studies have showed that copper NPs have antimicrobial efficiency higher than 8% against phytopathogenic fungi of vines comparing to the same dose of commonly used non-nano-chemicals of cupric hydroxide as Cuprozin and Spiess-Urania Chemicals (Gogos et al. 2012). Interestingly, copper nanoparticles and nanocomposites were considered more effective than commercially known copper chemical products in suppressing tomato fungal disease caused by *Phytophthora infestans* (Giannousi et al. 2013). Cu-chitosan NPs adopt an inhibitory effect on the spore germination and growth of plant pathogenic fungi as *Alternaria alternata*, *Macrophomina phaseolina*, and *Rhizoctonia solani* (Saharan et al. 2013). In addition, Kanhed et al. (2014) investigated the significant antifungal activity was performed by Cu NPs of 3–10 nm size against known fungal strains of *Phoma destructiva*, *C. lunata*, *A. alternata*, and *F. oxysporum* considered as superior effect exceeding that of the commercially known fungicide Bavistin.

According to Van et al. (2014), copper nanoparticles were shown to exhibit strong fungicidal activity against *Corticium salmonicolor* when applied to the infected rubber trees, causing inhibition of the development and spread of fungal pink disease. Both silver and copper nanoparticles and their combination (Ag/Cu NPs) showed the ability to perform fungicidal activity against plant pathogenic fungi *Alternaria alternata* and *Botrytis cinerea* where silver nanoparticles (Ag-NPs) have shown the maximum inhibitory efficiency (Ouda 2014). Through testing the antifungal activity possessed by Cu NPs using Kirby-Bauer disk diffusion protocol, *Fusarium culmorum* was found to be the most sensitive fungus to the treatment of Cu NPs, followed by *F. oxysporum* and *F. graminearum* (Shende et al. 2015).

Zabrieski et al. (2015) have studied the significant sensitivity shown by isolates of soilborne phytopathogen *P. aphanidermatum* under treatment with NPs aiming to investigate the potential of metal NPs as pesticides in a study where authors found that CuO NPs could cause inhibition in the enzymatic activity of ferric reductase at the mycelial surfaces. In a study conducted by Bramhanwade et al. (2016), authors performed chemical synthesis of copper nanoparticles using cetyltrimethylammonium bromide and copper nitrate. The resulting Cu-NPs were tested for the antimicrobial activity where they could significantly inhibit the fungal growth of *Fusarium* spp., i.e., *F. oxysporum*, *F. culmorum*, and *F. equiseti*. Another study to investigate the antifungal activity adopted by copper nanoparticles has been conducted by Ponmurugan et al. (2016) where authors synthesized copper nanoparticles through treating *Actinomycetes* spp., namely, *Streptomyces griseus* biomass, with the aqueous solution of copper sulfate. The extracellularly formed copper nanoparticles have been introduced to naturally infected plants to determine their fungicidal activity against *Poria hypolateritia*, the causal agent of red root-rot disease in tea plants. Within the study conducted by Ponmurugan et al. (2016), copper nanoparticles have been applied on the infected tea plants in the field experiments as sprays at different concentration rates of 1–2.5 ppm for each 1.5 L/bush. The study results have shown reduction in the incidence of red root-rot disease in response to the application of Cu NPs. The maximum reduction in the disease accounting for 52.7% was achieved at copper nanoparticle dosage of 2.5 ppm concentration.

As proven by Viet et al. (2016), Cu NPs of 450 ppm concentration showed fungicidal activity enabling them to inhibit the fungal growth of *Fusarium* sp. by 94% under 9-day incubation. Banik and Pérez-de-Luque (2017) have investigated the antimicrobial activity of copper NPs against fungi, bacteria, beneficial microbes such as *Trichoderma harzianum* and *Rhizobium* spp., and wheat seeds under in vitro conditions. Moreover, mycelial development and sporulation of *A. alternata* and *Pseudomonas syringae* could be reduced by applying Cu NPs with volume of 200 mg/L along with copper oxychloride (Banik and Pérez-de-Luque 2017). However, the previous study showed the inductive effect of Cu-NPs at low concentrations promoting the fungal development of *Botrytis fabae*, *Fusarium oxysporum* f.sp. *ciceris*, *F. oxysporum* f.sp. *melonis*, *Alternaria alternate*, and *P. syringae* and sporulation of *T. harzianum.* Bouson et al. (2017) have declared the significant antifungal activity adopted by water-stable Cu-based benzenetricarboxylate (Cu-BTC) which could inhibit the fungal spores' development of *A. niger*, *A. oryzae*, and *F. oxysporum.* Under both in vitro and in vivo conditions, Cu NPs could inhibit the growth of pathogenic fungi since they completely hindered the fungal growth of *P. digitatum* and *F. solani* under concentrations of 20 and 40 μg/mL, respectively, in vitro while reducing the development of both green mold and Fusarium rot with concentrations of 20 and 60 μg/mL, respectively, under in vivo conditions (Youssef et al. 2017). From the aforementioned studies and investigations illustrating the fungicidal activity possessed by Cu NPs, we can come up with the possibility of involving copper NPs in the development of new fungicide formulations, aiming to generate eco-friendly methods for agricultural management and crop disease control.

3.3.3.2 Copper Nanocomposites as Antifungal

Adding to their potential as effective structures for generating agrochemicals, it is assumed that copper nanoparticles can enhance food quality and extend the shelf life of food products since Cu NP modified chitosan showed significant inhibitory impact on the development of two microbes affecting food quality (Cárdenaz et al. 2009; Cioffi et al. 2004 and 2005). Moreover, some reports have proved the inhibitory activity of polymer molecules combined with copper nanocomposites against plant pathogenic fungi (Cioffi et al. 2004). The use of Cu/Zn chitosan NPs was effective against many plant pathogenic bacteria and fungi and was found to be more beneficial for plants through enhancing their nutrition, causing acceleration in their development with protection against both biotic and abiotic stresses. Moreover, Cu/Zn chitosan NPs have showed a functional role in mobilizing food required for seed development and germination through stimulating the activity of amylase and protease enzymes, adding to other functions regarding the response to diseases, and activating plant defense system through activating defense enzymes. Most of the studies support the assumption that polymer/metal nanocomposites can exhibit their antimicrobial activity through releasing the metal ions then leaching the particle as a mechanism for inferring their action (Choudhary et al. 2017a).

With an attempt to generate effective copper NP-based agrochemicals, it was investigated that combining copper with chitosan nanogels is better than using chitosan solutions as a promising means for developing bio-fungicides. Copper-chitosan nanogels showed strong fungicidal activity against *Fusarium graminearum*, with potential advantages regarding the ease in handling and the long-term release of copper on leaves or in soil which in its turn can enhance the treatment distribution without loss or reduction in its antifungal function, guaranteeing their potential participation in the development of copper-based pesticides and fungicides with effective transporting systems (Brunel et al. 2013). In an in vitro study conducted by Saharan et al. (2015), authors could synthesize Cu-chitosan nanocomposites and test their effects when applied to tomato plants (Fig. 3.2). The study indicated that

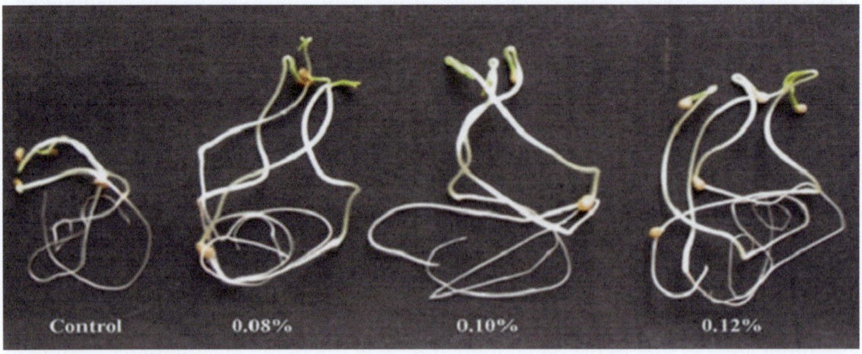

Fig. 3.2 Effect of Cu–chitosan NCs on growth of tomato seedling. (Reprinted from Saharan et al. 2015)

Fig. 3.3 Symptoms of *Curvularia lunata* disease on maize plant leaf in pot experiments (**a**) large necrotic lesion in control (**b**) micro lesions on Cu-chitosan NCs (0.16%)-treated leaf (Choudhary et al. 2017b)

treating tomato with synthesized Cu-chitosan CNs could enhance seed germination, seedling length, and both fresh and dry weight at different concentrations of 0.08%, 0.10%, and 0.12% (Fig. 3.1) with effective inhibitory effect on the spore germination of *Alternaria solani* and *Fusarium oxysporum* by 61.5% and 83.0% as well as on mycelial growth by 70.5% and 73.5%, respectively, at a concentration of 0.12% (Saharan et al. 2015).

It is assumed that Cu-chitosan NCs can help in controlling destructive diseases affecting crops of economic importance as they showed significant fungicidal activity against *Curvularia* leaf spot (CLS) disease in maize (Fig. 3.3). Hence, Cu-chitosan NPs are believed to be beneficial in inducing the systemic acquired resistance in maize as well as promoting its development (Choudhary et al. 2017b). In addition, exposing both *R. solani* and *S. rolfsii* to 100 mg of Cu-chitosan NPs could reduce their mycelial growth and inhibit their sclerotia formation. Such an activity can be explained by the ability of treatment solution to adhere to fungal hyphae and hinder their growth (Rubina et al. 2017). Various applications of Cu-based chitosan NCs in plant pathology are described in Table 3.1.

3.3.4 Antibacterial

Among their diverse antimicrobial activity, copper nanoparticles exhibit antibacterial activity against various bacterial strains, enabling them to be the promising alternatives for agrochemicals in plant protection and disease management with low cost and toxicity (Bogdanović et al. 2014; Young and Santra 2014). Moreover, Cu NPs showed inductive effect on the growth and nutrient accumulation of both *Vigna radiata* and maize (Pradhan et al. 2015; Saharan et al. 2016). However, it is a

Table 3.1 Applications of Cu-based chitosan NCs in plant pathology

Application	Nanocomposites	Targeted Pathogens	References
Antifungal activity	Cu–chitosan NCs	*Fusarium graminearum*	Brunel et al. (2013)
	Cu–chitosan NCs	*R. solani* and *S. rolfsii*	Rubina et al. (2017)
Plant growth and antifungal activity	Cu–chitosan NCs	Increased seed germination, seedling length, fresh and dry weight, and antifungal activity against *Alternaria solani* and *Fusarium oxysporum*	Saharan et al. (2015)
Plant growth promoting	Cu-chitosan hydrogel	Positive effects on tomato growth	Juarez-Maldonado et al. (2016)
Plant growth and antifungal activity	Cu–chitosan NCs	Increased seedling length, fresh and dry weight, and antifungal activity against *Rhizoctonia solani*	Abd-Elsalam et al. (2018)
Plant growth promoting activity	Cu–chitosan NCs	Increased seedling growth by upregulation of amylase and protease enzyme in maize seed germination	Saharan et al. (2015)
	Cu-chitosan hydrogel	Positive effects on tomato growth and quality	Juarez-Maldonado et al. (2016)

critical consideration for search for possible means to limit Cu NPs agglomeration which can reduce their antibacterial activity and efficiency (Karlsson et al. 2008). The global awareness about the bactericidal effect of copper nanoparticles has not been drawn until 2008 when copper was approved by the United States Environmental Protection Agency as an antimicrobial agent with inhibitory effects against harmful bacteria related to potentially deadly microbial infections (Theivasanthi and Alagar 2011). Gunawan et al. (2011) have illustrated that Cu NPs possess the ability for reducing the microbial growth of a wide range of microorganisms involving pathogenic bacteria. Moreover, Esteban-Tejeda et al. (2009) illustrated that adding copper NPs to the low melting point of soda-lime glass powder could inhibit the bacterial growth of both gram (+) and gram (−) bacteria, fungi, and yeast with a potential increase in its activity due to Ca^{+2} lixiviated from glass with some reports discussing their effectivity against *Xanthomonas* sp. such as limiting the spread of bacterial blight disease in rice caused by *Xanthomonas oryzae* and mung leaf spot causal agent *Xanthomonas campestris*.

Schlich and Hund-Rinke (2015) indicated that tests performed under two different soil types (sandy loam and sandy clay loam) for studying the effects of cuprous oxide and magnetite NPs have revealed their significant and various effects on the microbial community and bacterial activities in sandy loam soil. Copper nanoparticles showed bactericidal effect when applied on *Xanthomonas axonopodis* pv. *punicae* (Xap), causing inhibition to the water-soaked lesions caused by this bacteria under a concentration of 0.2 ppm which is lower by >10,000 times than that of recommended copper oxychloride dose (Mondal and Mani 2012). Treating tomato with copper NPs loaded on graphene oxide sheets (GO-Cu NPs) under in vivo experimental conditions has shown significant antibacterial activity through reducing the severity of bacterial speck diseases caused by *Pseudomonas syringae* pv. tomato (Pst). (Li et al. 2017).

3.4 Antimicrobial Mode of Actions

It is reported that copper is a redox-active metal which can produce reactive oxygen species (ROS) as hydroxyl radical through which copper-based fungicides can exhibit their inhibitory activity against microbes by destroying many crucial biomolecules and causing damages in their DNA, proteins, and lipids (Borkow and Gabbay 2005; Nimse and Pal 2015). Fenton reactions are responsible for copper toxicity through which free copper ions (Cu^{2+}) can undergo interconversion between Cu (I) and Cu (II) generating ROS that cause the degradation of lipids, proteins oxidation, and DNA damages (Suresh et al. 2013). Copper particles can destroy and kill bacterial cells through adhering to their membranes and produce ROS which on their turn can increase the cell permeability, causing bacterial cells to lose the control of managing CuO transporting through their cytoplasmic membranes (Subramanian et al. 2014). Moreover, Zhao et al. (2017) demonstrated the data regarding metabolomics showing the crucial role played by $Cu(OH)_2$ nanopesticides in activating the antioxidant defense and enhancing tolerance in both maize and cucumber. Therefore, a good knowledge and further investigations regarding Cu-based nanopesticides are essential to determine their potential impacts on crop development for better understanding and appliction of these novel materials.

3.5 Stimulate Crop Growth and Induce Resistance

Copper is one of the micronutrients essential for plant growth promotions and development. It has several functions in important physiological processes affecting the living organisms by being involved in protein and metabolic enzyme structures (Kasana et al. 2017). Different studies have discussed the possibility of nanoparticles to exhibit both negative and positive effects on crops, depending on the plant-nanoparticle interactions affected by some characteristics such as the plant species, size, structure, shape, concentration, and chemical properties of nanoparticles and their stability (Mirzajani et al. 2013; Rafique et al. 2014; Nhan et al. 2015; Tripathi et al. 2015, 2017; Costa and Sharma 2016; Wang et al. 2016a,b). While some studies showed the negative effects of NPs on plants represented in productivity reduction through hindering growth and affecting the content of pigments (Landa et al. 2016; Tripathi et al. 2017), others considered well-designed NPs in the form of nanopesticides, nanosensors, and growth regulators to possess positive effects on plants by stimulating their development and growth as well as increasing crop productivity

(Fraceto et al. 2016; Wang et al. 2016a,b). However, both *Landoltia* (*Spirodela*) *punctata* and *Zea mays* were found to be negatively affected by copper nanoparticles, which showed toxic effects on their growth and development (Shi et al. 2011; Wang et al. 2012). Depending on the postulates afforded by Wang et al. (2012), copper nanoparticles are consumed by different plant species which device their roots to uptake Cu-NPs from the surrounding, transporting them to the other plant parts through xylem tissues. However, the back translocation of Cu-NPs through the phloem can be accompanied by risks affecting the food safety since copper undergo reduction from Cu(II) to Cu(I). The effect of CuO nanoparticles of <50 nm size on germination and growth of seeds of *Glycine max* L. and *Cicer arietinum* L. showed that in both the crops, germination occurred up to 2000 ppm copper applied through copper oxide-nanoparticles, but the root growth was stopped above 500 ppm copper (Adhikari et al. 2012). Metal- and metal oxide-based nanomaterials have been shown to act as mediators of DNA damage in mammalian cells, organisms, and even in bacteria, but the molecular mechanisms through which this occurs are poorly understood (Atha et al. 2012). Along with the promising Cu NPs antifungal efficacy against *Phytophthora infestans* on tomato, it was also found that copper-based nanoparticles did not induce any permanent damage/deleterious effect to the plants (Giannousi et al. 2013). Furthermore, on the contrary findings concerning the negative effect of Cu NPs application on the soil nutrients, the current report has shown that applying Cu-NPs was accompanied by minor variation in the degree of soil pH and electrical conductivity (Karunakaran et al. 2013). Additionally, considerable variations were observed in the soil content of organic matter and macronutrients of NPK. However, the authors concluded that these results verify that application of copper nanoparticles has influences on the soil but in a considerable way. The application of CuNPs (10–50 ppm) to soil in pots caused a considerable increase of growth and yield of wheat, whereby treatment with 30 ppm CuNPs led also to a significantly higher Chl content, leaf area, number of spikes/pot, number of grains/spike, 1000-grain weight, and grain yield as compared with the control (Hafeez et al. 2015). The number of grains per spike and 1000-grain weight were improved in wheat varieties with Cu NPs exposure (Yasmeen et al. 2017). CuO NPs (<50 nm) supplied in solution culture as well as in the form of spray enhanced the growth of *Z. mays* plant by 51% compared to the control and significantly affected the activity of glucose 6-phosphate dehydrogenase and the pentose phosphate pathway of maize plant (Adhikari et al. 2016). Very recently, positive effects on tomato plant growth and fruit quality were observed by copper nanoparticles absorbed in chitosan hydrogel (Juarez-Maldonado et al. 2016). The influence of Cu and CuO NPs on the soil microbial community which was calculated using culture-dependent and culture-independent procedures showed that nanoparticles changed the microbial community structure. Some bacterial strains isolated from agricultural soil were used to evaluate the cytotoxicity of copper oxide nanoparticles, and strains like *Brevibacillus laterosporus*, *Chryseobacterium indoltheticum*, and *Pantoea ananatis* showed high sensitivity (Guerrero et al. 2014). *Pseudomonas syringae* was inhibited at 200 mg/L of Cu NPs. Cu NPs were not significantly biocidal against *Rhizobium* spp. and *Trichoderma harzianum* compared to CoC (Banik and

Pérez-de-Luque 2017). The effect of chitosan-PVA and Cu NPs absorbed on chitosan-PVA on growth, antioxidant capacity, mineral content, and saline stress in *Solanum lycopersicum* plants was investigated. Chitosan-PVA + Cu NPs promoted the vegetative and reproductive growth of plants and increased the quality of tomato fruit (Hernández-Hernández et al. 2018). Chitosan was reported as an excellent metal chelating or encapsulating agent for micro- or nanosize particles for Cu individually (Saharan et al. 2013; Rajasekaran and Santra 2015). So, Cu-based chitosan NPs can exhibit a dual role in plant as antimicrobial agents and plant growth enhancers (Du et al. 2009; Saharan et al. 2013, 2015). To date, fundamental studies have been performed to induce the plant innate system for plant defense and subsequent higher growth and yield by NP applications, thus, need further study of Cu-chitosan NCs for its effect on plant growth and protection for its comprehensive application in crop (Choudhary et al. 2017a,b). Dimkpa et al. (2017) briefed data, by no means exhaustive, from soil-based studies with micronutrient nanomaterials, in which the bioactivity of the nanomaterial is regulated based on soil property, in comparison with their bulk or ionic equivalents. Notably, Table 3.2 shows clearly that crop responses are more often than not positive at low doses compared to doses of traditional micronutrients used in real settings but negative at high doses, many of which are too high for the crop requirements of the respective nutrients. The increasing concentration of different metal nanoparticles in the agroecosystem may cause a negative impact on agriculture in the future (Kasana et al. 2017).

3.6 Phytotoxicity

Phytotoxicity of Cu and Cu ions and Cu/CuO NPs was carefully evaluated in details (Yruela 2005; Dimkpa et al. 2013; Rastogi et al. 2017). Many plant species are known to absorb NPs by the roots and translocate them in stems and leaves, depending on the physicochemical features of NPs, the type of plant, and the growth medium. For example, CuO NPs gathered on the plant epidermis of conventional cotton leaves, whereas it entered transgenic cotton leaf cells by means of endocytosis (Le Van et al. 2016). Spraying of Indian mustard (*Brassica juncea* L.) with CuO NPs (0, 20, 50, 100, 200, and 500 mg/L) suppressed both root and shoot boom in a dose-established manner (Nair and Chung 2015). At 500 mg/L, CuO NPs was inhibited plant growth pramaters including; (plant height, fresh weight, and leaf surface) of maize (*Zea mays* L.) (Adhikari et al. 2016). The phytotoxic effect of Cu NPs was investigated at the developmental increase of *Raphanus sativus*, *Lolium perenne*, and *Lolium rigidum* with the aid of inhibiting their practical structures and degradation their DNA by using introducing mutations and inducing DNA lesions (Atha et al. 2012). Instead, remedy of plants with excess Cu concentrations effects in plant growth inhibition and insufficiency of vital cellular processes (Yruela 2005). The root morphology was observed to be harmfully affected with Cu and CuO NP, with nearly complete inhibition with a high dose of NP (Lee et al. 2008; Adhikari et al. 2012; Perreault et al. 2014; Shaw et al. 2014; Song et al. 2016; Adams et al. 2017).

Table 3.2 Effects of nanocopper micronutrients on crops in soil-based systems, without and with comparisons to bulk-scale or ionic micronutrients

Nanofertilizer	Comparison with	Concentration (mg/kg or mg/L)	Main soil test condition and treatment application	Main agronomic findings	Effect compared to non-nano equivalent	References
Bare CuO or ZnO	N/A	10–100–1000	pH 8.2; loamy soil; soil applied	Spinach growth inhibited at 1000 mg/L but stimulated at lower rates	N/A[a]	Singh and Kumar (2016)
Bare CuO	N/A	100–300	pH 8.3 vs 4.8; applied in soil	At high pH, Cu uptake less and toxicity negated in wheat at all doses; at low pH and high dose, elevated Cu uptake corresponded with root growth inhibition	N/A	Anderson et al. (2017)
Bare CuO	Bulk CuO	50–200	Foliar applied	No reduction in cucumber yield compared to control; compromised fruit quality	Nano increased yield at 200 mg/L; bulk increased yield at 50 mg/L; nano reduced photosynthesis and transpiration at 200 mg/L	Hong et al. (2016)
Bare Cu, CuO	Bulk Cu, bulk CuO, $CuCl_2$	20–80	Soil application	Reduced cilantro germination, stimulated root and shoot biomass	Bulk CuO reduced biomass at 80 mg/kg	Zuverza-Mena et al. (2015)
Nanocomposite: ZnO + CuO + B_2O_3	$ZnSO_4$ + $CuSO_4$ + H_3BO_3	2.8 + 0.6 + 1.3	Sandy loam soil; pH 6.87; foliar applied	Increased soybean growth, yield; and grain N, K, Zn, and B accumulation under drought	Similar effect as salt formulation	Dimkpa et al. (2017)

Bare CuO, MnO, or ZnO	Bulk CuO, MnO, or ZnO; Cu, Mn, or zinc sulfate	100–1000	Sandy loam; pH 6.1; soil infested with *Verticillium* wilt fungus; shoot dipping and transplanting	Stimulation of plant defense against fungal disease effect; yield stimulation of eggplant and tomato by CuO	Effects greater than bulk and salt equivalents for Cu and Mn in eggplant; Zn salt better than nano for tomato	Elmer and White (2016)
Bacteria-synthesized Cu	Bulk Cu	1–2.5	pH 4.5–5.3; root-rot infestation; foliar application	Reduction of disease incidence; enhancement of yield in tea under disease condition	Effect better than bulk Cu at 2.5 mg/L	Ponmurugan et al. (2016)

[a]N/A, not applicable; comparisons were not made with conventional nutrients. (Modified and reprinted from Dimkpa et al. (2017)

Cu NPs inhibited increase rate of mung bean (*Phaseolus radiatus*) and wheat (*Triticum aestivum*), which is one of a symptom of toxicity (Lee et al. 2008). Phytotoxicity of CuO nanoparticles was evaluated in wheat plants, which were treated, these nanoparticles found to be toxic to wheat and causing a reduction in root length (Dimkpa et al. 2013). It was discovered that Cu-NPs have robust phytotoxic results when applied as colloidal solution at *Allium cepa* roots (Konotop et al. 2014). CuO NP was evaluated to negatively affect the photosynthetic activity through inactivating PS II response centers, and inflicting a lower in electron shipping, thylakoid range in line with grana, photosynthetic rate, photosynthetic pigments, transpiration rate, stomatal conductance (Perreault et al. 2014; Costa and Sharma 2016). A combined treatment on the plant with ultraviolet radiation and CuO NP was investigated to significantly increase the phytotoxic impact of CuO NP (Regier et al. 2015). It was found out in the assessment of the effect of CuNP on wheat that the charge of germination of wheat seeds becomes higher in the presence of CuNPs and CoC compared to control. Germination index, root length, shoot dry weight, and seed metabolic performance of wheat had been negatively affected (Banik and Pérez-de-Luque 2017). The growing release of CuO NPs from each intended and unintended resources into plant surroundings may additionally pose hazards to rice plant life, thereby lowering the quality or amount of this staple grain in the human food regimen. Few studies have addressed CuO NP phytotoxicity to rice, and the interactions of CuO NPs with As are poorly described (Liu et al. 2018). Phytotoxicity assays are executed in two stages of plant development: (1) all through germination, when the germination percent is measured, where the seeds ought to be exposed to the take a look at solution during germination (ideally as a minimum 4 days), and (2) at some point of seedling growth, wherein root/shoot elongation and dry weight are regularly used variables to evaluate the effects of plant exposure to dangerous materials (Wang and Freemark 1995). Genotoxic and phytotoxic consequences of various nanoparticles inside the roots of *Fagopyrum esculentum* seedlings was estimated the usage of the RAPD assay (Lee et al. 2013). It was found that the seedling growth was affected due to remarkable regulations in root morphology and root length as nanoparticles have been found to be accumulated, within the root epidermis. High doses of CuO NPs disturbed mobile homeostatic balance after altering intracellular signalling pathways which led to genotoxic effects for plant cells. Adverse results of different Cu/Cu NPs on special plant species were defined in Table 3.3.

3.7 Conclusion and Future Prospective

Research during the last years has documented the toxic effects of Cu NPs, both during the germination stage or with regard to the shoot-root length, while few others have explored the opportunities of using them as agrochemicals such as fungicides, bactericides, fertilizers, and nanosensors. Particularly, improved yield might not only be directly related to the decreased presence of pathogenic organisms but also to the potential nutritional value of the nanoparticles themselves, mainly for the essential

Table 3.3 Negative effects of different Cu/Cu NPs on different plant species

Copper NMs	Plant	Negative effect	References
Cu NPs	*Allium cepa*	Inhibited the root growth of onion	Konotop et al. (2014)
Cu NPs	*Cucumis sativus*	Decrease in root length, reduction of root biomass, bioaccumulation mainly in roots, a little in stems	Zhao et al. (2017)
CuO NPs	*Cucurbita pepo*	Reduced emerging root length	Stampoulis et al. (2009)
CuO NPs	*Fagopyrum esculentum*	Inhibition of root length, genomic DNA damaging, altered gene expression	Lee et al. (2013)
CuO NPs	*Brassica juncea*	Inhibited both root and shoot growth	Nair and Chung (2015)
CuO NPs	*Zea mays*	Inhibited plant height, fresh weight, leaf surface area	Adhikari et al. (2016)
CuO NPs	*Lactuca sativa*	Reduction of seed germination and root elongation	Liu and Lal (2015)
Core–shell NPs Cu/CuO	*Lactuca sativa*	Reduction of water content, root length, and dry biomass of the plant, alteration of the nutritional quality of lettuce	Liu and Lal (2015)
CuO NPs	*Raphanus sativus, Lolium perenne, Lolium rigidum*	Oxidative damage to plant DNA, inhibition of seedling growth (root and shoot growth)	Atha et al. (2012)
Cu NPs and CoC	*Triticum aestivum*	Germination vigor index, root length, shoot dry weight and seed metabolic efficiency of wheat were negatively affected	Banik and Pérez-de-Luque (2017)
CuO NPs	*Elsholtzia splendens*	No effect on seed germination, reduction of root length, accumulation of CuO NPs in root and leaf cells	Dimkpa et al. (2013)
CuO NPs	*Zea mays, Oryza sativa*	Inhibited root elongation at 2000 mg/L (95.73% for maize and 97.28% for rice) of two crop plants and reduced shoot length of maize by 30.98%	Yang et al. (2015)
CuO NPs	*Zea mays*	Did not affect the germination of maize, but suppressed root elongation.	Wang et al. (2012)
CuO NPs	*Oryza sativa*	Reduced the fresh weights and root length	Shaw and Hossain (2013)

micronutrients important for host defense. Many researchers have proven the effect of nanoparticles on plant growth and their accumulation in food source; however there are necessities for research to recognize the molecular mechanism of plant nanoparticle interaction. There are few researches displaying the useful role of metal and metallic oxide nanoparticles in agriculture, but the mechanism at huge extent are not understood, and the studies are in its primary stages. Therefore, screening and assay of nanoparticles is requested for enforcing them in specific fields alongside with agriculture. Copper oxides or composite NPs together with Cu/Cu_2O protected by means of polymers or embedded in matrices appear to be an effective nano-metal fungicide utility as inexpensive eco-fungicide for agriculture.

Acknowledgment This research was supported by the Science and Technology Development Fund (STDF), Joint Egypt (STDF)–South Africa (NRF) Scientific Cooperation, Grant ID. 27837 to Kamel Abd-Elsalam.

References

Abd-Elsalam KA, Vasil'kov AY, Said-Galiev EE, Rubina MS, Khokhlov AR, Naumkin AV, Shtykova EV, Alghuthaymi MA (2018) Bimetallic and chitosan nanocomposites hybrid with trichoderma: novel antifungal agent against cotton soil–borne fungi. Eur J Plant Pathol 151:57–72. https://doi.org/10.1007/s10658-017-1349-8

Adams J, Wright M, Wagner H, Valiente J, Britt D, Anderson A (2017) Cu from dissolution of CuO nanoparticles signals changes in root morphology. Plant Physiol Biochem 110:108–117

Adhikari T, Kundu S, Biswas AK, Tarafdar JC, Rao AS (2012) Effect of copper oxide nano particle on seed germination of selected crops. J Agric Sci Technol A 2:815–823

Adhikari T, Sarkar D, Mashayekhi H, Xing BS (2016) Growth and enzymatic activity of maize (*Zea mays* L.) plant: solution culture test for copper dioxide nano particles. J Plant Nutr 39:102–118

Ahamed M, Alhadlaq HA, Khan MM, Karuppiah P, Aldhabi NA (2014) Synthesis, characterization and antimicrobial activity of copper oxide nanoparticles. J Nano Mater 2014:1–4. https://doi.org/10.1155/2014/637858

Anderson A, McLean J, McManus P, Britt D (2017) Soil chemistry influences the phytotoxicity of metal oxide nanoparticles. Int J Nanotechnol 14(1–6):15–21

Atha DH, Wang H, Petersen EJ, Cleveland D, Holbrook RD, Jaruga P, Dizdaroglu M, Xing B, Nelson BC (2012) Copper oxide nanoparticle mediated DNA damage in terrestrial plant models. Environ Sci Technol 46(3):1819–1827

Banik S, Pérez-de-Luque A (2017) In vitro effects of copper nanoparticles on plant pathogens, beneficial microbes and crop plants. Spanish J Agric Res 15(2):1005. https://doi.org/10.5424/sjar/2017152-10305

Boehm AL, Martinon I, Zerrouk R, Rump E, Fessi H (2003) Nanoprecipitation technique for the encapsulation of agrochemical active ingredients. J Microencapsul 20:433–441

Bogdanović U, Lazić V, Vodnik V, Budimir M, Marković Z, Dimitrijević S (2014) Copper nanoparticles with high antimicrobial activity. Mater Lett 128:75–78

Boonham N, Glover R, Tomlinson J, Mumford R (2008) Exploiting generic platform technologies for the detection and identification of plant pathogens. Eur J Plant Pathol 121:355–363

Borkow G, Gabbay J (2005) Copper as a biocidal tool. Curr Med Chem 12:2163–2175

Borkow G, Gabbay J (2009) Copper, an ancient remedt returning to fight microbial, fungal and viral infections. Curr Chem Biol 3:272–278

Bouson S, Krittayavathananon A, Phattharasupakun N, Siwayaprahm P, Sawangphruk M (2017) Antifungal activity of water–stable copper–containing metal–organic frameworks. R Soc Open Sci 4:170654 https://doi.org/10.1098/rsos.170654

Bramhanwade K, Shende S, Bonde S, Gade A, Rai M (2016) Fungicidal activity of Cu nanoparticles against *Fusarium* causing crop diseases. Environ Chem Lett 14(2):229–235

Brunel F, ElGueddari NE, Moerschbacher BM (2013) Complexation of copper (II) with chitosan nanogels: toward control of microbial growth. Carbohydr Polym 92:1348–1356

Cárdenaz G, Díaz JV, Meléndrez MF, Cruzat CC, Cancino AG (2009) Colloidal Cu nanoparticles/chitosan composite film obtained by microwave heating for food package applications. Polym Bull 62:511–524

Carmen IU, Chithra P, Huang Q, Takhistov P, Liu S, Kokini JL (2003) Nanotechnology: a new frontier in food science. Food Technol 57:24–29

Chatterjee AK, Sarkar RK, Chattopadhyay AP, Aich P, Chakraborty R, Basu T (2012) A simple robust method for synthesis of metallic copper nanoparticles of high antibacterial potency against. Nanotechnology 23 (8):085103

Chen S, Sommers JM (2001) Alkanethiolate–protected copper nanoparticles: spectroscopy, electrochemistry, and solid–state morphological evolution. J Phys Chem B 105:816–8820

Choudhary RC, Kumaraswamy RV, Kumari S, Pal A, Raliya R, Biswas P, Saharan V (2017a) Synthesis, characterization, and application of chitosan nanomaterials loaded with zinc and copper for plant growth and protection. In: Prasad R, Kumar V, Kumar M (eds) Nanotechnology: food and environmental paradigm. Springer Nature Singapore Pte Ltd, Singapore, pp 227–248

Choudhary RC, Kumaraswamy RV, Kumari S, Sharma SS, Pal A, Raliya R, Biswas P, Saharan V (2017b) Cu-chitosan nanoparticle boost defense responses and plant growth in maize (*Zea mays* L.). Sci Rep 7:9754. https://doi.org/10.1038/s41598-017-08571-0

Cioffi N, Rai M (2012) Nano-antimicrobials: progress and prospects. Springer-Verlag, Berlin, Heidelberg

Cioffi N, Torsi L, Ditaranto N (2004) Antifungal activity of polymer–based copper nanocomposite coatings. Appl Phys Lett 85(12):2417–2419

Cioffi N, Torsi L, Ditaranto N, Tantillo G, Ghibelli L, Sabbatini L, Traversa E (2005) Copper nanoparticle/polymer composites with antifungal and bacteriostatic properties. Chem Mater 17(21):5255–5262

Costa MVJD, Sharma PK (2016) Effect of copper oxide nanoparticles on growth, morphology, photosynthesis, and antioxidant response in *Oryza sativa*. Photosynthetica 54:110–119

Dang V P, Vo TKL, Nguyen TKL, Nguyen ND, Nguyen DC, Bui DD, Bui DC, Nguyen QH (2010) Synthesis and antimicrobial effects of colloidal silver nanoparticles in chitosan by -irradiation. J Exp Nanosci 5(2):169–179

Dimkpa CO, McLean GE, Britt DW, Anderson AJ (2013) Antifungal activity of ZnO nanoparticles and their interactive effect with a biocontrol bacterium on growth antagonism of the plant pathogen *Fusarium graminearum*. Bio Metals 26(6):913–924

Dimkpa C, Bindraban P, Fugice J, Agyin-Birikorang S, Singh U, Hellums D (2017) Composite micronutrient nanoparticles and salts decrease drought stress in soybean. Agron Sustain Dev 37:5

Du WL, Niu SS, Xu YL, Xu ZR, Fan CL (2009) Antibacterial activity of chitosan tripolyphosphate nanoparticles loaded with various metal ions. Carbohydr Polym 75:385–389

Eastman JA, Choi S, Li S, Yu W, Thompson L (2001) Anomalously increased effective thermal conductivities of ethylene glycol–based nanofluids containing copper nanoparticles. Appl Phys Lett 78(6):718–720

Elmer W, White JC (2016) The use of metallic oxide nanoparticles to enhance growth of tomatoes and eggplants in disease infested soil or soilless medium. Environ Sci Nano 3:1072–1079

El-Sayed NR (2003) Effect of catalysis on the stability of metallic nanoparticles: Suzuki reaction catalyzed by PVP–palladium nanoparticles. J Am Chem Soc 125:8340–8347

Esteban-Tejeda L, Malpartida F, Esteban-Cubillo A, Pecharromn C, Moya JS (2009) Antibacterial and antifungal activity of a soda–lime glass containing copper nanoparticles. Nanotechnology 20(50):505701

Etefagh R, Azhir E, Shahtahmasebi N (2013) Synthesis of CuO nanoparticles and fabrication of nanostructural layer biosensors for detecting *Aspergillus niger* fungi. Sci Iran 20(3):1055–1058

Fraceto LF, Grillo R, de Medeiros GA, Scognamiglio V, Rea G, Bartolucci C (2016) Nanotechnology in agriculture: which innovation potential does it have? Front Environ Sci 4:20. https://doi.org/10.3389/fenvs.2016.00020

Garcıa VN, Gonzalez A, Fuentes M, Aviles M, Rios MY, Zepeda G, Rojas MG (2003) Antifungal activities of nine traditional Mexican medicinal plants. Jethnopharmacol 87(1):85–88.

Garcia M, Forbe T, Gonzalez E (2010) Potential applications of nanotechnology in the agro–food sector. Ciênc Tecnol Aliment 30:573–581

Ghasemian E, Naghoni A, Tabaraie B, Tabaraie T (2012) In vitro susceptibility of filamentous fungi to copper nanoparticles assessed by rapid XTT colorimetry and agar dilution method. J Mycol Med 22:322–328

Ghormade V, Deshpande MV, Paknikar KM (2011) Perspectives for nano–biotechnology enabled protection and nutrition of plants. Biotechnol Adv 29:792–803

Giannousi K, Avramidis I, Dendrinou-Samara C (2013) Synthesis, characterization and evaluation of copper based nanoparticles as agrochemicals against *Phytophthora infestans*. RSC Adv 3:21743–21752

Giannousi K, Pantazaki A, Dendrinou-Samara C (2017) Copper based nanoparticles as antimicrobials. In: Ficai A, Grumezescu AM (eds) Nanostructures for antimicrobial therapy. Elsevier, Amsterdam, pp 515–527

Gogos A, Knauer K, Bucheli TD (2012) Nanomaterials in plant protection and fertilization: current state, foreseen applications, and research priorities. J Agric Food Chem 60(39):9781–9792

Guerrero SIC, Brito EMS, Castillo HAP, Rivero SHT, Caretta CA, Velasco AL, Duran R, Borunda EO (2014) Effect of CuO nanoparticles over isolated bacterial strains from agricultural soil. J Nanomater 2014:1–13

Gunawan C, Teoh WY, Marquis CP, Amal R(2011) Cytotoxic Origin of Copper(II) Oxide Nanoparticles: Comparative Studies with Micron-Sized Particles, Leachate, and Metal Salts. ACS Nano 5 (9):7214–7225

Hafeez A, Razzaq A, Mahmood T, Jhanzab HM (2015) Potential of copper nanoparticles to increase growth and yield of wheat. J Nanosci Adv Technol 1:6–11

Hernández-Hernández H, González-Morales S, Benavides-Mendoza A, Ortega-Ortiz H, Cadenas-Pliego G, Juárez-Maldonado A (2018) Effects of chitosan–PVA and Cu nanoparticles on the growth and antioxidant capacity of tomato under saline stress. Molecules 23(1):178

Hirsh S, Schiefer J, Gschwandtner A, Hartmann M (2014) The determinants of firm profitability differences in EU food processing. J Agric Econ 65:703–721

Honary H, Barabadi H, Gharaei-Fathabad E, Naghibi F (2012) Green synthesis of copper oxide nanoparticles using *Penicillium aurantiogriseum*, *Penicillium citrinum* and *Penicillium waksmanii*. Dig J Nanomater Biostruct 7(3):999–1005

Hong J, Wang L, Sun Y, Zhao L, Niu G, Tan W, Gardea-Torresdey JL (2016) Foliar applied nanoscale and microscale CeO_2 and CuO alter cucumber (*Cucumis sativus*) fruit quality. Sci Total Environ 563:904–911

Hooley G, Piercy NF, Nicoulaud B (2014) Marketing strategy and competitive positioning. Prentice Hall/Financial Times, London (ISBN 9780273740933)

Ingle AP, Duran N, Rai M (2014) Bioactivity, mechanism of action, and cytotoxicity of copper–based nanoparticles: a review. Appl Microbiol Biotechnol 98:1001–1009

Jeong S, Woo K, Kim D, Lim S, Kim JS, Shin H, Moon J (2008) Controlling the thickness of the surface oxide layer on Cu nanoparticles for the fabrication of conductive structures by ink–jet printing. Adv Funct Mater 18(5):679–686

Juarez-Maldonado A, Ortega-Ortiz H, Perez-Labrada F, Cadenas-Pliego G, Benavides-Mendoza A (2016) Cu nanoparticles absorbed on chitosan hydrogels positively alter morphological, production, and quality characteristics of tomato. J Appl Bot Food Qual 89:183–189

Kah M, Hofmann T (2014) Nanopesticide research: current trends and future priorities. Environ Int 63:224–235

Kanhed P, Birla S, Gaikwad S, Gade A, Seabra AB, Rubilar O, Duran N, Rai M (2014) In vitro antifungal efficacy of copper nanoparticles against selected crop pathogenic fungi. Mater Lett 115:13–17

Karlsson HL, Cronholm P, Gustafsson J, Moller L (2008) Copper oxide nanoparticles are highly toxic: a comparison between metal oxide nanoparticles and carbon nanotubes. Chem Res Toxicol 21:1726–1732

Karunakaran G, Suriyaprabha R, Manivasakan P, Yuvakkumar R, Rajendran V, Kannan N (2013) Impact of nano and bulk ZrO_2, TiO_2 particles on soil nutrient contents and PGPR. J Nanosci Nanotechnol 13(1):678–685

Kasana RC, Panwar NR, Kaul RK, Kumar P (2017) Biosynthesis and effects of copper nanoparticles on plants. Environ Chem Lett 15:233–240
Khiyami MA, Almoammar H, Awad YM, Alghuthaymi MA, Abd-Elsalam KA (2014) Plant pathogen nanodiagnostic techniques: forthcoming changes? Biotechnol Biotechnol Equip 28:775–785
Kim MH, Lim B, Lee EP, Xia Y (2008) Polyol synthesis of Cu_2O nanoparticles: use of chloride to promote the formation of a cubic morphology. J Mater Chem 18:4069–4073
Konotop YO, Kovalenko MS, Ulynets VZ, Meleshko AO, Batsmanova LM, Taran NY (2014) Phytotoxicity of colloidal solutions of metal–containing nanoparticles. Cytol Genet 48:99–102
Kumar CSSR (2009) Metallic nanomaterials. Wiley-VCH Verlag GmbH and Co. KGaA, Weinheim
Landa P, Cyrusova T, Jerabkova J, Drabek O, Vanek T, Podlipna R (2016) Effect of metal oxides on plant germination: phytotoxicity of nanoparticles, bulk materials, and metal ions. Water Air Soil Pollut 227:448. https://doi.org/10.1007/s11270–016–3156–9
Lee WM, An YJ, Yoon H, Kweon HS (2008) Toxicity and bioavailability of copper nanoparticles to the terrestrial plants mung bean (*Phaseolus radiatus*) and wheat (*Triticum aestivum*) plant agar test for water–insoluble nanoparticles. Environ Toxicol Chem 27:1915–1921
Lee S, Chung H, Kim S, Lee I (2013) The genotoxic effect of ZnO and CuO nanoparticles on early growth of buckwheat, *Fagopyrum esculentum*. Water Air Soil Pollut 224(9):1668. https://doi.org/10.1007/S11270-013-1668-0
Le Van N, Ma C, Shang J, Rui Y, Liu S, Xing B(2016) Effects of CuO nanoparticles on insecticidal activity and phytotoxicity in conventional and transgenic cotton. Chemosphere 144:661–670
Li Y, Yang D, Cui J (2017) Graphene oxide loaded with copper oxide nanoparticles as an antibacterial agent against *Pseudomonas syringae* pv. tomato. RSC Adv 7:38853–38860
Liu R, Lal R (2015) Potentials of engineered nanoparticles as fertilizers for increasing agronomic productions. Sci Total Environ 514:131–139
Liu J, Dhungana B, Cobb GP (2018) Environmental behavior, potential phytotoxicity, and accumulation of copper oxide nanoparticles and arsenic in rice plants. Environ Toxicol Chem 37 (1):11–20
Mirzajani F, Askari H, Hamzelou S, Farzaneh M, Ghassempour A (2013) Effect of silver nanoparticles on Oryza sativa L and its rhizosphere bacteria. Ecotoxicol Environ Saf 88:48–54
Mohan R, Shanmugharaj AM, Hun RS (2011) An efficient growth of silver and copper nanoparticles on multiwalled carbon nanotube with enhanced antimicrobial activity. J Biomed Mater Res B 96:119–126
Mondal KK, Mani C (2012) Investigation of the antibacterial properties of nanocopper against *Xanthomonas axonopodis* pv. *punicae*, the incitant of pomegranate bacterial blight. Ann Microbiol 62(2):889–893
Montag J, Schreiber L, Schönherr J (2006) An in vitro study on the postinfection activities of copper hydroxide and copper sulfate against conidia of Venturia inaequalis. J Agric Food Chem 54(3):893–899
Nair PMG, Chung IM (2015) Study on the correlation between copper oxide nanoparticles induced growth suppression and enhanced lignification in Indian mustard (*Brassica juncea* L.). Ecotoxicol Environ Saf 113:302–313
Navarro E, Baun A, Behra R, Hartmann NB, Filser J, Miao A-J, Quigg A, Santschi PH, Sigg L (2008) Environmental behavior and ecotoxicity of engineered nanoparticles to algae, plants, and fungi. Ecotoxicology 17 (5):372–386
Nelson SC (2008) Late blight of tomato (*Phytophthora infestans*). Honolulu (HI): University of Hawaii. 10 p. (Plant Disease; PD–45).
Nhan LV, Ma C, Rui Y, Liu S, Li X, Xing B, Liu L (2015) Phytotoxic mechanism of nanoparticles: destruction of chloroplasts and vascular bundles and alteration of nutrient absorption. Sci Rep 5:11618. https://doi.org/10.1038/srep11618
Nimse SB, Pal D (2015) Free radicals, natural antioxidants, and their reaction mechanisms. RSC Adv 5(35):27986–28006

Ouda SM (2014) Antifungal activity of silver and copper nanoparticles on two plant pathogens, *Alternaria alternata* and *Botrytis cinerea*. Res J Microbiol 9:34–42
Park HJ, Kim SH, Kim HJ, Choi SH (2006) A new composition of nanosized silica–silver for control of various plant diseases. Plant Pathol J 22:295–302
Patolsky F, Zheng G, Lieber CM (2006) Nanowire sensors for medicine and life sciences. Nanomedicine 1:51–65
Pérez-de-Luque A, Hermosín MC (2013) Nanotechnology and its use in agriculture. Wiley-Blackwell, Chichester, pp 299–405
Perreault F, Samadani M, Dewez D (2014) Effect of soluble copper released from copper oxide nanoparticles solubilisation on growth and photosynthetic processes of *Lemna gibba* L. Nanotoxicology 8:374–382
Ponmurugan P, Manjukarunambika K, Elango V, Gnanamangai BM (2016) Antifungal activity of biosynthesized copper nanoparticles evaluated against red root–rot disease in tea plants. J Exp Nanosci 11(13):1019–1031
Pradhan S, Patra P, Mitra S, Dey KK, Basu S, Chandra S, Palit P, Goswami A (2015) Copper nanoparticle (CuNP) nanochain arrays with a reduced toxicity response: a biophysical and biochemical outlook on *Vigna radiata*. J Agric Food Chem 63:2606–2617
Prasad R (2014) Synthesis of silver nanoparticles in photosynthetic plants. J Nanopart 2014:1–8 https://doi.org/10.1155/2014/963961
Prasad R, Kumar V, Prasad KS (2014) Nanotechnology in sustainable agriculture: present concerns and future aspects. Afr J Biotechnol 13(6):705–713
Prasad R, Bhattacharyya A, Nguyen QD (2017a) Nanotechnology in sustainable agriculture: recent developments, challenges, and perspectives. Front Microbiol 8:1014. https://doi.org/10.3389/fmicb.2017.01014
Prasad R, Gupta N, Kumar M, Kumar V, Wang S, Abd-Elsalam KA (2017b) Nanomaterials act as plant defense mechanism. In: Prasad R, Kumar V, Kumar M (eds) Nanotechnology. Springer Nature Singapore Pte Ltd, Singapore, pp 253–269
Rafique R, Arshad M, Khokhar MF, Qazi IA, Hamza A, Virk N (2014) Growth response of wheat to titania nanoparticles application. NUST J Eng Sci 7:42–46
Rai M, Kratosova G (2015) Management of phytopathogens by application of green nanobiotechnology: emerging trends and challenges. J Agric Sci 66:15–22
Rajasekaran P, Santra S (2015) Hydrothermally treated chitosan hydrogel loaded with copper and zinc particles as a potential micronutrient–based antimicrobial feed additive. Front Vet Sci 2:62
Ramyadevi J, Jeyasubramanian K, Marikani A, Rajakumar G, Rahuman AA (2012) Synthesis and antimicrobial activity of copper nanoparticles. Mater Lett 71:114–116
Rastogi A, Zivcak M, Sytar O, Kalaji HM, He X, Mbarki S, Brestic M (2017) Impact of metal and metal oxide nanoparticles on plant: a critical review. Front Chem 5:78. https://doi.org/10.3389/fchem.2017.00078
Regier N, Cosio C, von Moos N, Slaveykova VI (2015) Effects of copper–oxide nanoparticles, dissolved copper and ultraviolet radiation on copper bioaccumulation, photosynthesis and oxidative stress in the aquatic macrophyte *Elodea nuttallii*. Chemosphere 128:56–61
Rubina RS, Vasil'kov AY, Naumkin AV, Shtykova EV, Abramchuk SS, Alghuthaymi MA, Abd-Elsalam KA (2017) Synthesis and characterization of chitosan–copper nanocomposites and their fungicidal activity against two sclerotia–forming plant pathogenic fungi. J Nanostruct Chem 7:249–258. https://doi.org/10.1007/s40097–017–0235–4
Saharan V, Mehrotra A, Khatik R, Rawal P, Sharma SS, Pal A (2013) Synthesis of chitosan based nanoparticles and their in vitro evaluation against phytopathogenic fungi. Int J Biol Macromol 62:677–683
Saharan V, Sharma G, Yadav M, Choudhary MK, Sharma SS, Pal A, Biswas P (2015) Synthesis and in vitro antifungal efficacy of Cu–chitosan nanoparticles against pathogenic fungi of tomato. Int J Biol Macromol 75:346–353

Saharan V, Kumaraswamy RV, Choudhary RC, Kumari S, Pal A, Raliya R, Biswas P (2016) Cu-chitosan nanoparticle mediated sustainable approach to enhance seedling growth in maize by mobilizing reserved food. J Agric Food Chem 64(31):6148–6155

Salavati-Niasari M, Davar F, Mir N (2008) Synthesis and characterization of metallic copper nanoparticles via thermal decomposition. Polyhedron 27 (17):3514–3518

Salzemann C, Lisiecki I, Urban J, Pileni MP (2004). Anisotropic copper nanocrystals synthesized in a supersaturated medium: Nanocrystal growth. Langmuir 20(26): 11772–11777.

Sangeetha J, Thangadurai D, Hospet R, Purushotham P, Manowade KR, Mujeeb MA, Mundaragi AC, Jogaiah S, David M, Thimmappa SC, Prasad R, Harish ER (2017a) Production of bionanomaterials from agricultural wastes. In: Prasad R, Kumar M, Kumar V (eds) Nanotechnology. Springer Nature Singapore Pte Ltd, Singapore, pp 33–58

Sangeetha J, Thangadurai D, Hospet R, Harish ER, Purushotham P, Mujeeb MA, Shrinivas J, David M, Mundaragi AC, Thimmappa AC, Arakera SB, Prasad R (2017b) Nanoagrotechnology for soil quality, crop performance and environmental management. In: Prasad R, Kumar M, Kumar V (eds) Nanotechnology. Springer Nature Singapore Pte Ltd, Singapore, pp 73–97

Schlich K, Hund-Rinke K (2015) Influence of soil properties on the effect of silver nanomaterials on microbial activity in five soils. Environ Pollut 196:321–330

Scrinis G, Lyons K (2007) The emerging nano-corporate paradigm: nanotechnology and the transformation of nature, food and agri-food systems. Int J Sociol Agric Food 15:22–44

Sekhon BB (2014) Nanotechnology in agri-food production: an overview. Nanotechnology, Nanotechnol Sci Appl. 7:31–53

Servin A, Elmer W, Mukherjee A, Torre-Roche RD, Hamdi H, White JC, Bindraban P, Dimkpa C (2015) A review of the use of engineered nanomaterials to suppress plant disease and enhance crop yield. J Nanopart Res 17:92 https://doi.org/10.1007/s11051-015-2907-7

Shaw AK, Hossain Z (2013) Impact of nano-CuO stress on rice (*Oryza sativa L*) seedlings. Chemosphere 93:906–915

Shaw AK, Ghosh S, Kalaji HM, Bosa K, Brestic M, Zivcak M, Hossain Z (2014) Nano-CuO stress induced modulation of antioxidative defense and photosynthetic performance of Syrian barley (*Hordeum vulgare* L). Environ Exp Bot 102:37–47

Shende S, Ingle AP, Gade A, Rai M (2015) Green synthesis of copper nanoparticles by *Citrus medica* Linn. (Idilimbu) juice and its antimicrobial activity. World J Microbiol Biotechnol 31:865–873

Shi J, Abid AD, Kennedy IM, Hristova KR, Silk WK (2011) To duckweeds (Landoltia punctata), nanoparticulate copper oxide is more inhibitory than the soluble copper in the bulk solution. Environ Poll 159 (5):1277–1282

Singh D, Kumar A (2016) Impact of irrigation using water containing CuO and ZnO nanoparticles on Spinacia oleracea grown in soil media. Bull Environ Contam Toxicol 97(4):548–553

Sodano V, Verneau F (2014) Competition policy and food sector in the European Union. J Int Food Agribusiness Mark 26:155–172

Somers E (1959) The preparation of bordeaux mixture. J Sci Food Agric 10 (1):68–72

Sonkaria S, Ahn SH, Khare V (2012) Nanotechnology and its impact on food and nutrition: a review. Recent Pat Food Nutr Agric 4(1):8–18.

Song G, Hou W, Gao Y, Wang Y, Lin L, Zhang Z, Wang H (2016) Effects of CuO nanoparticles on *Lemna minor*. Bot Stud 57:3. https://doi.org/10.1186/s40529-016-0118-x

Stampoulis D, Sinha SK, White JC (2009) Assay-dependent phytotoxicity of nanoparticles to plants. Environ Sci Technol 43(24):9473–9479.

Subramanian B, Anu Priya K, Thanka Rajan S, Dhandapani P, Jayachandran M (2014) Antimicrobial activity of sputtered nanocrystalline CuO impregnated fabrics. Mater Lett 128:1–4

Suresh AK, Pelletier DA, Doktycz MJ (2013) Relating nanomaterial properties and microbial toxicity. Nanoscale 5(2):463–474

Theivasanthi T, Alagar M (2011) Studies of copper nanoparticles effects on micro-organisms. arXiv preprint arXiv:1110: 1372.

Tripathi DK, Singh VP, Prasad SM, Chauhan DK, Dubey NK (2015) Silicon nanoparticles (SiNp) alleviate chromium (VI) phytotoxicity in *Pisum sativum* (L) seedlings. Plant Physiol Biochem 96:189–198

Van Acker H, Van Dijck P, Coenye T (2014) Molecular mechanisms of antimicrobial tolerance and resistance in bacterial and fungal biofilms. Trends Microbiol 22 (6):326–333

Tripathi DK, Singh S, Singh S, Srivastava PK, Singh VP, Singh S, Chauhan DK (2017) Nitric oxide alleviates silver nanoparticles (AgNps)–induced phytotoxicity in *Pisum sativum* seedlings. Plant Physiol Biochem 110:167–177

Viet PV, Nguyen HT, Cao TM, Hieu LV (2016) Fusarium antifungal activities of copper nanoparticles synthesized by a chemical reduction method. J Nanomater 2016:1–7 https://doi.org/10.1155/2016/1957612

Wang WC, Freemark K (1995) The Use of Plants for Environmental Monitoring and Assessment. Ecotoxicol Environ Safe 30 (3):289–301

Wang Z, Xie X, Zhao J, Liu X, Feng W, White JC, Xing B (2012) Xylem- and phloem-based transport of CuO nanoparticles in maize (*Zea mays* L). Environ Sci Technol 46:4434–4441

Wang P, Lombi E, Zhao FJ, Kopittke PM (2016a) Nanotechnology: a new opportunity in plant sciences. Trends Plant Sci 21:699–712

Wang Z, Xu L, Zhao J, Wang X, White JC, Xing B (2016b) CuO nanoparticle interaction with *Arabidopsis thaliana*: toxicity, parent-progeny transfer, and gene expression. Environ Sci Technol 50:6008–6016

Wang L, Liu Y, Liu J, Zhang Y, Zhang X, Pan H (2016c) The Gene Is Required for Apothecial Development . Phytopathology 106 (5):484–490

Wani IA, Ahmad T (2013) Size and shape dependent antifungal activity of gold nanoparticles: a case study of Candida. Colloids Surf B 101:162–170

Wei TY, Huang CT, Hansen BJ, Lin YF, Chen L J, Lu SY, Wang ZL (2010) Large enhancement in photon detection sensitivity via Schottky-gated CdS nanowire nanosensors. Appl Physics Lett 96(1):013508.

Weir E, Lawlor A, Whelan A, Regan F (2008) The use of nanoparticles in antimicrobial materials and their characterization. Analyst 133:835–845

Whitesides GM (2003) The "right" size in nanobiotechnology. Nat Biotechnol 21:1161–1165

Yang Z, Chen J, Dou R, Gao X, Mao C, Wang L(2015) Assessment of the Phytotoxicity of Metal Oxide Nanoparticles on Two Crop Plants, Maize (*Zea mays* L.) and Rice (*Oryza sativa* L.). Int J Environ Rese Public Health 12 (12):15100–15109

Yasmeen F, Raja NI, Razzaq A, Komatsu S (2017) Proteomic and physiological analyses of wheat seeds exposed to copper and iron nanoparticles. Biochim Biophys Acta 1865:28–42

Young M, Santra S (2014) Copper (Cu)–Silica nanocomposite containing valence-engineered Cu: a new strategy for improving the antimicrobial efficacy of Cu biocides. J Agric Food Chem 62:6043–6052

Youssef K, Hashim AF, Rubina RS, Alghuthaymi MA, Abd-Elsalam KA (2017) Fungicidal efficacy of chemically–produced copper nanoparticles against *Penicillium digitatum* and *Fusarium solani* on citrus fruit. Philipp Agric Sci 100:69–78

Yruela I (2005) Copper in plants. Braz J Plant Physiol 17:145–156

Zabrieski Z, Morrell E, Hortin J, Dimkpa C, McLean J, Britt D, Anderson A (2015) Pesticidal activity of metal oxide nanoparticles on plant pathogenic isolates of *Pythium*. Ecotoxicology 24(6):1305–1314

Zhao L, Huang Y, Keller AA (2017) Comparative metabolic response between cucumber (*Cucumis sativus*) and corn (*Zea mays*) to a $Cu(OH)_2$ nanopesticide. J Agric Food Chem. https://doi.org/10.1021/acs.jafc.7b01306

Zuverza-Mena N, Medina-Velo IA, Barrios AC, Tan W, Peralta-Videa JR, Gardea-Torresdey JL (2015) Copper nanoparticles/compounds impact agronomic and physiological parameters in cilantro (*Coriandrum sativum*). Environ Sci Process Impacts 17(10):1783–1793

Chapter 4
Nanoantimicrobials for Plant Pathogens Control: Potential Applications and Mechanistic Aspects

Mohamed A. Mohamed and Kamel A. Abd–Elsalam

4.1 Introduction

Nanotechnology is an emerging field of applied science and cutting-edge technology that utilizes the physicochemical properties of the bulk material as a means to minimize and control their size, shape, and surface area, in order to generate different nanosized materials with new unique physical, chemical, and biological properties (Prasad et al. 2016). Nanotechnology is directly linked with physics, chemistry, biology, material science, and medicine. Up to now, nanotechnology finds a deep accurate application in multiple aspects of research and in everyday life such as electronics, agriculture, medicine, pharmaceutical research, and new material design. However, their use particularly in medical treatment research as antimicrobial or antiviral agents compared to agricultural research is probably one of the fastest growing areas in which the functional mechanisms of nanoparticles and especially metal-based nanoparticles are just beginning to be exploited.

Implementation of new technology like nanoscience in agriculture sector is of extreme importance, particularly in dealing with major problems facing this sector like plant growth, climate change, pest management, and nutrient shortage (Yaseen et al. 2018). The growing risk offered by viral and bacterial infections regarding the danger of annual food production loss is considered the scope of interest by many scientists nowadays; this is why it represents the subject of the present chapter. There are many more important factors, with single aim to solve: how to introduce a reliable safe and cost-effective solution in controlling those microbial plant pathogens and protect or improve the crop production?

M. A. Mohamed (✉) · K. A. Abd–Elsalam
Plant Pathology Research Institute, Agricultural Research Center (ARC), Giza, Egypt

K. A. Abd-Elsalam, R. Prasad (eds.), *Nanobiotechnology Applications in Plant Protection*, Nanotechnology in the Life Sciences,
https://doi.org/10.1007/978-3-319-91161-8_4

Nanotechnology may play an effective role for this purpose. Already, there are many important developments on this aspect like enhancement of nutrients absorption by plants, protection of plants, nanoformulated food ingredients, and water treatment processes. However, protection of plants with metal-based nanomaterials having a much greater surface area to volume ratio and unique antimicrobial agents compared to their bulk materials represents one of the best recent applied solution offering successful results in controlling microbes particularly bacterial and viral plant microbial pathogens (Prasad et al. 2014, 2017a,b). The most simple and obvious way is direct application of nanoparticles in the soil, on seeds, or on foliage to protect plants from pathogen invasion. In this way, the nanoparticles may suppress the pathogen in a way comparable to chemical pesticides. When nanoparticles are to be applied directly in soil, their effects on nontarget organism especially the mineral fixing/solubilizing microorganisms will be of great significance (Sangeetha et al. 2017). Secondly, the nanomaterials, carbon tubes, cups, etc., can also be used as a carrier of some fancy chemicals such as pheromones, polyamine, synthesis inhibitors, or even concentrated active ingredients of pesticides for their controlled release especially under flooded conditions (Khan et al. 2013).

Metal-based nanomaterials have a much greater surface area to volume ratio, which can exhibit completely new or improved and unique properties compared to their bulk counterparts, can be wisely exploited, and are found to be interesting candidates for various varied biological applications (Raveendran et al. 2003). In this regard, different types of the nanoparticles have entered into the arena of controlling plant diseases. Carbon, silver, zinc, silica, and aluminosilicates are best examples for nanoparticles that showed excellent antimicrobial activity against many plant pathogens.

In consequence, the present chapter aims at a description of the reported bacterial and antiviral activities of different nanomaterials for plant pathogens control and their proposed modes of action in the agricultural systems.

4.2 Activities of Engineered Nanomaterials against Plant Pathogens

Different types of nanomaterials like copper, zinc, titanium, magnesium, gold (Zhao et al. 2010), alginate (Ahmad et al. 2009), and silver (Bhattacharyya et al. 2016) have come up in recent years, and most of them have proven to be effective against diverse microbial pathogens. The present chapter aims at a description of the reported antibacterial and antiviral activities of metal nanoparticles and their modes of action, with particular regard to the metal nanoparticles (Gholami–Shabani et al. 2017).

4.2.1 Carbon Nanotubes

Carbon nanotubes (CNTs) are nanostructures, which are allotropes of carbon with extraordinary unique physical, chemical, optical, thermal, electrical, and mechanical properties. Two classes of CNTs are described, namely, single-walled carbon nanotubes (SWCNTs), which consist of a single graphite sheet impeccably wrapped into a cylindrical tube with a diameter between 0.4 and 2.5 nm, and multi-walled carbon nanotubes (MWCNTs), which comprise more layers of graphite sheet with different diameters of up to 100 nm. A nanotube can be visualized as a hexagonal network of hexagonal carbon atoms which upon rolling give a cylindrical structure. The unique structure of CNTs offers excellent physical and chemical properties that allow a wide range of applications including microbial control (Zhang et al. 2010).

Their elevated loading capacity and ability to readily penetrate membranes represent a potential for antimicrobial delivery. One of the best successful trials in this regard was done in 2009 by Khodakovskaya and Biris who used CNTs to deliver some antimicrobial molecules to provide the target plants protection against diseases during their seed germination stage (Khodakovskaya and Biris 2009). In the past few years, Chen and his co-workers indicated that the antimicrobial potentiality of CNTs depends on many factors like the length and type of functional groups on CNTs surface together with the bacterial shape needed to lyse their walls. Moreover, the antibacterial action of CNTs was mainly dependent on the length and diameter of CNTs used in penetrating the microbial cells and lysis of their walls, membranes, inducing release of intracellular components and loss of bacterial cell membrane functionality, resulting in a complete bacterial cell death.

On the other hand, it was indicated that SWCNTs with unique fine thin and rigid wall properties show more effective wall/membrane penetration on the spherical bacteria cell than MWCNTs. In addition, the long MWCNT may be covering around the bacterial cells, offering a high surface area to contact with the bacterial wall. Thus, CNTs may be broad-spectrum antibacterial agents in the gut, and selective application of CNTs could reduce the potential hazard to probiotic bacteria. Also, Pangule et al. (2010) prepared a nanocomposite films embedded with antimicrobial lysostaphin-carbon nanotube and revealed the high antibacterial activity to reduce the viable bacterial counts of methicillin-resistant *S. aureus*, for more than 99% in only 2 hours. In fact, considering this unique strategy with the CNTs, properties may have important application for controlling bacterial contamination in preserved foods, preventing the risk of serious bacterial infection and biofouling of surfaces.

4.2.2 Nano Silver

Silver (Ag) has been widely used throughout history for its antibacterial, antiviral, and antifungal properties (Castellano et al. 2007; Rai et al. 2009; Fabrega et al. 2011; Aziz et al. 2014, 2015, 2016). The first scientist who actually connects the use of silver to its action as an antimicrobial agent is believed to be Vonnaegele (Alexander 2009). He attributed the effectiveness of silver as an antibacterial agent to the silver ion (Ag+), opening up the use of silver as an effective antimicrobial agent until the discovery of penicillin. While the golden era of antibiotics flourished, the use of silver as antimicrobial agent is minimized. In the past decade, however, almost most known antibiotic has given rise to resistant strains of infectious bacteria like multidrug-resistant (MDR) bacteria. In the quest of finding potent and non-microbial-resistant alternatives to antibiotics, the use of the Ag has reemerged again and is coming back into mainstream as an antimicrobial and anti-infective agent (Rai et al. 2012). Since then the antimicrobial properties of silver have been investigated and exploited more extensively than any other molecule showing antibacterial potentiality in treatment of diverse diseases caused by bacterial pathogens (Aziz et al. 2015, 2016; Joshi et al. 2018).

Consequently, AgNPs together with their nontoxicity to living tissues at low concentrations make them more attractive to be utilized as antimicrobial agents in different biological fields particularly plant disease control. According to literatures, silver nanoparticles are the most popular inorganic nanoparticles used as antimicrobial agents (Zinjarde 2012; Singh et al. 2018); however, only a few studies were done on microbial plant pathogens (Krishnaraj et al. 2012).

The antimicrobial effect of different controlled sized silver nanoparticles to control different soilborne fungi as well as *Bipolaris sorokiniana* and *Magnaporthe grisea* was investigated (Jo et al. 2009). Interestingly, their results showed that silver nanoparticles with the size (20 to 30 nm) could better penetrate and colonize within the plant tissue than larger ones. The results also indicated that Ag nanoparticles had a great potential for use in controlling spore-producing fungal plant pathogens. They suggested that these nanoparticles might be less toxic than synthetic fungicides (Jo et al. 2009). In the other study, Mishra and his co-workers also indicated a strong antifungal activity exhibited against the phytopathogenic fungus *Bipolaris sorokiniana*, causing spot blotch disease of wheat crop, by using small-sized silver nanoparticles (Mishra et al. 2014).

On the other side, other reports assayed the antibacterial activity of a well-formed silver nanoparticles with 19 nm size against different eight microorganisms using the disk-diffusion method. Their results strongly indicated that the used silver nanoparticles have a noticeable antibacterial activity against all used bacterial species. Those findings strongly suggest to exploit those AgNPs as commercial antibacterial agents. On the other hand, Ali and his co-workers revealed that the exposure of the snails and soil matrix to silver nanoparticles in a laboratory experiment reduced the activity and the viability of the land snail (20% of silver nanoparticles-treated snails died) as well as the frequency of fungal population in the surrounding soil (Ali et al. 2015). Spherical-shaped silver nanoparticles in size range of ~10 to 20 nm using culture supernatant of *Serratia* sp. BHU-S4 and their effective application for the management of spot blotch disease in wheat have been experimented. The experimental observations indicated that AgNPs have several mechanisms in its antimicrobial activity where it can bind not only membrane-bound targets but also penetrate the bacterial cell. Also, it can strongly conjugate with the electron donor groups containing N,O, and S and exhibit an elevated binding constant for thiol-containing proteins.

On the other hand, it was revealed that the antimicrobial efficacy of silver nanoparticles has been confirmed to be shape-dependent in studies that utilized differentially shaped nanoparticles and measured the inhibition of bacterial growth, suggesting that the most effective geometry is truncated triangular silver nanoparticles (Morones et al. 2005; Pal et al. 2007).

In this study it was shown that the triangular geometry needed silver content >1 μg to exert bactericidal properties, while spherical- and rod-shaped nanoparticles needed 12.5 and 50–100 μg, respectively. Moreover, it has been shown that in order to function effectively as bactericidal agents, AgNPs have to fall with a narrow size range. Those findings backed to the high-atom density facets ({111}) that are found in triangular plates when compared to spheres and rods ({100}). Those conclusions were also in agreement with that exposed by Morones and his co-workers (Morones et al. 2005). Recently, the International Center for Technology Assessment (ICTA) has submitted a petition to the Environmental Protection Agency (USA) requesting that it regulates nanosilver used in products as a pesticide under the Federal Insecticide, Fungicide, and Rodenticide Act (FIFRA). Silver is now globally accepted as agrochemical replacement. Silver that exhibits excellent qualities should be tasteless, strong nontoxic disinfectant, and good growth stimulator. In 2013, Ocsoy and his co-workers indicated that DNA-directed silver nanoparticles grown on graphene oxide (GO) composites effectively decrease cell viability in culture and on plants of *Xanthomonas perforans* causing bacterial spot of tomatoes (*Solanum lycopersicum*) in Florida while the pathogen has developed resistance to Cu fungicides (Ocsoy et al. 2013). These compounds (Ag@dsDNA@GO) show excellent antibacterial activity in culture at a very low concentration of 16 ppm with higher adsorption rate. Severity of tomato bacterial spot is significantly reduced by application of Ag@dsDNA@GO at 100 ppm in greenhouse when compared to untreated and showed no phytotoxicity.

4.2.3 Nanosilica-Silver Composite

Different reports revealed the potentiality of using silica solution in controlling microbial plant pathogens causing powdery and downy mildew (Mao et al. 2001; Brecht et al. 2004). Moreover, it was suggested that using silica in plant disease control may induce disease and stress resistance and promote the growth of plants (Mao et al. 2001). However, the biological effect of silica significantly varies with the physiological environment, and thus, it is not registered as an agricultural chemical-controlling agent. As mentioned above, the antimicrobial activity of the metallic silver is well-known as a powerful disinfecting agent killing bacteria by inactivating enzymes having metabolic cell functions (Kim et al. 2009). In 2006, the scientist Part and his co-workers introduced a unique "nanosized silica-silver formula composed of nanosilver combined with silica molecules and water-soluble polymer, prepared by exposing a solution, capable of controlling many plant diseases (Park et al. 2006).

Park et al. studied the effect of different concentrations from this formula in controlling different many microbial plant pathogens (Park et al. 2006). Their findings revealed the potent antimicrobial activity (100% growth suppression) of this formula against crops' fungal pathogens particularly *Magnaporthe grisea, Colletotrichum gloeosporioides*, *Rhizoctonia solani*, *Pythium ultimum*, and *Botrytis cinerea*, when applied at 10 ppm, while with bacterial pathogens like *Azotobacter chrococuum, Bacillus subtilis, Pseudomonas syringae, Rhizobium tropici*, and *Xanthomonas campestris pv. Vesicatoria*, a complete bacterial growth suppression was reported at 100 ppm. On the other hand, their findings indicated that using high concentrations from the nanosized silica-silver caused chemical contact injuries on cucumber and pansy testing plants.

4.2.4 Titanium Dioxide Nanoparticles

Titanium dioxide nanoparticles (TiO_2NPs) are another form of the metallic form of titanium (Ti) in the environment. There has been an increasing amount of attention in the literature regarding effects of TiO_2NPs on plant performance and their potential as antimicrobial agents (Roy et al. 2010; Haghighi et al. 2013; Lyu et al. 2017). Cui and his co-workers indicated that TiO_2 NP abled to reduce *P. syringae pv. lachrymans* and *P. cubensis* infection of cucumber by 69 and 91%, respectively (Cui et al. 2009). Their report also revealed an increase in the photosynthetic activity by (30%). On the other hand, Paret et al. (2013a,b) also showed that after TiO_2 NP photoactivation, the bacterial spot (*Xanthomonas* sp.) control on roses and tomato plants was equivalent to or better than other classical control options.

The antimicrobial property of TiO_2 is suggested to be related to its crystal structure, shape, and size (Roy et al. 2010). Furthermore, different reports proposed that oxidative stress via the generation of ROS may be particularly important mechanism for their antimicrobial activity, where the produced ROS may cause site-specific nucleic acid DNA damage (Roy et al. 2010; Cioffi and Rai 2012). Reports by Roy and his co-workers revealed that TiO_2 nanoparticles decorated with different antimicrobial agents abled to control different resistant bacteria (Roy et al. 2010). Furthermore, their experiments indicated the potentiality of TiO_2 nanoparticles to improve the antimicrobial effect of different antimicrobial agents against many bacterial-resistant species and the noticeable decrease of bacterial resistance toward different antimicrobial agents. On the other hand, different reports indicated the antifungal activity of TiO_2 nanoparticles (Haghighi et al. 2013). The unique photocatalytic properties of the TiO_2 nanoparticles may play a core efficient role in the eradication of the microbial cells as a result of lipid peroxidation that cause to enhance membrane fluidity and disrupt and destroy the microbial cell integrity (Carré et al. 2014). Although TiO_2 faced major precaution in their biological use due to their toxicity (Shah et al. 2017), however, conjugation of TiO_2 nanoparticles with eco-friendly polymers may be an acceptable approach for others in a way to avoid their toxicity issue (Allahverdiyev et al. 2011)

4.2.5 Nanoaluminosilicate

Many reports proved that using nanoparticles in nanoformulated products makes them highly efficient, biologically active, and environmentally safe against broad spectrum of microbial pathogen. Thus, many factories recently started to follow this strategy in producing new nanoformulated products as better alternatives to the nanoparticles alone. Nowadays many factories started to produce and utilize nanoparticles in a new nanoformulated products, to improve its effect. Aluminosilicate nanotubes with active ingredients are one such type of the best example of nanoformulation strategy, which has proved a real advantage when sprayed on plant surfaces because of its easy pick by microbes and insect hairs, and lastly these nanotubes get consumed by them leading to their rapid direct death.

4.2.6 Zinc Nanoparticles

Zinc nanoparticles were also evaluated as antiviral agents by El-Sawy and his co-workers to control *Cucumber mosaic virus* (CMV) in eggplant in comparison with 2-nitromethyl phenol and seaweeds extract (El–Sawy et al. 2017). The nanozinc was sized in 100 nm and used as spray solution at three different concentrations of 50, 100, and 200 μgmL^{-1}. The eggplants were sprayed with 2-nitromethyl phenol, zinc nanoparticles, and seaweed extract at 1 day before CMV inoculation under greenhouse conditions. In contrast to the potent antiviral activity proved by silver nanoparticles in many reports, the results clearly indicated that seaweed extract treatment exhibited the highest reduction of CMV infection at a concentration of 3 gmL^{-1} followed by 2-nitromethyl phenol and zinc nanoparticles which was also significantly higher than the control. Moreover, all treatments showed significant increase in morphological and physiological characters of eggplant plants such as plant height, leaf area, number of leaves and number of leaves branches, numbers of flowers and fruits, and fruit weight per plant (El–Sharkawy et al. 2017). The authors concluded that although zinc nanoparticles showed the lowest activity against CMV under greenhouse conditions, but it was similarly like other treatments (2-nitromethyl phenol, seaweed extract) that induced the activation of free and total phenols which resulted in enhanced resistance against the CMV infection.

4.2.7 Copper Nanoparticles

Copper is considered an essential micronutrient which plays a vital role in different biological processes for all living organisms. This may come from its incorporation into an array of proteins and metalloenzymes required for performing various metabolic functions by the plant cells/organisms. More interestingly, copper in its metallic form has been used as an antimicrobial agent for hundreds of years. In ancient Egypt (2000 BCE), copper was used to heal wounds and sterilize water from microbes. During the Roman Empire, the copper cooking utensils were ordinary used to prevent the spread of microbial contaminants and their infectious diseases. Furthermore, in the Second World War, Japanese soldiers put small copper pieces in their water bottles to prevent dysentery. However, copper ions and copper compounds can be toxic to microorganisms, humans, and the environment (Georgopoulos et al. 2001). In this respect, copper nanoparticles have attracted the attention of researches for using it as an essential component to act as a reservoir for the controlled release of copper ions and thus inhibit their toxicity. Also, many studies have suggested that copper nanoparticles can be directly used as antimicrobial agents against an array of phytopathogenic microbes including fungi and bacteria (Kanhed et al. 2014; Banik and Pérez–de–Luque 2017; Viet et al. 2016).

The scientist Kanhed and his co-workers as example indicated the broad-spectrum antifungal activity of a chemically synthesized copper nanoparticles against different phytopathogenic fungi including *Fusarium oxysporum* (MTCC 1755), *Curvularia lunata* (MTCC 2030), *Phoma destructiva* (DBT66), and *Alternaria alternata* (MTCC 6572) and used disk-diffusion method. Copper nanoparticles showed remarkable activity against all the mentioned plant pathogenic fungi with maximum antifungal activity against *Curvularia lunata* MTCC 2030 followed by *Alternaria alternata* MTCC 6572 and minimum activity against *Phoma* destructive DBT 66 (Kanhed et al. 2014). More interestingly, the experimental results reported by Mondal and Mani on their studies on copper nanoparticles indicated that nanoform of copper (Cu) showed four-order higher activity against bacterial blight on pomegranate at 10000 times less concentration of recommended Cu (Mondal and Mani 2012). Similarly, it was reported that the green synthesized copper nanoparticles showed a significant bactericidal activity against numerous bacterial pathogens as well as against pathogenic fungi including *Fusarium culmorum*, *Fusarium oxysporum*, and *Fusarium graminearum* (Shende et al. 2015). Also, about 21 bacterial strains of *Brevibacillus laterosporus*, *Chryseobacterium indoltheticum*, and *Pantoea ananatis* isolated from agricultural soil showed strong sensitivity toward copper oxide nanoparticles at low concentrations (Guerrero et al. 2014).

Keeping in mind the above findings and showing a unique antimicrobial activity of CuNPs against different phytopathogenic bacteria and fungi, this suggests the possibility of using CuNPs antimicrobial agent in a large-scale safe mode, but under controlled concentrations. On the other side, copper nanoparticles with antimicrobial activity can also be employed for the production of a broad range of polymer nanocomposites, by embedding in a polymerated matrix. This can also be employed in pesticides and the food-packaging field for delay of deterioration, shelf-life extension, maintaining quality, and safety of packaged food.

4.2.8 Mesoporous Silica Nanoparticles

Mesoporous silica nanoparticles (MSNs) are thermally stable nanoparticles with tractable porosity and channels, having the tendency to deliver biomolecules like bio-antimicrobial agents, chemicals, pesticides, and nucleic acids into plant tissues by targeted release (Wang et al. 2010). Mesoporous silica nanoparticles were firstly developed by the scientist Trewyn and his co-workers in 2007 (Trewyn et al. 2007). In this system, the authors developed a honeycomb-like structure having arrays of independently formed channels which can be used as a nanocarrier to fill with the bioagents needed to be delivered. Interestingly, this type of nanomaterials has a unique capping properties that permit to hold and protect the carried material inside the channels without wasting it. Interestingly, the caps of this system can be chemically activated to pop open and release the cargo inside the cells where it is delivered. Furthermore, the delivery efficiency of MSNs can also be improved by optimizing and controlling the time of release. This system is successfully tested using different modeling plants like tobacco (*Nicotiana tabacum*), Arabidopsis (*Arabidopsis thaliana*) and maize (*Zea mays*), plants to inject nucleic acid (DNA) and other bioactive compounds and desired results were obtained.

4.2.9 Molybdenum Nanoparticles

Molybdenum nanoparticles (MoNPs) are important element of nitrogen-fixation system in plants. Many scientists have been working on development of Mo nanoparticle. In this context Taran et al. (2014) reported that Mo nanoparticle with nitrogen-fixing bacteria treatment to chickpea seed showed enhanced growth two to three times in comparison to water, only Mo nanoparticle, and only nitrogen-fixing bacterial incubation treatment and concluded that the combination could be the optimal treatment for better plant nutrition, and Mo nanoparticle increases the microbial activity.

4.2.10 Chitosan Nanoparticles

Chitosan is a naturally occurring compound that has broad-spectrum antimicrobial and antiviral activities (El Hadrami et al. 2010; Choudhary et al. 2017). The efficiency of chitosan nanoparticles as natural antimicrobial agent mainly depends on its type (native or modified), the polymerization degree, the host, the substrate composition, and also environmental conditions (Rabea et al. 2003; Kulikov et al. 2006). As example, it was reported that oligomeric chitosans (pentamers and heptamers) are more potent against pathogenic fungi compared to the larger ones (Rabea et al. 2003). In another studies, it was reported that the antimicrobial activity is directly increased with the chitosan molecular weight (Kulikov et al. 2006) and interestingly seems to be faster on fungi and algae than on bacteria (Savard et al. 2002). Chitosan nanoparticles were shown to inhibit the systemic propagation of viruses and viroids throughout the plant and to enhance the host's hypersensitive response to infection (Faoro et al. 2001; Chirkov 2002). The level of viral infection suppression directly depends on the chitosan molecular weight (Kulikov et al. 2006). Different reports also confirmed this issue on different viral pathogens including the potato virus X, tobacco mosaic and necrosis viruses, alfalfa mosaic virus, peanut stunt virus, and cucumber mosaic virus (Chirkov 2002; Struszczyk 2002). On the other hand, other reports indicated the antibacterial activity of chitosan against a wide range of bacterial species (El Hadrami et al. 2010).

Although the exact mechanism of action of chitosan nanoparticles as unique antimicrobial agents in controlling plant disease is not well understood till now, however, some reports suggested their toxicity to the microbial pathogens may be backed to the biopolymer properties of chitosan itself that can form physical natural barriers around the penetration sites of the microbial pathogens, preventing them from spreading and infecting the healthy tissues. On the other side, chitosan is also known to induce or improve reactions locally and systemically that involve signaling cascades and the activation of some defense-related antimicrobial compounds including proteins and also their accumulation inside the plant cells (El Hadrami et al. 2010).

4.3 Antiviral Activity of Nanomaterials

Metal nanoparticles have been studied for their antimicrobial potential and have proven to be lead antibacterial (Morones et al. 2005; Kim et al. 2007; Shahverdi et al. 2007; Rai et al. 2009) and antifungal (Kim et al. 2009) agents. However, despite the financial interest, there are very few reports on the effectiveness of nano-based materials against plant viruses (Elbeshehy et al. 2015; Cordero et al. 2017; El–Sawy et al. 2017). Theoretically, any metal could be analyzed for antiviral activity; however, little effort has been done as mentioned above to determine the interactions of metal nanoparticles with viruses in all biological fields. Mainly, only three types of metallic nanoparticles are used, silver, gold, and zinc nanoparticles, in testing the antiviral activity of nanomaterials. In medicine, few studies are recently emerged to study the antiviral activity of some gold and silver nanoparticles against different human viruses like monkeypox virus (Rogers et al. 2008), herpes simplex virus (Baram–Pinto et al. 2009), HIV (Elechiguerra et al. 2005; Sun et al. 2005; Lara et al. 2010), and hepatitis B virus (Lu et al. 2008), respiratory syncytial virus, influenza virus (Papp et al. 2010), and Tacaribe virus (Speshock et al. 2010).

Similarly, very few reports were published on the effectiveness of nanomaterials particularly silver and zinc nanoparticles against plant viruses like *Bean yellow mosaic virus* (BYMV) (Elbeshehy et al. 2015), *Potato virus Y* (PVY) (Cordero

Fig. 4.1 Effect of mycosynthesized silver nanoparticles on potato Y virus (PVY–Ros1) infectivity: Pictures of representative leaves of tobacco plants mock-treated with different concentrations of silver nanoparticles at 100 and 1000 ppm, as indicated. Two days after treatment, the leaves were mechanically inoculated with PVY–Ros1. Pictures were taken at 6 dpi with PVY–Ros1

2017), and *Cucumber mosaic virus* (CMV) (El–Sawy et al. 2017) and their use as antiviral agents in agricultural system (Cordero 2017; El–Sawy et al. 2017) (Fig. 4.1). It is interesting to note that in the case of report published by Elbeshehy and his co-workers in controlling CMV, the authors did not detect any beneficial effect of silver nanoparticles in pre-infection treatment, and only remarkable positive results were observed in postinfection treatment (Elbeshehy et al. 2015), in contrast to reported study published by Cordero and her co-workers that clearly observed a significant beneficial effect of silver nanoparticles in treatments 48 h before and ahead of viral challenging as well.

4.4 Mechanistic Aspects of Nanomaterials as Antibacterial Agents

Although different reports revealed the antimicrobial action of nanomaterials against microorganisms, however the mechanism of action in controlling the microbes is not properly understood. Many reports backed the efficient antimicrobial activity of nanoparticles as compared to their salts due to their extremely large surface area. This enables the particles at the nanoscale to achieve better attachment with the microorganisms, consequently get adsorbed onto the bacterial cell membrane causing major structural and morphological changes, and also penetrate inside the bacterial cell, distributing in the cytoplasm and the respiration process. While the nanomaterials enter with the respiration process, they interact with the thiolated functional group which is found in the respiratory enzymes of the bacterial cells.

In case of silver nanomaterials as example, the metallic silver itself has the potentiality to interact by inhibiting the uptake of phosphate and releasing carbohydrate compounds like succinate, mannitol, and amino acids like proline and glutamine and phosphate from the bacterial cells (Yamanaka et al. 2005). Also, different studies indicated the effect of nanoparticles on the microorganism nucleic acid (DNA), causing complete deformation and lysis for the DNA strands (Hackenberg et al. 2011; Ahn et al. 2014; Kasyanenko et al. 2016). This was clarified by following the pathway action of the silver nanoparticles (Fig. 4.2), when they penetrate inside the bacterial cell, and observing their interaction with sulfur-containing proteins present in the cell membrane as well as with the phosphorous-containing compounds like DNA; thus the DNA molecule is reduced into condensed form and loses its replication ability leading to cell death (Feng et al. 2000).

The bactericidal activity of nanoparticles can also be enhanced compared to their bulk materials as a result of releasing their ions released inside the bacterial cell and interacting with the cell components (Feng et al. 2000; Morones et al. 2005; Rai et al. 2009), resulting in a range of effects from inhibition of growth and loss of infectivity to cell death. The degree of these effects mainly varied from microorganism to another one depending on the nanoparticles used regarding their shape (Pal et al. 2007), size (Yen et al. 2009), concentration (Asharani et al. 2009), and the sensitivity of the microbial species to those nanomaterials (Kim et al. 2007; Illingworth et al. 2000; Yamanaka et al. 2005; Lu et al. 2008; Mehrbod et al. 2009; Ruparelia et al. 2008).

On the other hand, in the case of silver nanoparticles, several studies demonstrated that the positive charge on the Ag + ion is crucial for its antimicrobial activity through the electrostatic attraction between the negative charges on the microbial cell membrane surface and the positive charges on the nanoparticles surface (Kim et al. 2007; Prasad and Swamy 2013; Swamy and Prasad 2012; Aziz et al. 2015, 2016). In contrast, Sondi and Salopek-Sondi suggested that the broad-spectrum antimicrobial activity of AgNPs on Gram-negative bacteria depends on the concentration of AgNPs used and is closely responsible for the formation of morphological changes like pores and pits on the bacterial cell wall surface (Sondi and Salopekk–Sondi 2004); consequently, AgNPs accumulated in the bacterial membrane disturbing the membrane permeability, resulting in bacterial cell death. As this assumption is not sufficient enough to explain the antimicrobial action more accurately, scientists theorized that there are another possible mechanisms. Hence, in 2000 the scientist Amro and his co-workers suggested that their antimicrobial activity may be backed to the metal depletion causing the formation of irregularly shaped pits in the outer membrane and change membrane permeability, which is caused by the progressive release of lipopolysaccharide molecules and membrane proteins (Amro et al. 2000).

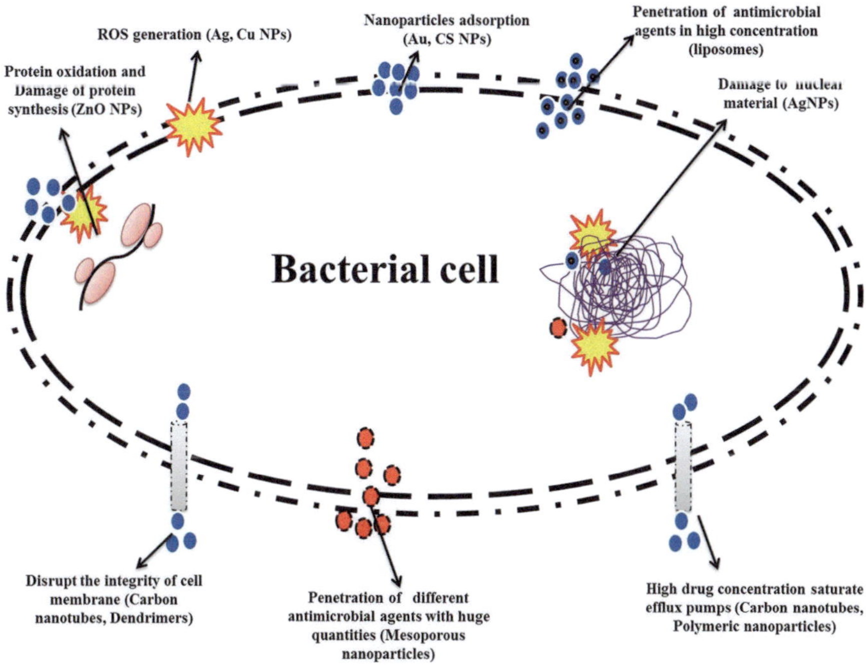

Fig. 4.2 A summary of the mechanisms associated with the antimicrobial behavior of metal nanoparticles

Similarly, Sondi and Salopek-Sondi also supposed the same mechanism after observing a noticeable cell membrane degradation of in the bacterial cells after challenging with a prepared solution of AgNPs (Sondi and Salopek–Sondi 2004). Although the assumption of nanoparticles-cell membrane components binding is revealed in different studies, but the mechanism of the binding and interaction process is still unclear. Recently, Danilczuk and co-workers reported that the nanomaterials when interact with the living cell generate free radicals derived from the nanomaterial surface may be responsible for the antimicrobial activity (Danilczuk et al. 2006). In contrast to Danilczuk study, another mechanism was proposed that bactericidal activity of nanomaterials mainly based on inhibition of the cell wall synthesis, protein synthesis mediated by the 30s ribosomal subunit, and nucleic acid synthesis (Lara et al. 2010). Similarly, the data recovered from proteomic analyses reported in 2006 by Lok et al. stated that a short exposure of the bacterial cells to a very low concentration of AgNPs resulted in an accumulation of envelope protein precursors, indicative of the dissipation of proton motive force (Lok et al. 2006). In parallel with those findings, silver nanoparticles were also shown to destroy the stability of the outer membrane, collapse the plasma membrane potential, and finally deplete the intracellular ATP level inside the bacterial cell (Dibrov et al. 2002).

4.5 Mechanistic Aspects of Nanomaterials as Antiviral Agents

The interaction of antimicrobial nanoparticles with viral particles is still an unexplored field. However, the mechanism of action of AgNPs as an antiviral and virucidal has been studied against different enveloped plant viruses causing sever crop loss (Elbeshehy et al. 2015; Cordero et al. 2017; El Sawy et al. 2017). In 2015, Elbeshehy and his co-workers studied the effect of silver nanoparticles on *Bean yellow mosaic virus* and suggested that antiviral activity of nanoparticles comes from their ability to bind with a viral envelope glycoprotein and inhibit the virus by binding to the disulfide bond regions of the CD4-binding domain within the yellow mosaic viral envelope glycoprotein gp120 (Elbeshehy et al. 2015). Besides the immediate interaction with glycoprotein of the virus surface, AgNPs may also enter the cell and fulfill their antiviral activity through interactions with the viral nucleic acids.

Continually, this strategy concluded in fusion inhibition was later elegantly demonstrated by Lara and her co-workers in their report (Lara et al. 2010). Their obtained results introduced an evidence for the high binding affinity of the used nanoparticles for DNA of the virus and extracellular virions with different sizes (10 and 50 nm). Moreover, the authors also revealed that the silver nanoparticles also able to inhibit the production of the viral RNA and extracellular virions in vitro, which was determined using a UV-vis absorption titration assay.

In another study, the scientist Sun and his co-workers indicated that AgNPs were superior to gold nanoparticles for cytoprotective activities toward virus. It is generally understood that Ag, in various forms, inactivates viruses by denaturing enzymes via reactions with sulfhydra, amino, carboxyl, phosphate, and imidazole groups

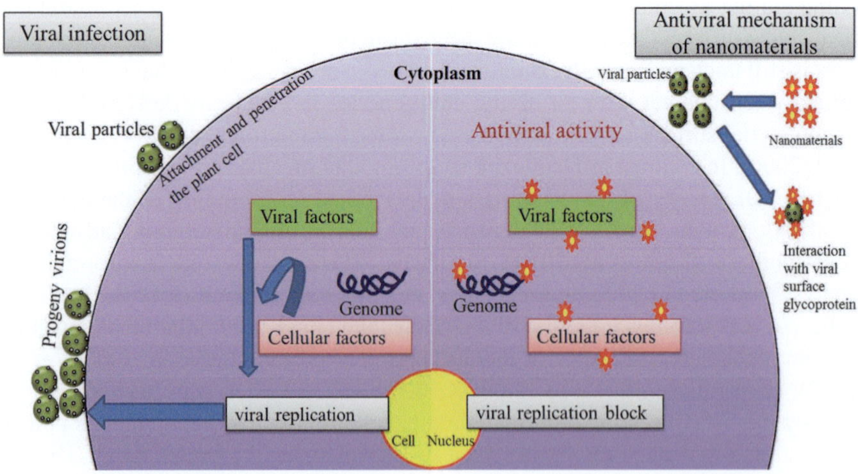

Fig. 4.3 Schematic model of virus particles infecting eukaryotic living cell and antiviral mechanism of metallic nanomaterials

(Borkow and Gabbay 2004, 2009; Baker et al. 2005; Ruparelia et al. 2008; Rai et al. 2009). Also, silver nanoparticles were indicated for their ability to interfere with the fusion of the viral membrane, inhibiting viral penetration into the host cell (Mehrbod et al. 2009) (Fig. 4.3). It has also been suggested that the antiviral activity of AgNPs depends on the particle size, as well as on the distribution of interacting ligand/receptor molecules (Lu et al. 2009; Papp et al. 2010). However, it is necessary to design studies in vivo to increase their application benefits and minimize adverse effects.

4.6 Factors Affecting Antimicrobial/Antiviral Activity of Nanomaterials

Different studies indicated that the biological activity of any nanomaterial is governed by their physical properties regarding their size, shape, and surface area (Kreibig and Vollmer 1995; Mulvaney 1996). The large surface area of the smaller-sized nanoparticles permits to come in contact with the bacterial surface cells covering large area on their surface cells, and therefore, it has higher efficiency than bigger particles and better chance to kill (Morones et al. 2005). Different reports revealed that AgNPs with size less than 10 nm have more antibacterial affinity compared to larger ones, proving that the antibacterial activity is a size-dependent manner (Raimondi et al. 2005; Morones et al. 2005). Furthermore, it was also indicated that the antimicrobial activity of nanoparticles is varied depending on the nanoparticle shape (spherical, rod-shaped, nanoshells, nanocages, nanowires, triangular, dimensional, etc.). Pal and his co-workers indicated that the content of the metallic silver content in the nanoparticles is varied depending on the AgNPs shape (triangular =1 μg, spherical =12.5 μg, rod, = 50–100 μg); consequently different effects on bacterial cell inhibition based on the AgNPs were observed.

In another study, Pan et al. (2007) studied the cytotoxic effect of different sized gold nanoparticles (1.2 nm, 1.8 nm, and 15 nm) on four cell lines (HeLa, Sk-Mel-28, L929, J774A1). According to the MTT assay results, it was found that the larger nanogold particles (15 nm sizes) showed less cytotoxic activity as compared to those ones with 2 and 1 nm size, respectively, which can irreversibly bind to the nucleic acid (DNA) and possibly other crucial molecules as well (Pan et al. 2007). Similarly, Dasgupta and his co-workers studied the size-dependent toxicological behavior of AgNPs prepared by thermal co-reduction approach. Those findings confirmed that although both AgNPs with 85 and 60 nm size had high antimicrobial and cytotoxic potential against the bacterial cells and cancerous cell lines, respectively, but the antimicrobial and cytotoxic behavior of AgNPs with 60 nm size was higher than that with 85 nm (Dasgupta et al. 2015). The surface plasmon resonance also plays a major role in the determination of optical density of metal nanoparticles. Particle size is proportional to the wavelength. The large surface area of nanoparticles will result in a higher intensity of interaction than larger particles size in micrometers or more. Thus, it is corroborated that the antibacterial effect of silver nanoparticles is size-dependent.

4.7 Demerits and Limits of Nanoparticles in Plant Disease Management

As the kinetics associated with the nanomaterials is very fast and is highly reactive, they inherently interact with contaminants and impurities. Furthermore, it is noticeable that most experimental studies with nanoparticles have been carried out with a degree of aggregates/agglomerates. Hence, retaining a high purified nanoparticles with 100% can become a challenge hard to overcome. This has significant repercussions on the biokinetics of the material. Consequently, many questions can be raised: What is the degree of nanoparticles stability and the size distribution of the aggregates/agglomerates, and what is the portion of the formed nanoparticles present as a mono-dispersed material?

On the other hand, current research work revealed that the uptake, translocation, and accumulation of nanoparticles depend on different factors including the microbial pathogen that needs to be controlled, the plant part which will be sprayed, the species of plant, and the size, chemical composition, functionalization, and stability of the nanoparticles used (Kole and Vittal 2013; Raliya et al. 2015). Among the carbon-based nanoparticles, only the fullerene C70 and fullerols were shown to get readily accumulated in plants (Rico et al. 2011; Nair et al. 2012). Most of the data corresponds to the germination stage and cell culture, because the protocols for quantification of nanoparticles within tissues are not well-defined yet. The discussion of the current research is more oriented to the effect of the nanoparticles on plants. A very few of the nanoparticles to the next generation of plants exposed to nanoparticles are unknown.

On the other hand, although the broad-spectrum antimicrobial activity of nanoparticles against different microbial plant pathogens is a well-known and accepted fact worldwide, their impact on soil biota is still less documented, consequently attracting noteworthy attention from scientists. Many gaps need to be filled in our knowledge regarding the toxicity of those nanomaterials on the environment and the ecological systems. Therefore, when talking about the application of nanomaterials in agroecosystems, their major interaction with residing soil biota cannot be ignored. The negative effect of nanoparticles is found to be more pronounced on denitrifying bacteria, disrupting the process of denitrification in soil (VandeVoort and Arai 2012). As a result, nanoparticles in soil have been used as a model system to evaluate the dose-dependent effects of metal itself (Throbäck et al. 2007). In this viewpoint, Yang et al. (2013) studied the interaction of a carbon coated AgNPs with 35 nm in size and Ag + (provided as $AgNO_3$) with *Pseudomonas stutzeri* (denitrifier), *Azotobacter vinelandii* (nitrogen fixer), and *Nitrosomonas europaea* (nitrifier).

They concluded lower toxicity of AgNPs toward these bacteria compared to 20–48 times higher toxicity exerted by the Ag + ions. Conversely, low and sub-lethal concentrations of Ag + and AgNPs (20–25 μg/L) yielded no significant impact on the expression pattern of denitrifying genes and nitrogen-fixing genes but showed 2.1- to 3.3-fold upregulation in nitrifying genes, indicating the sensitivity of the nitrification process toward silver. Another study representing the impact of nanosilver on aerobic denitrification process by Shahrokh et al. (2014) advocated that a low dose of AgNPs had no adverse effect on nitrate reductase activity of *Rhizobium* and *Azotobacter*. Interestingly, 0.2 ppm of AgNP treatment enhanced the process of nitrate reduction in *Azotobacter*. Also and in another study, when nanosized silica-silver particles were applied under in vivo condition to control the fungal disease cucurbit powdery mildew, 100% control was achieved after 25 days (Park et al. 2006).

These nanoparticles were found to be phytotoxic only at a very high dose of 3200 ppm when tested in cucumber and pansy plants. Taken together, these findings clearly indicate a dose-dependent effect of AgNPs on the microbial process of nitrogen cycle, giving us clue that entering an optimum concentration of AgNPs into the environment could be favorable for microbial processes with no hindrance in beneficial plant-microbe interactions in agroecosystems. On the other hand, size-dependent toxicity of AgNPs has also been evidenced by Choi and Hu (2008), where they found that AgNPs of size less than 5 nm were more toxic to nitrification bacteria. So, beside the need to understand the possible benefits of applying nanotechnology to agriculture, there is also an urgent need to feel secure about nanomaterials phytotoxicity when applied on crop plants. In consequence, the first step should be to analyze penetration of those particles and the pathway of their transport inside plant cells. Formulation stability is also an important aspect of the biosafety of nanomaterials. Liu et al. (2008) successfully formulated a stable nanopesticide (bifenthrin) using polymer stabilizer such as poly(acrylic acid)-b-poly(butyl acrylate) (PAA-b-PBA), polyvinylpyrrolidone (PVP), and polyvinyl alcohol (PVOH). A flash nanoprecipitation technique was used to prepare 60–200 nm bifenthrin particles. While using such techniques commercially, the stability profile of the polymers over an extended time period needs to be firstly considered.

4.8 Conclusion and Future Road Map

The growing challenges facing the humanity and food security demonstrate clearly that there is a growing demand and need in the agriculture field to produce more output with less input. The application of large amounts of microbial pesticides together with the existence of new microbial-resistant strains is one of the main critical problems facing the sector of food safety. Hence, we are on the edge of time where we have to adopt modern agriculture techniques and new innovative technologies to control those threats more wisely and in accurate way, as conventional agricultural practices will not be sufficiently able to control those threats without putting the human health under critical risk. Among the most recent technical approaches in the field of agriculture and plant disease control, nanotechnology holds an eminent position in remodeling agriculture and fighting plant microbes which directly reflects positively on the level of a healthy food production and fulfill the demands in an efficient and cost-effective way. Nanotechnology being studied since the last few decades is still in its premature phase of development. However, the whole course of action is very broad and being popularized day by day.

Nanotechnology in combination with biotechnology has led to the rapid development of marketable formulations involving deployment of artificially designed nanoparticles for crop protection. To restrict the indiscriminate use of excess pesticides and pesticides in plants, nanoparticles are proved to be a gifted tool of this age. Different nanomaterials have been proved to have beneficial and functional role with potent antimicrobial/antiviral efficiency against plant pathogens, whereas some nanoparticles have been reported to have a deleterious role regarding reactivity and phytotoxicity in plants. Hence, we are supposed to be very careful during screening and selection of the type of nanomaterial needed to be used regarding their physical, chemical, and accumulation behavior, concentration, and toxicity.

Otherwise, they could become the source of potential serious threat toward the whole ecosystem. Effect of nanoparticles depends on multiple factors, and their results can easily and completely change, if the condition varies. Such aspects would include the type of nanomaterials used and their concentrations, the chemical/biological coating surface agent, the plant type regarding its age, and the target part tissue. Nowadays, nanobiotechnology industries are growing very rapidly, and new generated antimicrobial nano-based systems were produced; however, there is an urgent need to perform profound research studies regarding this matter in order to develop comparatively economic, safe, and eco-friendly stabilized nanomaterials in the long run. The widespread assessment of these nanomaterials particularly in food production and agri-food sectors should also be carried out more wisely for public acceptance to avoid the unlike challenges as were faced by genetically modified organisms worldwide. In conclusion, the unique impact of nanotechnology and their applied applications in farmers' field is just in its infancy, but the expectations hoped from this science to erase challenges associated with plant disease control, food production, environment sustainability, and even fossil fuel are still very high.

References

Ahmad A, Rafatullah M, Sulaiman O, Ibrahim MK, Chii YY, Siddique BM (2009) Removal of Cu(II) and Pb(II) ions from aqueous solutions by adsorption on sawdust of Meranti wood. Desalination 247:636–646

Ahn JM, Eom HJ, Yang X, Meyer JN, Choi J (2014) Comparative toxicity of silver nanoparticles on oxidative stress and DNA damage in the nematode, *Caenorhabditis elegans*. Chemosphere 108:343–352

Ali SM, Yousef NM, Nafady NA (2015) Application of biosynthesized silver nanoparticles for the control of land snail *Eobania vermiculata* and some plant pathogenic fungi. J Nanomater 1:3

Alexander JW (2009) History of the medical use of silver. Surg Infect 10(3):289–292

Allahverdiyev AM, Abamor ES, Bagirova M, Rafailovich M (2011) Antimicrobial effects of TiO2 and Ag2O nanoparticles against drug–resistant bacteria and *leishmania parasites*. Future Microbiol 6(8):933–940

Amro NA, Kotra LP, Wadu–Mesthrige K, Bulychev A, Mobashery S, Liu G (eds) (2000) High–resolution atomic force microscopy studies of the *Escherichia coli* outer membrane: structural basis for permeability. Langmuir 16:2789–2796

Asharani PV, Hande MP, Valiyaveettil S (2009) Anti–proliferative activity of silver nanoparticles. BMC Cell Biol 10(65):65

Aziz N, Pandey R, Barman I, Prasad R (2016) Leveraging the attributes of Mucor hiemalis-derived silver nanoparticles for a synergistic broad-spectrum antimicrobial platform. Front Microbiol 7:1984. https://doi.org/10.3389/fmicb.2016.01984

Aziz N, Fatma T, Varma A, Prasad R (2014) Biogenic synthesis of silver nanoparticles using Scenedesmus abundans and evaluation of their antibacterial activity. J Nanoparticles 2014 https://doi.org/10.1155/2014/689419

Aziz N, Faraz M, Pandey R, Sakir M, Fatma T, Varma A, Barman I, Prasad R (2015) Facile algae-derived route to biogenic silver nanoparticles: synthesis, antibacterial and photocatalytic properties. Langmuir 31:11605–11612. https://doi.org/10.1021/acs.langmuir.5b03081

Baker C, Pradhan A, Pakstis L, Pochan DJ, Shah SI (2005) Synthesis and antibacterial properties of silver nanoparticles. J Nanosci Nanotechnol 5:244–249

Banik S, Pérez–de–Luque A (2017) In vitro effects of copper nanoparticles on plant pathogens, beneficial microbes and crop plants. Span J Agric Res 15(2):1005

Baram–Pinto D, Shukla S, Perkas N, Gedanken A, Sarid R (2009) Inhibition of herpes simplex virus type 1 infection by silver nanoparticles capped with mercaptoethane sulfonate. Bioconjug Chem 20:1497–1502

Bhattacharyya A, Duraisamy P, Govindarajan M, Buhroo AA, Prasad R (2016) Nano-biofungicides: Emerging trend in insect pest control. In: Prasad R (ed) Advances and Applications through Fungal Nanobiotechnology. Springer International Publishing, Switzerland, pp 307–319

Borkow G, Gabbay J (2004) Putting copper into action: copper–impregnated products with potent biocidal activities. FASEB J 18:1728–1730

Borkow G, Gabbay J (2009) Copper, an ancient remedy returning to fight microbial, fungal and viral infections. ACS Chem Biol 3(3):272–278

Brecht MO, Datnoff LE, Kucharek TA, Nagata RT (2004) Influence of silicon and chlorothalonil on the suppression of gray leaf spot and increase plant growth in St. Augustine grass. Plant Dis 88:338–344

Carré G, Hamon E, Ennahar S, Estner M, Lett MC, Horvatovich P, Andre P (2014) TiO2 photocatalysis damages lipids and proteins in *Escherichia coli*. Appl Environ Microbiol 80(8):2573–2581

Castellano JJ, Shafii SM, Ko F, Donate G, Wright TE, Mannari RJ et al (2007) Comparative evaluation of silver–containing antimicrobial dressings and drugs. Int Wound J 4(2):114–122

Chirkov SN (2002) The antiviral activity of chitosan (review). Appl Biochem Microbiol 38:1–8

Choi O, Hu Z (2008) Size dependent and reactive oxygen species related nanosilver toxicity to nitrifying bacteria. Environ Sci Technol 42:4583–4588

Choudhary RC, Kumaraswamy RV, Kumari S, Sharma SS, Pal A, Raliya R, Saharan V (2017) Cu–chitosan nanoparticle boost defense responses and plant growth in maize (*Zea mays* L.). Sci. Rep 7(1):9754

Cioffi N, Rai M (eds) (2012) Nano–antimicrobials: progress and prospects. Springer Science & Business Media, Heidelberg, Dordrecht, London, New York

Cordero T, Mohamed MA, López-Moya J.-J, Daròs J.-A (2017) A Recombinant Potato virus Y infectious clone tagged with the rosea1 visual marker (pvy–ros1) facilitates the analysis of viral infectivity and allows the production of large amounts of Anthocyanins in plants. Front Microbiol 8:611. https://doi.org/10.3389/fmicb.2017.00611

Cui X, Kim HM, Kawashita M, Wang L, Xiong T, Kokubo T, Nakamura T (2009) Preparation of bioactive titania films on titanium metal via anodic oxidation. Dent Mater 25(1):80–86

Danilczuk M, Lund A, Sadlo J, Yamada H, Michalik J (2006) Conduction electron spin resonance of small silver particles. Spectrochim Acta A Mol Biomol Spectrosc 63:189–191

Dasgupta J, Sikder J, Chakraborty S, Curcio S, Drioli E (2015) Remediation of textile effluents by membrane based treatment techniques: a state of the art review. J Environ Manag 147:55–72

Dibrov P, Dzioba J, Gosink KK, Hase CC (2002) Chemiosmotic mechanism of antimicrobial activity of ag(+) in Vibrio cholerae. Antimicrob Agents Chemother 46:2668–2670

El Hadrami A, Adam LR, El Hadrami I, Daayf F (2010) Chitosan in plant protection. Mar Drugs 8(4):968–987

Elbeshehy EK, Elazzazy AM, Aggelis G (2015) Silver nanoparticles synthesis mediated by new isolates of *Bacillus* spp., nanoparticle characterization and their activity against Bean Yellow Mosaic Virus and human pathogens. Front Microbiol 6:453

El–Sawy MM, Elsharkawy MM, Abass JM, Kasem MH (2017) Antiviral activity of 2–nitromethyl phenol, zinc nanoparticles and seaweed extract against cucumber mosaic virus (CMV) in Eggplant. J Virol Antivir Res 6:2. https://doi.org/10.4172/2324–8955.1000173

El–Sharkawy MS, El–Beshsbeshy TR, Mahmoud EK, Abdelkader NI, Al– Shal RM, Missaoui AM (2017) Response of alfalfa under salt stress to the application of potassium sulfate nanoparticles. Am J Plant Sci 8(08):1751

Fabrega J, Luoma SN, Tyler CR, Galloway TS, Lead JR (2011) Silver nanoparticles: behavior and effects in the aquatic environment. Environ Int 37:517–531

Faoro, F, Sant S, Iriti M, Appiano A (2001) Chitosan–elicited resistance to plant viruses: a histochemical and cytochemical study. In Muzzarelli, R.A.A (ed)Chitin Enzymology. Atec: Grottammare, Italy, 2001, pp. 57–62

Feng QL, Wu J, Chen GQ, Cui FZ, Kim TN, Kim JO (2000) A mechanistic study of the antibacterial effect of silver ions on Escherichia coli and Staphylococcus aureus. J Biomed Mater Res B 52(4):662–668

Georgopoulos G, Roy A, MJ Y–L, Opiekun RE, Lioy P (2001) Environmental copper: its dynamics and human exposure issues. J Toxicol Environ Health B Crit Rev 4(4):341–394

Gholami–Shabani M, Gholami–Shabani Z, Shams–Ghahfarokhi M, Jamzivar F, Razzaghi–Abyaneh M (2017) Green nanotechnology: biomimetic synthesis of metal nanoparticles using plants and their application in agriculture and forestry. In: Nanotechnology. Springer, Singapore, pp 133–175

Guerrero SIC, Brito EMS, Castillo HAP, Rivero SHT, Caretta CA, Velasco AL Duran R, Borunda EO (2014) Effect of CuO nanoparticles over isolated bacterial strains from agricultural soil. J Nanomater 2014:1–13. https://doi.org/10.1155/2014/148743

Hackenberg S, Scherzed A, Kessler M, Hummel S, Technau A, Froelich K, Kleinsasser N (2011) Silver nanoparticles: evaluation of DNA damage, toxicity and functional impairment in human mesenchymal stem cells. Toxicol Lett 201(1):27–33

Haghighi F, Roudbar Mohammadi S, Mohammadi P, Hosseinkhani S, Shipour R (2013) Antifungal activity of TiO2 nanoparticles and EDTA on Candida albicans biofilms. Infection Epidemiol Microbiol 1(1):33–38

Illingworth B, Bianco RW, Weisberg S (2000) In vivo efficacy of silver–coated fabric against Staphylococcus epidermidis. J Heart Valve Dis 9:135–141

Elechiguerra JL, Burt JL, Morones JR, Camacho– Bragado A, Gao X, Lara HH, Yacaman MJ (2005) Interaction of silver nanoparticles with HIV–1. J Nanobiotechnology 3(1):6

Jo YK, Kim BH, Jung G (2009) Antifungal activity of silver ions and nanoparticles on phytopathogenic fungi. Plant Dis 93(10):1037–1043

Joshi N, Jain N, Pathak A, Singh J, Prasad R, Upadhyaya CP (2018) Biosynthesis of silver nanoparticles using Carissa carandas berries and its potential antibacterial activities. J Sol-Gel Sci Techn. https://doi.org/10.1007/s10971-018-4666-2

Kanhed P, Birla S, Gaikwad S, Gade A, Seabra AB, Rubilar O, Duran N, Rai M (2014) In vitro antifungal efficacy of copper nanoparticles against selected crop pathogenic fungi. Mater Lett 115:13–17

Kasyanenko N, Varshavskii M, Ikonnikov E, Tolstyko E, Belykh R, Sokolov P, Lopatko K (2016) DNA modified with metal nanoparticles: preparation and characterization of ordered metal–DNA nanostructures in a solution and on a substrate. J Nanomater 2016:1–12. https://doi.org/10.1155/2016/3237250

Khan SA, Mulvaney RL, Ellsworth TR (2013) The potassium paradox: implications for soil fertility, crop production and human health. Renew Agri Food Syst 29:3–27. https://doi.org/10.1017/S1742170513000318

Khodakovskaya MV, Biris AS (2009) Method of using carbon nanotubes to affect seed germination and plant growth. WO 2011059507 A1—patent application

Kim SW, Kim KS, Lamsal K, Kim YJ, Kim SB, Jung M, Sim SJ, Kim HS, Chang SJ, Kim JK, Lee YS (2009) An in vitro study of the antifungal effect of silver nanoparticles on oak wilt pathogen *Raffaelea* sp. J Microbiol Biotechnol 19:760–764
Kim JS, Kuk E, Yu KN, Kim JH, Park SJ (2007) Antimicrobial effect of silver nanoparticles. Nanomedicine 3:95–101. https://doi.org/10.1016/j.nano.2006.12.001
Kole GK, Vittal JJ (2013) Solid–state reactivity and structural transformations involving coordination polymers. Chem Soc Rev 42(4):1755–1775
Kreibig U, Vollmer M (1995) In "Optical Properties of Metal *Clusters*," Springer Series in *Mater Sci* 532
Krishnaraj C, Jagan EG, Ramachandran R, Abirami SM, Mohan N, Kalaichelvan PT (2012) Effect of biologically synthesized silver nanoparticles on Bacopa monnieri (Linn.) Wettst. Plant growth metabolism. Process Biochem 47(4):51–658
Kulikov SN, Chirkov SN, Il'ina AV, Lopatin SA, Varlamov VP (2006) Effect of the molecular weight of chitosan on its antiviral activity in plants. Prik Biokhim Mikrobiol 42(2):224–228
Lara HH, NV A–N, Ixtepan–Turrent L, Rodriguez–Padilla C (2010) Bactericidal effect of silver nanoparticles against multidrug–resistant bacteria. World J Microbiol Biotechnol 26:615–621
Liu S, Yuan L, Yue X, Zheng Z, Tang Z (2008) Recent advances in nanosensors for organophosphate pesticide detection. Adv Powder Technol 19:419–441
Lok CN, Ho CM, Chen R, He QY, Yu WY, Sun H, Tam PK, Chiu JF, Che CM (2006) Proteomic analysis of the mode of antibacterial action of silver nanoparticles. J Proteome Res 5:916–924
Lu JM, Wang X, Muller CM, Wang H, Lin PH, Qizhi Y et al (2009) Current advances in research and clinical applications of PLGA based Nanotechnology. Expert Rev Mol Diagn 9:325–341. https://doi.org/10.1586/erm.09.15
Lu L, Sun RW, Chen R, Hui CK, Ho CM, Luk JM, Lau GK, Che CM (2008) Silver nanoparticles inhibit hepatitis B virus replication. Antivir Ther 13:253–262
Lyu S, Wei X, Chen J, Wang C, Wang X, Pan D (2017) Titanium as a beneficial element for crop production. Front Plant Sci 8:597
Mao HQ, Roy K, VL T–L, Janes KA, Lin KY, Wang Y, August JT, Leong KW (2001) Chitosan–DNA nanoparticles as gene carriers: synthesis, characterization and transfection efficiency. J Control Release 70:399–421
Mehrbod P, Motamed N, Tabatabaian M, Soleimani Estyar R, Amini E, Shahidi M (2009) In Vitro Antiviral Effect of "Nanosilver" on Influenza Virus. DARU 17:88–93
Mishra V, Mishra RK, Dikshit A, Pandey AC (2014) Interactions of nanoparticles with plants: an emerging prospective in the agriculture industry. In: Ahmad P, Rasool S (eds) Emerging technologies and management of crop stress tolerance: Biolog Techniq, vol 1. Academic Press, Elsevier, USA, pp 159–180
Mondal KK, Mani C (2012) Investigation of the antibacterial properties of nanocopper against *Xanthomonas axonopodis pv. punicae*, the incitant of pomegranate bacterial blight. Ann Microbiol 62(2):889–893
Morones JR, Elechiguerra JL, Camacho A, Holt K, Kouri JB, Ramirez JT, Yacaman MJ (2005) The bactericidal effect of silver nanoparticles. Nanotechnology 16:2346–2353
Mulvaney P (1996) Surface plasmon spectroscopy of nanosized metal particles. Langmuir 12(3):788–800
Nair RR, Wu HA, Jayaram PN, Grigorieva IV, Geim AK (2012) Unimpeded permeation of water through helium–leak–tight graphene–based membranes. Science 335(6067):442–444
Ocsoy I, Paret ML, Ocsoy MA, Kunwar S, Chen T, You M, Tan W (2013) Nanotechnology in plant disease management: DNA–directed silver nanoparticles on graphene oxide as an antibacterial against *Xanthomonas perforans*. ACS Nano 7:8972–8980
Pal S, Tak YK, Song JM (2007) Does the antibacterial activity of silver nanoparticles depend on the shape of the nanoparticle? A study of the gram–negative bacterium Escherichia coli. Appl Environ Microbiol 73:1712–1720
Pan Y, Neuss S, Leifert A, Fischler M, Wen F, Simon U, Jahnen-Dechent W (2007) Size-dependent cytotoxicity of gold nanoparticles. Small 3(11):1941–1949
Pangule RC, Brooks SJ, Dinu CZ, Bale SS, Salmon SL, Zhu G, Dordick JS (2010) Antistaphylococcal nanocomposite films based on enzyme– nanotube conjugates. ACS Nano 4(7):3993–4000

Papp I, Sieben C, Ludwig K, Roskamp M, Böttcher C, Schlecht S (2010) Inhibition of influenza virus infection by multivalent sialic–acid functionalized gold nanoparticles. Small 6:2900–2906

Paret M, Palmateer A, Knox G (2013a) Evaluation of a lightactivated nanoparticle formulation of TiO2/Zn for management of bacterial leaf spot on Rosa 'Noare. Hortscience 48(2):189–192

Paret ML, Vallad GE, Averett DR, Jones JB, Olson SM (2013b) Photocatalysis: effect of light–activated nanoscale formulations of TiO2 on *Xanthomonas perforans* and control of bacterial spot of tomato. Phytopathology 103(3):228–236

Park HJ, Kim SH, Kim HJ, Choi SH (2006) A new composition of nanosized silica–silver for control of various plant diseases. J Plant Pathol 22:295–302

Prasad R, Bhattacharyya A, Nguyen QD (2017a) Nanotechnology in sustainable agriculture: recent developments, challenges, and perspectives. Front Microbiol 8:1014. https://doi.org/10.3389/fmicb.2017.01014

Prasad R, Gupta N, Kumar M, Kumar V, Wang S, Abd-Elsalam KA (2017b) Nanomaterials act as plant defense mechanism. In: Prasad R, Kumar V, Kumar M (eds) Nanotechnology. Springer, Singapore, pp 253–269

Prasad R, Pandey R, Barman I (2016) Engineering tailored nanoparticles with microbes: quo vadis. WIREs Nanomed Nanobiotechnol 8:316–330. https://doi.org/10.1002/wnan.1363

Prasad R, Kumar V, Prasad KS (2014) Nanotechnology in sustainable agriculture: present concerns and future aspects. Afr J Biotechnol 13(6):705–713

Prasad R, Swamy VS (2013) Antibacterial activity of silver nanoparticles synthesized by bark extract of *Syzygium cumini*. Journal of Nanoparticles 2013:1. https://doi.org/10.1155/2013/431218

Rabea EI, El Badawy MT, Stevens CV, Smagghe G, Steurbaut W (2003) Chitosan as antimicrobial agent: applications and mode of action. Biomacromolecules 4:1457–1465

Rai M, Yadav A, Gade A (2009) Silver nanoparticles as a new generation of antimicrobials. Biotechnol Adv 27:76–83

Rai V, Acharya S, Dey N (2012) Implications of nanobiosensors in agriculture. J Biomater Nanobiotechnol 3:315–324

Raimondi F, Scherer GG, Kötz R, Wokaun A (2005) Nanoparticles in energy technology: examples from electrochemistry and catalysis. Angew Chem Int Ed 44(15):2190–2209

Raliya R, Nair R, Chavalmane S, Wang WN, Biswas P (2015) Mechanistic evaluation of translocation and physiological impact of titanium dioxide and zinc oxide nanoparticles on the tomato (*Solanum lycopersicum* L.) plant. Metallomics 7(12):1584–1594

Raveendran P, Fu J, Wallen SL (2003) Completely "Green" synthesis and stabilization of metal nanoparticles. J Am Chem Soc 125(46):13940–13941

Rico CM, Majumdar S, Duarte–Gardea M, Peralta–Videa JR, Gardea–Torresdey JL (2011) Interaction of nanoparticles with edible plants and their possible implications in the food chain. J Agric Food Chem 59:3485–3498

Rogers JV, Parkinson CV, Choi YW, Speshock JL, Hussain SM (2008) A preliminary assessment of silver nanoparticle inhibition of monkeypox virus plaque formation. Nanoscale Res Lett 3:129

Roy AS, Parveen A, Koppalkar AR, Prasad MA (2010) Effect of nano–titanium dioxide with different antibiotics against methicillin–resistant Staphylococcus aureus. J Biomater Nanobiotechnol 1:37

Ruparelia JP, Chatterjee AK, Duttagupta SP, Mukherji S (2008) Strain specificity in antimicrobial activity of silver and copper nanoparticles. Acta Biomater 4:707–716

Savard T, Beaulieu C, Boucher I, Champagne CP (2002) Antimicrobial action of hydrolyzed chitosan against spoilage yeasts and lactic acid bacteria of fermented vegetables. J Food Prot 65:828–833

Sangeetha J, Thangadurai D, Hospet R, Harish ER, Purushotham P, Mujeeb MA, Shrinivas J, David M, Mundaragi AC, Thimmappa AC, Arakera SB, Prasad R (2017) Nanoagrotechnology for soil quality, crop performance and environmental management. In: Prasad R, Kumar M, Kumar V (eds) Nanotechnology. Springer Nature Singapore Pte Ltd, Singapore, pp 73–97

Shah SNA, Shah Z, Hussain M, Khan M (2017) Hazardous effects of titanium dioxide nanoparticles in ecosystem. Bioinorg Chem Appl 2017:1. https://doi.org/10.1155/2017/4101735

Shahrokh S, Hosseinkhani B, Emtiazi G (2014) The impact of nano–silver on bacterial aerobic nitrate reductase. J Bioprocess Biotechnol 4:162
Shahverdi AR, Fakhimi A, Shahverdi HR, Minaian S (2007) Synthesis and effect of silver nanoparticles on the antibacterial activity of different antibiotics against Staphylococcus aureus and *Escherichia coli*. Nanomedicine 3:168–171
Shende S, Ingle AP, Gade A, Rai M (2015) Green synthesis of copper nanoparticles by Citrus medica Linn.(Idilimbu) juice and its antimicrobial activity. World J Microbiol Biotechnol 31:865–873
Singh A, Sharma B, Deswal R (2018) Green silver nanoparticles from novel Brassicaceae cultivars with enhanced antimicrobial potential than earlier reported Brassicaceae members. J Trace Elem Med Biol 47:1. https://doi.org/10.1016/j.jtemb.2018.01.001
Sondi I, Salopek–Sondi B (2004) Silver nanoparticles as antimicrobial agent: a case study on *E coli* as a model for Gram negative bacteria. J Colloid Interf Sci 275:177–182
Speshock JL, Murdock RC, LK B–S, Schrand AM, Hussain SM (2010) Interaction of silver nanoparticles with Tacaribe virus. J Nanobiotechnol 8:19
Struszczyk MH (2002) Chitin and chitosan – Part II. Applications of chitosan Polimery 47:396–403
Sun RW, Chen R, Chung NP, Ho CM, Lin CL, Che CM (2005) Silver nanoparticles fabricated in Hepes buffer exhibit cytoprotective activities toward HIV–1 infected cells. Chem Commun (Camb) 40:5059–5061
Swamy VS, Prasad R (2012) Green synthesis of silver nanoparticles from the leaf extract of Santalum album and its antimicrobial activity. J Optoelectronic and Biomedical Materials 4(3):53–59
Taran NY, Gonchar OM, Lopatko KG, Batsmanova LM, Patyka MV, Volkogon MV (2014) The effect of colloidal solution of molybdenum nanoparticles on the microbial composition in rhizosphere of Cicer arietinum L. Nanoscale Res Lett 9:289
Throbäck IN, Johansson M, Rosenquist M, Pell M, Hansson M, Hallin S (2007) Silver (Ag +) reduces denitrification and induces enrichment of novel nirK genotypes in soil. FEMS Microbiol Lett 270:189–194
Trewyn BG, Slowing II, Giri S, Chen HT, Lin VSY (2007) Synthesis and functionalization of a mesoporous silican based on the Sol–Gel process and applications in controlled release. Acc Chem Res 40:846–853. https://doi.org/10.1021/ar600032u
VandeVoort AR, Arai Y (2012) Effect of silver nanoparticles on soil denitrification kinetics. Ind Biotechnol 8(6):358–364
Viet PV, Nguyen HT, Cao TM, Hieu LV (2016) Fusarium antifungal activities of copper nanoparticles synthesized by a chemical reduction method. J Nanomater 6:7
Wang Z, Wei F, Liu SY, Xu Q, Huang JY (2010) Electrocatalytic oxidation of phytohormone salicylic acid at copper nanoparticles–modified gold electrode and its detection in oilseed rape infected with fungal pathogen *Sclerotinia sclerotiorum*. Talanta 80:1277–1281
Yamanaka M, Hara K, Kudo J (2005) Bactericidal actions of a silver ion solution on Escherichia coli, studied by energy–filtering transmission electron microscopy and proteomic analysis. Appl Environ Microbiol 71:7589–7593
Yang H, Wang Y, Huang H, Gell L, Lehtovaara L, Malola S, Zheng N (2013) All–thiol–stabilized Ag44 and Au12Ag32 nanoparticles with single–crystal structures. Nat Commun 4:2422
Yen HJ, Hsu SH, Tsai CL (2009) Cytotoxicity and immunological response of gold and silver nanoparticles of different sizes. Small 5:1553–1561
Yaseen T, Pu H, Sun DW (2018) Functionalization techniques for improving SERS substrates and their applications in food safety evaluation: a review of recent research trends. Trends Food Sci Technol 72:162–174
Zhang F, Ali Z, Amin F, Feltz A, Oheim M, Parak WJ (2010) Ion and pH sensing with colloidal nanoparticles: influence of surface charge on sensing and colloidal properties. Chem Phys Chem 11:730–735
Zhao X, Wang J, Wu F, Wang T, Cai Y, Shi Y, Jiang G (2010) Removal of fluoride from aqueous media by Fe 3 O 4 @Al(OH) 3 magnetic nanoparticles. J Hazard Mater 173:102–109
Zinjarde S (2012) Bio–inspired nanomaterials and their applications as antimicrobial agents. Chron Young Sci 3(1):74–74

Chapter 5
Role of Microbes in Plant Protection Using Intersection of Nanotechnology and Biology

Manoj Kaushal

5.1 Introduction

Plant diseases have escalated internationally which results in about 50% estimated loss caused by insect pests, and the value of this crop disaster was assessed to be US $2000 billion per year (van Lenteren and Martin 1999). Pathogens and pests trim growth of plants due to several adverse soil and climatic factors and thus lead to diminish the overall crop productivity. Habitually, worldwide the crops are protected from pathogens and pests with pesticide operations. This extensive usage of pesticides raises the resistance of crop pathogens and pests (Dzhavakhiya et al. 2012; Alghuthaymi et al. 2015), recedes nitrogen fixation, and scales down soil biodiversity which later contributes to the bioaggregation of pesticides in environment. Moreover, maximum share of the pesticides and fertilizers that are utilized are lost to air during operational hours and some as runoff which results in ecotoxicity (Chen et al. 2015; Vu et al. 2015). Also, concerning the global food security and climate change objections, crop protection is benefited from technological modernizations such as synthetic chemicals and hybrid varieties, but researchers are now focusing on green technology in particular. To tackle these situations, attempts have been made for secure crop productions with best management practices under varying circumstances (Leake et al. 2002). Biocontrol management of phytopathogens utilizing microbes or their derived products is useful as some of them are natural antagonists of pathogens with various inherent challenges (Frampton et al. 2012). Worldwide molecular techniques involving enzymes and primers are also widely used in laboratories for identification and control of specific plant pathogens due to their high degree of specificity (harmless to nontarget organisms), environmental compatibility, and applicability in integrated pest management. Still demerits exist for on-site detection of phytopathogens, shorter shelf

M. Kaushal (✉)
International Institute of Tropical Agriculture, Ibadan, Oyo State, Nigeria

K. A. Abd-Elsalam, R. Prasad (eds.), *Nanobiotechnology Applications in Plant Protection*, Nanotechnology in the Life Sciences,
https://doi.org/10.1007/978-3-319-91161-8_5

life, and high diagnostic price especially in developing countries which caps the application of these conventional molecular approaches. Therefore, the crop up nanobiotechnology sounds to be of preeminent attention for the detection and control of phytopathogens especially in prior detection of disease and smart delivery of biocontrol agents to crop (Prasad et al. 2014, 2017a, b). In general, nanobiotechnology refers to the intersection of nanotechnology and biology which involves bio-fabrication, manipulation, and utilization of submicron or nano-objects (≤100 nm). Ample of products are used to construct nanoparticles, such as lipids, polymers, ceramics, quantum dots, silicates, dendrimers, and metal oxides (silver, gold, platinum, silica, selenium, titania, zirconia, etc.) (Oskam 2006; Puoci et al. 2008). Usually nanoparticles are synthesized chemically which is curbed by high cost, toxicity, and complexity of the technique. Therefore, microbes are considered as the ideal architect of these diversified nanostructures due to their small size, control culturability, genetic manipulability, regeneration, high biosorption capacity, wide diversity, selective adsorption of metal ions, and indeed allowing an easier downstream purification procedure. Accordingly, the links between nanotechnologies and microbiology are bolstering in the last few years with continuous revealing the potentials of microbes as nanoparticle industries. Also, higher surface area to volume ratio and unique optical properties of nanoparticles enables more interaction chances with pathogens and destroys them. Nanoparticle biosynthesis by microbes can be classified into intracellular and extracellular synthesis according to the site (Simkiss and Wilbur 1989; Mann 2001; Prasad et al. 2016). Transportation of ions occurs into microbial cell to form nanoparticles in the existence of enzymes during intracellular method. Trapping of metal ions on cell surface and reducing ions again in the companionship of enzymes appear during extracellular synthesis of nanoparticles (Zhang et al. 2011; Prasad et al. 2016). Known fact of bacteria is to use an enzyme and metabolize oxygen for sustaining life. Silver ions stop this oxygen metabolization by debilitating the enzyme which results in suffocation to fungi and bacteria and ultimately death (Puebla et al. 2004). If the concentrations of nanoparticles exceed a limit, some bacterial genera such as *Geobacter sulfurreducens* and *Shewanella oneidensis* reduce and precipitate the cations as nanoparticles that are counterpart of metabolic processes enabling nanoparticle synthesis as a protective mechanism (Tanzil et al. 2016). Magnetotactic bacteria (*Magneto spirillum*) that are known to generate natural magnetic nanoparticles due to their size, homogeneity, and stabilization have also been used for bioremediation, sensor development, and agrochemical degradation (Jacob and Suthindhiran 2016). Also, polyhydroxyalkanoate (PHA) nanobead biosynthesis proved to be a less costly technique due to accessible downstream procedure of PHA intracellular granules (Dinjaski and Prieto 2015). There are the evidences which provide in vitro efficacy of distinct nanoparticles (Guo et al. 2015) and biodegradable nanoparticles that can be opted for plant disease management (Chowdappa and Gowda 2013). Unique nanosize of nanoparticles makes them suitable in gene transfer (Rai et al. 2012a, b) and management of pathogens in crop production (Mishra and Singh 2015). Efficacy of nano-based materials provides the quick on-site detection of phytopathogens, addresses toxicity issues to manage different plant diseases, and allows only minor doses to be used (Prasad et al. 2017b). Many researchers over the globe reported development of nanoparticles for the diseases detection and phytopathogen management employing bacteria (Tiwari et al. 2014), fungi (Potara et al. 2015), actinomycetes (Golinska et al. 2014), algae (El-Kassas and El-Sheekh 2014), and plants (Mallikarjuna et al. 2015).

This chapter provides a brief overview of the concepts and current state of the nanobiotechnology applications in disease management and plant protection involving nanoparticles that either incorporate biological elements or are entirely biological in origin. Next, biosynthesis mechanisms along with examples and current applications of diversified nanoparticles biosynthesized in the agricultural fields are presented. The chapter concludes with discussions on the current limitations and prospects of nanobiotechnology in sustainable agriculture.

5.2 Synthesis and Application of Nanoparticles

Nanoparticles are synthesized by a variety of physical, chemical, and biological methods which involve size reduction, high-pressure homogenization, sonication, reactive precipitation, and solvent displacement (Sasson et al. 2007). Nanoparticles generated from various natural resources such as plant extracts (pectin, cellulose, and chitin) are more expedient in terms of biodegradability, environmental safety, and biodegradation. Microbiocidal nanoparticles such as AgO and MgO due to their high stability, versatility, and biocompatibility are employed for agrochemical degradation, soil remediation, detection, and control of food spoilage (Baruah and Dutta 2009). Immunomolecules labeled ultrasensitive QD bioconjugates are developed as sensors for disease detection crops due to their potential for recognition of specific antibodies or antigens (Vinayaka et al. 2009). Other inorganic materials such as montmorillonite and other clay nanoparticles have a structure of stacked platelets with one dimension of the platelet in the nanoscale. Several investigations also demonstrated the potential use of nanoclays in the deployment of agrochemicals such as plant growth promoters, pesticides, and fertilizers (Bin et al. 2009).

5.2.1 Microbial Synthesis of Nanoparticles

Development of efficient and green nanoparticles is a crucial facet of nanobiotechnology. To achieve this objective, soil microorganisms and natural bioresources (plant–/marine-based material) are important options. However, soil microorganisms have gained preferences over plant material due to their natural ability of biocontrol agents. Also, the macro matter can be reformed into nanomaterials with the use of microbes through either intracellular or extracellular routes (Table 5.1). A diverse range of microbes have been found to produce nanoparticles in the substrate (El-rafie et al. 2012).

Table 5.1 Synthesis of nanoparticles by various microbes

Mode	Microbe	Nanoparticle
Extracellular	**Bacteria**	
	Penicillium sp.	Ag
	Klebsiella pneumonia	Ag
	Pseudomonas aeruginosa	Au
	Thermomonospora spp.	Au
	Rhodopseudomonas capsulate	Au
	Bacillus subtilis	Ag, Au
	Shewanella oneidensis	U
	Clostridium thermoaceticum	CdS
	Fungi	
	Aspergillus fumigates	Ag
	Aspergillus flavus	Ag
	Phanerochaete chrysosporium	Ag
	Phoma sp.	Ag
	Verticillium	Ag
	Trichoderma asperellum	Ag
	Fusarium oxysporum	Ag, magnetite
Intracellular	**Bacteria**	
	Bacillus licheniformis	Ag
	Pseudomonas stutzeri	Ag
	Streptomyces albidoflavus	Ag
	Bacillus subtilis	Ag, Au
	Lactobacillus sp.	Ag, Au
	Escherichia coli	Au, CdS
	Clostridium thermoaceticum	CdS
	Shewanella oneidensis	Magnetite
	Yeast	
	Pichia jadinii	Au
	Candida glabrata	CdS
	Schizosaccharomyces pombe	CdS
	Torulopsis	CdS

Various microbes (e.g., culture supernatant of *E.coli*) reduce the Ag^+ ions to form silver nanoparticles, most of which are found to be spherical particles (Mukherjee et al. 2001; Ahmad et al. 2003a, b; Fayaz et al. 2010). Further purification is done by using sucrose density gradient centrifugation, and purified sample was characterized by fluorescence spectroscopy, TEM, and UV-vis spectra. *Pseudomonas stutzeri* AG259 isolated from silver mines, when placed in a concentrated aqueous solution of silver nitrate, reduce Ag^+ ions and form silver nanoparticles (AgNPs) of well-defined size within the periplasmic space of the bacteria (Klaus et al. 1999). *E. coli* was used for the extracellular biosynthesis of AgNPs and characterized by UV-vis spectra, FTIR, and SEM of nanoparticles (Manonmani and Juliet 2011). Some researchers also focused on optimum reaction conditions which resulted in the reduction of particle size and maximize the synthesis of AgNPs (Gurunathan et al. 2009) using different medium (nitrate medium as the most efficient one), varying concentrations (5 mM

$AgNO_3$), pH (10.00), and reaction temperatures (60 °C). Utilizing these ideal conditions, 95% conversion was obtained with culture supernatant of *E. coli* in a time interval of 30 min. Mourato et al. (2011) investigated the biosynthesis of AgNPs by utilizing extremophilic yeast strain isolated from acid drainage mine. A mechanism involving c-type cytochromes reduces Ag(I) as insoluble AgCl or Ag^+ ions, precipitating extracellular nanoscale silver by *Geobacter sulfurreducens* (Law et al. 2008). Silver nanoparticles are also isolated from *Morganella morganii*, Gram-positive bacteria (Abd et al. 2013). Fungi are considered as sometimes advantageous in the synthesis of nanoparticles due to their ease of handling in laboratory and production of large quantities of enzymes (Mandal et al. 2006; Mohanpuria et al. 2007). When nanoparticles are synthesized outside the cell (extracellularly), it becomes easy to purify and can be used as direct applications (Mukherjee et al. 2008; Gaikwad et al. 2013). Also, fungal mycelial mesh can withstand flow pressure in bioreactors in resemblance to bacteria or plant extract (Narayanan and Sakthivel 2010). AgNPs were synthesized in film or solution (intracellulary) or accumulated on the cell surface (extracellulary) when fungi, *Verticillium*, *Fusarium oxysporum*, *Aspergillus flavus*, and *Penicillium*, were employed (Senapati et al. 2004; Bhainsa and D'Souza 2006; Vigneshwaran et al. 2007; Jain et al. 2011). Extracellular biosynthesis of AgNPs has been also achieved by using bacterium *Bacillus licheniformis* (Kalimuthu et al. 2008) and fungus *Aspergillus niger* (Gade et al. 2008). *F. oxysporum* and *Verticillium* sp. synthesized magnetite nanoparticles in the presence of ferric and ferrous salts. *F. oxysporum* f. sp. *lycopersici* was found to be efficiently producer of inter- and extracellular platinum nanoparticles and AgNPs using nitrate reductase-mediated technique. Biosynthesis of spherical and silver colloidal NPs was also investigated using *F. oxysporum* (Ahmad et al. 2003a, b). Biosynthesis of AgNPs was obtained by *Penicillium fellutanum* using $AgNO_3$ from coastal mangrove sediment as a substrate (Shaligram et al. 2009). Culture supernatants of *A. niger*, *A. fumigatus*, *A. clavatus*, and *A. terreus* were utilized for extracellular biosynthesis of AgNPs (Bhainsa and D'Souza 2006; Saravanan and Nanda 2010; Verma et al. 2010; Jaidev and Narasimha 2010; Li et al. 2012) that effectively suppress different plant pathogens. Stable forms of AgNPs were synthesized extracellularly with aqueous silver nitrate solution employing *Phoma sorghina*; *Phanerochaete chrysosporium*, the white rot fungus (Birla et al. 2009; Gade et al. 2013); and *Mucor* (Aziz et al. 2016). Single or aggregated AgNPs (8–60 nm) with round and uniform shape were produced by the biocontrol agent, *Trichoderma asperellum*, and other five *Trichoderma* species, viz., *T. asperellum*, *T. harzianum*, *T. longibrachiatum*, *T. pseudokoningii*, and *T. virens*, which remain stable for several months (Devi et al. 2013). $AgNO_3$ were reduced to polydispersed and spherical-shaped AgNPs with extracellular solution of *Cladosporium cladosporioides* revealed by TEM analysis (Balaji et al. 2009). Biosynthesis of AgNPs was also investigated using *Pleurotus* sp. (Gade et al. 2007) and *Pleurotus sajor-caju* (Nithya and Ragunathan 2009) with antimicrobial activities.

A superb quality of gold nanoparticles was synthesized extracellularly by fungus *Fusarium oxysporum* and extremophilic bacteria *Thermomonospora* sp., respectively (Ahmad et al. 2003a, b), and intracellularly by fungus *Verticillium* sp. Mesophilic bacterium, *Shewanella*, was utilized for synthesis of gold nanoparticles using H_2 as an electron donor. Monodisperse gold nanoparticles of size 10–20 nm have been synthesized by using *Rhodopseudomonas capsulata* (He et al. 2007) and alkalotolerant *Rhodococcus* sp. under intense alkaline and exalted temperature conditions. Synthesis of different shapes (spherical, cubic, and octahedral), gold nanostructures, and analysis of their formation mechanisms by filamentous cyanobacteria from Au(I)-thiosulfate and Au(III)-chloride complexes were identified by various researchers (Lengke et al. 2006). Intracellular synthesis of gold nanoparticles of various morphologies and sizes were obtained by fungi *V. luteoalbum* and *Penicillium* sp. (Kathiresan et al. 2009). Particle formation rate and size can be manipulated by controlling various factors such as concentration, pH, temperature, and exposure time to $AuCl^{4-}$. *Hormoconis resinae* also proved as an excellent source with high stability for extracellular synthesis of gold NPs (Mishra et al. 2010).

On the other hand, Au-Ag alloy nanoparticles were synthesized via an extracellular approach as marked by TEM and fluorescence microscopy. Some workers reported the synthesis of bimetallic Au-Ag alloy by fungus *F. oxysporum* and contend that the secreted cofactor NADH plays an influential role in determining the configuration of Au-Ag alloy nanoparticles (Senapati et al. 2005). Biosynthesis of Au-Ag alloy nanoparticles by yeast cells were also studied by Zheng et al. (2010a, b). Sawle et al. (2008) also proved the synthesis of core-shell Au-Ag alloy nanoparticles from fungal strains *Fusarium semitectum* and revealed that the nanoparticle suspensions remain highly stable for weeks. Au, Ag, and Au-AgNPs as reducing and protecting agents were also synthesized by a newly developed extracellular synthesis method utilizing an extract from edible mushroom, *Volvariella volvacea* (Philip 2009).

Several studies were conducted successfully to achieve the nanoparticles using heavy metals. Many heavy metals such as U(VI), Tc(VII), Cr(VI), Co(III), and Mn(IV) were reported to reduce by *Pyrobaculum islandicum*, an anaerobic hyperthermophilic bacterium. Platinum nanoparticles were accomplished using the metal ion-reducing bacterium *Shewanella algae* (Konishi et al. 2007) which reduce aqueous $PtCl_6^{2-}$ ions into elemental platinum at room temperature and neutral pH within 60 min with lactate as the electron donor. Uniform-sized (2–5 nm) and monodispersed intracellular mercury nanoparticles can also be synthesized by *Enterobacter* sp. cells (Sinha and Khare 2011) at pH 8.0 and lower concentration of mercury. Sulfate-reducing bacterium, *Desulfovibrio desulfuricans*, and metal ion-reducing bacterium, *S. oneidensis*, enable the synthesis of palladium nanoparticles (DeWindt et al. 2005). Long-term studies immobilized fungus *Coriolus versicolor* which bioremediates cadmium and is utilized for synthesizing stable CdS NPs in continuous column mode (Sanghi and Verma 2009). Mono- and bimetallic Au/AgNPs are synthesized from a biological agent *Neurospora crassa*, a filamentous fungus (Castro-Longoria et al. 2011). Mycelia-free culture filtrates of *Nigrospora oryzae* with gold chloride were used for mycosynthesis of AuNPs of size 6–18 nm (Kar et al. 2014).

Many magnetotactic bacteria also played a crucial role in the synthesis of oxide nanoparticles due to their super paramagnetic and high coercive force properties and unique micro-configurations. Magnetic nanoparticles such as Fe_3O_4 (magnetite) and Fe_2O_3 (maghemite) are found to be highly biocompatible, synthesized intracellularly by magnetotactic bacteria, and are referred to as bacterial magnetic particles (BacMPs) (Arakaki et al. 2008). BacMPs consist of phospholipids and proteins and thus smoothly spread in aqueous solutions. BacMPs aligned in chains within the bacterium, which enable the bacterium to shift along oxygen gradients in aquatic environments, with the clout of the Earth's geomagnetic field. Furthermore, magnetotactic bacteria are observed to possess various morphological types (ovoid bacteria, rod-shaped bacteria, cocci, vibrios, spirilla) and to inhabit various aquatic environments such as aquatic sediments, salt marshes, freshwater sediments, etc. Already established magnetic isolation techniques and different growth medium resulted in the isolation of a considerable number of the magnetotactic bacteria including *Magnetospirillum magnetotacticum* strain MS-1, *Magnetospirillum gryphiswaldense* strain MSR-1, and *Magnetospirillum magneticum* AMB-1 (Arakaki et al. 2008), and various others are uncultured. Among these isolated magnetotactic bacteria, the cultured ones are mostly mesophilic in nature and only a few reports describing thermophilic magnetotactic bacteria. Magnetotactic bacteria HSMV-1 were found in samples from springs that can grow at temperatures ranged from 32 to 63 °C (Lefevre et al. 2010). Besides magnetic oxide nanoparticles, a green low-cost and reproducible *Saccharomyces cerevisiae*-mediated biosynthesis of Sb_2O_3 nanoparticles was also investigated with a spherical aggregate of 2–10 nm size (Jha et al. 2009). Several other workers prepared tetragonal $BaTiO_3$ (4–5 nm) and quasispherical ZrO_2 nanoparticles (3–11 nm) from *F. oxysporum* (Bansal et al. 2006).

In addition to nanoparticles mentioned above, $SrCO_3$ crystals were retrieved with *Fusarium oxysporum* when incubated with aqueous Sr^{2+} ions and zinc phosphate nanopowders with yeasts as biotemplates (Pandian et al. 2009). Other researchers showed that highly luminescent CdSe quantum dots can be generated by *F. oxysporum* at room temperature (Kumar et al. 2007a, b). *Clostridium thermoaceticum* precipitate CdS on the surface of cell in existence of cysteine hydrochloride in growth medium where cysteine acts as the source of sulfide. 20–200 nm CdS was found on the cell surface when *Klebsiella pneumoniae* was exposed to Cd^{2+} ions in the growth medium. When *Escherichia coli* is incubated with $CdCl_2$ and Na_2SO_4 intracellular CdS nanocrystals are formed which poised a wurtzite crystal phase (Sweeney et al. 2004). *S. pombe* and *C. glabrata* (yeasts) have also successfully utilized to produce intracellular CdS nanoparticles with cadmium salt solution. ZnS and PbS were the other successfully synthesized nanoparticles by different microbes. ZnS nanoparticles were generated from *Rhodobacter sphaeroides* and *Desulfobacteraceae* intracellularly with 8 nm and 2–5 nm size, respectively (Bai et al. 2006; Bai and Zhang 2009). Eukaryotic organisms such as fungi (*F. oxysporum*) were also found for the synthesis of metal sulfide nanoparticles by extracellular means when exposed to aqueous solution of metal sulfate.

5.2.2 *Mechanism of Nanoparticle Synthesis by Microbes*

In general, nanoparticles are synthesized by microbes through entrapment of metal ions on the surface of cell (extracellular) or inside the cell (intracellular) followed by reduction with the help of enzymes (Fig. 5.1). Absorption and reduction of these metal ions are done with the help of fungal cell wall and cell wall sugars. However,

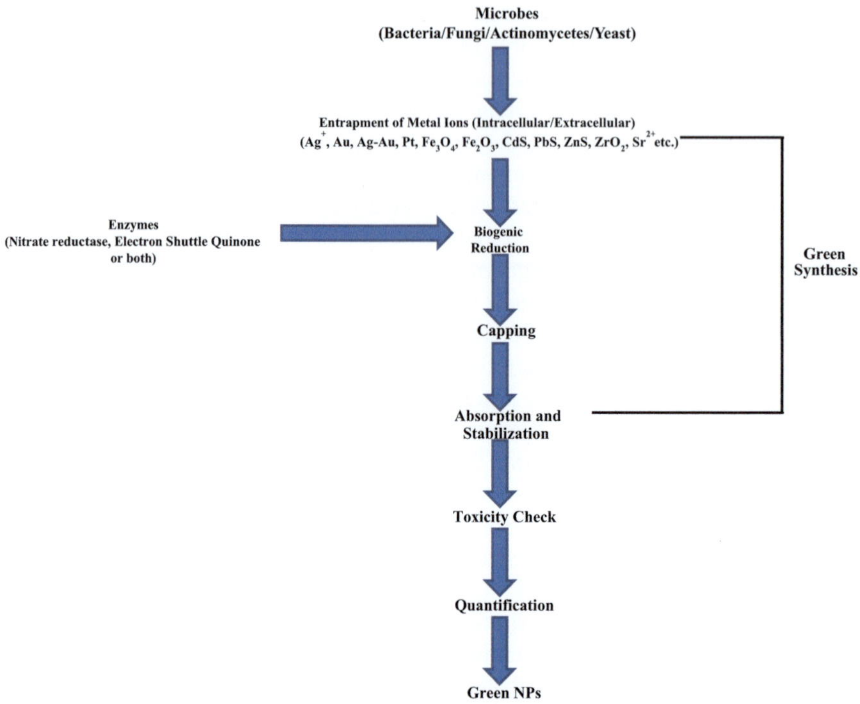

Fig. 5.1 Microbial synthesis of nanoparticles

mechanisms of nanoparticles synthesis vary with different microorganisms. For instance, synthesis of nanoparticles by extracellular means includes three possible mechanisms, viz., nitrate reductase action, electron shuttle quinones, or both. Nanoparticle synthesis was initiated by nitrate reductase utilizing *Penicillium* species and many other fungal species (Perez-de-Luque et al. 2008; Deepa and Panda 2014). The nitrate reductase assay was performed by the reaction of nitrite with 2,3-diaminophthalene (Duran et al. 2005; Kumar et al. 2007a, b). AgNP synthesis for *F. oxysporum* was involved with extracellular shuttle quinone, a-NADPH-dependent reductases, and nitrate-dependent reductases. Investigation also revealed that AgNP synthesis for *A. flavus* takes place earlier with a "33 kDa" protein and then by a protein (cysteine and free amine groups) which forms a capping agent and stabilizes the NPs (Soni and Prakash 2011).

During intracellular synthesis, metal ions were first entrapped at the cell surface of fungi via electrostatic interaction which further reduced by the enzymes within the cell wall, which leads to the accumulation and development of NPs (Singh et al. 2014). Synthesis of silver nanoparticles in *B. licheniformis* was involved with nitrate reductase enzyme (Kalishwaralal et al. 2008). *B. licheniformis* secretes the NADH and NADH-dependent enzymes that are responsible for the bioreduction of Ag^+ and the subsequent formation of AgNPs (Husseiny et al. 2007). Reduction of Ag^+ involves electron shuttle enzymatic metal reduction mechanism convinced with nitrate ions and dwindle silver ions to metallic silver.

Heavy metal (Co^{2+}, $CrO_4{}^{2+}$, Pb^{2+}, Zn^{2+}, Hg^{2+}, and Cd^{2+}) nanoparticles are synthesized by the microbes that developed genetic and proteomic responses to strictly regulate metal homeostasis and counter the toxic effects (Reith et al. 2007). On the other hand, synthesis of magnetite using *Shewanella oneidensis* involves both passive and active mechanisms. Initially, when bacteria utilize ferrihydrite, pH value rises due to amino acid metabolism, and active production of Fe^{2+} occurs followed by passive concentration of Fe^{2+} and Fe^{3+} which enabled the magnetite phase to precipitate. Investigation was done for the synthesis of CdS NPs through disulfide (cysteine) bridges which causes cleavage of S–H bond and generation of a new bond (Cd–S–CH_2COOH) complex on nanoparticle (Sanghi and Verma 2009). The –COOH groups from cadmium–thiolate complex react with hydrogen bond resulted in the capping of CdS nanoparticles bonding to –NH_2 groups (Tang et al. 2005). One oxygen atoms of carboxylic group (–COOH) formed coordinate bond between oxygen atom and Cd^{2+} ions, which competes with thiol group to accumulate on the CdS nanoparticles surfaces (Li et al. 2011).

5.2.3 Microbial-Based Nanoparticle Applications for Crop Protection

Nanobiotechnology can have ample of utilizations such as the identification and control of phytopathogens, protecting food from pathogens through nanosensors fabrications, biofertilizers, and biopesticide delivery in agriculture and highly competent gene transfer approach (Fig. 5.2). NP application is considered as successful for crop protection if it remains active in extreme environments (such as temperature fluctuations); penetrates the target pathogen; resists defense of phytopathogens; has low cost to formulate, preferably with advanced mode of action; and caters to social benefits and economic returns (Smith et al. 2008). NPs are also playing a dominant act for heighten the efficacy and stability of cells and enzymes used. The integration of biomolecules (enzymes, metabolites, etc.) or whole cells with nanomaterials leads to hybrid systems with numerous applications in agriculture (Bailey et al. 2010). Nanoparticles embedded with microbes offered the advantage of improved biological efficacy, easy attachment through larger surface area, higher solubility and mobility, lower toxicity, and enhanced mass delivery systems. After the entrapment of NPs and bonding of the nanomaterials, controlled release of the active ingredient is accomplished. The use of NP assisted delivery will require a targeted delivery approach focused on the behavior of phytopathogens and environmental conditions. For instance, for bombardment of plant cells and tissues to achieve gene transfer, DNA-coated AuNPs were used as bullets in "gene gun" system (Vijayakumar et al. 2010). Microbes (bacteria, fungi) and their products (enzymes, inhibitors, antibiotics, and toxins) can serve as biocontrol agents from decades for crop protection from phytopathogens as well as productivity boost. Coating of polymeric NPs offered advanced paths to enhance the efficiency and stability of these biocontrol agents as biofertilizer preparations to yield

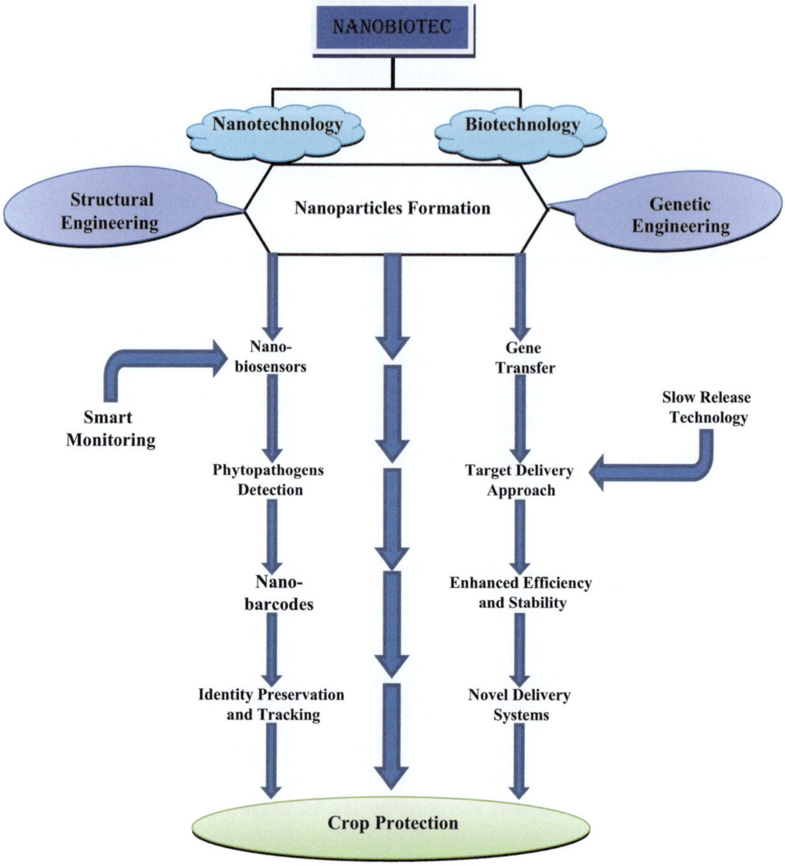

Fig. 5.2 Overview of nanoparticles impact on crop protection

formulations with directed delivery mechanism toward targeted pathogens. Furthermore, nanomaterial entrapped products would promote plant growth and soil health (Peteu et al. 2010).

Fungal biocontrol agents are relatively specific, act by contact without ingestion, and are easy for mass production. Some fungal genera (*Nomuraea*, *Beauveria*, and *Verticillium*) spread infection through conidia that require the moisture for their germination to activate host pathogenesis (Kulkarni et al. 2008). In oil emulsion formulation of *L. giganteum* (a water mold), inclusion of hydrophobic silica NPs (7–14 nm) to water in oil mycelium diminishes the desiccation, which reduced cell sedimentation and imparted N95% efficiency (Vandergheynst et al. 2007). A saprophytic fungus (*Myrothecium verrucaria*) produced endochitinase that assassinates mosquito larvae within 48 h of *Aedes aegypti* (Chavan 2009). Nanoformulation was prepared using chitosan and montmorillonite clay NPs to stabilize *Myrothecium* enzyme complex and observed for the biocontrol activity against *Fusarium* spp. and *Phenacoccus gossypiphilous* (cotton mealy bug) with slow discharge of the enzymes. Plants of *Curcuma longa* treated with chitosan nanoparticles induce antifungal hydrolases and increased chitinases and chitosanases enzymes responsible for the defense of host plants which enabled them to become resistant to *Pythium aphanidermatum*, the causal organism of rhizome rot of turmeric (Anusuya and Sathiyabama 2013). Sensitive detection of phytopathogens (even for a single bacterial cell) was demonstrated with silica-based NPs (60 nm) filled with a fluorescent dye and conjugated to an antibody specific to a surface antigen of the microbe of interest. With the assistance of endomycorrhizal fungi, the common wetland plants (*Phragmites australis* and *Iris pseudacorus*) transform copper into metallic NPs when grown in contaminated soil (Manceau et al. 2008) to prevent copper biomagnification. Ag_2S nanocrystals and $ZnTiO_3$ exhibited higher growth inhibition efficacy against *A. niger* (Fateixa et al. 2009; Jo et al. 2009; Ruffolo et al. 2010). Higher resistance against *Fusarium oxysporum* and *Aspergillus niger* was demonstrated in maize plants treated with silica NPs of 20–40 nm (Suriyaprabha et al. 2014). Titania NPs increased bacterial population (*Bacillus amyloliquefaciens*) in the rhizosphere of *Brassica napus* and provides protection against *Alternaria brassicae* (Palmqvist et al. 2015). ZnO NPs proved for the inhibition of conidiophores and conidia of *Penicillium expansum* resulting in the disappearance of fungal mats (He et al. 2010). A highly specific and more sensitive method employs magnetic nanoparticle-based reverse. Transcription loop-mediated isothermal amplification (RTLAMP) was developed for the prompt detection of *Prunus necrotic ringspot virus* (Zong et al. 2014).

Highly fluorescent CdSe QDs and CdTe QDs were biosynthesized with *F. oxysporum* when incubated with a mixture of $CdCl_2 + SeCl_4$ and $CdCl_2 + TeCl_2$, respectively, at ambient conditions by transmission electron microscopy (TEM) and selected area electron diffraction (SAED) (Shaligram et al. 2009). Yeast cells were also utilized for biosynthesis of biocompatible cadmium telluride (CdTe) QDs with tunable fluorescence emission (Nayak et al. 2010). Nucleic acid probe attached with the surface of a quartz crystal microbalance biosensor can also be combined with fast PCR protocols to reduce the time for specific detection of phytopathogens (Maliszewska et al. 2013).

Recently, nanobiotechnology received a greater effectiveness against various phytopathogens with the use of AgNPs. Contacts of AgNPs with microbes increase due to their larger surface area-to-volume ratio and thus more ability to permeate (Kim et al. 2008). When aqueous silver (Ag^+) ions exposed to a filtrate of *Trichoderma viride*, it reduced in solution resulted in the formation of extremely stable AgNPs with 5–40 nm size (Fayaz et al. 2010). When evaluated for antimicrobial activities, it was observed that combination of antibiotics with AgNPs has better antimicrobial effects against various Gram-positive and Gram-negative bacteria (Aziz et al. 2014, 2015, 2016). Pathogenic bacterial (*Staphylococcus aureus*) infection was minimized into textile fabrics with extracellularly produced AgNPs incorporating *Fusarium oxysporum* (Duran et al. 2007). AgNPs were also demonstrated for highest-disease inhibition rate against six species of *Colletotrichum* (*C. acutatum*, *C.dematium*, *C. gloeosporioides*, *C. higginsianum*, *C. nigrum*, and *C. orbiculare*) and powdery mildew in cucumbers and pumpkins (Lamsal et al. 2011).

Leaf-spot disease caused by *Xanthomonas perforans* can be suppressed by DNA-directed AgNPs (Ocsoy et al. 2013). In another study, biogenically synthesized silver nanoparticles impregnated antibiotic discs (chloramphenicol) and reported excellent antibacterial activity against two bacteria (*Citrobacter freundii* and *Erwinia cacticida*) causing diseases on *Abelmoschus esculentus* and *Citrullus lanatus* (Paulkumar et al. 2014). AgNPs using *Serratia* spp. were biosynthesized and demonstrated for remarkable antifungal activity against spot blotch disease in wheat caused by *Bipolaris sorokiniana* (Mishra et al. 2014). *Xanthomonas axonopodis* pv. *vesicatoria* causing bacterial spot disease in tomatoes and peppers was successfully detected by fluorescent silica nanoparticles (FSNP) combined with antibody molecules (Mishra et al. 2010). The nanoparticles embedded with antibodies are used for detection of *Xanthomonas axonopodis* (Yao et al. 2009). AgNPs as antimicrobial agents are rapidly growing attraction toward researchers worldwide to make their production more economical and profitable for the control of plant diseases. Various studies also revealed strong impact of AgNPs when used in consortium with other nanocrystals. *Botrytis cinerea* was diminished by strong antifungal effect when used with Ag-SiO_2 NPs (Oh et al. 2006). AgNPs were evaluated for antifungal activity along with fluconazole by disc diffusion technique against *Phoma glomerata*, *Phoma herbarum*, *F. semitectum*, *Trichoderma* spp., and *C. albicans* (Gajbhiye et al. 2009). *Colletotrichum gloeosporioides* (responsible for anthracnose), *Bipolaris sorokiniana*, *Magnaporthe grisea*, *Sclerotium cepivorum*, and sclerotium-forming phytopathogenic fungi were significantly suppressed in a dose-dependent manner with the existence of Ag NPs (Jung et al. 2010; Aguilar-Mendez et al. 2010). Fungistatic and fungicidal effect of the AgNPs was assayed against ambrosia fungus *Raffaelea* spp., *F. culmorum*, and certain pathogenic yeasts (*Candida tropicalis*, *C. albicans*, and *C. parapsilosis*) (Kasprowicz et al. 2010). It was also demonstrated that 15 mg concentration of AgNPs showed excellent inhibitory activity against *Alternaria alternata*, *Sclerotinia sclerotiorum*, *Macrophomina phaseolina*, *Rhizoctonia solani*, *B. cinerea*, and *Curvularia lunata* (Krishnaraj et al. 2012).

5.3 Microbial-Based Nanoparticles Delivery Systems

With the utilization of NPs, nanofibers and nanocapsules, nanobiotechnology offers a novel set of procedures to multiply genes and upgrade plant resistance (Rai et al. 2012a, b). Delivery of genetic material (DNA and small interfering RNA) is important for the development of phytopathogen-resistant strains of plants by alteration of gene expression (Price and Gatehouse 2008). Gene expressions confirmed the successful insertion and integration of plasmid DNA in plant genome (Filipenko et al. 2007) and thus have more potential to develop disease resistance embedded with genetic modifications. The methods (microinjection, *Agrobacterium*-mediated transformation, and microprojectile bombardment) applied for gene delivery systems have very low efficiency (0.01%–20%) and were mostly applied for dicotyledonous plant transformation (Sivamani et al. 2009). When NPs were employed, these technologies expand to both dicotyledonous and monocotyledonous plants with tissue specificity. During transformation of plant cells, delivery of genetic material (DNA) was employed with AuNPs (5–25 nm) embedded carbon matrices in both monocotyledonous *Oryza sativa* and dicotyledonous *Nicotiana tabacum*. This resulted in the easy access to plant cell due to its increased size transformation efficiency. In comparison to the commercial micrometer-sized gold particles, NPs also reduced plasmid and gold requirement (Vijayakumar et al. 2010) with minimal plant-cell damage and enhanced plant regeneration. Gene expressions were also triggered in the plants when uncapped with AuNPs by disulfide reduction in cellular environment under controlled release conditions (Torney et al. 2007). Starch NPs (50–100 nm) embedded with fluorescent material Tris-(2,2′-bipyridine) ruthenium-(Ru (bpy)3)2^+ were used to transfer plasmid DNA through *Dioscorea* spp. plant cell wall, cell membrane, and nuclear membrane (Liu et al. 2008).

Polymer/DNA complexes are more stable for the protection against nuclease degradation. Also, cationic polymers (chitosan) are used to condense and deliver DNA both in vitro and in vivo (Kim et al. 2007) because of biocompatibility and low cytotoxicity. Biocontrol of crops from the phytopathogens that feed upon the double-stranded RNA (dsRNA)-producing plants was improved by RNA-based silencing (Auer and Frederick 2009). Due to the properties of binding with RNA in a highly efficient manner, safety, and ability to invade the cell membrane, chitosan NPs can prove to be competent in dsRNA delivery. In *C. elegans*, delivery of dsRNA was followed by the feeding of bacteria-expressing dsRNA or even soaking in dsRNA solution (Tomoyasu and Denell 2004). For siRNA delivery vehicles, chitosan NPs (<500 nm) entrapped with siRNA performed well as vectors (Katas and Alpar 2006). Studies also demonstrated that chitosan NPs (100–200 nm) stabilized and delivered dsRNA (against chitin synthase genes) resulted in enhanced larval susceptibilities, in existence of chitin synthase inhibitors (Zhang et al. 2010). Nucleic acid sensors were fabricated via immobilization of ssDNA onto a chitosan nanobiocomposite film containing Fe_3O_4 NPs (22 nm) for cypermethrin and permethrin detection at 0.0025 ppm (Kaushik et al. 2009).

5.3.1 Nanosensors for Crop Protection

In addition to crop protection by NPs, nanosensors help farmers in maintaining farm with precise control, crop monitoring, accurate analysis of nutrients, report of plant needs (Mousavi and Rezaei 2011), as well as detection of phytopathogens to foster a smart agriculture (Prasad et al. 2014, 2017a, b; Bhattacharyya et al. 2016; Sangeetha et al. 2017a, b). More accurately, nanosensors can be used to monitor presence and identification of phytopathogens in order to carry out remedial activities timely for crop protection. Nanosensors located in cultivated fields will provide a real-time and comprehensive monitoring of the crop growth with high-quality data for best management practices (ElBeyrouthya and ElAzzi 2014). Some of the strategies such as antibody–antigen, adhesion receptor, antibiotic, and complementary DNA sequence recognitions are used for a specific detection between target phytopathogenic cells and bio-functionalized nanomaterials (Sastry et al. 2010; Duran et al. 2010). Gold nanoparticles are used in biosensors to develop biomolecular detection with DNA or protein-functionalized gold nanoparticles as the target-specific probes (Khosravi and Shojaosadati 2009). *Bacillus subtilis* were known to form spherical selenium nanoparticles with diameters of 50–400 nm employed for building HRP (*horse radish peroxidase*) biosensor (Wang et al. 2010). Yeast cells were used for biosynthesis of Au-Ag alloy nanoparticles to fabricate a sensitive electrochemical vanillin sensor (Zheng et al. (2010a, b). AuNP-based glucose oxidase (GOx) biosensors were synthesized for the enhancement of enzyme activity of GOx (Zheng et al. (2010a, b). Reaction rates in microbiology have also been enhanced using magnetic nanoparticles. Magnetic (Fe_3O_4)-coated nanoparticles with cells of *Pseudomonas delafieldii* were utilized to fulfill desulfurization of dibenzothiophene (Shan et al. 2005). Various researchers also emphasize on the incorporation of nanobiotechnology to emerging biotechnological methods for mycotoxin bioassays (Babu and Gunasekaran 2009) and utilization of nanosensors to reduce the fungal pathogen detection time with high sensitivity. *Escherichia coli* also rapidly determined using flow-injection system approach in which electrochemical measurement of $K_3Fe(CN)_6$ was shortened by microbial metabolism and thus allowed the quantitative determination of bacteria within 20 min. A new biosensor system utilizing equal quantities of two different microbes was developed for the rapid diagnosis of soil-borne diseases in which each microbe is individually immobilized on an electrode (Perez-Gonzalez et al. 2010). Copper oxide (CuO) nanoparticles and nanostructural layer biosensors fabricated by sol-gel and spray pyrolysis methods, respectively, were employed for detecting the *A. niger* fungi (Bao et al. 2003).

5.4 Nanotoxicity: A Major Obstacle

Although nanoparticles proved to have very huge applicability for the biocontrol of phytopathogens and thus management of various plant diseases, this increased application emerges the possibilities of getting nanoparticle accumulation in the environment which results in the toxicity of soil ecosystem (with more effect on soil microflora) leading to various harmful impacts. Activities of soil enzymes (*protease*, *catalase*, and *peroxidase*) were inhibited by TiO_2 and ZnO nanoparticles, thereby affecting the soil quality and health, and also have negative impact on the biomass of wheat growth (Du et al. 2011). Root elongation of *Zea mays*, *Cucumis sativus, Glycine max*, *Brassica oleracea*, and *Daucus carota* was affected with Al_2O_3, Al, Zn, and ZnO when conjugated with and without phenanthrene (Yang and Watts 2005) which resulted in suppression of plant germination. TiO_2 was observed for the reduction of the water usage in *Z. mays* which changes the path of apoplast (Asli and Neumann 2009). Accumulations of Fe_2O_3 and Pd nanoparticles were found in tissue of pumpkin and leaves of barley, respectively (Battke et al. 2008; Zhu et al. 2008). One hundred percent control was achieved for powdery mildew diseases of cucurbits with silica-silver nanoparticles but become phytotoxic at 3200 ppm (Park et al. 2006). ZnO and TiO_2 nanoparticles also affected diversity of soil microbial community and biomass. Soybean exposed to ZnO and TiO_2 nanoparticles impacted directly on biomass or plant-microbe interactions, including N_2-fixing symbiosis (Ge et al. 2011). ZnO NP toxic effects on rhizospheric population might be reduced with soil–plant interactive system (Lee et al. 2012). NPs also have an impact on plant sensitivity to bacterial infection, bacterial growth and stress resistance, and the interactions occurring between host plants and associated bacterial community (Degrassi et al. 2012).

5.5 Advantages of Microbial-Based Nanoparticles

Biosynthesis of microbial-based nanoparticles is considered as nontoxic, clean, and environmentally green. Biosynthesis mechanism by microbes (bacteria, fungi, and yeast) can be classified according to the location where NPs are formed, i.e., intracellular or extracellular. Various parameters (pH, substrate concentration, reaction temperature, etc.) influence the rate of NP biosynthesis and thus can be manipulated as per needs.

NP biosynthesis can also be manipulated through gene expression by genetic delivery materials (DNA and small interfering RNA). Below are the some of the advantages of microbial NPs in crop protection: highly effective and on time response for integrated plant disease management, quick and reliable response on phytopathogens with antifungal and bactericidal effects, smart delivery along with modification as genomic and proteomic levels, development of nanobiosensor for biodetection and management of plant pathogens in field, advanced nanosystem procedures for sampling of soil and plant, and enhancement of microbiological reaction rates.

5.6 Current Prospects and Way Forward for more Efficient Microbial-Based Nanoparticles

Till date nanotechnology has a wide range of potential applications in agriculture, enabling intense research at academic and industrial levels (Chen and Yada 2011; Dasgupta et al. 2015; Parisi et al. 2015). Besides the numerous advantages of NPs, still, there are many issues that remain to be resolved for this technology to make significant contributions in sustainable agriculture. Some aspects require immediate attention such as mode of uptake, impact of size, agglomeration, penetration, transport and stability of nanoparticles on crops, and consequences of their exposure to the environment. These factors indicate the promises of NPs for suppression/inhibition to various phytopathogens. Phytotoxicity study of NPs to determine root length, germination effect, and adsorption into the plant systems needs to be addressed (Kumari et al. 2012). Biocompatibility and biodegradability studies of nanoparticles are also desirable. Attention may also be given on the development of specific hybrid carriers for delivering active agents in order to maximize their efficiency (DeOliveira et al. 2014), upscaling of technology, risk assessment, and advancement in the regulations for the use of nanoparticles (Amenta et al. 2015). Nanosensors developed are also very specific, fast, and inexpensive, but their commercialization is again a major concern. Mechanisms of nanoparticles should be taken care to understand the interactions with target (Oliveira et al. 2015a) and nontarget organisms (Oliveira et al. 2015b) as well as risk-assessment analysis (Sadiko et al. 2014). Quantification at different concentrations in the localized environment (Kah et al. 2014) and advances in life cycle assessment (Kookana et al. 2014) are other two important factors to consider while developing any new nanoparticles. Isolation and characterization of the compounds responsible for reduction of NPs can be carried out at genomic and proteomic levels to expect shorter reaction time and high biosynthesis efficiency. Efforts are also required in defining the interactions of nanoparticles with microbes, plants, and soil for both constructive and catastrophic impacts on plant and soil health. In this context, expertise among the researchers from multidisciplinary fields would be imperative (Malysheva et al. 2015).

5.7 Conclusion

Nanobiotechnology has potential applications in the multiple detection and control of phytopathogens utilizing microbial-based nanoparticles, nanosensors, and quantum dots which greatly contribute to address the issue of sustainability. Biosynthesis of nanoparticles by microbes is eco-friendly, green, and economical viable, and issue of phytotoxicity and microbes resistant to microbicides or fungicides is also resolved. Moreover, nano-based diagnostic kits enhanced the rate of specific pathogen detection in field with high diagnosis accuracy. Additionally, nanosensors are able to detect phytopathogens at early stages of plant disease before these can be observed by farmers. Apart from detection, nanobiotechnology can be applied to know the interaction between plant and phytopathogens as well as solutions for soil remediation. Thus, application of green nanobiotechnology can greatly contribute to the management of phytopathogens and appropriate disease management action. However, experimental field trials at large scale are required to figure out the host and phytopathogen synergy, infection steps, and disease interpretation which will help in developing modern disease management approach to make agriculture as smart systems.

References

Abd FG, Al-Kawaz AJAH, HOM A-D (2013) Phenotypic and genotypic investigations on silver nanoparticles of *Morganella morganii* recovered from (cauti). Iraq Inter J Med Pharma Sci 3:29–38

Aguilar-Mendez MA, San Martin-Martinez E, Ortega-Arroyo L, Cobian-Portillo G, Sanchez-Espindola E (2010) Synthesis and characterization of silver nanoparticles: effect on phytopathogen *Colletotrichum gloesporioides*. J Nanopart Res 13:2525–2532

Ahmad A, Mukherjee P, Senapati S, Mandal D, Khan MI, Kumar R et al (2003a) Extracellular biosynthesis of silver nanoparticles using the fungus *Fusarium oxysporum*. Colloids Surfaces B 28(4):313–318

Ahmad A, Senapati S, Khan MI, Kumar R, Sastry M (2003b) Extracellular biosynthesis of monodisperse gold nanoparticles by a novel extremophilic actinomycete, *Thermomonospora* sp. Langmuir 19(8):3550–3553

Alghuthaymi MA, Almoammar H, Rai M, Said-Galiev E, Abd-Elsalam KA (2015) Myconanoparticles: synthesis and their role in phytopathogens management. Biotechnol Biotechnol Equ 29(2):221–222

Amenta V, Aschberger K, Arena M, Bouwmeester H, Moniz FB, Brandhoff P et al (2015) Regulatory aspects of nanotechnology in the Agri/feed/food sector in EU and non-EU countries. Regul Toxicol Pharmacol 73:463–476

Anusuya S, Sathiyabama M (2013) Effect of chitosan on rhizome rot disease of turmeric caused by *Pythium aphanidermatum*. ISRN Biotechnol 305349:1–5

Arakaki A, Nakazawa H, Nemoto M, Mori T, Matsunaga T (2008) Formation of magnetite by bacteria and its application. J Royal Soc Inter 5(26):977–999

Asli S, Neumann PM (2009) Colloidal suspensions of clay or titanium dioxide nanoparticles can inhibit leaf growth and transpiration via physical effects on root water transport. Plant Cell Environ 32:577–584

Auer C, Frederick R (2009) Crop improvement using small RNAs: applications and predictive ecological risk assessments. Trends Biotechnol 27:644–651

Aziz N, Fatma T, Varma A, Prasad R (2014) Biogenic synthesis of silver nanoparticles using *Scenedesmus abundans* and evaluation of their antibacterial activity. J Nanoparticles 689419. https://doi.org/10.1155/2014/689419

Aziz N, Faraz M, Pandey R, Sakir M, Fatma T, Varma A, Barman I, Prasad R (2015) Facile algae-derived route to biogenic silver nanoparticles: synthesis, antibacterial and photocatalytic properties. Langmuir 31:11605–11612. https://doi.org/10.1021/acs.langmuir.5b03081

Aziz N, Pandey R, Barman I, Prasad R (2016) Leveraging the attributes of Mucor hiemalis-derived silver nanoparticles for a synergistic broad-spectrum antimicrobial platform. Front Microbiol 7:1984. https://doi.org/10.3389/fmicb.2016.01984

Babu MMG, Gunasekaran P (2009) Production and structural characterization of crystalline silver nanoparticles from *Bacillus cereus* isolate. Colloids Surfaces B 74(1):191–195

Bai HJ, Zhang ZM (2009) Microbial synthesis of semiconductor lead sulfide nanoparticles using immobilized *Rhodobacter sphaeroides*. Mater Lett 63(9–10):764–766

Bai HJ, Zhang ZM, Gong J (2006) Biological synthesis of semiconductor zinc sulfide nanoparticles by immobilized *Rhodobacter sphaeroides*. Biotechnol Lett 28(14):1135–1139

Bailey KL, Boyetchko SM, Langle T (2010) Social and economic drivers shaping the future of biological control: a Canadian perspective on the factors affecting the development and use of microbial biopesticides. Biol Control 52:221–229

Balaji DS, Basavaraja S, Deshpande R, Mahesh DB, Prabhakar BK, Venkataraman A (2009) Extracellular biosynthesis of functionalized silver nanoparticles by strains of *Cladosporium cladosporioides* fungus. Colloids Surfaces B 68:88–92

Bansal V, Poddar P, Ahmad A, Sastry M (2006) Room temperature biosynthesis of ferroelectric barium titanate nanoparticles. J Amer Chem Soc 128(36):11958–11963

Bao H, Hao N, Yang Y, Zhao D (2003) Biosynthesis of biocompatible cadmium telluride quantum dots using yeast cells. Nano Res 3:491–498

Baruah S, Dutta J (2009) Nanotechnology applications in pollution sensing and degradation in agriculture: a review. Environ Chem Lett 7:161–204

Battke F, Leopold K, Maier M, Schmidhalter U, Schuster M (2008) Palladium exposure of barley: uptake and effects. Plant Biol 10:272–276

Bhainsa KC, D'Souza SF (2006) Extracellular biosynthesis of silver nanoparticles using the fungus *Aspergillus fumigatus*. Colloids Surfaces B 47(2):160–164

Bhattacharyya A, Duraisamy P, Govindarajan M, Buhroo AA, Prasad R (2016) Nano-biofungicides: emerging trend in insect pest control. In: Prasad R (ed) Advances and applications through fungal Nanobiotechnology. Springer International Publishing, Switzerland, pp 307–319

Bin Hussein MZ, Hashim N, Yahaya AH, Zainal Z (2009) Controlled release formulation of agrochemical pesticide based on 4-(2,4-dichlorophenoxy) butyrate nanohybrid. J Nanosci Nanotechnol 9:2140–2147

Birla SS, Tiwari VV, Gade AK, Ingle AP, Yadav AP, Rai MK (2009) Fabrication of silver nanoparticles by *Phoma glomerata* and its combined effect against *Escherichia coli*, *Pseudomonas aeruginosa* and *Staphylococcus aureus*. Lett Appl Microbiol 48:173–179

Castro-Longoria E, Vilchis-Nestor AR, Avalos-Borja M (2011) Biosynthesis of silver, gold and bimetallic nanoparticles using the filamentous fungus *Neurospora crassa*. Colloids Surfaces B 83:42–48

Chavan S (2009) Biocontrol of insect pests in agriculture using chitinolytic enzyme complex *of Myrothecium verrucaria* [Thesis]. University of Pune, Pune

Chen HD, Yada R (2011) Nanotechnologies in agriculture: new tools for sustainable development. Trends Food Sci Technol 22:585–594

Chen L, Song Y, Tang B, Song X, Yang H, Li B, Zhao Y, Huang C, Han X, Wang S, Li Z (2015) Aquatic risk assessment of a novel strobilurin fungicide: a microcosm study compared with the species sensitivity distribution approach. Ecotoxic Environ Safety. https://doi.org/10.1016/j.ecoenv.2015.06.027

Chowdappa P, Gowda S (2013) Nanotechnology in crop protection: status and scope. Pest Manage Hort Ecosys 19(2):131–151

Dasgupta N, Ranjan S, Mundekkad D, Ramalingam C, Shanker R, Kumar A (2015) Nanotechnology in agro-food: from field to plate. Food Res Int 69:381–400

Deepa K, Panda T (2014) Synthesis of gold nanoparticles from different cellular fractions of *Fusarium oxysporum*. J Nanosci Nanotechnol 14:3455–3463

Degrassi G, Bertani I, Devescovi G, Fabrizi A, Gatti A, Venturi V (2012) Response of plant-bacteria interaction models to nanoparticles. EQA 8:39–50

DeOliveira JL, Campos EVR, Bakshi M, Abhilash PC, Fraceto LF (2014) Application of nanotechnology for the encapsulation of botanical insecticides for sustainable agriculture: prospects and promises. Biotechnol Adv 32:1550–1561

Devi TP, Kulanthaivel S, Kamil D, Borah JL, Prabhakaran N, Srinivasa N (2013) Biosynthesis of silver nanoparticles from *Trichoderma* species. Indian J Exp Biol 51:543–547

DeWindt W, Aelterman P, Verstraete W (2005) Bioreductive deposition of palladium (0) nanoparticles on *Shewanella oneidensis* with catalytic activity towards reductive dechlorination of polychlorinated biphenyls. Environ Microbiol 7(3):314–225

Dinjaski N, Prieto MA (2015) Smart polyhydroxyalkanoate nanobeads by protein based functionalization. Nanomedicine 11:885–899

Du W, Sun Y, Ji R, Zhu J, Wub J, Guo H (2011) TiO_2 and ZnO nanoparticles negatively affect wheat growth and soil enzyme activities in agricultural soil. J Environ Monitor 13:822–828

Duran N, Marcato PD, Alves OL, DeSouza G, Esposito E (2005) Mechanistic aspects of biosynthesis of silver nanoparticles by several *Fusarium oxysporum* strains. J Nanobiotechnol 3:1–8

Duran N, Marcato PD, De Souza GIH, Alves OL, Esposito E (2007) Antibacterial effect of silver nanoparticles produced by fungal process on textile fabrics and their effluent treatment. J Biomed Nanotech 3(2):203–208

Duran N, Marcato PD, Alves OL, Da Silva JPS, De Souza GIH, Rodrigues FA, Esposito E (2010) Ecosystem protection by effluent bioremediation, silver nanoparticles impregnation in a textile fabrics process. J Nanopart Res 12:285–292

Dzhavakhiya V, Shcherbakova L, Semina Y, Zhemchuzhina N, Campbell B (2012) Chemosensitization of plant pathogenic fungi to agricultural fungicides. Front Microbiology 3:1–9

ElBeyrouthya M, ElAzzi D (2014) Nanotechnologies: novel solutions for sustainable agriculture. Adv Crop Sci Technol 2:e118

El-Kassas HY, El-Sheekh MM (2014) Cytotoxic activity of biosynthesized gold nanoparticles with an extract of the red seaweed *Corallina officinalis* on the MCF-7 human breast cancer cell line. Asian Pacific J Cancer Preven 15(10):4311–4317

El-rafie MH, Shaheen TI, Mohamed AA, Hebeish A (2012) Biosynthesis and applications of silver nanoparticles onto cotton fabrics. Carbohydr Polym 90:915–920

Fateixa S, Neves MC, Almeida A, Oliveira J, Trindade T (2009) Anti-fungal activity of SiO_2/Ag_2S nanocomposites against *Aspergillus niger*. Colloids Surfaces B 74:304–308

Fayaz AM, Balaji K, Girilal M, Yadav R, Kalaichelvan PT, Venketesan R (2010) Biogenic synthesis of silver nanoparticles and their synergistic effect with antibiotics: a study against gram-positive and gram-negative bacteria. Nanomedicine: nanotech. Biol Med 6(1):e103–e109

Filipenko EA, Filipenko ML, Deineko EV, Shumnyi VK (2007) Analysis of integration sites of T-DNA insertions in transgenic tobacco plants. Cytol Genet 41:199–203

Frampton RA, Pitman AR, Fineran PC (2012) Advances in bacteriophage-mediated control of plant pathogens. Inter J Microbiol. https://doi.org/10.1155/2012/326452

Gade A, Rai M, Karwa A, Bonde P, Ingle A (2007) Extracellular biosynthesis of silver nanoparticles by *Pleurotus* species. Int J Med Mushroom Res 9(3–4):298–299

Gade AK, Bonde P, Ingle AP, Marcato PD, Duran N, Rai MK (2008) Exploitation of *Aspergillus niger* for the synthesis of silver nanoparticles. J Biobased Mater Bioener 2:243–247

Gade A, Gaikwad S, Duran N, Rai M (2013) Screening of different species of *Phoma* for synthesis of silver nanoparticles. Biotechnol Appl Biochem 60(5):482–493

Gaikwad S, Birla SS, Ingle AP, Gade AK, Marcato PD, Rai MK, Duran D (2013) Screening of different *Fusarium* species to select potential species for the synthesis of silver nanoparticles. J Braz Chem Soc 24:1974–1982

Gajbhiye M, Kesharwani J, Ingle A, Gade A, Rai M (2009) Fungus mediated synthesis of silver nanoparticles and their activity against pathogenic fungi in combination with fluconazole. Nanomedicine 5:382–386

Ge Y, Schimel JP, Holden PA (2011) Evidence for negative effects of TiO_2 and ZnO nanoparticles on soil bacterial communities. Environ Sci Technol 45:1659–1664

Golinska P, Wypij M, Ingle AP, Gupta I, Dahm H, Rai M (2014) Biogenic synthesis of metal nanoparticles from Actinomycetes: biomedical applications and cytotoxicity. Appl Microbiol Biotech 98:8083–8097

Guo BL, Han P, Guo LC, Cao YQ, Li AD, Kong JZ, Zhai HF, Wu D (2015) The antibacterial activity of Ta-doped ZnO nanoparticles. Nanoscale Res Lett 10(1):1047

Gurunathan S, Kalishwaralal K, Vaidyanathan R, Venkataraman D, Pandian SRK, Muniyandi J et al (2009) Biosynthesis, purification and characterization of silver nanoparticles using *Escherichia coli*. Colloids Surfaces B 74(1):328–335

He SY, Guo ZR, Zhang Y, Zhang S, Wang J, Gu N (2007) Biosynthesis of gold nanoparticles using the bacteria *Rhodopseudomonas capsulate*. Mater Letter 61:3984–3987

He L, Liu Y, Mustapha A, Lin M (2010) Antifungal activity of zinc oxide nanoparticles against *Botrytis cinerea* and *Penicillium expansum*. Microbiol Res 166:207–215

Husseiny MI, El-Aziz MA, Badr Y, Mahmoud MA (2007) Biosynthesis of gold nanoparticles using *Pseudomonas aeruginosa*. Spectrochim Acta A 67(3–4):1003–1006

Jacob JJ, Suthindhiran K (2016) Magnetotactic bacteria and magnetosomes – scope and challenges. Mater Sci Eng C 68:919–928

Jaidev LR, Narasimha G (2010) Fungal mediated biosynthesis of silver nanoparticles, characterization and antimicrobial activity. Colloids Surfaces B 81:430–433

Jain N, Bhargava A, Majumdar S, Tarafdar JC, Panwar J (2011) Extracellular biosynthesis and characterization of silver nanoparticles using *Aspergillus flavus* NJP08: a mechanism perspective. Nanoscale 3(2):635–641

Jha AK, Prasad K, Prasad K (2009) A green low-cost biosynthesis of Sb_2O_3 nanoparticles. Biochem Eng J 43(3):303–306

Jo YK, Kim BH, Jung G (2009) Antifungal activity of silver ions and nanoparticles on phytopathogenic fungi. Plant Dis 93:1037–1043

Jung JH, Kim SW, Min JS, Kim YJ, Lamsal K, Kim KS (2010) The effect of nano-silver liquid against the white rot of the green onion caused by *Sclerotium cepivorum*. Mycobiology 38(1):39–45

Kah M, Machinski P, Koerner P, Tiede K, Grillo R, Fraceto LF et al (2014) Analysing the fate of nanopesticides in soil and the applicability of regulatory protocols using a polymer-based nanoformulation of atrazine. Environ Sci Pollut Res Int 21:11699–11707

Kalimuthu K, Babu RS, Venkataraman D, Bilal M, Gurunathan S (2008) Biosynthesis of silver nanocrystals by *Bacillus licheniformis*. Colloids Surface B: Biointerfaces 65:150–153

Kalishwaralal K, Deepak V, Ramkumarpandian S, Nellaiah H, Sangiliyandi G (2008) Extracellular biosynthesis of silver nanoparticles by the culture supernatant of *Bacillus licheniformis*. Mater Lett 62(29):4411–4413

Kar PK, Murmu S, Saha S, Tandon V, Acharya K (2014) Anthelmintic efficacy of gold nanoparticles derived from a phytopathogenic fungus, *Nigrospora oryzae*. PLoS One 9(1):e84693

Kasprowicz MJ, Kozio M, Gorczyca A (2010) The effect of silver nanoparticles on phytopathogenic spores of *Fusarium culmorum*. Can J Microbiol 56:247–253

Katas H, Alpar HO (2006) Development and characterisation of chitosan nanoparticles for siRNA delivery. J Control Release 115:216–225

Kathiresan K, Manivannan S, Nabeel MA, Dhivya B (2009) Studies on silver nanoparticles synthesized by a marine fungus, *Penicillium fellutanum* isolated from coastal mangrove sediment. Colloids Surfaces B 7:133–137

Kaushik A, Solanki PR, Ansarib AA, Malhotra BD, Ahmad S (2009) Iron oxide-chitosan hybrid nanobiocomposite based nucleic acid sensor for pyrethroid detection. Biochem Engineer J 46:132–140

Khosravi A, Shojaosadati SA (2009) Evaluation of silver nanoparticles produced by fungus *Fusarium oxysporum*. Int J Nanotechnol 6:973–983

Kim TH, Jiang HL, Jere D, Parka IK, Cho MH, Nah JW et al (2007) Chemical modification of chitosan as a gene carrier in vitro and in vivo. Prog Polym Sci 32:726–753

Kim KJ, Sung WS, Moon SK, Choi JS, Kim JG, Lee DG (2008) Antifungal effect of silver nanoparticles on dermatophytes. J Microbiol Biotechnol 18:1482–1484

Klaus T, Joerger R, Olsson E, Granqvist CG (1999) Silver based crystalline nanoparticles, microbially fabricated. Proc Nat Acad Sci USA 96(24):13611–13614

Konishi Y, Ohno K, Saitoh N, Nomura T, Nagamine S, Hishida H et al (2007) Bioreductive deposition of platinum nanoparticles on the bacterium *Shewanella algae*. J Biotech 128(3):648–653

Kookana RS, Boxall AB, Reeves PT, Ashauer R, Beulke S, Chaudhry Q et al (2014) Nanopesticides: guiding principles for regulatory evaluation of environmental risks. J Agric Food Chem 62:4227–4240

Krishnaraj C, Ramachandran R, Mohan K, Kalaichelvan PT (2012) Optimization for rapid synthesis of silver nanoparticles and its effect on phytopathogenic fungi. Spectrochim Acta A 93:95–99

Kulkarni SA, Ghormade V, Kulkarni G, Kapoor M, Chavan SB, Rajendran A et al (2008) Comparison of Metarhizium isolates for biocontrol of *Helicoverpa armigera* (Lepidoptera: Noctuidae) in chickpea. Biocontrol Sci Tech 18:809–828

Kumar SA, Abyaneh MK, Gosavi SW, Kulkarni SK, Pasricha R, Ahmad A, Khan MI (2007a) Nitrate reductase-mediated synthesis of silver nanoparticles from $AgNO_3$. Biotechnol Lett 29:439–445

Kumar SA, Ansary AA, Abroad A, Khan MI (2007b) Extracellular biosynthesis of CdSe quantum dots by the fungus, *Fusarium oxysporum*. J Biomed Nanotech 3(2):190–194

Kumari M, Ernest V, Mukherjee A, Chandrasekaran N (2012) In vivo nanotoxicity assays in plant models. Methods Mol Biol 926:399–410

Lamsal K, Kim SW, Jung JH, Kim YS, Kim KS, Lee YS (2011) Application of silver nanoparticles for the control of *Colletotrichum* species *in vitro* and pepper anthracnose disease in field. Mycobiology 39(3):194–199

Law N, Ansari S, Livens FR, Renshaw JC, Lloyd JR (2008) Formation of nanoscale elemental silver particles via enzymatic reduction by *Geobacter sulfurreducens*. Appl Environ Microbiol 74:7090–7093

Leake JR, Donnelly DP, Boddy L (2002) Interactions between ectomycorrhizal fungi and saprotrophic fungi. In: der Heijden MGA V, Sanders IR (eds) Mycorrhizal ecology. Ecological Studies Springer Verlag, Heidelberg, p 157

Lee S, Kim S, Kim S, Lee I (2012) Effects of soil-plant interactive system on response to exposure to ZnO nanoparticles. J Microbiol Biotechnol 22:1264–1270

Lefevre CT, Abreu F, Schmidt ML, Lins U, Frankel RB, Hedlund BP et al (2010) Moderately thermophilic magnetotactic bacteria from hot springs in Nevada. Appl Environ Microbiol 76(11):3740–3743

Lengke MF, Ravel B, Fleet ME, Wanger G, Gordon RA, Southam G (2006) Mechanisms of gold bioaccumulation by *filamentous cyanobacteria* from gold (III)-chloride complex. Environ Sci Technol 40(20):6304–6309

van Lenteren JC, Martin NA (1999) Biol Control of whitefly. In: Albajes R, Gullino ML, van Lenteren JC, Elad Y (eds) Integrated pest and Dis Manag in greenhouse Crops Dordrecht. Kluwer Academic Publishers, Netherlands, pp 202–216

Li X, Huizhong X, Zhe-Sheng C, Guofang C (2011) Biosynthesis of nanoparticles by microorganisms and their applications. J Nanomater. https://doi.org/10.1155/2011/270974

Li G, He D, Qian Y, Guan B, Gao S, Cui Y, Yokoyama K, Wang L (2012) Fungus-mediated green synthesis of silver nanoparticles using *Aspergillus terreus*. Int J Mol Sci 13:466–476

Liu J, Wang FH, Wang LL, Xiao SY, Tong CY, Tang DY et al (2008) Preparation of fluorescence starch-nanoparticle and its application as plant transgenic vehicle. J Cent S Univ Technol 15:768–7737

Maliszewska I, Juraszek A, Bielska K (2013) Green synthesis and characterization of silver nanoparticles using ascomycota fungi *Penicillium nalgiovense* AJ12. J Clust Sci 25:989–1004

Mallikarjuna K, Narasimha G, John Sushma N, Dillip GR, Subba Reddy BV, Sreedhar B, Deva Prasad Raju B (2015) Biogenic preparation of gold nanostructures reduced from *Piper longum* leaf broth and their electrochemical studies. J Nanosci Nanotech 15(2):1280–1286

Malysheva A, Lombi E, Voelcker NH (2015) Bridging the divide between human and environmental nanotoxicology. Nat Nanotechnol 10:835–844

Manceau A, Nagy K, Marcus M, Lanson M, Geoffroy N, Jacquet T et al (2008) Formation of metallic copper nanoparticles at the soil-root Interface. Environ Sci Technol 42:1766–1772

Mandal D, Bolander ME, Mukhopadhyay D, Sarkar G, Mukherjee P (2006) The use of microorganisms for the formation of metal nanoparticles and their application. Appl Microbiol Biotechnol 69:485–492

Mann S (2001) Biomineralization: principles and concepts in bioinorganic materials chemistry. Oxford University Press, Oxford, UK

Manonmani V, Juliet V (2011) Biosynthesis of Ag nanoparticles for the detection of pathogenic bacteria. In: Proceedings of the 2nd international conference on innovation, management and service, Singapore, pp 307–311

Mishra S, Singh HB (2015) Biosynthesized silver nanoparticles as a nanoweapon against phytopathogens: exploring their scope and potential in agriculture. Appl Microbiol Biotech 99:1097–1107

Mishra AN, Bhadauria S, Gaur MS, Pasricha R (2010) Extracellular microbial synthesis of gold nanoparticles using fungus *Hormoconis resinae*. J Microbiol 62:45–48

Mishra S, Singh BR, Singh A, Keswani C, Naqvi AH, Singh HB (2014) Biofabricated silver nanoparticles act as a strong fungicide against *Bipolaris sorokiniana* causing spot blotch disease in wheat. PLoS One 9(5):e97881

Mohanpuria P, Rana NK, Yadav SK (2007) Biosynthesis of nanoparticles, technological concepts and future applications. J Nanopart Res 7:9275–9280

Mourato A, Gadanho M, Lino AR, Tenreiro R (2011) Biosynthesis of crystalline silver and gold nanoparticles by extremophilic yeasts. Bioinorganic Chem App doi. https://doi.org/10.1155/2011/546074

Mousavi SR, Rezaei M (2011) Nanotechnology in agriculture and food production. J Appl Environ Biol Sci 1:414–419

Mukherjee P, Ahmad A, Mandal D, Senapati S, Sainkar SR, Khan MI et al (2001) Fungus-mediated synthesis of silver nanoparticles and their immobilization in the mycelial matrix: a novel biological approach to nanoparticle synthesis. Nano Lett 1(10):515–519

Mukherjee P, Roy M, Mandal BP, Dey GK, Mukherjee PK, Ghatak J, Tyagi AK, Kale SP (2008) Green synthesis of highly stabilized nanocrystalline silver particles by a non-pathogenic and agriculturally important fungus *Trichoderma asperellum*. Nano 19:075103

Narayanan KB, Sakthivel N (2010) Biological synthesis of metal nanoparticles by microbes. Adv Colloid Interf Sci 156:1–13

Nayak RR, Pradhan N, Behera D, Pradhan KM, Mishra S, Sukla LB, Mishra BK (2010) Green synthesis of silver nanoparticle by *Penicillium purpurogenum* NPMF, the process and optimization. J Nanopart Res 13:3129–3137

Nithya R, Ragunathan R (2009) Synthesis of silver nanoparticle using *Pleurotus sajor caju* and its antimicrobial study. Digest J Nanomater Biostruct 4:623–629

Ocsoy I, Paret ML, Ocsoy MA, Kunwar S, Chen T, You M, Tan W (2013) Nanotechnology in plant disease management: DNA directed silver nanoparticles on graphene oxide as an antibacterial against *Xanthomonas perforans*. ACS Nano. https://doi.org/10.1021/nn4034794

Oh SD, Lee S, Choi SH, Lee IS, Lee YM, Chun JH, Park HJ (2006) Synthesis of Ag and Ag-SiO_2 nanoparticles by y-irradiation and their antibacterial and antifungal efficiency against *Salmonella enterica serovar Typhimurium* and *Botrytis cinerea*. Colloids Surfaces A 275:228–233

Oliveira HC, Stolf-Moreira R, Martinez CBR, Grillo R, DeJesus MB, Fraceto LF (2015a) Nanoencapsulation enhances the post-emergence herbicidal activity of atrazine against mustard plants. PLoS One 10:e0132971

Oliveira HC, Stolf-Moreira R, Martinez CBR, Sousa GFM, Grillo R, DeJesus MB et al (2015b) Evaluation of the side effects of poly (epsilon- caprolactone) nanocapsules containing atrazine toward maize plants. Front Chem 3:61

Oskam G (2006) Metal oxide nanoparticles: synthesis, characterization and application. J Sol-Gel Sci Technol 37:161–164

Palmqvist NGM, Bejai S, Meijer J, Seisenbaeva GA, Kessler VG (2015) Nano titania aided clustering and adhesion of beneficial bacteria to plant roots to enhance crop growth and stress management. Sci Rep 5:10146

Pandian SRK, Deepak V, Kalishwaralal K, Muniyandi J, Rameshkumar N, Gurunathan S (2009) Synthesis of PHB nanoparticles from optimized medium utilizing dairy industrial waste using *Brevibacterium casei* SRKP2: a green chemistry approach. Colloids Surfaces B 74(1):266–273

Parisi C, Vigani M, Rodriguez-Cerezo E (2015) Agricultural nanotechnologies: what are the current possibilities? NanoToday 10:124–127

Park HJ, Kim SH, Kim HJ, Choi SH (2006) A new composition of nanosized silica-silver for control of various plant diseases. Plant Pathol J 22(3):295–302

Paulkumar K, Gnanajobitha G, Vanaja M, Rajeshkumar S, Malarkodi C, Pandian K, Annadurai G (2014) *Piper nigrum* leaf and stem assisted green synthesis of silver nanoparticles and evaluation of its antibacterial activity against agricultural plant pathogens. The Scientific World J. https://doi.org/10.1155/2014/829894

Perez-de-Luque A, Rubiales D, Marquina CI, Ibarra MR, de la Fuente JM (2008) Nanoparticles in agriculture, development of smart delivery systems for plant research Nano-Spain. Braga-Portugal
Perez-Gonzalez T, Jimenez-Lopez C, Neal AL, Rull-Perez F, Rodriguez-Navarro A, Fernandez-Vivas A et al (2010) Magnetite biomineralization induced by *Shewanella oneidensis*. Geochimica Cosmochimica Acta 74(3):967–979
Peteu SF, Oancea F, Sicuia OA, Constantinescu F, Dinu S (2010) Responsive polymers for crop protection. Polymers 2:229–251
Philip D (2009) Biosynthesis of Au, Ag and Au-Ag nanoparticles using edible mushroom extract. Spectrochim Acta A 73:374–381
Potara M, Bawaskar M, Simon T, Gaikwad S, Licarete E, Ingle A, Banciu M, Vulpoi A, Astilean S, Rai M (2015) Biosynthesized silver nanoparticles performing as biogenic SERS-nanotags for investigation of C26 colon carcinoma cells. Colloids Surfaces B: Biointerfaces 133:296–303
Prasad R, Kumar V, Prasad KS (2014) Nanotechnology in sustainable agriculture: present concerns and future aspects. Afr J Biotechnol 13(6):705–713
Prasad R, Pandey R, Barman I (2016) Engineering tailored nanoparticles with microbes: quo vadis. WIREs Nanomed Nanobiotechnol 8:316–330. https://doi.org/10.1002/wnan.1363
Prasad R, Bhattacharyya A, Nguyen QD (2017a) Nanotechnology in sustainable agriculture: recent developments, challenges, and perspectives. Front Microbiol 8:1014. https://doi.org/10.3389/fmicb.2017.01014
Prasad R, Gupta N, Kumar M, Kumar V, Wang S, Abd-Elsalam KA (2017b) Nanomaterials act as plant defense mechanism. In: Prasad R, Kumar V, Kumar M (eds) Nanotechnology. Springer, Singapore, pp 253–269
Price DRG, Gatehouse JA (2008) RNAi-mediated crop protection against insects. Trends Biotechnol 26:393–399
Puebla RA, Dos Santos DS Jr, Aroca RF (2004) Surface enhanced Raman scattering for ultrasensitive chemical analysis of 1 and 2-naphthalenethiols. Analyst 129:1251–1256
Puoci F, Lemma F, Spizzirri UG, Cirillo G, Curcio M, Picci N (2008) Polymer in agriculture: a review. Am J Agri Biol Sci 3:299–314
Rai M, Deshmukh S, Gade A, Elsalam K-A (2012a) Strategic nanoparticles-mediated gene transfer in plants and animals - a novel approach. Curr Nano 8:170–179
Rai MK, Gade AK, Gaikwad S, Marcato PD, Duran N (2012b) Biomedical applications of nanobiosensors: the state-of-the-art. J Brazilian Chem Soc 23:1):14–1):24
Reith F, Lengke MF, Falconer D, Craw D, Southam G (2007) The geomicrobiology of gold. The ISME J 1(7):567–584
Ruffolo SA, La Russa MF, Malagodi M, Oliviero Rossi C, Palermo AM, Crisci GM (2010) ZnO and $ZnTiO_3$ nanopowders for antimicrobial stone coating. Appl Phys A Mater Sci Process 100:829–834
Sadiko A, Du N, Kariuki V, Okello V, Bushlyar V (2014) Current and emerging technologies for the characterization of nanomaterials. ACS Sustain Chem Eng 2:1707–1716
Sangeetha J, Thangadurai D, Hospet R, Purushotham P, Manowade KR, Mujeeb MA, Mundaragi AC, Jogaiah S, David M, Thimmappa SC, Prasad R, Harish ER (2017a) Production of bionanomaterials from agricultural wastes. In: Prasad R, Kumar M, Kumar V (eds) Nanotechnology. Springer Nature, Singapore, pp 33–58
Sangeetha J, Thangadurai D, Hospet R, Harish ER, Purushotham P, Mujeeb MA, Shrinivas J, David M, Mundaragi AC, Thimmappa AC, Arakera SB, Prasad R (2017b) Nanoagrotechnology for soil quality, crop performance and environmental management. In: Prasad R, Kumar M, Kumar V (eds) Nanotechnology. Springer Nature, Singapore, pp 73–97
Sanghi R, Verma P (2009) A facile green extracellular biosynthesis of CdS nanoparticles by immobilized fungus. Chem Eng J 155:886–891
Saravanan M, Nanda A (2010) Extracellular synthesis of silver bionanoparticles from *Aspergillus clavatus* and its antimicrobial activity against MRSA and MRSE. Colloids Surfaces B 77:214–218

Sasson Y, Levy-Ruso G, Toledano O, Ishaaya I (2007) Nanosuspensions: emerging novel agrochemical formulations. In: Ishaaya I, Nauen R, Horowitz AR (eds) Insecticides design using advanced technologies. Springer-Verlag, Netherlands, pp 1–32

Sastry RK, Rashmi HB, Rao NH, Ilyas SM (2010) Integrating nanotechnology (NT) into Agri-food systems research in India: a conceptual framework. Technol Forecasting Soc Change 77:639–648

Sawle BD, Salimath B, Deshpande R, Bedre MD, Prabhakar BK, Venkataraman A (2008) Biosynthesis and stabilization of Au and Au-Ag alloy nanoparticles by fungus, *Fusarium semitectum*. Sci Tech Adv Mater 9(3):1–6

Senapati S, Mandal D, Ahmad A (2004) Fungus mediated synthesis of silver nanoparticles: a novel biological approach. Ind J Phy A 78(1):101–105

Senapati S, Ahmad A, Khan MI, Sastry M, Kumar R (2005) Extracellular biosynthesis of bimetallic Au-Ag alloy nanoparticles. Small 1(5):517–520

Shaligram NS, Bule M, Bhambure R, Singhal RS, Singh SK, Szakacs G, Pandey A (2009) Biosynthesis of silver nanoparticles using aqueous extract from the compactin producing fungal strain. Process Biochem 44:939–943

Shan G, Xing J, Zhang H, Liu H (2005) Biodesulfurization of dibenzothiophene by microbial cells coated with magnetite nanoparticles. Appl Environ Microbiol 71(8):4497–4502

Simkiss K, Wilbur KM (1989) Biomineralization. Academic, New York, USA

Singh D, Rathod V, Ninganagouda S, Hiremath J, Singh AK, Mathew J (2014) Optimization and characterization of silver nanoparticle by endophytic fungi *Penicillium* sp. isolated from *Curcuma longa* (turmeric) and application studies against MDR *E. coli* and *S. aureus*. Bioinorg Chem Appl. https://doi.org/10.1155/2014/408021

Sinha A, Khare SK (2011) Mercury bioaccumulation and simultaneous nanoparticle synthesis by *Enterobacter* sp. cells. Bioresour Technol 102:4281–4284

Sivamani E, DeLong RK, Qu R (2009) Protamine-mediated DNA coating remarkably improves bombardment transformation efficiency in plant cells. Plant Cell Rep 28:213–221

Smith K, Evans DA, El-Hiti GA (2008) Role of modern chemistry in sustainable arable crop protection. Phil Trans R Soc B 363:623–637

Soni N, Prakash S (2011) Factors affecting the geometry of silver nanoparticles synthesis in *Chrysosporium tropicum* and *Fusarium oxysporum*. Am J Nanotechnol 2(1):112–121

Suriyaprabha R, Karunakaran G, Kavitha K, Yuvakkumar R, Rajendran V, Kannan N (2014) Application of silica nanoparticles in maize to enhance fungal resistance. IET Nanobiotechnol 8(3):133–137

Sweeney RY, Mao C, Gao X, Burt JL, Belcher AM, Georgiou G et al (2004) Bacterial biosynthesis of cadmium sulfide nanocrystals. Chem Biol 11(11):1553–1559

Tang H, Yan M, Zhang H, Xia M, Yang D (2005) Preparation and characterization of water-soluble CdS nanocrystals by surface modification of ethylene diamine. Mater Lett 59(8–9):1024–1027

Tanzil AH, Sultana ST, Saunders SR, Shi L, Marsili E, Beyenal H (2016) Biological synthesis of nanoparticles in biofilms. Enzym Microb Technol. https://doi.org/10.1016/j.enzmictec.2016.07.015

Tiwari M, Narayanan K, Thakar MB, Jagani HV, VenkataRao J (2014) Biosynthesis and wound healing activity of copper nanoparticles. IET Nanobiotechnol 8(4):230–237

Tomoyasu Y, Denell RE (2004) Larval RNAi in T*ribolium* (Coleoptera) for analyzing adult development. Dev Genes Evol 214:575–578

Torney F, Trewyn BG, Lin VS, Wang K (2007) Mesoporous silica nanoparticles deliver DNA and chemicals into plants. Nature Nanotechnol 2:295–300

Vandergheynst J, Scher H, Hy G, Schultz D (2007) Water-in-oil emulsions that improve the storage and delivery of the biolarvacide Lagenidium giganteum. BioControl 52:207–229

Verma VC, Kharwar RN, Gange AC (2010) Biosynthesis of antimicrobial silver nanoparticles by the endophytic fungus *Aspergillus clavatus*. Nanomedicine 5:33–40

Vigneshwaran N, Ashtaputre NM, Varadarajan PV, Nachane RP, Paralikar KM, Balasubramanya RH (2007) Biological synthesis of silver nanoparticles using the fungus *Aspergillus flavus*. Materials Lett 61(6):1413–1418

Vijayakumar PS, Abhilash OU, Khan BM, Prasad BLV (2010) Nanogold-loaded sharp-edged carbon bullets as plant-gene carriers. Adv Funct Mater 20:2416–2423

Vinayaka AC, Basheer S, Thakur MS (2009) Bioconjugation of CdTe quantum dot for the detection of 2,4-dichlorophenoxyacetic acid by competitive fluoroimmunoassay based biosensor. Biosens Bioelectron 24:1615–1620

Vu HT, Keough MJ, Long SM, Pettigrove VJ (2015) Effects of the boscalid fungicide Filan on the marine amphipod *Allorchestes compressa* at environmentally relevant concentrations. Environ Toxic Chem doi. https://doi.org/10.1002/etc.3247

Wang T, Yang L, Zhang B, Liu J (2010) Extracellular biosynthesis and transformation of selenium nanoparticles and application in H_2O_2 biosensor. Colloids Surfaces B 80(1):94–102

Yang L, Watts DJ (2005) Particle surface characteristics may play an important role in phytotoxicity of alumina nanoparticles. Toxicol Lett 158:122–132

Yao KS, Li SJ, Tzeng KC, Cheng TC, Chang CY, Chiu CY, Liao CY, Hsu JJ, Lin ZP (2009) Fluorescence silica nanoprobe as a biomarker for rapid detection of plant pathogens. Adv Mater Res 79:513–516

Zhang X, Zhang J, Zhu KY (2010) Chitosan/double-stranded RNA nanoparticle-mediated RNA interference to silence chitin synthase genes through larval feeding in the African malaria mosquito (*Anopheles gambiae*). Insect Mol Biol 19:683–693

Zhang X, Yan S, Tyagi RD, Surampalli RY (2011) Synthesis of nanoparticles by microorganisms and their application in enhancing microbiological reaction rates. Chemosphere 82(4):489–494

Zheng B, Qian L, Yuan H, Xiao D, Yang X, Paau MC et al (2010a) Preparation of gold nanoparticles on eggshell membrane and their biosensing application. Talanta 82(1):177–183

Zheng D, Hu C, Gan T, Dang X, Hu S (2010b) Preparation and application of a novel vanillin sensor based on biosynthesis of Au-Ag alloy nanoparticles. Sensors Actuators B Chem 148(1):247–252

Zhu H, Han J, Xiao JQ, Jin Y (2008) Uptake translocation and accumulation of manufactured iron oxide nanoparticles by pumpkin plants. J Environ Monitor 10:713–717

Zong X, Wang W, Wei H, Wang J, Chen X, Xu L, Zhu D, Tan Y, Liu Q (2014) Rapid detection of *Prunus necrotic ringspot virus* using magnetic nanoparticle-assisted reverse transcription loop-mediated isothermal amplification. J Virol Methods 208:85–89

Chapter 6
The Role of Nanoemulsions as Antimicrobial Agents in Plant Protection

Ayat F. Hashim, Khamis Youssef, and Kamel A. Abd-Elsalam

6.1 Advantages of Nanoemulsion

Nanoemulsions have several advantages and can be summarized as follows: (i) it may be used as substitute for liposomes and vesicles (Bouchemal et al. 2004); (ii) it is nontoxic and non-irritant in nature; (iii) it has improved physical stability; (iv) it has small-sized droplets having greater surface area providing greater absorption; (v) it can be formulated in variety of formulations such as foams, creams, liquids, and sprays; (vi) it provides better uptake of oil-soluble supplements in cell culture technology; (vii) it is helpful in taste masking; (viii) less amount of energy is required; and (ix) it improves the bioavailability of drug (Kim et al. 2001; Wagner et al. 1996).

6.2 Preparation of Nanoemulsions

Nanoemulsions are prepared using aqueous phase, oils, surfactants, and co-surfactants (Adnan et al. 2009) (Fig. 6.1). Oils used in nanoemulsions preparation include Captex 355, Captex 8000, Witepsol, Myritol 318, Isopropyl myristate, Capryol 90, triacetin, isopropyl myristate, castor oil, olive oil, etc. Surfactants for nanoemulsions stabilization may be nonionic, zwitterionic, cationic, and anionic. The surfactants may include Capryol 90; Gelucire 44/14; 50/13; Cremophor RH 40; Imwitor 191, 742, 780 k, 928, and 988; Labrafil CS; Lauroglycol 90; Tween 20,

A. F. Hashim
Food industries and Nutrition Division, National Research Center, Giza, Egypt

K. Youssef (✉) · K. A. Abd-Elsalam
Plant Pathology Research Institute, Agricultural Research Center (ARC), Giza, Egypt
e-mail: youssefeladawy@arc.sci.eg

K. A. Abd-Elsalam, R. Prasad (eds.), *Nanobiotechnology Applications in Plant Protection*, Nanotechnology in the Life Sciences,
https://doi.org/10.1007/978-3-319-91161-8_6

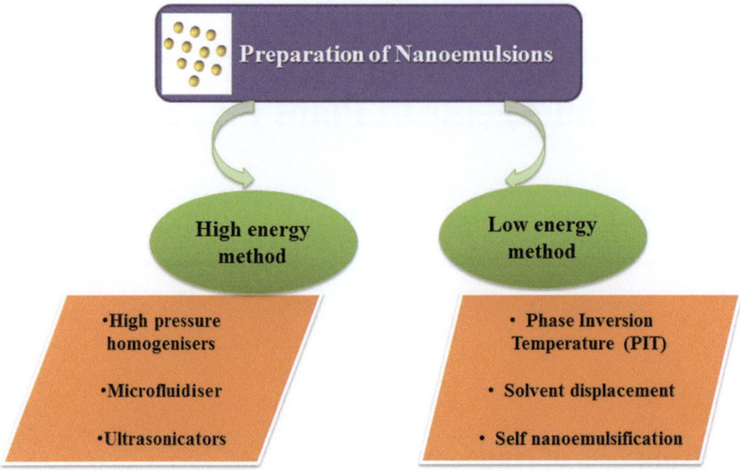

Fig. 6.1 Preparation of nanoemulsions

Tween 60, and Tween 80; etc. Co-surfactants are used to obtain nanoemulsion systems at low surfactant concentration (Kreilgaard et al. 2000). Short- to medium-chain length alcohols (C3–C8) are frequently added as co-surfactants, which further reduce the interfacial tension and increase the fluidity of the interface (Tenjarla 1999; Attwood 1994). They also increase the hydrocarbon tail mobility and allow greater penetration of the oil into this region. Alcohols may also increase the miscibility of the aqueous and oily phases due to its partitioning between these phases. Co-surfactants used in nanoemulsions include glycerin, ethylene glycol, ethanol, propanol, ethanol, isopropyl alcohol, Carbitol, propylene glycol, etc. Nanoemulsion area is used as the assessment criterion for the evaluation of co-surfactants. The larger the size of the nanoemulsion field, the greater the nanoemulsification efficiency of the system.

So these nanoemulsions are non-equilibrated systems (Ravi and Padma 2011; Mason et al. 2006), and so their preparation involves the input of a large amount of either energy or surfactants and in some cases a combination of both. As a result, high-energy or low-energy methods can be used in their formulation (Anton and Vandamme 2009).

6.2.1 *High-Energy Method*

This method uses devices that use very high mechanical energy to generate nanoemulsions with high kinetic energy. The high-energy method employs mechanical devices to produce intensely disruptive forces which break up the oil and water phases to form nano-sized droplets. This can be reached with high-pressure homogenizers, microfluidizer, and ultrasonicators (Mason et al. 2006; Jafari et al. 2007). Particle size will depend on the type of instruments used and their operating conditions like time and temperature along with sample properties and composition (Quin and Mc Clement 2011).

6.2.1.1 High-Pressure Homogenization

High-pressure homogenization is the most common method used for the production of nanoemulsions. During high-pressure homogenization, the coarse macro-emulsion is passed through a small orifice at an operating pressure in the range of 500–5000 psi. During this procedure, several forces, such as hydraulic shear, intense turbulence, and cavitation, act together to produce nanoemulsions with extremely small droplet size. The resulting product can be re-subjected to high-pressure homogenization until nanoemulsion with desired droplet size and polydispersity index is achieved (Mason et al. 2006).

6.2.1.2 Micro-fluidization

Micro-fluidization uses a high-pressure positive displacement pump operating at very high pressures, up to 20,000 psi. This pump forces macro-emulsion droplets through the interaction chamber consisting of a series of micro-channels. The macro-emulsion flowing through the micro-channels collides with high velocity on to an impingement area resulting in very fine nanoemulsions. The nanoemulsions with desired size range and dispersity can be achieved by changing the operating pressure and the number of passes through interaction chambers like high-pressure homogenization.

6.2.1.3 Ultrasonication

Ultrasonic emulsification utilizes a probe that emits ultrasonic waves to disintegrate the macro-emulsion by means of cavitation forces. The nanoemulsions with desired properties can be obtained by changing the ultrasonic energy input and time (Graves et al. 2005; Quin and Mc Clement 2011).

High-pressure homogenization and microfluidization can be used for fabrication of nanoemulsions at laboratory and industrial scale, while ultrasonic emulsification is mainly used at laboratory scale. Furthermore, high-energy methods need sophisticated instruments and extensive energy input, which considerably increases the cost of nanoemulsions production (Graves et al. 2005; Quin and Mc Clement 2011).

These methods can be employed to prepare both o/w and w/o nanoemulsions. High-energy methods allow for a greater control of particle size and a large choice of composition, which in turn controls the stability, rheology, and color of the emulsion. Although high-energy emulsification methods yield nanoemulsions with desired properties and have industrial scalability, they may not be suitable for macromolecules, including proteins, enzymes, and nucleic acids. Besides, these methods alone normally do not yield oil droplets (<100 nm).

6.2.2 *Low-Energy Methods (Condensation Method)*

Lower-energy method, also called the condensation method, is based on the phase transitions occurring during the emulsification process (Lamaallam et al. 2005; Solans et al. 2002). It involves low energy for the preparation of nanoemulsions. These methods are dependent on changing of interfacial phenomenon/phase transitions and intrinsic physicochemical properties of the surfactants, co-emulsifiers/co-surfactants, and oil to produce nano-sized emulsion droplets. These phase transitions result from modulation in the spontaneous curvature of the surfactant and can be achieved (i) at constant composition by changing the spontaneous curvature of nonionic surfactants with temperature, the well-known phase-inversion temperature (PIT), commonly used in industry (Izquierdo et al. 2005; Shinoda and Saito 1968), or (ii) at constant temperature by varying the composition of the system by the emulsion inversion point (EIP) method (Forgiarini et al. 2001; Porras et al. 2008).

Low-energy emulsification method was dependent on the phase behavior and properties of the constituents, to promote the formation of ultra-small droplets (Sonneville-Aubrun et al. 2004; Solans et al. 2005). These low-energy techniques include self-emulsification, phase-transition, and phase-inversion temperature methods (Wang et al. 2007). The low-energy method is interesting because it utilizes the stored energy of the system to form small droplets. This emulsification can be brought about by changing the parameters which would affect the hydrophilic-lipophilic balance (HLB) of the system like temperature, composition, etc. (Sole et al. 2006, 2010). The limitations include complexity, precise approach required, and use of synthetic surfactants. In a nutshell, the most commonly used low-energy emulsification methods include:

6.2.2.1 Phase-Inversion Temperature (PIT) Method

It employs temperature-dependent solubility of nonionic surfactants, such as polyethoxylated surfactants, to change their affinities for water and oil as a function of the temperature. Polyethoxylated surfactants become lipophilicon heating due to dehydration of polyoxyethylene groups. In this method, oil, water, and nonionic surfactants are mixed together at room temperature. The mixture contains o/w micro-emulsions coexisting with excess oil, and the surfactant monolayer shows positive curvature. When this macro-emulsion is heated gradually, the polyethoxylated surfactant becomes lipophilic. The surfactant gets completely solubilized in the oily phase, and the initial o/w emulsion undergoes phase inversion to w/o emulsion at higher temperatures. The surfactant monolayer has negative curvature at this stage (Izquierdo et al. 2005). The structure of the particles in a nanoemulsion is also very similar to those found in a micro-emulsion: the nonpolar tails of the surfactant molecules protrude into the hydrophobic core formed by the oil phase, while the polar head groups of the surfactant molecules protrude into the surrounding aqueous phase (Fig. 6.2).

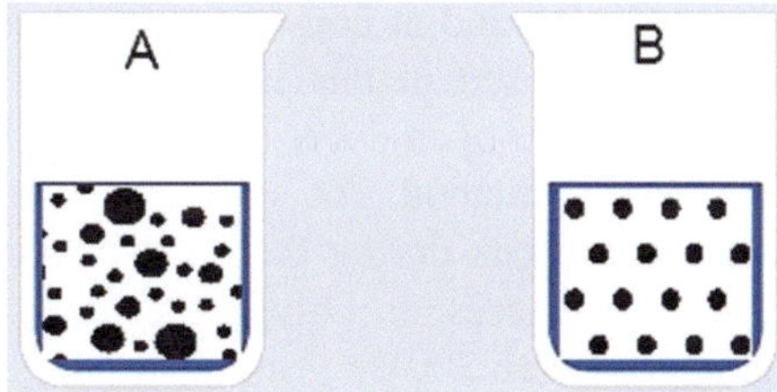

Fig. 6.2 Differences in the appearance of a conventional emulsion (**a**) and nanoemulsion (**b**) fabricated from oil, water, and surfactant

At HLB temperature (an intermediate temperature), the nonionic surfactant has similar affinity for aqueous and oily phase, and this ternary system has extremely low interfacial tension (in the order of 10^{-2}–10^{-5} mNm^{-1}), and spontaneous curvature typically reaches zero (Sole et al. 2006, 2010). The ternary system at this stage typically consists of a D-phase bicontinuous micro-emulsion or a mixture of a D-phase bicontinuous micro-emulsion and lamellar liquid crystalline phases. Rapid cooling of the single-phase or multiphase bicontinuous micro-emulsions maintained at either PIT or a temperature above PIT (transitional-phase inversion) can generate nanoemulsions with very small droplet size and polydispersity index (Shinoda and Saito 1968).

Nanoemulsion (o/w or w/o) can be obtained by rapidly diluting the single bicontinuous micro-emulsions with the aqueous or oil phase (catastrophic phase inversion). It has been detected that the nanoemulsion characteristics are mainly dependent on the structure of the surfactant at HLB temperature (bicontinuous or lamellar) and also on the surfactant/oil ratio. Initially, PIT method was believed to be useful for fabricating o/w nanoemulsions. However, in recent years, the application of the PIT method has been established for fabricating w/o emulsions and nanoemulsions. It is important to note that the use of lipophilic polyethoxylated surfactants and appropriate modifications in the typical PIT protocol are required for obtaining w/o nanoemulsions (Wang et al. 2007).

6.2.2.2 Solvent Displacement Method

This method for spontaneous nanoemulsion fabrication has been based on the nano-precipitation method for polymeric nanoparticles. In this method, oily phase is dissolved in water-miscible organic solvents (i.e., acetone, ethanol, and ethyl methyl ketone). Spontaneous nanoemulsion was prepared by pouring organic phase into an aqueous phase having surfactant through rapid diffusion of organic solvent. Suitable means (i.e., vacuum evaporation) are used to remove organic solvent from the nanoemulsion (Pey et al. 2006; Solans et al. 2005).

6.2.2.3 Phase-Inversion Composition Method (Self-Nanoemulsification Method)

It produces nanoemulsions at room temperature without heat and any organic solvent. Kinetically stable nanoemulsions with small droplet size (~50 nm) can be prepared by the stepwise adding of water into solution of surfactant in oil, with gentle stirring and at constant temperature (Forgiarini et al. 2001). Although the components used in this nanoemulsion were not of pharmaceutical grade, the research has to open doors to prepare pharmaceutically acceptable nanoemulsions via a similar method. The spontaneous nanoemulsification has been related to the phase transitions during the emulsification process and includes lamellar liquid crystalline phases or D-type bicontinuous micro-emulsion during the process (Forgiarini et al. 2001).

6.3 Formulation Factors that Affect the Stability of Nanoemulsions

Nanoemulsions stability studies are achieved by storing at room temperatures and refrigerator over a number of months. The viscosity, refractive index, and droplet size are determined during this period of storage. Insignificant changes in these parameters indicate formulation stability. Accelerated stability studies can also be performed on the nanoemulsions.

Factors to be considered during preparation of nanoemulsion include the following (Haritha et al. 2013):

- The main requirement in nanoemulsion production is that an ultra-low interfacial tension should be attained at the oil water interface, so surfactants must be carefully chosen.
- Concentration of surfactant must be high enough to provide the number of surfactant molecules needed to stabilize the nanodroplets.
- The interface must be flexible to promote the formation of nanoemulsion.

6.4 Characterization of Nanoemulsions

Characterization of nanoemulsions involves the physical and chemical tests as following.

6.4.1 Differential Scanning Calorimetry (DSC)

It provides information on the interactions of different components and polarization microscopy using crossed polarizers. It is employed to confirm isotropicity of the formulation (Chiesa et al. 2008).

6.4.2 Self-Diffusion Nuclear Magnetic Resonance (SD NMR): Small Angle X-Ray Scattering (SAXS)

SD NMR and SAXS have studied the structural features of nanoemulsions.

6.4.3 Freeze Fracture Electron Microscopy

It has used to study nanoemulsion structure; however, extremely rapid cooling of the sample is required in order to maintain the structure and minimize the possibility of artifacts (Debnath et al. 2011; Karthikeyan et al. 2012).

6.4.4 Scanning Electron Microscopy (SEM)

It provides a three-dimensional image of the drops (Kayes 1999). The samples are scanned at suitable accelerating voltage, usually 20 kV, at different magnifications. A good analysis of surface morphology of disperse phase in the formulation is achieved through SEM. Image analysis software may be employed to obtain an automatic analysis result of the shape and surface morphology (Barea et al. 2010).

6.4.5 Transmission Electron Microscopy (TEM)

In TEM, higher-resolution images of the disperse phase are acquired. The sample is negatively stained with 1% aqueous solution of phosphotungstic acid or by dropping 2% uranyl acetate solution onto a 200 μm mesh size Pioloform™-coated copper grid or a microscopic carbon-coated grid using a micropipette and the sample examined at appropriate voltage. A digital image processing program can make qualitative measurements of sizes and size distribution of TEM micrographs.

6.4.6 Dynamic Light Scattering (DLS)/Photon Correlation Spectroscopy (PCS)

It is used to assess nanoemulsion droplet size, polydispersity, and zeta potential using a particle size analyzer. This instrument also measures polydispersity index, which is a measure of the broadness of the size distribution derived from the cumulative analysis of dynamic light scattering. The polydispersity index indicates the quality or homogeneity of the dispersion (Li et al. 2011). PCS gives z-average particle diameter.

6.4.7 Laser Diffraction

It is another technique for measuring particle size. The fundamental particle size distribution derived by this technique is volume based and is expressed in terms of the volume of equivalent spheres (*DN%*) and weighted mean of the volume distribution (mass mean diameter). Since the laser diffraction system is used for this analysis, a rough equivalent of particle polydispersity could be given by two factors/values, namely, uniformity (how symmetrical the distribution is around the median point) and span (the width of the distribution). The span value is defined by the expression: Span = $(D90\%–D10\%)/D50\%$ (2) where $DN\%$ (N = 10%, 50%, 90%) means that the volume percentage of particles with diameters up to $DN\%$ equals to $N\%$. The smaller the span value, the narrower the particle size distribution.

6.4.8 *Viscosity Measurement*

This carried out using a viscometer. Viscosity measurement can indicate the presence of rod-like or worm-like reverse micelles, and conductivity measurements provide the means of determining whether a nanoemulsion is oil-continuous or water-continuous, as well as providing a means of monitoring phase-inversion phenomena (Chiesa et al. 2008). The viscosity of nanoemulsions is a function of the surfactant, water and oil components, and their concentrations. Increasing the water content lowers the viscosity, while decreasing the amount of surfactant and co-surfactant increases interfacial tension between water and oil resulting in increased viscosity. Monitoring of viscosity change is a method of assessing stability of liquid and semi-solid preparations including nanoemulsion formulations (Chiesa et al. 2008).

6.4.9 *Surface Tension Analysis*

Surface tension was measured by using the Du Nouy ring method through Kruss K6 tension meter (Kruss, UK) equipped with a platinum plate. Prior to measurements, calibration was conducted using deionized water with a surface tension of 70.2 mN/m. Sufficient time was allowed to reach equilibrium until no significant modifications were detected. The ring was washed with methanol and acetone and finally flamed prior to the next measurement. Each run was repeated three times. All measurements were carried out at 25 °C.

6.5 Stability Test

Selected formulations were centrifuged at 3500 rpm for 30 min and kept at room temperature for 4 weeks (Shafiq et al. 2007). The formulations were then assessed for the ability to keep transparent one-phase appearance after 4 weeks in order to indicate the presence of a nanoemulsion.

6.5.1 *Thermostability Test*

The selected formulations were stored at room temperature (25 °C) for 3 months and kept at 54 °C for 14 days, prescribed by the Food and Agricultural Organization (FAO) as a standard evaluation of agrochemical products to show stability in the tropical climate (Chen et al. 2000). The change in physical appearance of samples was visually detected.

6.5.1.1 Turbidity Measurement

It is used to determine the rapid equilibrium reached by the dispersion and reproducibility of this process.

6.6 Prevention of Food Spoilage

Essential oils are natural organic compounds and have antibacterial, antifungal, and antiviral properties. Some of the essential oils used in food industry are carvacrol, eugenol, carvone, citral, geraniol, terpineol, thymol, vanillin, and cinnamaldehyde (Burt 2004). Essential oils, including clove essential oil, cinnamon oil, mandarin essential oil, lime oil, cinnamon oil, and basil oil, are commonly used ingredients as natural antibacterial agents. They can endow the nanoemulsion powerful antimicrobial activity due to the components, including pinene, benzaldehyde, carvacrol, carvone, eugenol, eugenyl acetate, geraniol, limonene, menthol, terpineol, thymol, and vanillin (Ghosh et al. 2013b). Nanoemulsions containing essential oils are thus emerging as alternative food preservation method to control the growth of pathogens on food products. It has been confirmed that nanoemulsions possess a wide antimicrobial activity against various food pathogens (McClements and Rao 2011; Karthikeyan et al. 2011; Hamouda et al. 2001; Sugumar et al. 2013), including bacteria (*Lactobacillus delbrueckii*, *E. coil*, *S. aureus*, *Vibrio cholera*, and so on), viruses (herpes simplex, influenza A, vaccinia viruses, and so on), fungi (*Candida albicans*, *Dermatophytes*, and so on), and spores (*Bacillus anthracis*, *Bacillus cereus*. Eucalyptus oil nanoemulsions were formulated and characterized by Sugumar et al. (2013) in order to test their antimicrobial activity against food-borne pathogens. Also, nisin is a kind of natural antibacterial compound commonly contained in nanoemulsions to combine with some essential oils. It was found that the combination with D-limonene displayed the positive effect (Zhang et al. 2014).

6.7 Antimicrobial Nanoemulsions

O/w nanoemulsions for antimicrobial applications contain droplets of size from 200 to 600 nm. The droplets are stabilized by surfactants and medium-chain alcohols, which act as co-surfactant. The co-surfactant concentration in these formulations is generally low, and it helps to reduce the interfacial tension promoting the formation of small droplets. The nanoemulsion is active against a broad spectrum of bacteria, viruses, fungi, and spores.

Anise oil and its (nano and coarse) emulsion antimicrobial activities were tested by the minimum inhibitory concentration (MIC) and time kill assay. It was showed that bulk anise oil reduced the population of *Escherichia coli* O157:H7 and *Listeria monocytogenes* by 1.48 and 0.47 log CFU/ml, respectively, after 6 h of contact time. However, anise oil nanoemulsion (AO75) reduced *Escherichia coli* O157 : H7 and *Listeria monocytogenes* count by 2.51 and 1.64 log CFU/ml, respectively, under the same condition. Microbial analysis and physicochemical showed that both nano- and coarse emulsions of anise oil showed better and long-term physicochemical stability and antimicrobial activity compared to bulk anise oil (Topuz et al. 2016).

A terpenes mixture extracted from *Melaleuca alternifolia* and D-limonene were encapsulated into nanoemulsions based on food-grade ingredients, prepared by high-pressure homogenization (300 MPa). The antimicrobial activity of terpenes was examined by determining the minimum inhibitory concentration and minimum bactericidal concentration (MBC) for three different types of microorganisms (*Lactobacillus delbrueckii*, *Saccharomyces cerevisiae*, *Escherichia coli*). The increase of the antimicrobial activity depended on the formulation and mean diameter of the delivery systems and on the microorganism type. Furthermore, gas chromatography-mass spectrometry (GC-MS) analysis has shown that high-intensity processing for nanoemulsion preparation may affect the chemical stability of several active compounds. The antimicrobial activity of nanocapsules was tested in pear and orange juices inoculated with *Lactobacillus delbrueckii*. The higher the antimicrobial activity of the nano-encapsulated mixtures, the lower antimicrobial concentrations are needed under accelerated aging for a bactericidal action, with a minimal alteration of the organoleptic properties of the juice. The encapsulation into nanoemulsion-based delivery systems of two essential oils, a terpenes mixture and D-limonene, was examined as a process to improve the safety and quality of foods through the addition of natural preservatives. The carrier system lecithin-based nanoemulsion for the terpenes mixture is highly efficient, while D-limonene was successfully nano-encapsulated pure or in a blend with palm oil using as emulsifier modified starch and soy lecithin, respectively. The nano-encapsulated terpenes (minimum inhibitory concentration and minimum bactericidal concentration) values, tested on three different microorganisms (*Lactobacillus delbrueckii*, *Saccharomyces cerevisiae*, *Escherichia coli*) resulted in equal or lower than the unencapsulated mixture values. On the other hand, the nano-encapsulation of D-limonene was able to reduce only the minimum inhibitory concentration values, without any significant variation of the minimum bactericidal concentration values in comparison to the unencapsulated D-limonene. The terpenes nanocapsules were tested in orange and pear juices, inoculated with *Lactobacillus delbrueckii*. The nano-encapsulated terpenes at low concentrations were able to delay the microbial growth (1.0 g/l terpenes) or completely inactivate the microorganisms (5.0 g/l terpenes) while minimally altering the organoleptic properties of the fruit juices (Donsì et al. 2011).

Some micro-/nanoemulsions may be effective as anti-biofilm agents and have antimicrobial activities. It was designated TEOP and BCTP, to inactivate suspensions of vegetative cells of *Salmonella* spp., *Escherichia coli* 0157 : H7 (VT-), *Pseudomonas aeruginosa*, *Staphylococcus aureus*, and *Listeria monocytogenes*. TEOP is an O/W micro-emulsion of ethyl oleate with Tween 80 as emulsifier and n-pentanol as a co-emulsifier, while BCTP is an O/W nanoemulsion of soybean oil and tri-*n*-butyl phosphate emulsified with Triton X-100. BCTP was effective in reducing the cell numbers of *Listeria monocytogenes*, while TEOP was effective against all five bacteria examined. The abilities of these emulsions to reduce preformed biofilms of the above bacteria were also studied. With the exception of the biofilm formed by *Listeria monocytogenes*, which surprisingly was not significantly affected by BCTP, all biofilms were inhibited by both BCTP and TEOP (Teixeira et al. 2007).

Different concentrations of whey protein isolate (WPI) (e.g. 0.1%, 0.5%, 1.0%, 2.5%, and 5.0% (w/v), having 1.0% (w/v) eugenol, were prepared by high-speed homogenization to formulate nanoemulsions (NEs) and to examine their antimicrobial activity. It was shown that particle size decreased with increases in whey protein isolate concentration. Also, the potential value was reduced to a negative charge when using whey protein isolate concentrations >0.1% (w/v). In contrast, no significant differences in particle size were detected during 1 month of storage, except for the 0.1% (w/v) whey protein isolate nanoemulsion. The potential value depended on the increase in whey protein isolate concentration and storage duration, except for NE1 and NE5, suggesting that a low or high concentration of emulsifier was not effective for maintaining the droplet form of the eugenol NE. It was indicated that the growth of *Escherichia coli* was inhibited based on an increase in eugenol concentration in all NE formulations. Moreover, a membrane permeability study showed that total leakage content increased according to incubation time (Bejrapha et al. 2011).

Geraniol and linalool have been found to be effective against food-borne microorganisms in vitro. However, due to their hydrophobic nature, it is difficult to achieve an even dispersion in foods with high-water content resulting in dramatic loss of activity. It was fabricated with geraniol or linalool nanoemulsions, and their effect was investigated against *Escherichia coli* K12, *Listeria innocua*, and *Pseudomonas lundensis* in a meat simulation medium. The agar diffusion assay revealed that geraniol and linalool had a potent antimicrobial activity against all tested bacteria. Dynamic light scattering showed that geraniol and linalool nanoemulsions had a mean diameter of 68.22 ± 2.46 and 173.59 ± 4.15 nm, respectively. Killing assay results showed that both nanoemulsions were able to significantly reduce *Escherichia coli* and *Listeria innocua* counts by approx. 3 log CFU/ml. *Pseudomonas lundensis* proved to be more resistant to both nanoemulsions showing a reduction of approx. 1.2 log CFU/ml. Overall, it was shown that nanoemulsions loaded with geraniol or linalool represent a promising antimicrobial system to improve food preservation and food safety (Balta et al. 2017).

Huanglongbing is the most serious disease affecting the citrus industry worldwide to date. Nanoemulsion formulation (water/oil) may provide a useful model for the effective delivery of chemical compounds into citrus phloem via a foliar spray for controlling citrus Huanglongbing. The causal agent, *Candidatus Liberibacter asiaticus*, lives in citrus phloem, which makes it difficult to successfully treat with chemical compounds. A transcuticular nanoemulsion formulation was developed to enhance the permeation of an effective antimicrobial compound (ampicillin) against Huanglongbing disease through the citrus cuticle into the phloem via a foliar spray. It was demonstrated that efficiency of cuticle isolation using an enzymatic method (pectinase and cellulase) was dependent on the citrus cultivar and Las-infection, and it was more difficult to isolate cuticles from Valencia orange (*Citrus sinensis*) and Huanglongbing symptomatic leaves. Of eight adjuvants tested, Brij 35 provided the maximum increase in permeability of the Huanglongbing-affected cuticle with a 3.33-fold enhancement of cuticular permeability over water control. An in vitro assay using *Bacillus subtilis* showed that nanoemulsion formulations containing ampicillin (droplets size = 5.26 ± 0.04 nm and 94 ± 1.48 nm) coupled with Brij 35 resulted in greater inhibitory zone diameters (5.75 and 6.66 mm) compared to those of Brij 35 (4.34 mm) and ampicillin solution (2.83 mm) alone. Additionally, the nanoemulsion formulations removed *Candidatus Liberibacter asiaticus* bacteria in Huanglongbing-affected citrus *in planta* more efficiently than controls (Yang et al. 2015).

Biodegradable antibacterial agent clove oil is vital oil in food products. Emulsions are used to stabilize and increase the antimicrobial efficacy of oils in aqueous solutions. Smaller droplets provide better distribution of the oil. Clove oil was formulated in aqueous solution by nanoemulsion. The parameters key in the formulation of stable clove oil nanoemulsion were optimized to obtain emulsions with less than 50 nm droplet size. The evaluated parameters include the concentration of clove oil, concentration of emulsifier, hydrophilic-lipophilic balance number, and ultrasonication time. It was determined the following parameters: droplet size, polydispersity index, and zeta potential, and to achieve <50 nm droplet size with low polydispersity index and a highly negative zeta potential. The optimization and evaluation was obtained via Taguchi method. Taguchi method identified the concentration of clove oil and the hydrophilic-lipophilic balance as the most affecting factors in the preparation of nanoemulsions. The optimum conditions for the preparation of nanoemulsion such as mean droplet size, polydispersity index, and zeta potential were 50 nm, 0.49, and −40.7 mV, respectively. This formulation of clove oil resulted in minimum inhibitory concentration against *Escherichia coli* and *Bacillus cereus* of 16 and 32 μg/ml, respectively, and minimal bactericidal concentrations of 16 and 64 μg/ml, respectively (Shahavi et al. 2016). Recently, the citral nanoemulsions showed significantly different antimicrobial activities against *Listeria monocytogenes* and *Staphylococcus aureus* (Lu et al. 2018).

6.8 Nanoemulsion Against Plant Pathogens

The application of nanoemulsion as an antimicrobial agent is a promising and new innovation (nano.med.umich.edu). The nanoemulsion has a broad spectrum activity against fungi (e.g., *Ganoderma* sp., *Fusarium oxysporum*, *Colletotrichum* sp.), bacteria (e.g., *Escherichia coli*, *Bacillus cereus*, *Saccharomyces cerevisiae*, *Staphylococcus aureus*), and *Candidatus Liberibacter asiaticus*, the causal agent of Huanglongbing, the most serious disease affecting citrus industry worldwide.

Nanoemulsions have attracted much attention due to their applications especially in plant protection field. It was developed and characterized oil-in-water (O/W) phenazine extract nanoemulsions for controlling *Ganoderma boninense* (as the major pathogen that affects the oil palm tree) PER71. A phase diagram (PD) was fabricated based on nonionic surfactant Tween 80, oil carrier E2126 and water, and 30% crude phenazine extract as active ingredient by low-energy method. As primary screening, six formulations were examined for its stability and thermostability over time after constructing PD. Four formulations were selected to proceed in physiochemical characterizations. The characterized formulations have a mean droplet size ranging from 130.54 to 309.9 nm with polydispersity index varied between 0.32 and 0.97. The larger drop size (309.9 nm) shifted to a smaller size of 130.54 nm with decrease in the concentration of oil carrier Emereen 2126. The zeta potential of all formulation is yet stable with the value ranged from −11.8 to −16 mV. The surface tension was around 30.82–30.88 mN/m. The fungicidal effect of the phenazine nanoemulsion was tested against *Ganoderma boninense* PER71. Nanoemulsion with 174.43 nm size was obtained at a ratio 5:5:90, and it was found to be stable in terms of polydispersity index (0.6), zeta potential (−16.0 mV), and surface tension (30.88 mN/m) and effective in controlling *Ganoderma boninense* PER71 at 70.74%. This is the first time that a phenazine extract nanoemulsion has been reported. The results obtained might corroborate to the application of phenazine extract nanoemulsion as potential candidate for controlling *Ganoderma boninense* PER71. The nanoemulsion with active ingredient phenazine extracts containing Tween 80, E2126, and deionized water was effectively improved by the low-energy method. A nano-sized droplet of 174.43 nm was obtained. Phenazine nanoemulsion with this droplet size was found to be more effective in controlling *Ganoderma boninense* compared with any larger or smaller droplet size. Phenazine nanoemulsion is the green innovation alternative to chemical fungicide in agricultural field and provides a lower cost, nontoxic, and effective agent for development as a biofungicide for basal stem rot disease (Lee et al. 2016).

Antifungal activity of eugenol oil nanoemulsion against *Fusarium* wilt of cotton was evaluated in in vitro and in vivo tests (Fig. 6.3). The inhibitory effect of eugenol oil nanoemulsion on growth of *Fusarium oxysporum* f. sp. *vasinfectum* was confirmed. The biobased oil in water nanoemulsion was characterized by particle size analyzer, stability test, TEM, and thin-layer chromatography (TLC). Thin-layer chromatography assay confirmed the presence of eugenol as an active antifungal component in all tested concentration. Total protein SDS-PAGE assay further investigated the molecular weight of the decreased and stimulated proteins. Phytotoxic effects were determined on cottonseed treated with high concentration of eugenol oil nanoemulsion. The strong antagonistic activity against *Fusarium oxysporum* f. sp. *vasinfectum* recommends that eugenol oil nanoemulsion could be used as an efficient nanofungicide in plant disease control (Abd-Elsalam and Khokhlov 2015).

Fig. 6.3 Effectiveness of EON in controlling *Fusarium* wilt under greenhouse conditions. (**a**) Cotton seedlings treated with eugenol oil nanoemulsion. (**b**) Severely affected seedlings after planting in artificially infested soil in a greenhouse test (untreated cottonseed) (Vein discoloration began at the margin of cotyledonary leaf). (Cited from Abd-Elsalam and Khokhlov 2015)

It was showed that chitosan in the form of nanoemulsions displayed better antifungal activity as compared to its conventional form. The effect of chitosan in the form of nanoemulsions against *Colletotrichum musae* and *Colletotrichum gloeosporioides*, important pathogens of fresh fruits and vegetables in the tropics, was studied. Therefore, it could be used as a potential alternative treatment to synthetic chemicals for controlling postharvest anthracnose of tropical fruits and vegetables. Also, it was suggest that instead of applying chitosan in a conventional form, chitosan in the form of nanoemulsions could be more effective as a biofungicide for controlling anthracnose of fresh fruits. Moreover, it could be cost-effective as the amount of chemical used is reduced when applied in the form of nanoemulsions (Zahid et al. 2012).

6.9 Future Perspectives in Nanoemulsions Applications

Further studies in this area are clearly needed to improve the nanoemulsions production and to obtain formulations that add high stability, higher shelf life, and satisfactory physical characteristics. Toxicological studies should be also conducted to ensure that the new technologies are safe for widespread use in food and agricultural sectors especially in plant protection applications.

Acknowledgments This research was supported by the International Foundation for Science, Stockholm, Sweden, through a grant to Hashim Ayat (F5853).

References

Abd-Elsalam KA, Khokhlov AR (2015) Eugenol oil nanoemulsion: antifungal activity against *Fusarium oxysporum* f. sp. *vasinfectum* and phytotoxicity on cottonseeds. Appl Nanosci 5:255–265

Adnan A, Mohammad R, Farhan JA, Zeenat I, Roop KK, Aqil M, Sushama T (2009) Nanoemulsion components screening and selection: a technical note. AAPS PharmSciTech 10(1):69–76

Anton N, Vandamme T (2009) The universality of low-energy nano-emulsification. Int J Pharm 377(1–2):142–147

Attwood D (1994) Microemulsions. In: Kreuer J (ed) Colloidal drug delivery systems. Marcel Dekker, New York, pp 31–71

Balta I, Brinzan L, Stratakos A, Linton M, Kelly C, Pinkerton L, Corcionivoschi N (2017) Geraniol and linalool loaded nanoemulsions and their antimicrobial activity. Bull UASVM Anim Sci Biotechnol 74(2):157–161

Barea MJ, Jekins MJ, Gaber MH (2010) Evaluation of liposomes coated with a pH responsive polymer. Int J Pharm 402(1–2):89–94

Bejrapha P, Choi M, Surassmo S, Chun J, Min S (2011) Formulation and antimicrobial activity on *Escherichia coli* of nanoemulsion coated with whey protein isolate. Korean J Food Sci Anim Resour 31:543–550

Bouchemal K, Briancon S, Perrier E, Fessi H (2004) Nano-emulsion formulation using spontaneous emulsification: solvent, oil and surfactant optimization. Int J Pharm 280:241–251

Burt S (2004) Essential oils: their antibacterial properties and potential applications in foods—a review. Int J Food Microbiol 94:223–253

Chen F, Wang Y, Zheng F, Wu Y, Liang W (2000) Studies on cloud point of agrochemical microemulsions. Colloids Surf A 175:257–262

Chiesa M, Garg J, Kang YT, Chen G (2008) Thermal conductivity and viscosity of water in oil nanoemulsions. Colloids Surf A 326:67–72

Debnath S, Satayanarayana P, Gampa VK (2011) Nanoemulsion-a method to improve the solubility of lipophilic drugs. Int J Adv Pharm Sci 2(2–3):72–83

Donsì F, Annunziata M, Sessa M, Ferrari G (2011) Nanoencapsulation of essential oils to enhance their antimicrobial activity in foods. Food Sci Technol 44:1908–1914

Forgiarini A, Esquena J, Gonzalez C, Solans C (2001) Formation of nano-emulsions by low-energy emulsification methods at constant temperature. Langmuir 17:2076–2083

Ghosh V, Saranya S, Mukherjee A, Chandrasekaran N (2013) Cinnamon oil nanoemulsion formulation by ultrasonic emulsification: investigation of its bactericidal activity. J Nanosci Nanotechnol 13:114–122

Graves S, Meleson K, Wilking J (2005) Structure of concentrated nanoemulsions. J Chem Phys 122(13):134703

Hamouda T, Myc A, Donovan B, Shih AY, Reuter JD, Baker JR (2001) A novel surfactant nanoemulsion with a unique non-irritant topical antimicrobial activity against bacteria, enveloped viruses and fungi. Microbiol Res 156:1–7

Haritha A, Syed PB, Koteswara RP, Chakravarthi V (2013) A brief introduction to methods of preparation, applications and characterization of nanoemulsion drug delivery systems. Indian J Res Pharm Biotechnol 1(1):25–28

Izquierdo P, Feng J, Esquena J, Tadros TF, Dederen JC, Garcia MJ (2005) The influence of surfactant mixing ratio on nano-emulsion formation by the pit method. J Colloid Interface Sci 285(1):388–394

Jafari SM, He Y, Bhandari B (2007) Optimization of nanoemulsion production by microfluidization. Eur Food Res Technol 225:733–741
Karthikeyan R, Amaechi BT, Rawls HR, Lee VA (2011) Antimicrobial activity of nanoemulsion on cariogenic *Streptococcus mutans*. Arch Oral Biol 56:437–445
Karthikeyan S, Jeeva PA, Jerobin J, Mukherjee A, Chandrasekaran N (2012) Formulation and characterization of nanoemulsion coatings from *Azadirachta indica*. Int J ChemTech Res 4(4):1566–1570
Kayes JB (1999) Disperse systems. In: Aulton ME (ed) Pharmaceutics the science of dosage form design. Churchill Livingstone, Edinburgh, pp 81–118 571
Kim CK, Cho YJ, Gao ZG (2001) Preparation and evaluation of biphenyl dimethyl dicarboxylate microemulsions for oral delivery. J Control Release 70:149–155
Kreilgaard M, Pedersen EJ, Jaroszewski JW (2000) NMR characterization and transdermal drug delivery potential of microemulsion systems. J Control Release 69:421–433
Lamaallam S, Bataller H, Dicharry C, Lachaise J (2005) Formation and stability of mini-emulsions produced by dispersion of water/oil/surfactants concentrates in a large amount of water. Colloids Surf A 270–271(1–3):44–51
Lee KW, Omar D, Abdan K, Wong MY (2016) Physiochemical characterization of nanoemulsion formulation of phenazine and their antifungal efficacy against *Ganoderma boninense* PER71 *in vitro*. Res J Pharm Biol Chem Sci 7(6):3056–3066
Li X, Anton N, Ta TMC, Zhao M, Messaddeq N, Vandamme TF (2011) Microencapsulation of nano-emulsions: novel Trojan particles for bioactive lipid molecule delivery. Int J Nanomedicine 6:1313–1325
Lu W, Huang D, Wang C, Yeh C, Tsai J, Huang Y, Li P (2018) Preparation, characterization, and antimicrobial activity of nanoemulsions incorporating citral essential oil. J Food Drug Anal 26:82–89 https://doi.org/10.1016/j.jfda.2016.12.018
Mason TG, Graves SM, Wilking JN, Lin MY (2006) Extreme emulsification: formation and structure of nanoemulsions. Condens Matter Phys 9(1):193–199
McClements DJ, Rao J (2011) Food-grade nanoemulsions: formulation, fabrication, properties, performance, biological fate, and potential toxicity. Crit Rev Food Sci Nutr 51:285–330
Pey CM, Maestro A, Solé I, González C, Solans C, Gutiérrez JM (2006) Optimization of nano-emulsions prepared by low-energy emulsification methods at constant temperature using a factorial design study. Colloids Surf A 288(1–3):144–150
Porras M, Solans C, González C, Gutiérrez JM (2008) Properties of water-in-oil (W/O) nano-emulsions prepared by a low-energy emulsification method. Colloids Surf A 324(1–3):181–188
Quin C, Mc Clement DJ (2011) Formation of nanoemulsions stabilized by model food grade emulsifiers using high pressure homogenization: factors affecting particle size. Food Hydrocoll 25(5):1000–1008
Ravi TPU, Padma T (2011) Nanoemulsions for drug delivery through different routes. Res Biotechnol 2(3):1–13
Shafiq S, Shakeel F, Talegaonkar S, Ahmad FJ, Khar RK, Ali M (2007) Development and bioavailability assessment of ramipril nanoemulsion formulation. Euro J Pharm Biopharm 66(2):227–243
Shahavi MH, Hosseini M, Jahanshahi M, Meyer RL, Darzi GN (2016) Clove oil nanoemulsion as an effective antibacterial agent: Taguchi optimization method. Desalin Water Treat 57(39):18379–18390
Shinoda K, Saito H (1968) The effect of temperature on the phase equilibria and the types of dispersions of the ternary system composed of water, cyclohexane, and nonionic surfactant. J Colloid Interface Sci 26(1):70–74
Solans C, Esquena J, Forgiarini AM, Uson N, Morales D, Izquierdo P (2002) Nanoemulsions: formation and properties. In: Mittal KL, Shah DO (eds) Surfactants in solution: fundamentals and applications. Marcel Dekker, New York, p 525

Solans C, Izquierdo P, Nolla J, Azemar N, Garcia-Celma MJ (2005) Nano-emulsions. Curr Opin Colloid Interface Sci 10(3–4):102–110
Sole I, Maestro A, Pey CM, Gonzalez C, Solans C, Gutierrez JM (2006) Nanoemulsions preparation by low energy methods in an ionic surfactant system. Colloids Surf A 288:138–143
Sole I, Pey CM, Maestro A, Gonzalez C, Porras M, Solans C, Gutierrez JM (2010) Nanoemulsions prepared by phase inversion composition method: preparation variables and scale up. J Colloid Interface Sci 344(2):417–423
Sonneville-Aubrun O, Simonnet JT, L'Alloret F (2004) Nanoemulsions: a new vehicle for skincare products. Adv Colloid Interface Sci 108–109:145–149
Sugumar S, Nirmala J, Ghosh V, Anjali H, Mukherjee A, Chandrasekaran N (2013) Bio-based nanoemulsion formulation, characterization and antibacterial activity against food-borne pathogens. J Basic Microbiol 53(8):677–685
Teixeira PC, Leite GM, Domingues RJ, Silva J, Gibbs PA, Ferreira JP (2007) Antimicrobial effects of a microemulsion and a nanoemulsion on enteric and other pathogens and biofilms. Int J Food Microbiol 118(1):15–19
Tenjarla S (1999) Microemulsions: an overview and pharmaceutical applications. Crit Rev Ther Drug Carrier Syst 16:461–521
Topuz OK, Özvural EB, Zhao Q, Huang Q, Chikindas M, Gölükçü M (2016) Physical and antimicrobial properties of anise oil loaded nanoemulsions on the survival of foodborne pathogens. Food Chem 203:117–123
Wagner JG, Gerrard ES, Kaiser DG (1996) The effect of the dosage form on serum levels of indoxole. Clin Pharmacol Ther 7:610–619
Wang L, Li X, Zhang G, Dong J, Eastoe J (2007) Oil-in-water nanoemulsions for pesticide formulations. J Colloid Interface Sci 314(1):230–235
Yang C, Powell CA, Duan Y, Shatters R, Zhang M (2015) Antimicrobial nanoemulsion formulation with improved penetration of foliar spray through citrus leaf cuticles to control citrus huanglongbing. PLoS One 10(7):e0133826
Zahid N, Ali A, Manickam S, Siddiqui Y, Maqboo M (2012) Potential of chitosan-loaded nanoemulsions to control different *Colletotrichum* spp. and maintain quality of tropical fruits during cold storage. J Appl Microbiol 113:925–939
Zhang Z, Vriesekoop F, Yuan Q, Liang H (2014) Effects of nisin on the antimicrobial activity of D-limonene and its nanoemulsion. Food Chem 150:307–312

Chapter 7
Nano-carbon: Plant Growth Promotion and Protection

Mohamed A. Mohamed, Ayat F. Hashim, Mousa A. Alghuthaymi, and Kamel A. Abd-Elsalam

7.1 Introduction

Carbon nanomaterials are materials of diverse structure and size such as fullerenes, nano-onions, nano-cones, nanohorns, carbon dots, carbon nanotubes, nano-beads, nano-fibers, nano-diamonds, and graphene (Sharon et al. 2010; Chai et al. 2013). Carbon-based nanomaterials were utilized by scientists in various environmental applications such as in solar cells, for the production of renewable energy, soil remediation, and contaminant degradation, and in the detection as sensors for pollutants (Mauter and Elimelech 2008; Rasool and Lee 2015). In agriculture, carbon-based nanomaterials contribute to approximately 40% of the total engineered nanoparticles used and are mainly used either as additives or as active components (Gogos et al. 2012). For example, fungicides encapsulated in multi-walled carbon nanotubes were more toxic to *Alternaria alternata* compared to bulk pesticides which were not capsulated (Sarlak et al. 2014). In the case of fertilizer application, for slow and efficient release, encapsulation with graphene oxide films was found to be effective (Zhang et al. 2014). For example, Zhang et al. (2014) reported that encapsulation of potassium nitrate in graphene oxide prolonged the release into the soil thereby making the availability of potassium nitrate more efficiently to the

M. A. Mohamed
Plant Pathology Research Institute, Agricultural Research Center (ARC), Giza, Egypt

A. F. Hashim
Food industries and Nutrition Division, National Research Center, Giza, Egypt

M. A. Alghuthaymi
Department of Biology, Science and Humanities College, Shaqra University, Alquwayiyah, Saudi Arabia

K. A. Abd-Elsalam (✉)
Plant Pathology Research Institute, Agricultural Research Center (ARC), Giza, Egypt

K. A. Abd-Elsalam, R. Prasad (eds.), *Nanobiotechnology Applications in Plant Protection*, Nanotechnology in the Life Sciences,
https://doi.org/10.1007/978-3-319-91161-8_7

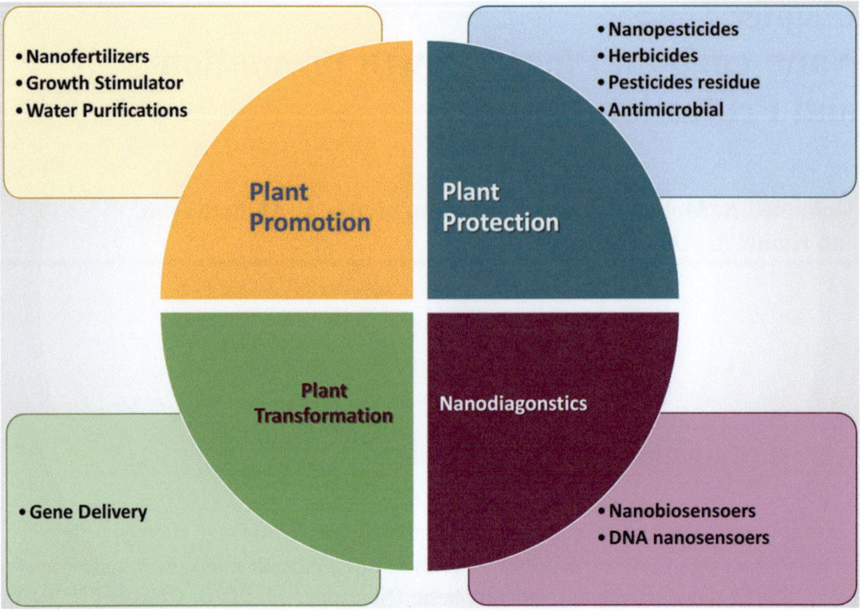

Fig. 7.1 Carbon nanomaterials applications in plant protection

plants. Carbon nanomaterials could be used as additives for the development of efficient fungicides due to antifungal properties (Wang et al. 2014a,b,c). In nano-biotechnology areas, the ability of carbon nanomaterials to penetrate and enter into cells could be used for the purpose of delivery of DNA molecules (Burlaka et al. 2015). In the past few years, it was proposed that carbon-based nanomaterials present about 40% of the total exploitation of nanotechnology in agriculture field, acting as additives as well as active components (Gogos et al. 2012). The diverse applications of carbon nanomaterials in plant crop protection are shown in Fig. 7.1.

7.2 Classification of Carbon-Based Nanomaterials

Carbon is one of the few elements known since antiquity with the ability to polymerize at the atomic level, thus forming very long carbon chains. It is nonmetallic and tetravalent—making four electrons available to form covalent chemical bonds. Carbon atoms have a valence of four due to the four electrons in the outer electron layer. The carbon atoms can bond together in different ways, termed allotropes of carbon. Until recently, only two natural carbon allotropes were known: diamond and graphite. In the meantime, various forms of new allotropic have been defined, including carbon nanomaterials. Generally, materials containing particles with at least one dimension between 1 and 100 nm in size are defined as nanomaterials (European Commission 2011). Carbon-based nanomaterials are composed of carbon atoms in their structure. Its classification is most commonly performed according to their geometrical structure. Particles of carbon nanostructures can be tube-shaped, horn-shaped, spherical, or ellipsoidal. Nanoparticles having horn-shaped particles are nanohorns and spheres or ellipsoids that belong to the group of fullerenes; the shape of tubes is termed as carbon nanotubes.

7.2.1 Fullerenes

In 1966, Harold W. Kroto, Robert F. Curl, and Richard E. Smalley have won Nobel Prize for their discovery in 1985 of fullerenes, in which the atoms are arranged in closed shells. Fullerenes consist of 20 hexagonal and 12 pentagonal rings as the basis of icosahedral symmetry closed-cage structure. Each carbon atom is bonded to three others and is sp^2 hybridized. Fullerenes are an allotropic carbon modification, often termed as a molecular form of carbon, or carbon molecules. The fullerene family includes a number of atomic Cn clusters ($n > 20$), composed of carbon atoms on a spherical surface. Carbon atoms are usually located on the surface of the sphere at the vertices of pentagons and hexagons. Fullerene C60 (is not super aromatic) is the most common and best-investigated fullerene. The spherical molecule is highly symmetric and consists of 60 carbon atoms, located at the vertices of 20 hexagons and 12 pentagons. The diameter of fullerene C60 is 0.7 nm (Yadav and Kumar 2008).

7.2.2 Carbon Nanotubes (CNTs)

In 1991, the Japanese researcher S. Iijima discovered CNTs. They are characterized by cylindrical structures with a diameter of several nanometers, consisting of rolled graphene sheets. Carbon nanotubes may vary in length, diameter, chirality (symmetry of the rolled graphite sheet), and the number of layers. CNTs may be classified into two main groups according to their structure: single-walled nanotubes (SWCNTs) and multi-walled nanotubes (MWCNTs). Some researchers additionally isolate a separate class of CNTs called double-walled carbon nanotubes (DWCNTs). Generally SWCNTs have a length of a few micrometers and a diameter around 1–3 nm. Multi-walled CNTs have a length around 10 μm and a diameter of 5–40 nm. Recently, synthesis of CNTs with a length of even 550 mm has been reported (Zhang et al. 2013). The valuable advantage of adding CNTs to polymeric membranes has been reported. These include increased oleophobicity and hydrophilicity (Zhang et al. 2016) and improved thermal, (Namasivayam and Shapter 2017) electrical, (Sarno et al. 2013) and mechanical properties of the composite membranes (Bai et al. 2017). CNTs as nanofiller in polymer matrix have attracted a great attention because of their easy functionalization, high specific surface area, proper compatibility, and chemical stability (Das et al. 2014a,b).

Bakajin et al. described a possible to modify CNTs pores to selectively rejections (Bakajin et al. 2009). Therefore, CNTs membrane can be used as a "gate keeper" for size-controlled separation of multiple pollutants. The fast and efficient method to separate emulsified oil from oily wastewater using functionalized multi-walled carbon nanotubes (F-MWCNTs) as a demulsifier was reported (Liu et al. 2016). Lately, CNTs could be covalently attached onto the surface of polymer to produce hybrid CNTs/polymer membranes for efficient separation of surfactant-stabilized oil/water emulsions (Zhang et al. 2016; Gu et al. 2016). Novel Ag/polyacrylic acid (PAA)–hybrid CNT membranes for treating oil/water/solid three-phase system have been developed (Gu et al. 2016). Also, the hybrid membranes exhibited an excellent antibacterial activity due to the carbon-based membrane that was integrated with AgNPs. Bacterial cells of *E. coli* were destroyed after incubating with the Ag/PAA–CNT membrane, indicating the antibacterial ability of the developed membrane.

7.2.3 Graphene

In 2004, the first graphene samples were defined by A. Geim (Dutch-British physicist) and K. Novoselov (Russian-British physicist), awarded with a Nobel Prize in 2010. It is a two-dimensional allotropic form of carbon, formed by single layers of carbon atoms. Carbon atoms show sp^2 hybridization linked by σ- and π-bonds in a two-dimensional hexagonal crystal lattice with a distance of 0.142 nm between carbon hexagons of closed atoms. Graphene represents a structural element of some other carbon allotropes, such as graphite, carbon nanotubes, and fullerenes. Graphene oxide (GO), the oxidation state of graphene nanosheets, could be an attractive candidate as a carbon filler because it contains epoxide, hydroxyl, and carboxylic acid groups.

7.3 Synthesis of Carbon-Based Nanomaterials

Since the discovery of carbon-based nanomaterials, their different methods for synthesis have been developed and outstanding properties have been intensively studied. The basic components for carbon nanomaterial production are carbon vapors.

7.3.1 Fullerenes

In 1990, W. Krätschmer and D.R. Huffman produced for the first time fullerenes by evaporation of graphite electrodes in a helium atmosphere (Kratschmer et al. 1990; Kratschmer 2011). Earlier, a reactor was changed by forming an electric arc between two graphite electrodes. The resulting soot condenses on the cold surface of the reactor and is collected and processed in boiling benzene, toluene, xylene, or other organic solvents. A black condensate is formed after evaporation of the solvents, having about 10–15% of C60 and C70 fullerenes, and small amounts of higher fullerenes. The ratio between the C60 and C70 fullerenes varies, depending on the synthesis parameters, but typically C60 represents the main fraction. The described method of arc discharge belongs to the large family of plasma methods which are most common and generally used compared to other techniques (Churilov 2008). However, the fullerenes are limited in practical use due to the low productivity of the methods and high costs currently available from their synthesis.

7.3.2 Carbon Nanotubes

Chemical vapor deposition (CVD), laser ablation, and arc discharge are basic techniques for CNT synthesis (Gore and Sane 2011). Presently, CVD is one of the most explored and mostly used methods for CNT production (Kumar and Ando 2010). It requires milder conditions and simpler equipment in temperature and pressure, making it more appropriate for the large-scale production of CNTs in contrast to two other methods (Zhang et al. 2011). CVD production is based on hydrocarbon decomposition to carbon, following synthesis of carbon nanostructures on various substrates having catalysts on which the nanotubes are rising. Nickel-, cobalt-, or iron-based nanoparticles are frequently utilized as catalysts. Their size strongly relates to the diameter of nanotubes synthesized on it (8–100 nm for MWCNTs, 0.5–5 nm for SWCNT synthesis).

Reactors for CVD synthesis usually consist of a reaction chamber and tubes filled with inert gas and hydrocarbon. Ethylene or acetylene is commonly used for MWCNTs, while methane for SWCNT synthesis. As a simplified method explanation, the substrate is heated up to 550–700 °C for MWCNT and up to 850–1000 °C in case of SWCNT production. Carbon is formed by thermal decomposition of hydrocarbons and dissolves in the metal nanoparticle catalyst. After reaching a certain concentration of carbon, it forms a semifullerene cap, as a starting structure for the growth of a cylindrical shell nanotube, formed by a continuous flow of carbon from the hydrocarbon source to the catalyst particle. Finally, the catalysts from the nanotubes tips are removed, and more purification is still under improvement and optimization to yield CNTs of a higher quality (Morsy et al. 2014; Matsuzawa et al. 2014).

In large scale, heterogeneous products are formed, containing impurities of amorphous carbon, carbon fiber, catalyst residuals, and other nanoparticles. So, production costs increase due to additional purification, and separation steps are needed. SWCNT production involves small metal catalyst particles to equal dispersion on the substrate and prevent aggregation. For example, when fine catalysts are sintering into larger particles, SWCNT diameter is increased or DWCNTs and MWCNTs are formed. A mixture of semiconductive and conductive SWCNTs is yielded requiring more extraction steps to get SWCNTs with defined chirality for specific applications. Vertically or horizontally aligned CNTs synthesis has numerous structural advantages as compared to bundles of agglomerated CNTs. Production of MWCNTs seems to be less complicated and expensive than SWCNTs. But, the controlled inner and outer diameters or defined numbers of walls are still major challenges.

7.3.3 Graphene

Gheim and Novoselov obtained graphene sheets using mechanical splitting of graphite with adhesive tape (Novoselov et al. 2004). Various methods for graphene production are available (Novoselov et al. 2012). These methods are based on splitting or cutting materials, such as graphite or nanotubes (Jiao et al. 2009), using a range of physical or chemical methods to obtain nanoscale graphene sheets. Graphene sheet synthesis by CVD synthesis or laser ablation methods is possible. The different methods are able to provide graphene or reduced graphene oxide sheets of different qualities, depending on the requirements of the corresponding applications. Moderate quality graphene for structural applications can be obtained in large quantities with relatively low production costs. High-quality graphene for electronic devices produced in smaller quantities is usually more expensive. The major methods suitable for mass production of graphene are liquid phase and thermal exfoliation of graphite, CVD synthesis (potentially most cost-effective), and synthesis on silicon carbide (Novoselov et al. 2012).

7.4 Chemical Functionalization of Carbon-Based Nanoparticles

Chemical functionalization can explain a wide variety of carbon-based nanomaterials. Nanoparticles are often functionalized by linking certain molecules to the nanoparticle surface, in order to modify the physical and chemical properties of the particles, which in turn greatly expands the field of applications (Hirsch and Vostrowsky 2005; Hernandez-Fernandez et al. 2010). One example for functionalization of carbon-based nanoparticles is an oxidation of CNTs. This process includes an ultrasonic treatment of nanotubes in a mixture of acids, leading to attachment of carboxylic functional groups (–COOH) on the sidewalls of the nanotubes. Oxidized CNTs attain solubility in aqueous solutions but retain their mechanical and electrical properties. Furthermore, carboxylic groups attached to the nanotube surface can serve as sites for further functionalization.

7.4.1 Properties of Carbon-Based Nanomaterials

Molecular manipulation of carbon-based nanomaterials involves controlling the conformation and structure of a material including size, length, chirality, and the number of layers.

7.4.1.1 Electronic, Optical, and Thermal Properties

CNTs possess a wide range of electrical and optical properties not only from their extended sp^2 carbon but also from their tunable physical properties (e.g., diameter, length, single-walled vs. multi-walled, surface functionalization, and chirality) (Saito et al. 1998). Diameter is a significant factor in determining the properties and applications of tube-shaped carbon nanostructures. Diameter of small single-walled carbon nanotubes is strongly related to synthesis technique (Andreas 2002). The diameter induces higher strain energies, mixing of σ and α bonds and electron orbital rehybridization. These bond structure modifications encourage essential modifications to the electronic, optical, mechanical, elastic, and thermal properties of SWCNTs.

CNTs display high thermal and electrical conductivity compared to other conductive materials. SWCNTs electrical properties depend on their chirality or hexagon orientation with respect to the tube axis. Accordingly, SWCNTs are classified into three subclasses: (1) armchair (electrical conductivity > copper), (2) zigzag (semiconductive properties), and (3) chiral (semiconductive properties). Due to variable chirality of MWCNTs, unusual mechanical properties instead of outstanding electrical characteristics can be revealed.

CNTs and graphene have similar electrical, optical, and thermal properties, but the electrical properties of graphene allotrope are basically different from the properties of three-dimensional materials due to its two-dimensional atomic sheet structure that allows more varied electronic characteristics.

The CNT structure leads to outstanding properties with a unique mixture of strength, rigidity, and elasticity compared with other fibrous materials. For example, CNTs display noticeably higher aspect ratios (length to diameter ratios) than other materials and larger aspect ratios for SWCNTs as compared with MWCNTs due to their smaller diameter. Graphene has several unique physical properties, for example, high thermal stability and extremely high mechanical rigidity.

7.4.1.2 Molecular Interaction and Sorption

The molecular interactions and sorption properties controlling carbon-based nanomaterials depend commonly on physical-chemical models and theories (electrostatics, hydrophobicity, adsorption, etc.). At the nanoscale molecular modeling can supply elucidations about physical-chemical processes. The possible energies of interaction between carbonaceous nanomaterials are already defined in the literature (Hunter 2001), considering both van der Waals attractive forces and Pauli repulsion rising from overlapping electron orbitals at short separation distances. The adsorption behavior and orientation of sorbates in microporous carbon, physic-sorption will be contributed by hydrophobicity and capillarity being the sorption dominant mechanism for not functionalized nanomaterials. Adsorption reports of high adsorption capacity, low sensitivity to pH value, rapid equilibrium rates, minimal hysteresis in dispersed nanoparticle (Hilding et al. 2001), and consistency with Langmuir were cited.

The sorptive capacity of traditional carbonaceous sorbents is limited by the surface-active site density, the slow kinetics, and the sorption nonequilibrium in heterogeneous systems. The large dimensions of conventional sorbents limit their transport through low-porosity environments and complicate the subsurface remediation. High surface area to volume ratio of carbonaceous nano-sorbent controlled pore size distribution, and their surface chemistry overcomes many of these intrinsic limitations.

7.5 Applications

7.5.1 Graphene-Based DNA Sensors

The exceptional electrical, optical, thermal, and mechanical properties of graphene oxide (GO) received worldwide attention since it was reported in 2004. Graphene with its large surface area reaching up to 2630 $m^2\ g^{-1}$ and the unique sp^2 (sp^2/sp^3)-bonded network and its other remarkable characteristics have led to its rapid development for use in the sensors making it an excellent promising candidate for biomolecule anchoring and detection (Suvarnaphaet and Pechprasarn 2017). On the other hand, the application of specific DNA sequences has been widely used for detection of bacteria, fungi, and genetically modified organisms having variety of applications such as clinical human disease detection, environmental horticulture, and food analysis. In consequence, new innovative types of nanosensor-based graphene using nucleic acid fragments as elements for pathogen detection have been developed. Although the nucleobase–graphene binding energy is slightly different via various strategies and equipments, one might be certain that ssDNA could be adsorbed on graphene sheet surface coupling crossover of several interaction forces and employing nucleobases as the anchors. This is also the root cause that ssDNA binds more strongly to graphene than dsDNA does in which nucleobases are entrapped and shielded by the phosphate–deoxyribose backbones. According to the binding affinity difference between ssDNA and dsDNA to graphene sheet, graphene oxide (GO) has been successfully adopted as a platform to discriminate DNA sequences. On the other hand and based on the specific nucleic acid hybridization of the immobilized DNA probe on the sensor and the analyte DNA sequence, DNA-based biosensor allows rapid, simple, and economical testing of genetic and infectious diseases. The most commonly adopted DNA probe is single-stranded DNA (ssDNA) on electrodes with electro-active indicators to measure hybridization between probe DNA and the complementary DNA analyte (Eun and Wong 2000).

There are four major types of DNA-based biosensors depending on their mode of transduction, namely, optical, piezoelectric, strip-type, and electrochemical DNA biosensors. Different reports revealed practically the potential effect of using this kind of nanosensors in detecting viral and bacterial pathogens due to their unique nucleic acid sequence, which can be specifically hybridized with the complementary DNA probe. The recognition of analyte DNA is dependent upon the formation of stable hydrogen bonds between the DNA probe and analyte DNA sequence. This is different from the antibody-based biosensors where hydrophobic, ionic, and hydrogen bonds play a role in the stabilization of antigen-antibody complex. Although the DNA-based nanosensors show a significant and promising rapid solutions for plant disease detection, however, polymerase chain reaction (PCR) technique may have to be performed prior to the probing process due to the small quantity of nucleic acid present in the bacterial cells (Ivnitski et al. 2000; Fang and Ramasamy 2015). We should clarify that there is still some limitations around this unique kind of sensing including the requirement for the synthesis of specific DNA probe, amplification of DNA, and high cost (DNA-based molecular beacons).

7.5.2 Graphene as a Biosensor

Construction of nanosensor-based materials for exploitation in agriculture is new and a promising direction of plant nanotechnology. Despite the material advancements, very scanty data is available where electrochemical biosensors have been applied for the detection of agricultural pathogens for their rapid and timely identification to prevent huge economic losses. The number of successful studies is still very limited, but nanosensors can be developed in the very near future. A great example is the recent building of single-walled carbon nanotubes (SWCNTs) radiometric sensors (for H_2O_2 and NO) performed by Giraldo et al. (2015) which proved the efficiency of radiometric nanosensing platform for detecting key compounds in plant tissues.

Carbon-based nanosensors were made of different types of carbon materials, metal-based nanomaterials, and screen-printed electrodes that generally utilize electrochemical mode of measurement and/or microfluidics-based system to achieve simple and compact analytical devices for detection of toxins and various applications in food, agriculture, and environmental monitoring. Use of single-walled carbon nanotubes (SWCNTs) with metal/metal oxide NPs were selected to deal agricultural by-products such as ammonia and nitrogen oxide as they are the most recent technologies in this issue. These techniques help in the development of nanosensor arrays which are of high density with a potential role in monitoring agricultural pollutants and their impact on biological and ecological health and thus in turn increasing crop productivity. On another side, the scientist Fernández-Baldo and his co-workers developed a faster screen-printed immunosensor assay modified with carbon nanotubes in a continuous-flow system for detecting the phytopathogenic fungus *Botrytis cinerea* with a detection limit reaching 0.02 μg mL^{-1} (Fernández-Baldo et al. 2009). This system was further modified by involving the use of micromagnetic beads coupled to the carbon-based screen-printed electrodes, apparently increasing in sensitivity to 0.008 μg mL^{-1} "pure" Botrytis antigen (Fernández-Baldo et al. 2010). Electrochemically functionalized single-walled carbon nanotube (SWCNT)-based nanosensors with metal/metal oxide nanoparticles or nanotubes for gases, viz., ammonia, nitrogen oxides, hydrogen sulfide, sulfur dioxide, and volatile organics have also a significant potential application in monitoring agricultural pollutants and assessment of impacts on living matter or health and in increase of crop productivity (Sekhon 2014). Also, the electrochemical DNA biosensor, graphene oxide/gold nanoparticles were also functionalized and used for rapid detection some microbial pathogens like *Helicobacter pylori* with a detection limit reaching 27 pM (Hajihosseini et al. 2016). More interestingly, a biosensor is reported for the screening of DNA-level modification and damage caused by chemotherapeutic drugs and a screen-printed gold electrode (SPGE) for the detection of *Cucumber mosaic virus* (CMV) (Ilkhani et al. 2016; Zulkifli et al. 2016).

7.5.3 Gene Delivery

Smart delivery systems of organic and inorganic agrochemicals to deliver the nucleic acids into the plant cells have a great attention recently (Zarei et al. 2018). This kind of technology is mainly based on the ability of nanomaterials to penetrate through cell walls and membranes of plant cells, solving major critical problems in agriculture sciences. Different types of nanomaterials with potential advantages were used as reliable vehicle systems to deliver therapeutic and agrochemical agents (i.e., liposomes, polymersomes, microspheres, and polymer conjugates). Carbon nanotubes are the best example showing safe interaction with biomacromolecules and a remarkable potential nano-vector to transfect plant cells with genes of interest (Wang et al. 2014a,b,c). Liu and his co-workers stated the ability of carbon-based nanomaterials particularly carbon nanotubes that showed a good potential to deliver nucleic acids (DNA or RNA) and other small molecules into tobacco plants (Liu et al. 2006a,b). Other studies also confirmed the ability of single-walled-CNTs (SWCNTs) act as nano-transporters for delivery of DNA and dye molecules into plant cells in more details (Srinivasan and Saraswathi 2010).

Different reports suggested that the multi-walled CNTs (MWCNTs) have a more magic ability to influence the seed germination and plant growth and work as a delivery system of DNA and chemicals to plant cells (Lahiani et al. 2015). On the other hand, scientists indicated that both carbon nanotubes (MWCNTs) and single-walled carbon nanohorns (SWCNHs) were documented by using Raman spectroscopy and transmission electron microscopy (TEM) (Lahiani et al. 2015). On the other hand, Serag and his research team observed that single-stranded DNA molecules wrapped around SWCNTs were able to target the cytoplasm of walled plant cells. This property could be used to introduce RNA pieces into the nucleus to activate or silence the genes of interest. Similarly, protoplast could be a target for delivering larger DNA molecules such as the delivery of plasmids into the plant cell genome. Thus, it is important to take into consideration all possible effects of carbon-based materials used as the DNA delivery machine on plant genome and proteome (Serag et al. 2015).

7.5.4 *Plant Growth Stimulators and Fertilizers*

Multi-walled carbon nanotubes (MWCNTs) are a kind of nanomaterial, and due to their unique nanostructures and extraordinary properties such as high electrical conductivity, large and special area, and significant thermal stability, they have been seriously taken into consideration in fundamental research and technological development (Milne et al. 2004).

7.5.4.1 Seed Germination

The growing consensus suggested that a high degree of the functionalized CNT leads to a dramatic reduction and toxic effects on different plant species based on CNT concentrations that have been increased daily. This may be backed up by the positive effects observed on the growth of *Cicer arietinum* L. plants after being treated by water-soluble MWCNT (wsMWCNT) (Tripathi and Shahi 2011), although some reports indicated that the germination rate of corn and rye plants treated with MWCNT was reduced, while the root length was increased (Begum and Fugetsu 2012). However, inhibitory effect of MWCNTs on plant growth has been reported by many researchers (Tiwari et al. 2014; Begum et al. 2014; Fathi et al. 2017). The MWCNTs had no effect on the germination of castor seeds and the length of plumule. But root length and wet weight in the treatment of 100 mg mL^{-1} seedlings were found improved in all factors related to the growth stage of seedlings (Fig. 7.2) (Fathi et al. 2017). Thus, the effect of NPs on plants varies from plant to plant, their growth stages, and the chemical composition of the used nanomaterial.

Other studies have also supported the positive influence of MWCNTs on seed germination and growth of six different crop species such as radish (*Raphanus sativus*), rapeseed (*Brassica napus*), rye, lettuce, maize, and cucumber (Lin and Xing 2007). Similarly, Ma et al. (2010) functionalized SWCNTs with poly-3-amino benzenesulfonic acid and studied the effects of both functionalized and non-functionalized SWCNTs on root growth of six crop plants, cabbage, carrot (*Daucus carota*), cucumber, lettuce, onion (*Allium cepa*), and tomato. Srinivasan and Saraswathi (2010) showed enhanced seed germination and growth rate in tomato seeds when exposed to CNTs. Highly maximized germination rate was observed in crop species such as tomato, hybrid Bt cotton (*Gossypium hirsutum*), Indian mustard (*Brassica juncea*), urd bean (*Vigna mungo*), and rice (*Oryza sativa*) with MWCNT treatment (Ghodake et al. 2010; Morla et al. 2011; Nalwade and Neharkar 2013). Interestingly, the effect of MWCNTs on the seed germination of *Brassica juncea* L. and *Phaseolus mungo* L. plants was 100%, which indicated their safety for seed germination (Ghodake et al. 2010).

The effects of engineered carbon nanomaterials of various dimensionalities on rice seed germination were studied, and an increase in germination rate with increased water uptake was observed for treated seeds than the control seeds. Surprisingly, the effects of MWCNTs and single-walled CNTs (SWCNTs) have

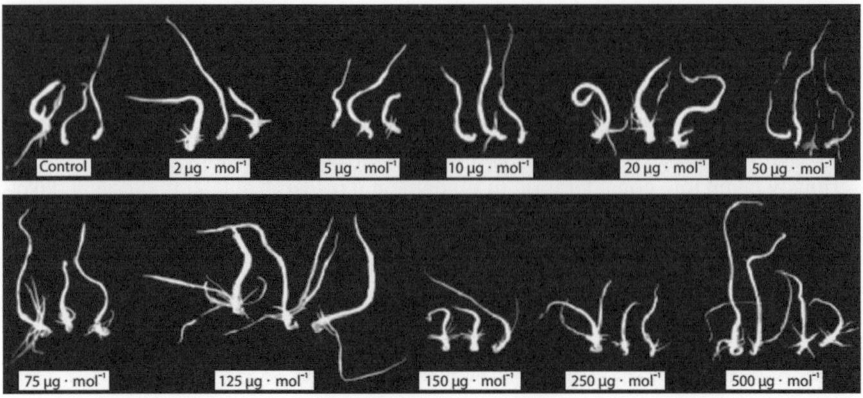

Fig. 7.2 Effects of multi-walled carbon nanotubes (MWCNTs) on root and shoot elongation of castor seedlings (Reprinted from Fathi et al. 2017)

been observed to vary. For example, zucchini plants exposed to MWCNTs did not show any detrimental effects on seed germination and root elongation, whereas a marked decrease in the biomass was recorded during further growth in the presence of SWCNTs (Stampoulis et al. 2009). Anjum et al. (2013) assessed the germinating faba bean (*Vicia faba* L.) seedling tolerance to different concentrations (0, 100, 200, 400, 800, and 1600 mgL^{-1}) of single-bilayer graphene oxide sheet (GO; size, 0.5–5 μm) and underlying potential mechanisms. Hu and Zhou (2014) reported a novel and biocompatible hydrated graphene ribbon (HGR) could promote aged (2 years) wheat seed germination, increase seed germination, and enhance resistance to oxidative stress *Onobrychis arenaria* (Kit.) DC seedlings treated with MWNTs showed an increased germination rate (Smirnova et al. 2012) that can be due to the positive role of carbon nanotubes in absorption of water (Kole et al. 2013). The growth of the roots and stems of *O. arenaria* seedlings is stimulated with MWNTs treatment (Smirnova et al. 2011). Similarly, it was observed that seed germination is activated with the sedimentation of MWCNTs on the surface of corn, barley, and soybean seeds (Lahiani et al. 2015). Synthesis of MWCNTs in low concentrations leads to increased corn seedling germination, while high concentrations lead to reduced growth. Increased growth mainly occurs via improvement of water transmission by carbon nanotubes. These compounds can be useful for the improvement of water flow, plant biomass, and essential concentrations of calcium and iron (Tiwari et al. 2014). Effects of carbon family-based nanomaterials on plant seed germination were shown in Table 7.1, Fig. 7.3.

7.5.4.2 Seedling Growth Promoters

Different studies conducted on CNMs evidently indicated their potential to enhance plant growth, nutrient uptake, seed germination, and fruit yield/quality (Khodakovskaya et al. 2009; Lahiani et al. 2015). Absorption of carbon nanotubes

Table 7.1 Effects of carbon family-based nanomaterials on the germination of plant seeds

Carbon NMs	Plant	Germination[a]	References
Graphene oxide	Faba bean	(+)	Anjum et al. (2013)
CNTs	Onion, rice, Indian mustard mung bean, tomato	(+)	Ghodake et al. (2010), Nair et al. (2010), Mondal et al. (2011), Morla et al. (2011)
	Garden cress, sorghum, tomato, radish, cucumber	(+)	Oleszczuk et al. (2011)
MWCNTs	Radish, rapeseed, rye, lettuce, maize, cucumber, zucchini	N	Lin and Xing (2007)
	Castor	N	Fathi et al. (2017)
Carbon nanohorns	Barley, maize, rice, soybean, switchgrass, tomato, and tobacco cell culture	(+)	Lahiani et al. (2015)

[a]Positive effect (+), negative effect (−), no effect (N)

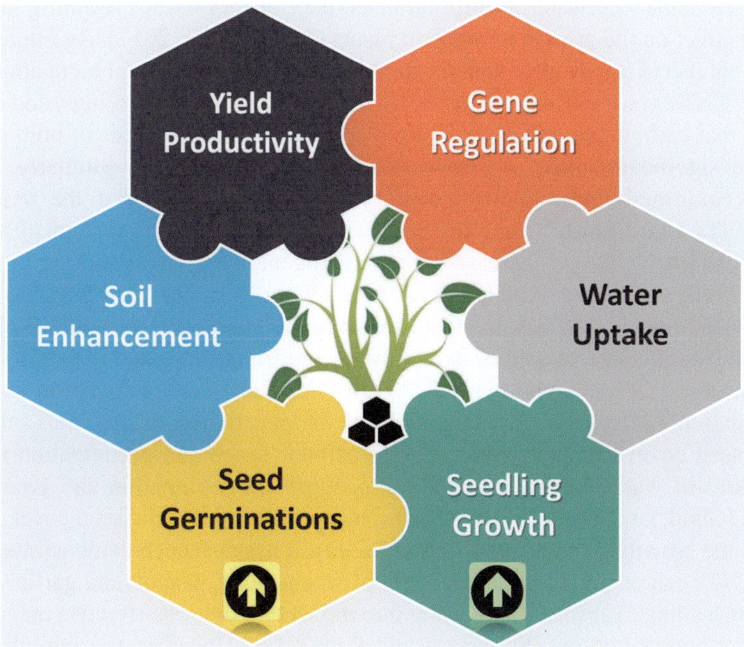

Fig. 7.3 The positive effect of nano-carbon in plants and soil

by plants is a new field in nano-agriculture. Carbon nanotubes can change the morphological and physiological properties of plant cells (Pourkhaloee et al. 2011; Lahiani et al. 2015) and are thought to regulate plant and seedling growth (Khodakovskaya et al. 2012; Haghighi and da Silva 2014). This stimulatory effect can have an important role on their application in the modeling of the plants' function. The improved root and stem growth on MWCNTs exposure may be due to the uptake and accumulation of MWCNTs by roots with their subsequent translocation to leaves (Smirnova et al. 2012). The presence of water-soluble CNTs inside wheat plants was evidenced by Tripathi and Sarkar (2015) with scanning electron and fluorescence microscope.

Furthermore, the authors linked this observation to the CNTs-induced root and shoot growth under both light and dark conditions. Interestingly, MWCNTs have been recognized to augment water retention; improve biomass, flowering, and fruit yield; and also enhance medicinal properties of plants (Khodakovskaya et al. 2013; Husen and Siddiqi 2014). In another study, it was indicated that the exposure of SWCNTs to maize seedlings promotes growth of seminal roots. Carbon-based fullerol [C60(OH)20] NPs treatment resulted in increases of up to 54% in biomass yield and 24% in water content in bitter melon (*Momordica charantia*). A 20% fruit length, 59% fruit number, and 70% fruit weight gain resulted in an overall improvement of up to 128% in fruit yield (Kole et al. 2013). The treatment of *Brassica juncea* seeds with oxidized MWCNTs increased moisture content of seeds and enhanced water absorption machinery of root tissues resulting in beneficial effect on the growth of mustard plants (Mondal et al. 2011). An interesting study published by Miralles and his co-workers indicated that root elongation due to MWCNT exposure was enhanced also in alfalfa and wheat seedlings, and it was found that carbon nanotubes were adsorbed onto the root surfaces of both plants without significant uptake or translocation (Miralles et al. 2012). Similarly, Wang et al. confirmed those findings, and their study indicated that the oxidized MWCNTs significantly promoted cell elongation in the root system of wheat (*Triticum aestivum*) and increased the dehydrogenase activity, resulting in faster root growth and higher biomass production (Wang et al. 2012). Later, the same group documented that tomato plants grown in soil supplemented with MWCNTs were able to produce two times more flowers and fruits compared to plants grown in control soil (Khodakovskaya et al. 2013). The effect of carbon nanotubes on the germination and growth of plants confirms the penetration of materials into the thick seed cover and their entrance into cellular space and participation in the processes of water absorption, leading to improved germination and growth of plants (Ghodake et al. 2010). MWCNTs, by penetrating into the seed cover, stimulated the growth of tomato and mustard seeds. It has also been demonstrated that MWCNTs can penetrate the thick cover of wheat, corn, peanut, and garlic seeds, thereby leading to absorption of water into the seeds – this was effective on growth at lower concentrations (Srivastava and Rao 2014). However, the inhibition of root elongation after application of CNTs on tomato was documented (Cañas et al. 2008). The inhibitory effects of MWCNTs on plant growth have also been reported by many researchers (Begum and Fugetsu 2012; Ikhtiar et al. 2013; Tiwari et al. 2014; Begum et al. 2014). In another study, the uptake, accumulation, and transmission of natural organic matter (NOM)-suspended MWCNTs in rice were reported (Lin et al. 2009). Some examples of CNTs influencing the yield and biomass of plants are compiled in Table 7.2.

Table 7.2 Positive and negative effects of different carbon nanomaterials on different plant species are summarized in the table

Carbon NMs	Plant	Seedling growth	References
CNTs	Alfalfa, wheat	Root elongation (+)	Miralles et al. (2012)
	Chickpea	Growth rate (+)	Tripathi and Shahi (2011)
	Tomato	Increased number of flowers and fruits (+)	Khodakovskaya et al. (2013)
	Tomato	Seedling growth (+)	Morla et al. (2011)
SWCNTs	Rice	Improved seed germination, water uptake, healthier seedlings (+)	Nair et al. (2012)
	Blackberry	Root elongations (+)	Flores et al. (2014)
	Arabidopsis, Rice protoplasts	Programmed cell death (−)	Shen et al. (2010)
	Maize	Enhanced root elongation (+)	Yan et al. (2013)
MWCNTs	Maize	Improved growth (+)	Tiwari et al. (2014)
	Rice	Improved seed germination, water uptake, healthier seedlings (+)	Nair et al. (2012)
	Red spinach	Exhibited growth inhibition, cell death (−)	Begum and Fugetsu (2012)
	Wheat	Root growth (+)	Wang et al. (2012)
	Tobacco	Growth rate (+)	Khodakovskaya et al. (2012)
	Tomato	Increase plant height and number of flowers (+)	Khodakovskaya et al. (2013)
	Castor	Improved growth (+)	Fathi et al. (2017)
	Zucchini	Reduced biomass (−)	Stampoulis et al. (2009)
	Lettuce	Reduced root length (−)	Lin and Xing (2007)
	Rye	Promoted root length	Lin and Xing (2007)
Poly-3-amino benzenesulfonic acid functionalized SWCNTs	Onion, cucumber	Enhanced root elongation (+)	Cañas et al. 2008
	Lettuce	Inhibited root elongation (−)	Cañas et al. 2008
Fullerol	Bitter melon	Increased biomass, fruit yield, and improved phytomedicines content (+)	Kole et al. (2013)
C60 fullerenes	Maize, soybean	Reduced biomass	Torre-Roche et al. (2013)

7.5.4.3 Plant Growth Promotion Mechanism

The possible mechanism for which the concentration-dependent MWCNTs affect the growth of castor seedlings needs to be clarified. Also, the mechanism behind the stimulatory effect of MWCNTs on the growth of castor seedlings is that the mentioned compounds can support the absorption of water by the seeds. Probably, carbon nanotubes can penetrate the thick seed coat and support the water uptake inside seeds. MWCNTs penetrate the cell wall and accumulate in the cells and tissues, and, via the vascular system, they are transferred from the root to the stem and to the leaf of the seedling (Srivastava and Rao 2014). MWCNTs induce the water and essential Ca and Fe nutrients uptake efficiency that could enhance the seed germination and plant growth and development (Villagarcia et al. 2012; Tiwari et al. 2014). The ability of SWCNTs to traverse across the plant cell wall and cell membrane was first reported by Liu et al. (2009). This has opened novel methods to deliver DNA and other molecules to intact plant cells. Liu and his group also studied changes in the cell wall of tobacco cells under the repression of water-soluble carboxy-fullerenes. Disruption in cell wall and cell membrane was observed on the adsorption of fullerenes which led to complete inhibition of cell growth (Liu et al. 2013a,b). Single-walled carbon nanotubes (SWCNTs) have been shown to exert adverse effects on *Arabidopsis* and rice leaf protoplasts through oxidative stress, leading to a certain amount of programmed cell death (PCD)/apoptosis, DNA damage, and chromatin condensation (Shen et al. 2010).

On the other hand, a noticeable increased glycosyl residue was observed in the cell wall of fullerene-treated plant cells with elevated levels of reactive oxygen species (ROS). Serag et al. (2011) investigated the ability of FITC-labeled MWCNTs to penetrate the cell membrane of periwinkle (*Catharanthus roseus*) protoplasts, and their internalization mechanism was studied with the help of confocal imaging and TEM techniques.

The direct penetration mode helped MWCNTs to bypass endosomes and hence opens new avenues in designing endosomes escaping nano-transporters for plant cells. MWCNTs have been shown to improve the peroxidase and dehydrogenase activity (Smirnova et al. 2012). Wang et al. (2012) reported oxidized MWCNTs significantly enhanced cell elongation in the root system and promoted dehydrogenase activity. In this regard, Zao and his co-workers noticed that graphene oxide (GO) exposure did not induce H_2O_2 production, formation of oxidative stress, increase in malondialdehyde content, or altered activities of antioxidant enzymes in *Arabidopsis* plants (Zhao et al. 2015). Although the role of CNTs in regulating cell division and plant growth and improvement of plant production is clear, however, the complex interactions of those nanomaterials with the environment and their inhabitants particularly terrestrial plant species need further investigation.

7.5.5 Nanofertilizers

Carbon-based nanomaterials significantly help in reducing losses of nutrients by leaching or evaporation caused by the classical working with common practice fertilizer application (e.g., foliar application by spraying) in conventional agriculture. On the contrary, carbon-based nanomaterials can provide a controlled-release technique for fertilizer delivery precisely functionalized to adapt the nutrient supply to the current demand of the target plant, extending the time of function, and inhibit losses by leaching with no risk of facing overdose effects. Using different cheap carbon nanomaterials like graphene oxide aids the process of commercial large-scale encapsulated fertilizer production with controlled-release technique to be possible with more economic cost (Zhang et al. 2014). Most of these fertilizers are based on amendments of mineral and organic fertilizers with nano-carbon, which in most cases acts as a fertilizer synergist with the aim of improving plant nutrient availability, reducing nutrient losses, and stimulating plant growth (Zaytseva and Neumann 2016). Several carbon-based nanomaterials have found applications in patents on nano-fertilizer formulations (Zhang and Liu 2010; Biris and Khodakovskaya 2011; Li and Guan 2011; Liu and Wangquan 2012; Xie and Liu 2012; Zhang and Chen 2012). Multi-walled carbon nanotubes (MWNTs) can be used as a carrier to improve nutrient uptake by plant cells (Torney et al. 2007; Serag et al. 2011). Upon successful delivery into the cells, the absorption and utilization of N by the plant can be increased leading to enhanced plant growth. Yatim et al. (2015) identified the significant process parameters to attach urea fertilizer (UF) onto MWNTs. The UF-MWNTs was then characterized by means of spectroscopic and microscopic analyses to confirm their bonding. Comparison study was also conducted between UF-MWNTs and UF-functionalized MWNTs on total N content bonding to the MWNTs. Their results indicated the importance of functionalization and amount of MWNTs for further optimization step in developing novel urea fertilizer and MWNTs for plant growth (Yatim et al. 2015).

On the other hand, carbon nanotubes could be used as a nutrient carrier for macro- and microelements and also as slow-release fertilizers that may reduce their higher concentrations which are usually used (Hasaneen et al. 2017). The nano-carbon is used as a coating material for slow-release fertilizer, and incorporation of nano-carbon into slow-release fertilizer is beneficial for reducing water pollution, especially the Ingzhengda slow-release fertilizer and nano-carbon (JSCU+C) (Wu 2013). To investigate the effects of combined application of nitrogen fertilizer and nano-carbon on nitrogen use of soil and rice yield, six treatments were applied by the scientist Fan and his co-workers. The utilization rate of nitrogen fertilizer increased after combined application of nano-carbon, which can save the N fertilizer in production practice. Therefore, combined treatment is suitable to application and dissemination in soda saline-alkali soil in the agriculture (Fan et al. 2012). More knowledge about nano-fertilizers in agriculture and the relationships between physicochemical characteristics of nanomaterials and biological interactions are still in need, but also more care about the risk with handling nanoparticle application in this important field is needed (Taha 2016). On the other side, much work should be done by scientists in a way to define the optimized concentrations of different types of nano-carbons in different plant species as organic fertilizer to reach for the positive impacts that are important for promoting plant growth and increasing the crop yields (Singh et al. 2017).

7.5.6 Antimicrobial

Generally, nanomaterials have an immense potential in plant disease control, providing unique novel cost-effective methodologies in crop protection with environment-friendly and ignorable limitations (Prasad et al. 2014, 2017a,b). Carbon-based nanomaterials particularly carbon nanotubes, fullerene, and graphene are best examples showing potent antimicrobial potential giving high possibility to be applied as novel fungicides and disinfection agents suitable for agricultural purposes (e.g., in plant protection) (Jung et al. 2011; Al-Hakami et al. 2013; Schmitt et al. 2015) that can be used to produce innovative nanocomposite materials. Also, Al-Hakami and his co-workers showed a good exploitation of using functionalized CNTs with the aliphatic alcohol 1-octadecanol ($C_{18}H_{38}O$) in water disinfection based on interactions of functionalized CNTs with microwaves. Based on their findings, they stated that the formed CNTs had outstanding antimicrobial properties since the long carbon chains contributed to a better absorption of the microwaves by CNTs (Al-Hakami et al. 2013). Zaytseva and Neumann reported interesting findings that revealed the broad-spectrum antifungal activity of those nanomaterials against two phytopathogenic fungi, namely, *Fusarium graminearum* and *Fusarium poae*, compared to other carbon-based nanomaterials like fullerenes and graphene oxide (Zaytseva and Neumann 2016). Also, it was indicated that CNTs displayed superior inactivation effects on the copper-resistant plant microns like *Ralstonia solanacearum*, *Fusarium graminearum*, and *F. oxysporum* (Wang et al. 2013, 2014a,b,c). Additionally, multi-walled carbon nanotubes (MWCNTs) have a noticeable strong antimicrobial activity against different microbial and viral and wastewater agents, where the MWCNTs were coated with different metallic nanomaterials like Ag NPs or Zn NPs in a good way to control the fungal growth of the phytopathogenic fungi *Aspergillus fumigatus* and *A. ochraceus* (Fosso-Kankeu et al. 2016). From the toxicological point of view, single-walled carbon nanotubes have higher antimicrobial properties than multi-walled carbon nanotubes (MWCNTs) (Oyelami and Semple 2015).

Although different published articles indicated the possible use of those nanomaterials as antimicrobial agents against bacteria and fungi (Upadhyayula et al. 2009; Das et al. 2014a,b; Pereira et al. 2014; Joshi et al. 2018), very few studies proposed the mechanism of action and suggested that the antimicrobial action may be backed up by either physical (cell wall damage and cytoplasm separation) or chemical effects (oxidative stress and ROS generation) (Wang et al. 2017). This assumption requires much more detailed investigations of the behavior of those nanomaterials, considering a wider range of phytopathogenic microbes to be tested under in vivo conditions.

7.5.7 Nano-pesticides

Nanomaterials have a great application in agriculture not only for diseases control but also for the enhancing growth effect in plants (Patel et al. 2014). Some of the nanoparticles that are widely used to control plant diseases are nanoforms of carbon, silver, silica, and aluminasilicates (Saurabh et al. 2015). Carbon nanomaterials can have different forms like hollow spheres, ellipsoids (fullerenes), or cylindrical [nanotubes such as single-walled carbon nanotube (SWCNT) and multi-walled carbon nanotube (MWCNT)] (Saurabh et al. 2015). Due to their antifungal properties, carbon-based nanomaterials are promising materials for the development of nano-antifungal pesticides. The effect of commercial multi-walled carbon nanotubes (MWCNTs) on conidia of the entomopathogenic fungus *Paecilomyces fumosoroseus* was investigated by Gorczyca et al. (2009). The CNTs strongly limited spore production of mycelium in applied concentrations. The polyethylene films with carbon nanotube nanocomposite base are prepared by solution casting from boiling xylene. This nanocomposite showed strong antifungal activity in packaged materials (Asgari et al. 2014). Six CNMs including SWCNTs, MWCNTs, graphene oxide (GO), reduced graphene oxide (rGO), fullerene (C60), and activated carbon (AC) were examined against two plant pathogenic fungi: *Fusarium graminearum* and *Fusarium poae* (Wang et al. 2014a,b,c).

The SWCNTs had the highest antifungal activity followed by MWCNTs, GO, and rGO. C60 and AC showed no significant antifungal activity. The antifungal activities of MWCNTs with different surface groups against *Fusarium graminearum* were explored by Wang et al. (2017). According to their findings, spore germination was remarkably inhibited by surface-modified MWCNTs, with germination rate being 18%, threefold lower than for pristine MWCNTs. Sarlak et al. (2014) have also used the multi-walled carbon nanotube (MWCNT) coated with poly(citric acid) as a carrier for the water-soluble dithiocarbamate fungicides. The encapsulation of pesticide was optimized in the pH range of 6–8 with a stirring time of 30–80 min. The pesticide loaded nanocarriers exhibited superior fungicidal activity against the leaf spot fungi *Alternaria alternata*, when compared to the bulk form of the fungicides.

7.5.8 Pesticide Monitoring Using Carbon Nanotube Sensors

Pesticides are organic compounds, which are developed to protect plant crops from pest invasion in agriculture. Although pesticides have a significant positive impact on the crop yield and food loss, however, intensification of agricultural practices can lead to an increase of the pesticide residues that can impose a serious risk to human health and the environment worldwide (Damalas and Eleftherohorinos 2011; Sangeetha et al. 2017a,b). Furthermore, their residual accumulation in the fresh and edible parts is considered a serious problem taken into account in global trade and countries particularly around the country borders in import-export exchange (Gholipour et al. 2012). On the other hand, transformation produced compounds released from the product active ingredients that may also be found in the final stage of commercial packages of pesticides which also are considered very toxic for human beings. Therefore, there is a need to develop a simple, sensitive, selective, cheap, and portable sensing platform for pesticide determination. Carbon nanotubes

(CNTs) are currently one of the most promising nanomaterials having unique properties including lightness, rigidity, high surface area, high mechanical strength in tension, good thermal conductivity, or resistance to mechanical damage. More interestingly, their unique properties put it an efficient alternative to other conventional analytical sorbents. Besides their analytical sorbent applications, carbon nanotubes are also widely enrolled in the development of different nanosensors to detect different biomolecules like drugs, dyes, ions, phenolic compounds, and also pesticide residues (Wang et al. 2014a,b,c; De Oliveira et al. 2015; Govindhan et al. 2015; Mani et al. 2015). In the past few years, electrochemical sensor-based nanotubes have been issued as a very sensitive electroanalytical method for detection of pesticide residues at very low quantities (Herrero et al. 2012).

Electrochemical sensors have been grouped into seven different carbon nanotube-based techniques including (1) carbon nanotube sensors, (2) phthalocyanine, (3) molecularly imprinted polymers (MIP), (4) ionic liquid, (5) metallic nanoparticles, (6) β-cyclodextrin, and (7) fullerene (C60)-functionalized carbon nanotubes (Wong et al. 2017). Some studies indicated that the possible combined use of carbon nanotube-based sensors showed a noticeable improvement in the detection analytical signal (Herrero et al. 2012). One of the most interesting studies in pesticide nanosensor-based electrode detection was reported by Mercan and his co-workers. They determined that the cyromazine insecticide using a multi-walled carbon nanotube paste electrode by square-wave adsorptive stripping voltammetry was assayed (Mercan et al. 2011).

Other studies used glassy carbon electrodes and a pyrolytic graphite electrode modified with multi-walled carbon nanotubes and iron (II) tetra-aminophthalocyanine (MPc) in detection of amitrole and different pesticides (Ribeiro et al. 2011). Molecularly imprinted polymers (MIP) are also used due to their biomimetic recognition of analytes similar to selective enzyme-substrate systems and/or antigen-antibody interactions and functionalized in selective sensing of certain pesticides (Yaqub et al. 2011). Also, metallic nanomaterials like gold (AU), silver (Ag), and palladium (Pd) nanoparticles combined with multi-walled carbon nanotubes show significant interesting use in the electroanalytical chemistry as electrode modifiers for a set of important advantages that can be achieved (Zhang et al. 2009). In 2013, Torre-Roche and his co-workers reported the positive impact of using such new detection technology in pesticide residue assessment in different crop plants like zucchini, corn, tomato, and soybean (Torre-Roche et al. 2013). Continually, Wong et al. reported the most recent work related to pesticide electrochemical monitoring using carbon nanotube-based electrochemical sensors and expressed that electrochemical sensors based on carbon nanotubes have an efficient potential to analyze very minute amounts of pesticide residues directly on plant surface; this for sure will have a significant great value in food production and industry (Wong et al. 2017).

7.5.9 Agricultural Wastewater Treatment

Remediation of pollutants in agricultural areas is considered one of the major global challenges up to date. Different protocols worked up in this issue either by improvement the classical and conventional methods or by introducing other innovative approaches in a way to improve the pollutant remediation strategies for environmental sustainability. In history, activated carbon (AC) has already been used as good organic sorbent in wastewater treatments due to its unique ability to adsorb a broad spectrum of organic and inorganic contaminants. However, AC has slow adsorption kinetics; it is a nonspecific adsorbent and its effectiveness against microbes is relatively low. On the other hand, in heavy wastewater that was contaminated with oils, greases, and hard solids, carbon is not well active and may often cause pore blockage in it. Furthermore, activated carbon itself is frequently removed together with the adsorbed pollutants and therefore needs to be replaced in regular intervals. In this regard, carbon-based nanomaterials offer an attractive and promising alternative tool in improving wastewater filtration systems with numerous examples in the available literature (Liu et al. 2013a,b; Das et al. 2014a,b; Smith and Rodrigues 2015). More interestingly, Husen and Siddiqi stated that carbon nanotubes and fullerene have the potential to increase the water-retaining capacity, biomass, and fruit yield in plants up to 118% (Husen and Siddiqi 2014). Carbon nanotubes (CNTs) are the best example of carbon nanomaterials that have received wide attention due to their unique properties, like large specific surface area with high absorption potential, high thermal stability, and high chemical stability. Numerous reports revealed the high adsorption capacity of CNTs toward microcystins (cyanobacterial toxins) (Yan et al. 2006), lead (Li et al. 2002), and copper (III) (Dichiara et al. 2015), which are even stronger than that of activated carbon. Furthermore, they can be used for sorption of herbicides or nitrogen (N) and phosphorus (Ph) elements in wastewater (Zheng et al. 2014). On the other hand, fullerenes as well as CNTs exhibit a mobilization potential for various organic pollutants (Mishra and Clark 2013), such as lindane (agricultural insecticide) (Srivastava et al. 2011) and persistent polychlorinated biphenyls (Wang et al. 2013). Despite the significant merits offered by carbon nanomaterials (CNMs) compared to activated carbon including the unique high surface area, mechanical and thermal stability, high chemical affinity for aromatic compounds (Yoo et al. 2011), potential antimicrobial properties, and ability to adsorb contaminants, filters based on CNMs can be recycled (Wang et al. 2014a,b,c). However, certain challenges related to the production cost, difficulties in obtaining of CNTs with certain morphological characters including the size shape, diameter distribution, uncertainties regarding the leaching potential of CNTs, as well as ecological safety and human health issues, are still limitations for commercial applications in wastewater cleanup technologies. The list of some environmental pollutants that can be remediated by different types of nano-carbon was shown in Table 7.3.

7.5.10 *Effect of Carbon Nanotubes on Soil Microorganisms*

Different reports indicated that fullerene NPs had no significant antimicrobial potential against microbial diversity like Eubacteria and Kinetoplastida (protozoans) community structure on DGGE profiles (20–30% of dissimilarity) (Tong et al. 2007; Johansen et al. 2008). In contrast, in 2011, Boonyanitipong and his

Table 7.3 List of some pollutants that can be removed by different types of nano-carbon

Pollutants	Initial concentration	Carbon nanotube dose	pH	Removal efficiency	References
1,2-Dichloro-benzene	20 mgL^{-1}	0.5 gL^{-1}	3–10	30.8 and 28.7 mg g^{-1} for as grown and graphitized CNT in 40 min	Peng et al. (2003)
Aniline	0.1–10,000 mgL^{-1}	/	7	114.8 mg g^{-1}	Peng et al. (2003)
Atrazine	5 mgL^{-1}	4 gL^{-1}	5	0.956 mg g^{-1}	El-Sheikh et al. (2008)
Cd^{+2}	2–15 mgL^{-1}	0.5 gL^{-1}	5	10.86 mg g^{-1}	Li et al. (2003a)
Cd^{+2}	9.5 mgL^{-1}	0.5 gL^{-1}	7	1.1 mg g^{-1}	Li et al. (2003b)
$CHBr_2$	2 mgL^{-1}	0.33 gL^{-1}	3–11	0.92 mg g^{-1}	Lu and Chiu (2006)
$CHCl_2Br$	2 mgL^{-1}	0.33 gL^{-1}	3–11	1.23 mg g^{-1}	Lu and Chiu (2006)
$CHCl_3$	2 mgL^{-1}	0.33 gL^{-1}	3–11	2.41 mg g^{-1}	Lu and Chiu (2006)
$CHClBr_2$	2 mgL^{-1}	0.33 gL^{-1}	3–11	1.08 mg g^{-1}	Lu and Chiu (2006)
Co^{+2}	10 mgL^{-1}	5 gL^{-1}	9	More than 90%	Pyrzynska and Bystrzejewski (2010)
Cu^{+2}	10 mgL^{-1}	0.5 gL^{-1}	9	Nearly 100%	Pyrzynska and Bystrzejewski (2010)
Cu^{+2}	5–30 mgL^{-1}	0.5 gL^{-1}	5	28.49 mg g^{-1}	Li et al. (2003a)
Fluoride	12 mgL^{-1}	2 gL^{-1}	5–9	14.4 mg g^{-1}	Li et al. (2003a)
Methidathion	5 mgL^{-1}	4 gL^{-1}	5	1.11 mg g^{-1}	El-Sheikh et al. (2008)
Ni^{+2}	6–20 mgL^{-1}	0.3 gL^{-1}	6.55	9.8 mg g^{-1}	Chen and Wang (2006)
Pb^{+2}	10–80 mgL^{-1}	0.5 gL^{-1}	5	97.08 mg g^{-1}	Li et al. (2003a)
Pb^{+2}	2–14 mgL^{-1}	0.5 gL^{-1}	7	1 mg g^{-1}	Li et al. (2002)
Phenol	0.1–100,000 mgL^{-1}	/	7	64.6 mg g^{-1}	Li et al. (2002)
Propoxur	5 mgL^{-1}	4 gL^{-1}	5	0.625 mg g^{-1}	El-Sheikh et al. (2008)
Th (IV)	32.32 μmol L^{-1}	0.2 gL^{-1}	1.9	65.8 μmol g^{-1}	El-Sheik et al. (2008)
U (VI)	10^{-7}–10^{-4} M	1 gL^{-1}	4	5.0 mmol g^{-1}	Schierz and Zanker (2009)
Zn^{+2}	10–80 mgL^{-1}	0.5 gL^{-1}	7	10.21–11.23	Liu et al. (2006a,b)
Zn^{+2}	60 mgL^{-1}	0.5 gL^{-1}	1–5	37.03–46.94 mg g^{-1} and 30.3–34.36 mg g^{-1} from 5 to 45 C	Lu and Chiu (2006)

co-workers revealed that about 10 g kg^{-1} soil from MWCNT have the potential to improve the degraders of recalcitrant contaminants (PAH) *Rhodococcus*. Also, their results suggested that the flourishment of their bacterial community and its diversity is passively affected by the MWCNT particularly at the highest concentrations. Interestingly, it was found that there is no observed effect on the bacterial community by MWCNT after only 1 week. The scientist attributed this early effect to the acidic nature of the MWCNT used, which caused a significant decrease in soil pH at higher exposure concentrations, and consequently the soil bacterial community and its chemistry totally changed (Boonyanitipong et al. 2011). Their studies demonstrated that the bacterial community particularly *Cellulomonas*, *Nocardioides*, and *Pseudomonas* was not changed, while the flourishment of some bacterial genera like *Derxia*, *Holophaga*, *Opitutus*, and *Waddlia* was noticeably changed and decreased (Shrestha et al. 2013).

More interestingly and by following a meta genomic comparative analysis for the bacterial populations and their diversity, Khodakovskaya et al. observed that the bacterial diversity and their richness was not significantly affected by MWCNTs, while a noticeable modification in the bacterial composition and their species was reported (Khodakovskaya et al. 2013; Jin et al. 2014). Those results are in agreement with that reported at the same year by Rodrigues and his co-workers, where a noticeable modification in the fungal community and bacterial community was reported after only 2 weeks of soil exposure to SWCNT. On the other hand, Gurunathan (2015) demonstrated the negative effect of graphene oxide nanoparticles (GO NPs) on isolated *Rhizobacteria* from the soil (five *Bacillus* species: *B. megaterium*, *B. cereus*, *B. subtilis*, *B. mycoides*, and *B. marisflavi*). They reported that after 2 weeks, the soil bacterial community composition was affected by the multi-walled carbon nanotubes at the highest concentrations (Kerfahi et al. 2015).

7.6 Phytotoxicity

An increasing number of studies outlined the environmental impacts and the safety profile of carbon nanomaterials (Lin and Xing 2007; Stampoulis et al. 2009; Shen et al. 2010; Kaphle et al. 2017). Most of the phytotoxicity studies have examined germination, cell cultures, and genetic effects (Kaphle et al. 2017). Although carbon nanomaterials were evidenced to penetrate through seed coats, enter into the plant cells, and translocate into different plant parts more efficiently, some studies indicated that exposure to carbon nanomaterials decreases seed germination and root growth and changes the root architecture (Begum and Fugetsu 2012; Kaphle et al. 2017). Carbon nanomaterials inhibit seedling growth and change morphological, physiological, biochemical, molecular, nutritional, and genetic levels in plants (Gopalakrishnan Nair 2018). Also, it was reported that carbon nanotubes and fullerene can cause damage to plants (Shen et al. 2010; Begum and Fugetsu 2012). In 2009, Lin and his co-workers revealed that blossoming of rice plants incubated with C70 fullerene was delayed by at least 1 month and their seed-setting rate was reduced by 4.6% compared to the controls (Lin et al. 2009). Also, water-soluble fullerene inhibits plant growth and causes shortening of seedling roots and loss of gravitropism. These adverse effects may be caused by auxin disruption, abnormal cell division, and microtubule disorganization. Furthermore, fullerene effect on fruit and crop production in edible plants and vegetables increases the water-retaining capacity, biomass, and fruit yield in plants up to ~118%. The internalization and accumulation of these compounds in plant roots and seedlings and their phytotoxicity have been investigated by Liu et al. (2013a,b).

Wang et al. (2016) investigated the bioaccumulation of fullerenol (water-soluble derivative of fullerene carbon nanomaterial) nanoparticles in wheat using ^{13}C-labeling methods. Also, it was revealed by Ghosh and his co-workers that the internalization of multi-walled carbon nanotubes inside the plant cells results in chromosomal aberrations, DNA fragmentation, and apoptosis in *Allium* root cells (Ghosh et al. 2015). Shen et al. (2010) found that certain amounts of single-walled carbon nanotubes can induce the production of reactive oxygen species (ROS), which eventually leads to programmed cell death, in *Arabidopsis* leaves, and protoplasts. In red spinach (*Amaranthus tricolor* L.), phytotoxicity of multi-walled carbon nanotubes causes growth inhibition and cell death; multi-walled carbon nanotubes also cause ROS production and hypersensitive response-type necrotic lesions of leaf cells and tissues (Begum and Fugetsu 2012). Although plant cells and mammalian cells have different structures, such as the thick and rigid plant cell wall, chloroplasts, and large central vacuoles, they show similar responses to fullerene. The bioaccumulation of dichloro-diphenyl-dichloro-ethylene, a persistent and estrogenic pollutant, in some food crops such as zucchini, soybean, and tomato, increased in the presence of C60 fullerene (De La Torre-Roche et al. 2012; Avanasi et al. 2014). Avanasi and his co-workers have measured plant uptake of C60 fullerene and found that ^{14}C-labeled C60 can be slowly absorbed by plants and will likely persist in soil for a long period (Avanasi et al. 2014). The modifications induced by carboxy-fullerenes to the cell wall of tobacco plants such as the mechanism of interaction and survival under repression were analyzed. The adsorption of this component led to the disruption of the cell wall and membrane with the consequent inhibition of cell growth. A further study evidenced the ability of functionalized carbon nanotubes to penetrate the plant cell wall and induce changes in specific organelles. Also, the diameter and length of single-walled carbon nanotubes are the major restraining features for their effective penetration into the plant cell wall (Serag et al. 2013). Toxicity of carbon nanomaterials was found to be largely dependent on their concentrations, growth/exposure conditions, and plant species. Due to their smaller size and altered physical, chemical, and structural properties, the absorption and translocation of different types of carbon nanomaterials raise serious concerns about their toxic effects on plants and soil microorganisms. In conclusion, different reports on CNT toxicity and safety were carried out, and it was demonstrated as a functionalization that reduces toxic effects, while pristine CNTs induce different phytotoxic effects, but further investigations are needed. Some examples on the mechanism of phytotoxic effects of different carbon nanomaterials on different plant species were shown in Table 7.4.

Table 7.4 The mechanism of phytotoxic effects of different carbon nanomaterials on different plant species

Carbon NMs	Plant	Mechanism	References
Graphene oxide (GO)	*Arabidopsis*	Fragmented nuclei, membrane damage, and mitochondrial dysfunction	Begum and Fugetsu (2012)
	Faba bean	Concentration-dependent decrease in oxidative enzyme activity	Anjum et al. (2013)
Water-soluble graphene oxide (ws-GO)	Lettuce, cabbage, red spinach, tomato	Reduced plant growth, biomass, the number and size of leaves, increased ROS along with necrotic symptoms	Anjum et al. (2013)
CNTs	Red spinach	Growth inhibition, changes to tissue structure	Begum and Fugetsu (2012)
MWCNTs	Rice cells	Increased ROS generation and decreased cell viability	Tan et al. (2009)
	Rice	Delayed flowering and seed setting. Reduced seed weight	Lin et al. (2009)
	Rice	Chromatin condensed inside the cytoplasm and caused cell death, plasma membrane detachment from cell wall and cell shrinkage	Tan et al. (2009)
	Zucchini	Negatively affected biomass production and transpiration	Stampoulis et al. (2009)
	Wheat	Enhanced the uptake of phenanthrene to the living cells	Wild and Jones (2009)
	Onion	Chromosomal aberrations, DNA fragmentation, and apoptosis in root cells	Ghosh et al. (2015)
	Onobrychis	Enhanced the POD activity	Smirnova et al. (2012)
ws-C70	Tobacco BY-2 cells	Cell boundary disruption and growth inhibition	Liu et al. (2013a,b)

7.7 Conclusion and Future Perspectives

Carbon nanotechnology has the potential to be exploited in different advanced applications in agriculture which finds use in biology, plant protection, medicine, and chemical engineering. Carbon-based nanomaterials are some of the best examples that find their application starting from increased crop yield, passing by organelle-targeted gene delivery, and ending with wastewater treatment and nanopesticide and antimicrobial pesticide extraction and detection. The antimicrobial effect of CNMs either on microbial pathogens or on soil microbiota together with the analytical-based information of the ecological interaction between those nanomaterials and organic/inorganic matter present in soil can introduce significant information in a more relevant context of a biological "realistic" scenario or will reduce the gap between experimental and environmental conditions.

On the other hand, carbon-based nanomaterials may also be helpful for the advancement of agriculture and plant protection. But in some cases, they may act a negative behavior in certain concentration that may be controlled within the permissible limit to prevent any damage. Furthermore, the insufficient information, their fate, and effects in the environment produce non-accurate assessment of their risk, which is necessary before their widespread application in agriculture generally and plant protection specifically. In some cases, carbon nanotubes were used in a modified form, by adding conductive polymers, metallic nanoparticles, phthalocyanines, porphyrins, ionic liquids, and graphene. These diverse modified nano-carbon composite electrodes are efficiently enrolled in different electroanalysis purposes, and future electrochemical investigations should be carried out on the sensing and biosensing of pesticides. As a result, more comprehensive investigations of chronic exposure under environmentally realistic scenarios will be needed at first to enable such efforts. In conclusion, we assume that carbon-based nanomaterials will find soon a fairly bright future in agriculture fields, introducing an economic solution for critical agronomic problems in a way to improve crop production and reduce the food loss.

Acknowledgment This research was supported by the Science and Technology Development Fund (STDF), **Joint Egypt (STDF)-South Africa (NRF) Scientific Cooperation**, Grant ID. 27837, to Kamel Abd-Elsalam. Also, this research was partly supported by the International Foundation for Science, Stockholm, Sweden, through a grant to Dr. Hashim Ayat (F5853).

References

Al-Hakami SM, Khalil AB, Laoui T, Atieh MA (2013) Fast disinfection of Escherichia coli bacteria using carbon nanotubes interaction with microwave radiation. Bioinorg Chem Appl 45:8943

Andreas H (2002) Functionalization of single-walled carbon nanotubes. Angew Chem Int Ed 41:1853–1859

Anjum NA, Gill SS, Duarte AC, Pereira E, Ahmad I (2013) Silver nanoparticles in soil–plant systems. J Nanopart Res 15(9):1896

Asgari P, Moradi O, Tajeddin B (2014) The effect of nanocomposite packaging carbon nanotube base on organoleptic and fungal growth of mazafati brand dates. Int Nano Lett 4:1–5

Avanasi R, Jackson WA, Sherwin B, Mudge JF, Anderson TA (2014) C60 fullerene soil sorption, biodegradation, and plant uptake. Environ Sci Technol 48(5):2792–2797

Bai L, Bossa N, Qu F, Winglee J, Li G, Sun K, Liang H, Wiesner MR (2017) Comparison of hydrophilicity and mechanical properties of nanocomposite membranes with cellulose nanocrystals and carbon nanotubes. Environ Sci Technol 51:253–262

Bakajin O, Noy A, Fornasiero F, Grigoropoulos CP, Holt JK, In JB, Kim S, Park HG (2009) Nanofluidics carbon nanotube membranes: applications for water purification and desalination. In: Savage NF (ed) Nanotechnology applications for clean water. William Andrew Inc., Norwich, NY, pp 77–93

Begum P, Fugetsu B (2012) Phytotoxicity of multi-walled carbon nanotubes on red spinach (*Amaranthus tricolor* L.) and the role of ascorbic acid as an antioxidant. J Hazard Mater 243:212–222

Begum P, Ikhtiari R, Fugetsu B (2014) Potential impact of multi-walled carbon nanotubes exposure to the seedling stage of selected plant species. Nanomaterials 4(2):203–221

Biris AS, Khodakovskaya M (2011) Method of using carbon nanotubes to affect seed germination and plant growth. WO 2011059507

Boonyanitipong P, Kositsup B, Kumar P, Baruah S, Dutta J (2011) Toxicity of ZnO and TiO_2 nanoparticles on germinating rice seed *Oryza sativa* L. Int J Biosci Biochem Bioinforma 1:282–285

Burlaka OM, Pirko YV, Yemets AI, Blume YB (2015) Plant genetic transformation using carbon nanotubes for DNA delivery. Cytol Genet 49:349–357

Cañas JE, Long M, Nations S, Vadan R, Dai L, Luo M, Ambikapathi R, Lee H, Olszyk D (2008) Effects of functionalized and nonfunctionalized single-walled carbon nanotubes on root elongation of select crop species. Environ Toxicol Chem 27:1922–1931

Chai M, Shi F, Li R, Liu L, Liu Y, Liu F (2013) Interactive effects of cadmium and carbon nanotubes on the growth and metal accumulation in a halophyte Spartina alterniflora (Poaceae). Plant Growth Regul 71:171–179

Chen C, Wang X (2006) Adsorption of Ni (II) from aqueous solution using oxidized multiwall carbon nanotubes. Ind Eng Chem Res 45:9144–9149

Churilov GN (2008) Synthesis of fullerenes and other nanomaterials in arc discharge. Fullerenes, Nanotubes, Carbon Nanostruct 16:395–403

Damalas CA, Eleftherohorinos IG (2011) Pesticide exposure safety issues, and risk assessment indicators. Int J Environ Res Public Health 8:1402–1419

Das R, Abd Hamid SB, Ali ME, Ismail AF, Annuar MSM, Ramakrishna S (2014a) Multifunctional carbon nanotubes in water treatment: the present, past and future. Desalination 354:160–179

Das R, Ali ME, Hamid SBA, Ramakrishna S, Chowdhury ZZ (2014b) Carbon nanotube membranes for water purification: a bright future in water desalination. Desalination 336:97–109

De La Torre-Roche R, Hawthorne J, Deng Y, Xing B, Cai W, Newman LA, Wang C, Ma X, White JC (2012) Fullerene-enhanced accumulation of p,p'-DDE in agricultural crop species. Environ Sci Technol 46(17):9315–9323

De Oliveira R, Hudari F, Franco J, Zanoni MVB (2015) Carbon nanotube-based electrochemical sensor for the determination of anthraquinone hair dyes in wastewaters. Chemosensors 3:22–35

Dichiara AB, Webber MR, Gorman WR, Rogers RE (2015) Removal of copper ions from aqueous solutions via adsorption on carbon nanocomposites. ACS Appl Mater Interface 7:15674–15680

El-Sheikh AH, Sweileh JA, Al-Degs YS, Insisi AA, Al-Rabady N (2008) Critical evaluation and comparison of enrichment efficiency of multi-walled carbon nanotubes, C18 silica and activated carbon towards some pesticides from environmental waters. Talanta 74:1675–1680

Eun AJC, Wong SM (2000) Molecular beacons: a new approach to plant virus detection. Phytopathology 90(3):269–275

European Commission (2011) European commission recommendations (2011) on the definition of nanomaterial. Off J Eur Union 54:38–40

Fan LL, Wang YH, Shao XW, Geng YQ, Wang ZC, Ma Y, Liu J (2012) Effects of combined nitrogen fertilizer and nano-carbon application on yield and nitrogen use of rice grown on saline alkali soil. J Food Agric Environ 10:558–562

Fang Y, Ramasamy RP (2015) Current and prospective methods for plant disease detection. Biosensors 5(3):537–561

Fathi Z, Nejad R-AK, Mahmoodzadeh H, Satari TS (2017) Investigating of a wide range of concentrations of multi-walled carbon nanotubes on germination and growth of castor seeds (*Ricinus communis* L.). J Plant Prot Res 57:228–236

Fernández-Baldo MA, Messina GA, Sanz MI, Raba J (2009) Screen-printed immunosensor modified with carbon nanotubes in a continuous-flow system for the *Botrytis cinerea* determination in apple tissues. Talanta 79(3):681–686

Fernández-Baldo MA, Messina GA, Sanz MI, Raba J (2010) Microfluidic immunosensor with micromagnetic beads coupled to carbon-based screen-printed electrodes (SPCEs) for determination of *Botrytis cinerea* in tissue of fruits. J Agric Food Chem 58(21):1201–11206

Flores D, Chacón R, Alvarado L, Schmidt A, Alvarado C, Chaves J (2014) Effect of using two different types of carbon nanotubes for blackberry (*Rubus adenotrichos*) in vitro plant rooting, growth and histology. Am J Plant Sci 5:3510–3518

Fosso-Kankeu E, De Klerk CM, Botha TA, Waanders F, Phoku J, Pandey S (2016) The antifungal activities of multi-walled carbon nanotubes decorated with silver, copper and zinc oxide particles. In: International conference on advances in science, engineering, technology and natural resources (ICASETNR-16), Parys, South Africa, 24–25 November 2016, pp 55–59

Ghodake G, Seo YD, Park D, Lee DS (2010) Phytotoxicity of carbon nanotubes assessed by *Brassica juncea* and *Phaseolus mungo*. J Nanoelectron Optoelectron 5(2):157–160
Gholipour Y, Erra-Balsells R, Nonami H (2012) Integrative analysis of physiological phenotype of plant cells by turgor measurement and metabolomics. Eng Lett 20(4):EL01
Ghosh M, Bhadra S, Adegoke A, Bandyopadhyay M, Mukherjee A (2015) MWCNT uptake in Allium cepa root cells induces cytotoxic and genotoxic responses and results in DNA hyper methylation. Mutat Res 774:49–58
Giraldo JP, Landry MP, Kwak S, Jain RM, Wong MH, Iverson NM, Ben-Naim M, Strano MS (2015) A ratiometric sensor using single chirality near infrared fluorescent carbon nanotubes: application to in vivo monitoring. Small 32:3973–3984
Gogos A, Knauer K, Bucheli TD (2012) Nanomaterials in plant protection and fertilization: current state, foreseen applications, and research priorities. J Agric Food Chem 60:9781–9792
Gopalakrishnan Nair PM (2018) Toxicological impact of carbon nanomaterials on plants. In: Gothandam K, Ranjan S, Dasgupta N, Ramalingam C, Lichtfouse E (eds) Nanotechnology, food security and water treatment. Environmental chemistry for a sustainable world. Springer, Cham
Gorczyca A, Kasprowicz MJ, Lemek T (2009) Physiological effect of multi–walled carbon nanotubes MWCNTs on conidia of the entomopathogenic fungus, *Paecilomyces fumosoroseus* Deuteromycotina, Hyphomycetes. J Environ Sci Health A 44(14):1592–1597
Gore JP, Sane A (2011) Flame synthesis of carbon nanotubes. INTECH Open Access Publisher, Rijeka
Govindhan M, Lafleur T, Adhikari BR, Chen A (2015) Electrochemical sensor based on carbon nanotubes for the simultaneous detection of phenolic pollutants. Electroanalysis 27(4):902–909
Gu J, Xiao P, Zhang L, Lu W, Zhang G, Huang Y, Zhang J, Chen T (2016) Construction of superhydrophilic and under-water superoleophobic carbon-based membranes for water purification. RSC Adv 6:73399–73403
Gurunathan S (2015) Cytotoxicity of graphene oxide nanoparticles on plant growth promoting rhizobacteria. J Ind Eng Chem 32:282–291
Haghighi M, da Silva TJA (2014) Effect of N-TiO_2 on tomato, onion and radish seed germination. J Crop Sci Biotechnol 17:221–227
Hajihosseini S, Nasirizadeh N, Hejazi MS, Yaghmaei P (2016) A sensitive DNA biosensor fabricated from gold nanoparticles and graphene oxide on a glassy carbon electrode. Mater Sci Eng C 61:506–515
Hasaneen MNA, Abdel-Aziz HMM, Omer AM (2017) Characterization of carbon nanotubes loaded with nitrogen, phosphorus and potassium fertilizers. Am J Nano Res Appl 5(2):12–18
Hernandez-Fernandez P, Montiel M, Ocón P, de la Fuente JLG, Garcia-Rodriguez S, Rojas S, Fierro JL (2010) Functionalization of multi-walled carbon nanotubes and application as supports for electrocatalysts in proton exchange membrane fuel cell. Appl Catal B 99:343–352
Herrero M, Simó C, García-Cañas V, Ibáñez E, Cifuentes A (2012) Foodomics: MS-based strategies in modern food science and nutrition. Mass Spectrom Rev 31:49–69
Hilding J, Grulke EA, Sinnot SB, Qian D, Andrews R, Jagtoyen M (2001) Sorption of butane on carbon multiwall nanotubes at room temperature. Langmuir 17:7540–7544
Hirsch A, Vostrowsky O (2005) Functionalization of carbon nanotubes. Top Curr Chem 245:193–237
Hu X, Zhou Q (2014) Novel hydrated graphene ribbon unexpectedly promotes aged seed germination and root differentiation. Sci Rep 4:3782
Hunter RJ (2001) Foundation of colloid science, 2nd edn. Oxford University Press, Oxford; New York
Husen A, Siddiqi KS (2014) Carbon and fullerene nanomaterials in plants system. J Nanobiotechnology 12:1–10
Iijima S (1991) Helical microtubules of graphitic carbon. Nature 354:56
Ikhtiar R, Begum P, Watari F, Fugetsu B (2013) Toxic effect of multiwalled carbon nanotubes on lettuce (*Lactuca sativa*). Nano Biomed 5:18–24
Ilkhani H, Hughes T, Li J, Zhong CJ, Hepel M (2016) Nanostructured SERS electrochemical biosensors for testing of anticancer drug interactions with DNA. Biosens Bioelectron 80:257–264

Ivnitski D, Abdel-Hamid I, Atanasov P, Wilkins E, Stricker S (2000) Application of electrochemical biosensors for detection of food pathogenic bacteria. Electroanalysis 12(5):317–325

Jiao L, Zhang L, Wang X, Diankov G, Dai H (2009) Narrow graphene nanoribbons from carbon nanotubes. Nature 458:877–880

Jin L, Son Y, DeForest JL, Kang YJ, Kim W, Chung H (2014) Single-walled carbon nanotubes alter soil microbial community composition. Sci Total Environ 446:533–538

Johansen A, Pedersen AL, Jensen KA, Karlson U et al (2008) Effects of C60 fullerene nanoparticles on soil bacteria and protozoans. Environ Toxicol Chem 27:1895–1903

Joshi N, Jain N, Pathak A, Singh J, Prasad R, Upadhyaya CP (2018) Biosynthesis of silver nanoparticles using Carissa carandas berries and its potential antibacterial activities. J Sol-Gel Sci Techn. https://doi.org/10.1007/s10971-018-4666-2

Jung JH, Hwang GB, Lee JE, Bae GN (2011) Preparation of airborne Ag/CNT hybrid nanoparticles using an aerosol process and their application to antimicrobial air filtration. Langmuir 27:10256–10264

Kaphle A, Navya PN, Umapathi A, Chopra M, Daima HK (2017) Nanomaterial impact, toxicity and regulation in agriculture, food and environment. In: Ranjan S et al (eds) Nanoscience in food and agriculture 5. Sustainable agriculture reviews, vol 26. Springer International Publishing AG, Cham, pp 205–242. https://doi.org/10.1007/978-3-319-58496-6_8

Kerfahi D, Tripathi BM, Singh D, Kim H, Lee S, Lee J, Adams JM (2015) Effects of functionalized and raw multi-walled carbon nanotubes on soil bacterial community composition. PLoS One 10(3):e0123042

Khodakovskaya M, Dervishi E, Mahmood M, Xu Y, Li Z, Watanabe F, Biris AS (2009) Carbon nanotubes are able to penetrate plant seed coat and dramatically affect seed germination and plant growth. Am Chem Soc 3(10):3221–3227

Khodakovskaya MV, de Silva K, Biris AS, Dervishi E, Villagarcia H (2012) Carbon nanotubes induce growth enhancement of tobacco cells. Am Chem Soc Nano 6(3):2128–2135

Khodakovskaya MV, Kim BS, Kim JN, Alimohammadi M, Dervishi E, Mustafa T, Cernigla CE (2013) Carbon nanotubes as plant growth regulators: effects on tomato growth, reproductive system, and soil microbial community. Small 9:115–123

Kole C, Kole P, Randunu KM, Choudhary P, Podila R, Ke PC, Rao AM, Marcus RK (2013) Nanobiotechnology can boost crop production and quality: first evidence from increased plant biomass, fruit yield and phytomedicine content in bitter melon (Momordica charantia). BMC Biotechnol 13(13):37

Kratschmer W (2011) The story of making fullerenes. Nanoscale 3:2485–2489

Kratschmer W, Lamb LD, Fostiropoulos K, Huffman DR (1990) Solid C60: a new form of carbon. Nature 347:354–358

Kroto HW, Heath JR, O'Brien SC, Curl RF, Smalley RE (1985) C60: buckminsterfullerene. Nature 318:162–163

Kumar M, Ando Y (2010) Chemical vapor deposition of carbon nanotubes: a review on growth mechanism and mass production. J Nanosci Nanotechnol 10:3739–3758

Lahiani MH, Chen J, Irin F, Puretzky AA, Green MJ, Khodakovskaya MV (2015) Interaction of carbon nanohorns with plants: uptake and biological effects. Carbon 81:607–619

Li H, Guan Y (2011) Foliar fertilizer containing carbon nanoparticles for plants under stress conditions. CN 102030595

Li Y, Wang S, Wei J, Zhang X, Xu C, Luan Z, Wu D, Wei B (2002) Lead adsorption on carbon nanotubes. Chem Phys Lett 357:263–266

Li YH, Ding J, Luan Z, Di Z, Zhu Y, Xu C, Wu D, Wei B (2003a) Competitive adsorption of Pb^{2+}, Cu^{2+} and Cd^{2+} ions from aqueous solutions by multiwalled carbon nanotubes. Carbon 41:2787–2792

Li YH, Wang S, Luan Z, Ding J, Xu C, Wu D (2003b) Adsorption of cadmium(II) from aqueous solution by surface oxidized carbon nanotubes. Carbon 41:1057–1062

Lin D, Xing B (2007) Phytotoxicity of nanoparticles: inhibition of seed germination and root growth. Environ Pollut 150(2):243–250

Lin S, Reppert J, Hu Q, Hudson JS, Reid ML, Ratnikova TA, Rao AM, Luo H, Ke PC (2009) Uptake, translocation, and transmission of carbon nanomaterials in rice plants. Small 5:1128

Liu Y, Wangquan T (2012) Special fertilizer for rapeseed base fertilizer. CN 102816021

Liu F, Wen LX, Li ZZ, Yu W, Sun HY, Chen JF (2006a) Porous hollow silica nanoparticles as controlled delivery system for water soluble pesticide. Mater Res Bull 41:2268–2275

Liu X, Feng Z, Zhang S, Zhang J, Xiao Q, Wang Y (2006b) Preparation and testing of cementing nano-subnano composites of slow- or controlled release of fertilizers. Sci Agric Sin 39:1598–1604

Liu Q, Chen B, Wang Q, Shi X, Xiao Z, Lin J, Fang X (2009) Carbon nanotubes as molecular transporters for walled plant cells. Nano Lett 9:1007–1010

Liu Q, Zhang X, Zhao Y, Lin J, Shu C, Wang C, Fang X (2013a) Fullerene-induced increase of glycosyl residue on living plant cell wall. Environ Sci Technol 47:7490–7498

Liu X, Wang M, Zhang S, Pan B (2013b) Application potential of carbon nanotubes in water treatment: a review. J Environ Sci (China) 25:1263–1280

Liu J, Li X, Jia W, Ding M, Zhang Y, Ren S (2016) Separation of emulsified oil from oily wastewater by functionalized multiwalled carbon nanotubes. J Dispers Sci Technol 37:1294–1302

Lu C, Chiu H (2006) Adsorption of zinc(II) from water with purified carbon nanotubes. Chem Eng Sci 61:1138–1145

Ma X, Geiser-Lee J, Deng Y, Kolmakov A (2010) Interactions between engineered nanoparticles (ENPs) and plants: phytotoxicity, uptake and accumulation. Sci Total Environ 408(16):3053–3061

Mani S, Cheemalapati S, Chen SM, Devadas B (2015) Anti-tuberculosis drug pyrazinamide determination at multiwalled carbon nanotubes/graphene oxide hybrid composite fabricated electrode. Int J Electrochem Sci 10:7049–7062

Matsuzawa Y, Takada Y, Kodaira T, Kihara H, Kataura H, Yoshida M (2014) Effective nondestructive purification of single-walled carbon nanotubes based on high-speed centrifugation with a photochemically removable dispersant. J Phys Chem C 118:5013–5019

Mauter M, Elimelech M (2008) Environmental applications of carbon-based nanomaterials. Environ Sci Technol 42:5843–5859

Mercan H, Inam R, Aboul-Enein HY (2011) Square wave adsorptive stripping voltammetric determination of cyromazine insecticide with multi-walled carbon nanotube paste electrode. Anal Lett 44:1392–1404

Milne WI, Teo KBK, Amaratunga GAJ, Legagneux P, Ganglof L, Schnell JP, Semet V, Binh VT, Groening O (2004) Carbon nanotubes as field emission sources. J Mater Chem 14:933–943

Miralles P, Johnson E, Church TL, Harris AT (2012) Multiwalled carbon nanotubes in alfalfa and wheat: toxicology and uptake. J R Soc Interface 9:3514–3527

Mishra A, Clark JH (2013) Green materials for sustainable water remediation and treatment. Royal Society of Chemistry, Cambridge

Mondal A, Basu R, Das S, Nandy P (2011) Beneficial role of carbon nanotubes on mustard plant growth: an agricultural prospect. J Nanopart Res 13(10):4519

Morla S, Ramachandra Rao CSV, Chakrapani R (2011) Factors affecting seed germination and seedling growth of tomato plants cultured in vitro conditions. J Chem Biol Phys Sci B 1:328

Morsy M, Helal M, El-Okr M, Ibrahim M (2014) Preparation, purification and characterization of high purity multi-wall carbon nanotube. Spectrochim Acta A Mol Biomol Spectrosc 132:594–598

Nair R, Varghese SH, Nair BG, Maekawa T, Yoshida Y, Kumar DS (2010) Nanoparticulate material delivery to plants. Plant Sci 179:154

Nair R, Mohamed SM, Gao W, Maekawa T, Yoshida Y, Ajayan PM, Kumar DS (2012) Effect of carbon nanomaterials on the germination and growth of rice plants. J Nanosci Nanotechnol 12:2212–2220

Nalwade AR, Neharkar SB (2013) Carbon nanotubes enhance the growth and yield of hybrid Bt cotton Var. ACH-177-2. Int J Adv Sci Technol Res 3:840

Namasivayam M, Shapter J (2017) Factors affecting carbon nanotube fillers towards enhancement of thermal conductivity in polymer nanocomposites: a review. J Compos Mater 51:3657–3668

Novoselov KS, Geim AK, Morozov SV, Jiang D, Zhang Y, Dubonos SV, Grigorieva IV, Firsov AA (2004) Electric field effect in atomically thin carbon films. Science 306:666–669
Novoselov KS, Fal'Ko VI, Colombo L, Gellert PR, Schwab MG, Kim K (2012) A roadmap for graphene. Nature 490:192–200
Oleszczuk P, Josko I, Xing BS (2011) The toxicity to plants of the sewage sludges containing multiwalled carbon nanotubes. J Hazard Mater 186:436–442
Oyelami AO, Semple KT (2015) Impact of carbon nanomaterials on microbial activity in soil. Soil Biol Biochem 86:172–180
Patel N, Desai P, Patel N, Jha A, Gautam HK (2014) Agronanotechnology for plant fungal disease management: a review. Int J Curr Microbiol Appl Sci 3(10):71–84
Peng X, Li Y, Luan Z, Di Z, Wang H, Tian B, Jia Z (2003) Adsorption of 1,2-dichlorobenzene from water to carbon nanotubes. Chem Phys Lett 376:154–158
Pereira A, Grillo R, Mello NF, Rosa AH, Fraceto LF (2014) Application of poly(epsilon-caprolactone) nanoparticles containing atrazine herbicide as an alternative technique to control weeds and reduce damage to the environment. J Hazard Mater 268:207–215
Pourkhaloee A, Haghighi M, Saharkhiz MJ, Jouzi H, Doroodmand MM (2011) Carbon nanotubes can promote seed germination via seed coat penetration. J Seed Technol 33(2):155–169
Prasad R, Kumar V, Prasad KS (2014) Nanotechnology in sustainable agriculture: present concerns and future aspects. Afr J Biotechnol 13(6):705–713
Prasad R, Bhattacharyya A, Nguyen QD (2017a) Nanotechnology in sustainable agriculture: recent developments, challenges, and perspectives. Front Microbiol 8:1014. https://doi.org/10.3389/fmicb.2017.01014
Prasad R, Gupta N, Kumar M, Kumar V, Wang S, Abd-Elsalam KA (2017b) Nanomaterials act as plant defense mechanism. In: Prasad R, Kumar V, Kumar M (eds) Nanotechnology. Springer, Singapore, pp 253–269
Pyrzyńska K, Bystrzejewski M (2010) Comparative study of heavy metal ions sorption onto activated carbon, carbon nanotubes, and carbon-encapsulated magnetic nanoparticles. Colloids Surf A 362:102–109
Rasool K, Lee DS (2015) Influence of multi-walled carbon nanotubes on anaerobic biological sulfate reduction processes. J Nanoelectron Optoelectron 10:485–489
Ribeiro WF, Selva TMG, Lopes IC, Coelho ECS, Lemos SG, de Abreu FC, do Nascimento VB, de Araújo MCU (2011) Electroanalytical determination of carbendazim by square wave adsorptive stripping voltammetry with a multiwalled carbon nanotubes modified electrode. Anal Methods 3:1202–1206
Saito R, Dresselhaus G, Dresselhaus MS (1998) Physical properties of carbon nanotubes. Imperial College Press, London
Sangeetha J, Thangadurai D, Hospet R, Purushotham P, Manowade KR, Mujeeb MA, Mundaragi AC, Jogaiah S, David M, Thimmappa SC, Prasad R, Harish ER (2017a) Production of bionanomaterials from agricultural wastes. In: Prasad R, Kumar M, Kumar V (eds) Nanotechnology. Springer Nature Singapore Pte Ltd, Singapore, pp 33–58
Sangeetha J, Thangadurai D, Hospet R, Harish ER, Purushotham P, Mujeeb MA, Shrinivas J, David M, Mundaragi AC, Thimmappa AC, Arakera SB, Prasad R (2017b) Nanoagrotechnology for soil quality, crop performance and environmental management. In: Prasad R, Kumar M, Kumar V (eds) Nanotechnology. Springer Nature Singapore Pte Ltd, Singapore, pp 73–97
Sarlak N, Taherifar A, Salehi F (2014) Synthesis of nanopesticides by encapsulating pesticide nanoparticles using functionalized carbon nanotubes and application of new nanocomposite for plant disease treatment. J Agric Food Chem 62:4833–4838
Sarno M, Tamburrano A, Arurault L, Fontorbes S, Pantani R, Datas L, Ciambelli P, Sarto MS (2013) Electrical conductivity of carbon nanotubes grown inside a mesoporous anodic aluminium oxide membrane. Carbon 55:10–22
Saurabh S, Bijendra KS, Yadav SM, Gupta AK (2015) Applications of nanotechnology in agricultural and their role in disease management. J Nanosci Nanotechnol 5:1–5

Schierz A, Zanker H (2009) Aqueous suspensions of carbon nanotubes: surface oxidation, colloidal stability and uranium sorption. Environ Pollut 157:1088–1094
Schmitt H, Creton N, Prashantha K, Soulestin J, Lacrampe MF, Krawczak P (2015) Melt-blended halloysite nanotubes/wheat starch nanocomposites as drug delivery system. Polym Eng Sci 55:573–580
Sekhon BS (2014) Nanotechnology in agri-food production: an overview. Nanotechnol Sci Appl 7:31–53
Serag MF, Kaji N, Gaillard C, Okamoto Y, Terasaka K, Jabasini M, Tokeshi M, Mizukami H, Bianco A, Baba Y (2011) Trafficking and subcellular localization of multiwalled carbon nanotubes in plant cells. ACS Nano 5:493–499
Serag MF, Kaji N, Habuchi S, Bianco A, Baba Y (2013) Nanobiotechnology meets plant cell biology: carbon nanotubes as organelle targeting nanocarriers. RSC Adv 3:4856–4862
Serag MF, Kaji N, Tokeshi M, Baba Y (2015) Carbon nanotubes and modern nanoagriculture. In: Siddiqui M, Al-Whaibi M, Mohammad F (eds) Nanotechnology and plant sciences. Springer, Cham, pp 183–201
Sharon M, Choudhary AK, Kumar R (2010) Nanotechnology in agricultural diseases and food safety. J Phytology 2(4):83–92
Shen CX, Zhang QF, Li J, Bi FC, Yao N (2010) Induction of programmed cell death in Arabidopsis and rice by single-wall carbon nanotubes. Am J Bot 97:1602–1609
Shrestha B, Acosta-Martinez V, Cox SB, Green MJ, Li S, Canas-Carrell JE (2013) An evaluation of the impact of multiwalled carbon nanotubes on soil microbial community structure and functioning. J Hazard Mater 261:188–197
Singh A, Bhati A, Gunture, Tripathi KM, Sonkar SM (2017) Nanocarbons in agricultural plants: can be a potential nanofertilizer? In: Hussain CM, Mishra AK (eds) Nanotechnology in environmental science, vol 2 Volumes. Wiley, Newark, NJ; Weinheim
Smirnova EA, Gusev AA, Zaitseva ON, Lazareva EM, Onishchenko GE, Kuznetsova EV, Tkachev AG, Feofanov AV, Kirpichnikov MP (2011) Multi-walled carbon nanotubes penetrate into plant cells and affect the growth of Onobrychis arenaria seedlings. Acta Nat 3(1):99–106
Smirnova E, Gusev A, Zaytseva O, Sheina O, Tkachev A, Kuznetsova E, Kirpichnikov M (2012) Uptake and accumulation of multiwalled carbon nanotubes change the morphometric and biochemical characteristics of *Onobrychis arenaria* seedlings. Front Chem Sci Eng 6(2):132–138
Smith SC, Rodrigues DF (2015) Carbon-based nanomaterials for removal of chemical and biological contaminants from water: a review of mechanisms and applications. Carbon 91:122–143
Srinivasan C, Saraswathi R (2010) Nano-agriculture-carbon nanotubes enhance tomato seed germination and plant growth. Curr Sci 99:273–275
Srivastava A, Rao DP (2014) Enhancement of seed germination and plant growth of whest, maize, peanut and garlic using multiwalled carbon nanotubes, enhancement of plant growth using multi-walled carbon nanotubes. Eur Chem Bull 3(5):502–504
Srivastava M, Abhilash PC, Singh N (2011) Remediation of lindane using engineered nanoparticles. J Biomed Nanotechnol 7:172–174
Stampoulis D, Sinha SK, White JC (2009) Assay dependent phytotoxicity of nanoparticles to plants. Environ Sci Technol 43:9473–9479
Suvarnaphaet P, Pechprasarn S (2017) Graphene-based materials for biosensors: a review. Sensors 17(10):2161
Taha RA (2016) Nano carbon applications for plant. Adv Plants Agric Res 5(2):00172
Tan XM, Lin C, Fugetsu B (2009) Studies on toxicity of multi-walled carbon nanotubes on suspension rice cells. Carbon 47:3479–3487
Tiwari DK, Dasgupta-Schubert N, Villasenor Cendejas LM, Villegas J, Carreto Montoya L, Borjas García SE (2014) Interfacing carbon nanotubes (CNT) with plants: enhancement of growth, water and ionic nutrient uptake in maize (Zea mays) and implications for nanoagriculture. Appl Nanosci 4(5):577–591
Tong Z, Bischoff M, Nies L, Applegate B, Turco RF (2007) Impact of fullerene (C60) on a soil microbial community. Environ Sci Technol 41:2985–2991
Torney F, Trewyn B, Lin VSY, Wang K (2007) Mesoporous silica nanoparticle deliver DNA and chemicals into plant. Nat Nanotechnol 2:295–300

Torre-Roche RDL, Hawthorne J, Deng Y, Xing B, Cai W, Newman LA, Wang Q, Ma X, Hamdi H, White JC (2013) Multiwalled carbon nanotubes and C60 fullerenes differentially impact the accumulation of weathered pesticides in four agricultural plants. Environ Sci Technol 47:12539–12547
Tripathi S, Sarkar S (2015) Influence of water soluble carbon dots on the growth of wheat plant. Appl Nanosci 5:609–619
Tripathi BP, Shahi VK (2011) Organic–inorganic nanocomposite polymer electrolyte membranes for fuel cell applications. Prog Polym Sci 36(7):945–979
Upadhyayula VKK, Deng S, Mitchell MC, Smith GB (2009) Application of carbon nanotube technology for removal of contaminants in drinking water: a review. Sci Total Environ 408:1–13
Villagarcia H, Dervishi E, de Silva K, Biris AS, Khodakovskaya MV (2012) Surface chemistry of carbon nanotubes impacts the growth and expression of water channel protein in tomato plants. Small 8(15):2328–2334
Wang YY, Hsu PK, Tsay YF (2012) Uptake, allocation and signaling of nitrate. Trends Plant Sci 17(8):458–467
Wang L, Fortner JD, Hou L, Zhang C, Kan AT, Tomson MB, Chen W (2013) Contaminant-mobilizing capability of fullerene nanoparticles (nC60): effect of solvent-exchange process in nC60 formation. Environ Toxicol Chem 32:329–336
Wang H, Ma H, Zheng W, An D, Na C (2014a) Multifunctional and recollectable carbon nanotube ponytails for water purification. ACS Appl Mater Interfaces 6:9426–9434
Wang T, Zhao D, Guo X, Correa J, Riehl BL, Heineman WR (2014b) Carbon nanotube-loaded nafion film electrochemical sensor for metal ions: europium. Anal Chem 86:4354–4361
Wang X, Liu X, Chen J, Han H, Yuan Z (2014c) Evaluation and mechanism of antifungal effects of carbon nanomaterials in controlling plant fungal pathogen. Carbon 68:798–806
Wang C, Zhang H, Ruan L, Chen L, Li H, Chang XL, Zhang X, Yang ST (2016) Bioaccumulation of 13C-fullerenol nanomaterials in wheat. Environ Sci Nano 3:799–805
Wang X, Zhou Z, Chen F (2017) Surface modification of carbon nanotubes with an enhanced antifungal activity for the control of plant fungal pathogen. Materials 10:1375. https://doi.org/10.3390/ma10121375
Wild E, Jones KC (2009) Novel method for the direct visualization of in vivo nanomaterials and chemical interactions in plants. Environ Sci Technol 43:5290–5294
Wong A, Silva TA, Caetano FR, Bergamini MF, Marcolino-Junior LH, Fatibello-Filho O, Janegitz BC (2017) An overview of pesticide monitoring at environmental samples using carbon nanotubes-based electrochemical sensors. J Carbon Res 3:8. https://doi.org/10.3390/c3010008
Wu M (2013) Effects of incorporation of nano-carbon into slow-released fertilizer on rice yield and nitrogen loss in surface water of paddy soil. Adv J Food Sci Technol 5:398–403 https://doi.org/10.1109/isdea.2012.161
Xie J, Liu J (2012) Nano-carbon synergism compound fertilizer for tobacco and preparation method thereof. CN 102718584
Yadav BC, Kumar R (2008) Structure, properties and applications of fullerenes. Int J Nanotechnol Appl 2:15–24
Yan H, Gong A, He H, Zhou J, Wei Y, Lv L (2006) Adsorption of microcystins by carbon nanotubes. Chemosphere 62:142–148
Yan S, Zhao L, Li H, Zhang Q, Tan J, Huang M, He S, Li L (2013) Single-walled carbon nanotubes selectively influence maize root tissue development accompanied by the change in the related gene expression. J Hazard Mater 246:110–118
Yaqub S, Latif U, Dickert FL (2011) Plastic antibodies as chemical sensor material for atrazine detection. Sensors Actuators B 160:227–233
Yatim NM, Azizah S, Fairuz DM, Faridah Y (2015) Statistical evaluation of the production of urea fertilizer-multiwalled carbon nanotubes using Plackett Burman experimental design. Procedia Soc Behav Sci 195:315–323
Yoo J, Ozawa H, Fujigaya T, Nakashima N (2011) Evaluation of affinity of molecules for carbon nanotubes. Nanoscale 3:2517–2522
Zarei F, Negahdari B, Eatemadi A (2018) Diabetic ulcer regeneration: stem cells, biomaterials, growth factors. Artif Cells Nanomed Biotechnol 46(1):26–32

Zaytseva O, Neumann G (2016) Carbon nanomaterials: production, impact on plant development, agricultural and environmental applications. Chem Biol Technol Agric 3:17. https://doi.org/10.1186/s40538-016-0070-8

Zhang Z, Chen J (2012) Method for preparation of compound organic fertilizer containing nanocarbon and sulfate radical organic fertilizer. CN 102816003

Zhang Z, Liu J (2010) Synergistic fertilizer containing nanometer carbon and rare earth and its preparation. CN 101633590

Zhang Y, Kang TF, Wan YW, Chen SY (2009) Gold nanoparticles-carbon nanotubes modified sensor for electrochemical determination of organophosphate pesticides. Microchim Acta 165:307–311

Zhang Q, Huang J, Zhao M, Qian W, Wei F (2011) Carbon nanotube mass production: principles and processes. ChemSusChem 4:864–889

Zhang R, Zhang Y, Zhang Q, Xie H, Qian W, Wei F (2013) Growth of half-meter long carbon nanotubes based on Schulz-Flory distribution. ACS Nano 7:6156–6161

Zhang M, Gao B, Chen J, Li Y, Creamer AE, Chen H (2014) Slow-release fertilizer encapsulated by graphene oxide films. Chem Eng J 255:107–113

Zhang L, Gu J, Song L, Chen L, Huang Y, Zhang J, Chen T (2016) Underwater superoleophobic carbon nanotubes/core–shell polystyrene@Au nanoparticles composite membrane for flow-through catalytic decomposition and oil/water separation. J Mater Chem A 4:10810–10815

Zhao S, Wang Q, Zhao Y, Rui Q, Wang D (2015) Toxicity and translocation of graphene oxide in Arabidopsis thaliana. Environ Toxicol Pharmacol 39:145

Zheng X, Su Y, Chen Y, Wei Y, Li M, Huang H (2014) The effects of carbon nanotubes on nitrogen and phosphorus removal from real wastewater in the activated sludge system. RSC Adv 4:45953–45959

Zulkifli H, Salam F, Saad SM, Rahman RA, Rani RM, Karim MSA, Ishak Z (2016) Preliminary study of electrochemical DNA sensor for cucumber mosaic virus. Procedia Chem 20:98–101

Chapter 8
Benefits and Potential Risks of Nanotechnology Applications in Crop Protection

Josef Jampílek and Katarína Kráľová

8.1 Introduction

Due to the increasing number of human population and changing climatic conditions, it is increasingly difficult to provide sufficient food for the population. With global hunger on the rise again, the Food and Agricultural Organization of the United Nations (FAO) has issued a sobering forecast on world food production. FAO says that if global population reaches 9.1 billion by 2050, the world food production will need to rise by 70%, and food production in the developing world will need to double. The FAO's forecast does not take into account any increase in agricultural production for biofuels. The projected 70% increase in food production will have to overcome rising energy prices, growing depletion of underground aquifers, the continuing loss of farmland to urbanization and increased drought and flooding resulting from climate changes (Population Institute 2017; FAO 2009). From the report of FAO, it results that crop yields would continue to grow but at a slower rate than in the past. Therefore, one of possible strategies is better protection of crops, although crop protection from pests and diseases can only reduce the amount lost after the potential for increased food production has been attained by proper utilization of all means possible. According to the data of FAO, every year the damage done to crops by pests and diseases constitutes ca. 20% of the potential world yield of food crops (FAO 2009). Crop protection becomes even more important in intensive agriculture, where increased fertilization, genetically uniform, high-yielding varieties, increased irrigation, and other methods are used. Crop losses due to

J. Jampílek (✉)
Department of Pharmaceutical Chemistry, Faculty of Pharmacy, Comenius University, Bratislava, Slovakia

K. Kráľová
Institute of Chemistry, Faculty of Natural Sciences, Comenius University, Bratislava, Slovakia

K. A. Abd-Elsalam, R. Prasad (eds.), *Nanobiotechnology Applications in Plant Protection*, Nanotechnology in the Life Sciences,
https://doi.org/10.1007/978-3-319-91161-8_8

diseases and pests not only affect national and world food supplies and economies but also affect individual farmers even more, whether they grow the crop for direct consumption or for sale. Because operating expenditures for the production of the crop remain the same in years of low or high disease incidence, harvests are lost due to diseases and pests lower the net return directly (Agrios 2005).

Crop protection can be defined as "the science and practice of managing invertebrate pests and vertebrate pests, plant diseases, weeds and other pest organisms that damage agricultural crops and forestry. Agricultural crops include field crops, vegetable crops and fruit and horticultural crops. Crop protection encompasses (*i*) pesticide-based approaches such as herbicides, fungicides and insecticides; (*ii*) biological pest control approaches such as cover crops, trap crops and beetle banks; (*iii*) barrier-based approaches such as agrotextiles and bird netting; (*iv*) animal psychology-based approaches such as bird scarers; and (*v*) biotechnology-based approaches such as plant breeding and genetic modification" (Crop Protection Definitions 2017). It is estimated that the discovery and development of a new agent costs about 150–200 million USD. A new product must be tested thoroughly for its action and its safety for the environment. It takes an average of 10–15 years to do this, so it is small wonder that worldwide, only about 12 agrochemicals are introduced each year. However, these chemicals are crucial for the efficient production of food (Essential Chemical Industry 2017).

As nanotechnology is one of the key technologies of the twenty-first century (Wennersten et al. 2008) that is able to provide "a new dimension", new properties to many current materials (Borm et al. 2006; Buzea et al. 2007; Jampílek et al. 2013, 2014, 2015; Vaculíková et al. 2016a, b; Jampílek and Kráľová 2017a, 2018a), it has been also widely used in food industry and for production of a new generation of agrochemicals (Chaudhry and Castle 2011; Rashidi and Khosravi-Darani 2011; Khot et al. 2012; Sekhon 2014; Parisi et al. 2015; Jampílek and Kráľová 2015, 2017b, c; Nuruzzaman et al. 2016; Fraceto et al. 2016). Thus, the use of nanotechnologies can significantly contribute to sustainable intensification of agricultural production (Garcia et al. 2010; Pérez-de-Luque and Hermosín 2013; Prasad et al. 2014, 2017; Sekhon 2014; Jampílek and Kráľová 2015), and vice versa, the agricultural production and food industry belong to important areas of nanotechnology application (Ghormade et al. 2011; Coles and Frewer 2013; Raliya et al. 2013; Chen et al. 2014a; Mukhopadhyay 2014).

Based on the definitions of the European Commission and/or US National Nanotechnology Initiative, nanomaterials/nanoparticles (NPs) can be generally classified as materials with a particle size less than 100 nm in at least one dimension (European Commission 2011; National Nanotechnology Initiative 2008). Pesticide nanosystems formulated into this particle size (similarly as nanoformulations of drugs) acquire enhanced bioavailability, targeted delivery, controlled release, protection against degradation and higher potency, and when currently applied and approved pesticides are used, a rapid and economically favourable solution is provided. Nanoformulations of pesticides can be classified either according to the nature of the nanocarrier, organic polymer-based formulations, lipid-based formulations, nanosized metals/metal oxides, metalloids, clay-based nanomaterials, etc. or according to various structures and morphologies of the nanosystem: nanocapsules, nanospheres, nanomicelles, nanogels, nanoemulsions, nanofibers, nanoliposomes, solid lipid nanoparticles, etc. (Balaure et al. 2017; Jampílek and Kráľová 2017a, b, 2018a). Alone pesticides, i.e. herbicides, fungicides and insecticides, can be divided into natural or synthetic and of inorganic or organic nature. In some cases, also a stabilizer/matrix of the nanosystem shows effectivity against phytopathogens, and thus it can be used alone or with a pesticide and amplify its potency.

In this chapter, advantageous effects of nanomaterials/nanoformulations of various herbicides, fungicides, bactericides and insecticides on weed and phytopathogens are discussed in detail, and special attention is devoted also to risks of applications of nanopesticides.

8.2 Nanoherbicides

Modern agriculture and land management uses chemical agents, i.e. herbicides for the control of unwanted vegetation. Although many compounds used for the control of unwanted vegetation were designed and applied, currently new agrochemicals more effective for specific weeds resulting in less damage of desirable vegetation, i.e. safer to human and the environment, are desirable. Since these herbicides of new generations should be affordable, attention is focused not only on nanoformulations of used current herbicides but also on nanosystems containing metals effective against weeds and their combinations. In addition, some polymers used in nanoformulations as excipients were found to potentiate effectivity and selectivity of herbicidal-effective organic compounds.

8.2.1 Synthetic Nanoherbicides

8.2.1.1 Nanoscale Phenoxyacetic Acid Herbicides

Nanohybrids of 2-chloro- (2-CPA) and 2,4,5-trichlorophenoxyacetic acids (2,4,5-T) prepared by hybridization of phenoxyacetic acid herbicides into zinc-aluminium-layered double hydroxide (Zn-Al-LDH) interlamellae, in which the successful intercalation of the herbicides into the layered double hydroxide inorganic interlayers was confirmed by basal spacing expansion from 8.9 Å in the layered double hydroxide to 18.5 and 26.2 Å, respectively, were reported by Sarijo et al. (2010a). The release process was found to be pH-dependent in the order of pH $12 > 3 > 6.25$, and longer release time estimated for 2,4,5-T compared to 2-CPA indicated stronger interaction of 2,4,5-T with the layered double hydroxide inorganic interlayer. The obtained results suggested that two-dimensional-type layered structure consisting of thin crystalline inorganic layers with a thickness of a few nanometers such as Zn-Al-LDH represents a suitable matrix for the controlled release formulation of agrochemicals based on halogen-substituted phenoxyacetic acid such as 2-CPA and 2,4,5-T. In another study, Sarijo et al. (2010b) investigated the release of chlorophenoxy herbicides, namely, 2-CPA, 4-chlorophenoxyacetic acid (4-CPA) and 2,4,5-T, from their nanohybrids into various aqueous solutions, carbonate, sulphate and chloride, whereby the release was found to be controlled by pseudo-second-order rate expression. The calculated $t_{1/2}$ values for 2-CPA were 71, 77 and 103 min in carbonate, sulphate and chloride aqueous solutions. The $t_{1/2}$ values of 79, 97 and 146 min and 210, 282 and 442 min were estimated for 4-CPA and 2,4,5-T for carbonate, sulphate and chloride, respectively, indicating that the percentage of the saturated amount of 4-CPA and 2,4,5,T released decreased in the following order: carbonate > sulphate > chloride. Thus, the release of phenoxyacetic acid herbicides into the media is preferred if the available anion in the media has higher affinity towards the Zn-Al-LDH inorganic interlayers, and, therefore, the exchangeable anions, either they are in the release media or in the nanohybrid, can be exploited as a means to tune the release properties. On the other hand, for all the media the percentage of saturated release decreased in the following order: 2-CPA > 4-CPA > 2,4,5-T. This can be connected with the fact that electrostatic forces with the host in the molecule of 2,4,5-T having three chlorine atoms attached to the benzene ring are stronger than in 2-CPA, which results in more difficult release of 2,4,5-T compared to 2-CPA; the release from the interlayer of the inorganic host could be also affected by the bulkier structure of 2,4,5-T. Easier release of 2-CPA compared to 4-CPA is connected with the fact that in 4-CPA, the chloride substituent in position 4 becomes more negatively charged and therefore held stronger in the interlayer. The inorganic Zn-Al-LDH was also used as a matrix for 2,4-dichlorophenoxyacetic acid (2,4-D) by Hussein et al. (2005) who found that the release rate of the 2,4-D anion from the interlamellae of the nanocomposite depended on the type of anion and its concentration in the release media, the release from the carbonate solution being more effective than from chloride solution or distilled water. Initially, the release of the guest 2,4-D into aqueous solutions containing chloride, carbonate and distilled water was rapid, followed by a more sustained release thereafter, and this behaviour was dependent on the type of anions and their concentrations in the release medium (aqueous solution). While in distilled water and NaCl aqueous solutions the layered structure of the nanohybrid was not destroyed by the release of 2,4-D anions for at least 24 h, in the presence of carbonate in aqueous solution, the release of 2,4-D ions from the nanohybrid resulted in the formation of two new phases, LDH and ZnO. This indicates higher affinity of carbonate towards the LDH inorganic interlamellae compared to chloride. However, independently on the structure of the resulting controlled release formulation, the release of 2,4-D anions from Zn-Al-LDH inorganic lamella was controlled by the first-order kinetic at least at the beginning of the deintercalation up to 12 h.

Using montmorillonite (MMT)-gelatin composites, Alromeed et al. (2015) prepared slow-release formulations of the (4-chloro-2-methylphenoxy)acetic acid (MCPA) herbicide, which could reduce the environmental risk associated with herbicide application by more effective reduction of leaching and improved bioactivity in the upper soil layer compared with a commercial product. MCPA was released much more slowly from the MMT-gelatin formulations prepared at lower pH than from those prepared at higher pH values. For all formulations, the herbicide was completely released after 48 h, and no fraction was bound irreversibly to the clay-gelatin matrix. The highest release was obtained for the formulation prepared at pH close to the isoelectric point of the protein (7.9–9.0). Increasing of pH results in the increased labile fraction of MCPA due to the reduction of strong electrostatic interactions involved in the retention of the herbicide in the clay-gelatin matrix at pH value exceeding the value of isoelectric point; this effect is counteracted by the presence of mostly exfoliated clay particles acting as a barrier to the diffusing out of herbicide molecules. By application of glycerol, the interaction of the herbicide within the clay-gelatin matrix can be modified by enhancing hydrogen bonding over stronger electrostatic interactions, which results in enhanced release of the herbicide. The slower release rate of 2,4-D in water and soil was estimated also from carboxymethyl cellulose gel formulation containing some modified bentonites prepared by intercalating inorganic or organic cations in interlayers of Na^+-saturated bentonite. The $t_{1/2}$ corresponding to the time when 50% of 2,4-D has been released in water varied from 8.8 to 19.8 h, and the largest value was shown by the formulation incorporating hydroxy-iron intercalated bentonite showing the highest sorption capacity to 2,4-D. Such gel formulations could also be used for controlled release of 2,4-D herbicide when applied to a thin soil layer (Li et al. 2009).

Formulations of herbicides 2,4-D and picloram which were anchored on porous gel of hexagonal mesoporous silica modified with carboxylic acid showing a nanometric structure with spheres <50 nm and porous diameter of 10 nm exhibited the controlled release of herbicides, which was lower for picloram than for 2,4-D (Prado et al. 2011).

Nanostructured liquid crystalline particles (NLCP) containing 18% (w/w) of phytantriol with the size of ~250 nm, polydispersity index of 0.22 and zeta potential of −15 mV, which are able safely interact with plant leaf cuticular surfaces with minimal impact on epicuticular waxes, were used to deliver 2,4-D to weeds, crops and model plants. In field trials, such nanoformulation used for the control of the invasive weed wild radish (*Raphanus raphanistrum* L.) in wheat was found to be effective at lower concentrations (0.03% and 0.06%) as compared with commercially available herbicide formulation (Estercide 800), while crop yield remained similar for nano- and commercial preparations. In a separate trial, the phytotoxicity on the crop *Hordeum vulgare* was assessed, along with the herbicidal effects on the weed *R. raphanistrum*, and the obtained results were consistent with earlier observations made on *Triticum aestivum*. High-concentration spray applications of 2,4-D NLCP resulted in greater epicuticular wax solubilization effects, and it was estimated that the area of epicuticular waxes was the highest for untreated controls and significantly decreased with the increase in the concentration of NLCP. This indicates that NLCP can reduce the risk of cuticle damage while still efficiently delivering the active ingredient, which can result in increased yield. The application of NLCP can also eliminate adverse environmental effects as well as negative effects on nontarget plants observed with the overuse of surfactants in agrochemical formulations (Nadiminti et al. 2016).

Sustained release and enhanced herbicidal activity against the tested target plant (*Brassica* sp.) were shown also by nanosized rice husk loaded with 2,4-D, while the nontarget plant *Zea mays* L. was not affected, and better herbicidal efficiency of this formulation as compared with that of the commercial 2,4-D could be connected with the reduced soil sorption or increased bioavailability of 2,4-D in the soil (Abigail et al. 2016).

8.2.1.2 Nanoscale Triazine Herbicides

Solid lipid NPs (SLNPs) prepared using glycerol tripalmitate and poly(vinyl alcohol) (PVA) containing atrazine (ATZ) and simazine (SMZ) showing the hydrodynamic diameter of 255 nm and encapsulation efficiency (EE) of 89.7 ± 0.02% of ATZ and 97.3 ± 0.05% of SMZ were prepared by de Oliveira et al. (2015). PVA used during the preparation of the formulations was adsorbed on the surface of the particles, creating a layer that provided steric stabilization. The SLNP formulations showed negative zeta potential values that were not affected by encapsulation of the herbicides, and after 30 days of storage, a mean value of −15 mV was estimated. The release of herbicides from SLNPs was slower compared to that of the free herbicides, which was reflected in significantly lower $t_{1/2}$ values corresponding to 50% release that were 2.5 h (ATZ) and 5.3 h (SMZ) compared to $t_{1/2}$ values estimated for free herbicides, namely, 52.9 h for ATZ and 51.1 h for SMZ. The values of the release constants showed that atrazine was released faster than simazine. The encapsulated herbicides showed decreased cytotoxicity when compared with the commercial formulation, and they were also investigated for pre- and postemergence treatments applied to a target species (*R. raphanistrum*) and a nontarget species (*Z. mays*) at concentrations equivalent to 0.3 and 3 kg/ha. SLNPs containing herbicides caused greater phytotoxic effects on both the aerial parts and roots of plants, compared to the commercial formulation, and they remained effective also at tenfold lower concentration than the recommended concentration. At postemergence treatment, SLNPs loaded with herbicides showed comparable phytotoxic effects on aerial parts and roots at both studied concentrations, and the interaction of SLNPs with *R. raphanistrum* was found to be species-specific, because no toxic effects of SLNPs were observed in assays with *Z. mays*.

Treatment with poly(ε-caprolactone) (PCL) nanocapsules containing ATZ induced faster and more severe development of toxicity symptoms, faster inhibition of photosystem (PS) II photochemistry and greater lipid peroxidation in *Brassica juncea* leaves compared with the commercial ATZ product, and it was very effective also when the tenfold diluted concentration of nanoformulation was used. The herbicidal effectiveness of nanocapsules containing ATZ could be connected (*i*) with the protection of the encapsulated active compound against physicochemical degradation; (*ii*) with the interaction of hydrophobic nanocapsules with the leaf cuticle resulting in increased delivery of herbicide to the plant tissues and decreased loss of the herbicide to the environment; and (*iii*) with slow release of ATZ from PCL nanocapsules, which can promote a gradual contact between the herbicide and the plant. Thus, PCL nanocapsules could be considered as an efficient carrier system for ATZ enabling the application of lower dosages of the herbicide and could be used as an effective tool in the postemergence control of weeds. Oliveira et al. (2015) evaluated also postemergence herbicidal activity of PCL nanocapsules containing ATZ with the average size of 240.7 ± 2.9 nm using mustard (*B. juncea*) as target plant species model. After 7 days also the leaves of the plants treated with tenfold diluted nanoformulation containing ATZ revealed similar symptoms of leaf wilt, yellowing, and necrosis as the commercial atrazine at the recommended dosage, and a strong reduction of the shoot dry weight was observed. Pereira et al. (2014) evaluated PCL NPs containing ATZ in terms of their herbicidal activity and genotoxicity and found that the encapsulation of the herbicide resulted in harmlessness to a nontarget organism (*Zea mays*), but it enhanced the effectiveness against a target organism (*Brassica* sp.), compared to the use of the free herbicide, which could be connected with increased herbicide bioavailability. At application of nanoencapsulated herbicide, the mobility of ATZ in the soil column was found to be increased because of reduced soil sorption, which led to better effectiveness of ATZ against the target organism. Moreover, the nanoformulations containing ATZ were less genotoxic, compared to the free herbicide, which would contribute to the improved level of safety in agricultural applications. Clemente et al. (2013) performed ecotoxicological evaluation of PCL nanocapsules containing ametryn and ATZ. The encapsulation of the herbicides in nanocapsules resulted in lower toxicity to the alga *Pseudokirchneriella subcapitata* and higher toxicity to the microcrustacean *Daphnia similis* compared to the herbicides alone. The cytogenetic tests employing human lymphocyte cultures showed that formulations of nanocapsules containing the herbicides were less toxic than the herbicides alone. The suitability of polymeric PCL nanocapsules containing three triazine herbicides (ametryn, atrazine and simazine) as controlled release systems that could reduce environmental impacts was studied also by Grillo et al. (2012), and the obtained results supported the previous findings that the use of PCL nanocapsules is a promising technique that could improve the behaviour of herbicides in environmental systems.

Controlled release formulations prepared by incorporation of ATZ in ethylcellulose, in which allophanic clays and nanoclays were incorporated as matrix-modifying agents, were designed by Cea et al. (2010), and their effect on the emergence and growth of field mustard (*Brassica campestris* L.) was evaluated under greenhouse conditions. The controlled release formulations effectively reduced the seedling emergence and caused greater death of seedlings than the commercial formulation, especially when nanoclays were added into the formulation, and they were characterized also by prolonged bio-efficiency enabling longer applications intervals and in this way minimizing the harmful impact of ATZ on the environment.

Metribuzine entrapped within a sepiolite-gel-based matrix with one of two proportions of clay/herbicide and used as either a gel or powder after freeze-drying remained active longer than commercial formulation, avoiding the need to use more frequently herbicide applications (Maqueda et al. 2009).

8.2.1.3 Other Nanoscale Aromatic-Type Herbicides

Paraquat encapsulated in the formulation of AgNPs in the chitosan (CS) matrix with the particle size of 100 nm and the entrapment efficiency of 90% exhibited steady release of herbicide in the early hours, and a total release of about 90% was estimated at 24 h. Surface treatment of the cut pieces of *Eichhornia crassipes* with 0.5, 10 and 25 μg/mL of this formulation resulted in greater necrotic lesions than at application of free paraquat at doses 10 and 25 μg/mL. The application of the nanoformulation did not affect soil physicochemical parameters and soil enzymes activity; nanoherbicide-treated seeds showed 90.1% seed germination, and no plant growth parameters of the nontarget plant *Vigna mungo* were adversely affected (Namasivayam et al. 2014). CS/tripolyphosphate (TPP) NPs loaded with paraquat showing 62.6 ± 0.7% association of the herbicide with the NPs exhibited delayed release of paraquat in laboratory conditions compared to the free herbicide (70% vs 90% within 350 min), and the diffusion and relaxation of the polymeric chain might be a factor affecting paraquat release. The encapsulation did not affect the herbicidal activity of paraquat in cultivations of maize (*Z. mays*) and mustard (*Brassica* sp.), and herbicide bound to NPs caused less chromosome damage compared to its free form (Grillo et al. 2014). Less chromosome damage in samples treated with nanoparaquat compared to conventional paraquat was estimated also by Nishisaka et al. (2014), indicating that the nanoformulation of paraquat loaded into NPs prepared from CS and TPP can be used to minimize damage caused by bulk herbicide and is suitable for safer control of weeds in agriculture. Silva et al. (2011) studied the release profile of paraquat from alginate (ALG)/CS NPs with particle size of 635 nm, zeta potential −22.8 ± 2.3 mV and entrapment efficiency of 74% and compared it with that of the free herbicide. They estimated that the complete herbicide release from NPs was extended by 2 h compared to free paraquat, which allows to reduce the amount of the herbicide resulting in lower environmental risk and lower energy costs. The release process was governed by mechanisms displaying non-Fickian kinetics, and it could be assumed that the release of paraquat from ALG/CS NPs is connected with the rupture of ionic bonds between paraquat and polymeric ALG chains. In another experiment, Silva et al. (2010) loaded clomazone herbicide into ALG/sodium bis(2-ethylhexyl) sulfosuccinate (AOT) or ALG/CS NPs and found that the association of the herbicide with the NPs prolonged the release time: in the time period of 240 min ca. 70% of clomazone was released, while from ALG/AOT or ALG/CS NPs within the same period, this amount was only 50% and 20%, respectively, indicating that ALG/AOT NPs have higher rates of association of the herbicide clomazone than ALG/CS NPs. The release of clomazone was also found to be governed by non-Ficknian kinetic processes, and the kinetic constant value (k) indicated a faster release for herbicide of the ALG/CS NPs ($k = 1.96\ \text{min}^{-1}$) compared to ALG/AOT NPs ($k = 1.12\ \text{min}^{-1}$).

Poly(butyl methacrylate-diacetone acrylamide)-based formulation used for controlled release of acetochlor showed improved herbicide incorporation and slower release, obviously due to potential interactions between the herbicide and the polymer (Guo et al. 2014). The evaluation of application of pretilachlor microemulsion and herbicide encapsulated monolithic dispersion with average particle size in the range of 1–100 nm against *Echinochloa crus-galli* in rice fields performed 30, 60 and 90 days after transplantation confirmed that tested nanoformulations were superior compared to the commercial pretilachlor formulation Rifit® 50 EC (Kumar et al. 2016).

Metsulfuron methyl-loaded pectin nanocapsules with particle size ranging from 50 to 90 nm, zeta potential value of −35.9 mV and 63 ± 2% EE, which applied on a weed (*Chenopodium album*) grown in a wheat crop were found to be more effective at a reduced dose than commercial formulation, showed less toxicity and longer lasting effects, while wheat crop was unaffected. The dry biomass of *C. album* treated with nanoformulation containing the herbicide was 5 g/m^2, while it reaches 48 g/m^2 at controls and 19 g/m^2 at application of the normal herbicide (Kumar et al. 2017). Subabul stem lignin was used as a matrix material in a controlled release nanoformulation of diuron with particle size ca. 166 nm and 74.3±4% EE. This nanoformulation exhibited a nonlinear biphasic release profile for diuron, and its application into soil caused earlier signs of leaf chlorosis and mortality in *Brassica rapa* seedlings compared to seedlings grown on soil supplemented with a commercial diuron preparation or bulk diuron (Yearla and Padmasree 2016). Isoproturon-loaded carboxymethyl starch/MMT composite microparticles showing about 75% EE demonstrated a significantly reduced release rate of herbicide than its commercial formulation, releasing 95% isoproturon after 700 h compared to 24 h estimated with the commercial formulation. Moreover, leaching in soil from composite formulations was relatively slower than release in water, which could positively affect the environmental pollution (Wilpiszewska et al. 2016).

Kanimozhi and Chinnamuthu (2012) fabricated manganese (II) carbonate core-shell NPs, in which the $MnCO_3$ core was coated with a single bilayer of the polyelectrolytes sodium polystyrene sulphonate and polyallylamine hydrochloride using a layer-by-layer method. The particle size distribution of the $MnCO_3$ core and core-shell was 126 and 250 nm, respectively. Then the NPs were treated with diluted hydrochloric acid to prepare inorganic/organic hollow spheres, which were subsequently loaded with pre-emergence herbicide pendimethalin programmed to release smartly upon requirements. Porous hollow-shell material could be considered as suitable also for loading of other active ingredients, e.g. fertilizers for conditional release.

The interlayer spaces of the methoxy-modified nanosized tubular halloysite (mHal) and platy kaolinite (mKaol) were found to be suitable for the effective intercalation of amitrole herbicide, which substantially promoted amitrole loading. The slow herbicide release from amitrole-loaded mKaol was connected with the restricted diffusion of the intercalated herbicide caused by the lamellar structure of mKaol as well as with the long diffusion path of the intercalated herbicide due to the large size of mKaol particles compared to mHAL particles (Tan et al. 2015).

ALG/CS and CS/TPP NPs with particle size <400 nm and zeta potentials of −30 and + 26 mV, respectively, were found to be suitable to encapsulate the herbicides imazapic and imazapyr with 60% EE. The treatment of target weed species, *Bidens pilosa* (blackjack), with a dose equal to that used in the field (400 g/ha) resulted in reduced growth compared to the control; the herbicides maintained adequate herbicidal activity, but their toxicity to nontarget organisms was reduced, and the researchers emphasized that the encapsulation of two herbicides in one carrier system could improve the activity and reduce the impacts on the environment (Maruyama et al. 2016). A natural smectite (SW) modified with CS or with Fe^{3+} cation was tested as an adsorbent or a carrier for controlled release formulations of imazamox, an herbicide used for the control of root-parasitic plants *Orobanche* spp. The herbicide release into water was inversely related to the strength of imazamox-clay interactions, whereby the herbicidal activity of the weak complex imazamox-SW modified with CS was comparable with that of commercial formulation, however showing a reduction in the total soil leaching losses (15%) and the peak maximum concentration in soil column leachates (40%) (Cabrera et al. 2016). The co-exposure of AgNPs (100 μM) and chiral herbicide imazethapyr (IM) (0.2 μM) to model plant *Arabidopsis thaliana* showed that the use of (*R*)-enantiomer led to preferential Ag uptake by plant roots, and also higher metal amount in shoots was estimated compared to co-exposure of AgNPs with (*S*)-enantiomer. A significant increase of free amino acids (except cysteine) following exposure to racemate IM, (*R*)-IM or their co-exposure with AgNPs resulted in increased release of Ag^+ due to formation of amino acid adducts with Ag^+ ions, which was then reflected in the toxicity enhancement under co-exposure of AgNPs and (*R*)-enantiomers. Treatment of roots with (*S*)-IM led to reduced production of reactive oxygen species (ROS) compared to the control, while the administration of (*R*)-IM and herbicide racemate as well as their co-exposure with AgNPs resulted in enhanced ROS formation compared to the control, indicating enantioselective ROS production (Wen et al. 2016).

8.2.1.4 Nanoscale Organophosphorus Herbicides

A nanoemulsion system consisting of long-chain fatty acid methyl esters (LFAMEs)/mixed surfactant (long-chain alkyl polyglucosides and ethoxylated 3-(3-hydroxypropyl)-heptamethyltrisiloxane (organosilicone))/water and glyphosate isopropylamine (IPA) herbicide was designed by Lim et al. (2012). The pre-formulation concentrate with less than 20% (w/w) of inerts (LFAMEs + mixed surfactant) appeared as a polymerized multi-connected network, and the dilution of the pre-formulation with water resulted in the destruction of the polymerized network and formation of dispersed NPs of nanoemulsion formulation. Because the emulsion particles had incorporated glyphosate IPA, the herbicide bioactivity, bioavailability and delivery efficiency were improved. Similar oil-in-water nanoemulsions incorporating glyphosate IPA with particle sizes of diameter <200 nm applied on narrow-leaved weed *Eleusine indica* showed lower ED_{50} (0.40 kg a.e./ha) compared to those estimated using Roundup® (0.48 kg a.e./ha), which indicates that the nanoemulsion system could increase penetration and uptake of glyphosate IPA (Jiang et al. 2012). The nanoemulsion formulations containing glyphosate IPA displayed a significantly lower spray deposition on creeping foxglove (2.9–3.5 ng/cm^2), slender button weed (2.6–2.9 ng/cm^2) and buffalo grass (1.8–2.4 ng/cm^2) than

Roundup® (3.7–5.1 ng/cm^2). At 3 and 7 days after treatment, the order of the mortality rates of the investigated weeds was buffalo grass > slender button weed > creeping foxglove, but the control rates were the same at the 14th day for the three weeds. Thus, the different cuticle permeability and foliar structures considerably affected the absorption rates of the herbicide and so its bioefficacy. Fourteen days after treatment with nanoformulation, the visible injury rates were comparable with that of Roundup® indicating the enhanced bioactivity of the nanoemulsion formulations (Lim et al. 2013).

8.2.2 *Metal-Based Nanoherbicides*

Adverse effects on plants are exhibited also by metal and metal oxide NPs because of stress or stimuli caused by the surface, size and/or shape of the particle, while inside the cells they might directly provoke alterations of membranes and other cell structures and molecules as well as protective mechanisms. The change of membrane permeability connected with the damage of cell membranes due to the production of ROS by metal NPs contributes to the enhanced probability of entry of NPs into the cell (Nel et al. 2006; Nair et al. 2010). Due to increasing environmental pollution with metals, numerous papers are devoted to the study of the negative effects of metal NPs on plants. On the other hand, some NPs of essential metals (e.g. Cu, Zn, Fe) used in appropriate concentration and also alumina and TiO_2 NPs were found to exhibit positive effects on plant growth. The beneficial and adverse effects of metal and metal oxide NPs were comprehensively reviewed by several researchers (e.g. Masarovičová and Kráľová 2013; Ma et al. 2015; Masarovičová et al. 2014; Du et al. 2017; Rizwan et al. 2017; Siddiqi and Husen 2017). However, metal NPs could be considered as non-selective herbicides, because they can damage not only undesired weeds but also crops, and therefore in agriculture selective herbicides targeting the weed without affecting nontarget crops are preferred.

8.3 Nanofungicides and Nanobactericides

There are approximately two million different species of fungi on Earth (Gauthier and Keller 2013). The vast majority of known fungal species are strict saprophytes (De Lucca 2007), but it is estimated that 270,000 fungal species can attack plants, such as genera *Botrytis*, *Sclerotinia*, *Aspergillus*, *Fusarium* and *Verticillium* (Sharon and Shlezinger 2013). Of the over 15,000 species of bacteria, about 200 species of phytopathogenic bacteria were identified, such as genera *Erwinia*, *Acidovorax*, *Pseudomonas*, *Ralstonia*, *Rhizobacter*, *Xanthomonas*, *Agrobacterium*, *Xylella*, *Arthrobacter*, *Clavibacter* and *Streptomyces* (Agrios 2005). Thus fungi and bacteria can cause crop losses worldwide (Gauthier and Keller 2013; Fisher et al. 2012; Carris et al. 2012; Jampílek 2016). Fungicides and bactericides are a specific type of pesticides that control fungal/bacterial diseases by specifically inhibiting or killing the fungus/bacteria that causes the disease (Jampílek 2016; Bhattacharyya et al. 2016; Ismail et al. 2017). As in other classes of pesticides, also dynamic development in the field of inorganic and organic nanofungicides and nanoscale bactericides can be recorded.

8.3.1 Natural and Synthetic Organic Nanoscale Fungicides and Bactericides

Zataria multiflora essential oil (ZEO)-loaded SLNPs with ca. particle size 255 nm, zeta potential approximately −37.8 ± 0.8 mV and EE 84 ± 0.92%, showed in vitro antifungal activity against pathogens such as *Aspergillus ochraceus* (MIC 200 ppm), *Aspergillus flavus* (MIC 200 ppm), *Alternaria solani* (MIC 100 ppm), *Rhizoctonia solani* (MIC 50 ppm) and *Rhizopus stolonifer* (MIC 50 ppm). These formulations showed higher potencies than those with pure essential oil (Nasseri et al. 2016). ZEO encapsulated in CS NPs with the mean particle size of 125–175 nm demonstrated a controlled and sustained release of ZEO for 40 days in vitro, and in vivo investigation showed that the encapsulated oil at 1500 ppm concentration considerably decreased both disease severity and incidence of *Botrytis*-inoculated strawberries during 7 days of storage at 4 °C followed by 2–3 more days at 20 °C. Increasing of the initial ZEO content in CS NPs led to a decrease of ZEO encapsulation and loading efficiency (Mohammadi et al. 2015). Encapsulation of thyme essential oils (TEO) in self-assembled polymer of CS and benzoic acid nanogel notably increased the half-life and the antifungal properties of TEO, and the estimated MIC of encapsulated TEO was 300 mg/L at unsealed and 500 mg/L at sealed condition compared to 400 mg/mL and 1000 mg/mL, respectively, determined for free TEO. Good antifungal effects of encapsulated TEO at concentrations >700 mg/L were confirmed also in in vivo experiment (Khalili et al. 2015). Similar results were obtained also with *Mentha piperita* essential oils encapsulated in CS-cinnamic acid nanogel showing MIC values of 500 ppm against *A. flavus* under sealed condition, while the corresponding MIC value for free oils was 4.2-fold higher. A test under non-sealed condition showed that treatment with 800 ppm of uncapsulated oil resulted in complete inhibition of fungal growth, while the same effect could be obtained only with 3000 ppm of free oils (Beyki et al. 2014). As environmentally friendly alternative products for postharvest disease control, polyethylene terephthalate punnets containing thyme oil and sealed with CS/boehmite nanocomposite lidding films were designed, which significantly reduced the incidence and severity of brown rot caused by *Monilinia laxa* in artificially inoculated peach fruits (cv. Kakawa) held at 25 °C for 5 days and caused considerable reduction of the brown rot incidence to 10% in naturally infected fruits stored at 0.5 °C and 90% relative humidity for 7 days and at simulated market shelf conditions at 15 °C for 3 days (Cindi et al. 2015). β-D-glucan (isolated from the cell wall of *Pythium aphanidermatum*) NPs prepared using sodium TPP, in which phosphoric groups of TPP were linked with

OH group of β-D-glucan with the size 20–50 nm showing spherical, smooth and almost homogenous structure, were found to inhibit the growth of *P. aphanidermatum*, suggesting that they could be used in crop protection against this devastating fungus (Anusuya and Sathiyabama 2014).

The CS NPs inhibited the growth of phytopathogens, namely, *Pyricularia grisea*, *A. solani* and *Fusarium oxysporum*, but they were able also to promote germination %, seed vigour index and vegetative biomass of chickpea seedlings. For example, CS NPs inhibited the radial growth of *P. grisea*, and their application delayed blast symptom expression on fingermillet leaves for 25 days compared to 15 days in control plants, which could be connected with the induction of ROS and the enhanced activity of peroxidase (reaching maximum at day 50) in leaves of fingermillet, which might be the reason for the delayed symptom (Sathiyabama and Manikandan 2016). ROS can directly act at the site of infection or function indirectly as second messengers (Arasimowicz and Floryszak-Wieczorek 2007), and H_2O_2, which could diffuse through the membrane, is considered to serve as a signal molecule under stress (Mittler 2002), while peroxidases, the scavengers of H_2O_2, are one of the pathogenesis-related proteins which are implicated in plant defence system against pathogenic fungi (Hiraga et al. 2001). Moreover, the disease incidence in CS NP-treated fingermillet plants was lower compared to control plants (Sathiyabama and Manikandan 2016), and CS NPs showed also potential in suppressing blast disease of rice, which can be used further under field conditions to protect rice plants from the devastating fungus (Manikandan and Sathiyabama 2016). Nanoemulsions prepared using 1.0% of low molecular weight CS showing 600 nm droplet size inhibited conidial germination and reduced dry weight of mycelia and sporulation of *Colletotrichum gloeosporioides* in vitro, and they could be used as biofungicide for controlling anthracnose of dragon fruit plants in the future (Zahid et al. 2013). CS and CS NPs characterized with low toxicity towards mammalia were also found to be effective for the control of *Fusarium* head blight disease in wheat (*Fusarium graminearum*), and greenhouse experiments showed that plants can be protected from the disease by spraying them at anthesis. CS and CS NPs showing polycationic properties can affect membrane permeability and leakage of cellular contents resulting in disorganized hyphae associated with inhibition of fungal growth. Moreover, application of CS to plant tissues often results in its agglutination around the penetration sites, and isolation of the penetration site through the formation of a physical barrier could prevent the pathogen from spreading and invading other healthy tissues (Kheiri et al. 2016).

The CS NPs prepared of CS having low (LMW) and high molecular weight (HMW) and *N*-trimethyl CS (TMCS) exhibited zeta potential ranging from +22 to +55 mV, and higher values of zeta potential were obtained when HMW CS was used. The CS NPs were tested against *Fusarium solani* and *Aspergillus niger*, and it was found that the smallest HMW CS NPs (CS concentration of 1 mg/mL) showed the best antifungal activity against *F. solani* (MIC = 0.5–1.2 mg/mL), the effect of particle size on the activity being higher than their surface charge. On the other hand, *A. niger* was found to be highly resistant to CS, and inhibition was observed only at treatment with CS solution (MIC = 3 mg/mL) and NPs prepared at high concentration (2 and 3 mg/mL) of HMW CS (MIC = 1.71–2.43 mg/mL). Unlike other types of CS NPs, TMCS NPs had no inhibitory activity against *F. solani* (Ing et al. 2012). Sulphonated CS showed antifungal activities against *Arthrinium sacchari* (MIC, 64.00 mg/mL) and *Botrytis cinerea* (MIC, 0.25 mg/mL), and it was found to damage and deform the structure of fungal hyphae (Sun et al. 2017). Oleoyl-CS NPs with the particle size of about 297 nm were tested against six plant pathogenic fungi in a mycelium growth experiment, and it was found that *Alternaria tenuissima*, *Botryosphaeria dothidea* and *Nigrospora sphaerica* were CS-sensitive in contrast to *Gibberella zeae* and *Fusarium culmorum*, which were CS-resistant. Increasing the NP concentration resulted in an increase of the antifungal index of CS-sensitive fungi, whereby their plasma membranes contained lower levels of unsaturated fatty acid than those of CS-resistant fungi (Xing et al. 2016).

In addition, CS can be also used as a matrix for loading and stabilizing various fungicides. For example, hexaconazole nanocapsules prepared using naturally occurring CS and TPP through ionotropic gelation showed 73% slowing down of the release of the active ingredient compared to a commercial preparation, and this effect was greater at pH 7 and pH 10 than at pH 4, and a release study in soil confirmed that this nanoformulation is suitable for alkaline soil. Also the antifungal activity of nanocapsules against *R. solani* exceeded that of the commercial preparation, and they showed lower toxicity on nontarget cell lines (Chauhan et al. 2017). The study of the effect of nanohexaconazole on the phenotype and pathogenicity of *R. solani* f. sp. *sasakii* causing banded leaf and sheath blight in maize showed that at the application of 1 ppm, it inhibited growth and sclerotial body formation similarly to commercial hexaconazole, while in vivo it exhibited notable restriction of lesion formation in insusceptible cultivar Vivek QPM-9 and also reduced the disease rating caused upon inoculation with the fungus *R. solani* exposed to 0.1 and 0.01 ppm of nanohexaconazole (Bheemaraya et al. 2014). Biodegradable CS-lactide copolymer (CS-PLA) NPs loaded with pyraclostrobin with particle sizes ranging from 77 to 128 nm prepared by varying the feed mass ratio of CS-PLA to fungicide from 50:1 to 5:1 exhibited an initial burst followed by sustained and pH-controlled pyraclostrobin release and better fungicidal activity against *Colletotrichum gossypii* Southw than 25% pyraclostrobin emulsifiable concentrate (Xu et al. 2014). Nanoformulations of carbendazim loaded into polymeric NPs (CS and pectin) with mean particle size of 70–90 nm applied at concentration 0.5 and 1.0 ppm caused complete inhibition of *F. oxysporum* and *Aspergillus parasiticus*, while the antifungal effectiveness of pure carbendazim was lower (80% and 97.2% inhibition at 0.5 and 1.0 ppm concentrations, respectively, against *F. oxysporum*; 86.0% and 100.0% inhibition at 0.5 and 1.0 ppm concentrations, respectively, against *A. parasiticus*), and even the inhibitory effect of commercial formulation WP 50 (50.5% and 70.0% inhibition at 0.5 and 1.0 ppm concentrations, respectively, against *F. oxysporum*; 42% and 58% inhibition at 0.5 and 1 ppm concentrations, respectively, against *A. parasiticus*) did not reach the effectiveness of carbendazim nanoformulations (Sandhya et al. 2017).

Besides CS, also other natural polysaccharides, synthetic polymers or inorganic materials have been used as stabilizers/matrices. Application of nanochitin suspension (0.001% (w/v)) exhibited synergistic effects on inhibition of tobacco root rot when mixed with metalaxyl mancozeb and thiophanate methyl fungicides indicating its protecting effects on tobacco plants from tobacco root rot diseases and suggesting that its co-administration could reduce the amount of chemical fungicides in tobacco plantations (Zhou et al. 2017). PEG 400 was used as the surface-stabilizing agent to prepare nanohexaconazole with a size of about 100 nm showing not only better fungicidal potential than the conventional registered formulation, but also it did not affect adversely the soil nitrifiers (Kumar et al. 2015a). Controlled release nanoformulations of carbendazim prepared using PEG-based functionalized amphiphilic copolymers released the fungicide between the 10th and the 35th day, while the release from a commercial preparation lasted only to the 7th day, and the half-release ($t_{1/2}$) values of the nanoformulation ranged between 9.47 and 24.20 days, showing increased release of the maximum amount of carbendazim with increasing PEG molecular weights. For antifungal activity of the most active formulations against *R. solani*, ED_{50} values ranging from 0.40 to 0.42 mg/mL were estimated (Koli et al. 2015). Azomethine-based nanofungicides with the particle size of 100 nm prepared using technically pure azomethines and PEG as a surface stabilizer exhibited twofold higher antifungal activity against *R. solani*, *R. bataticola* and *Sclerotium rolfsii* compared to bulk azomethines, and they were found to be better antifungal formulations than the conventional preparation also in pot experiments (Mondal et al. 2017).

NPs prepared by encapsulation of thiamine dilauryl sulphate (TDS), a vitamin B1 derivative, into lecithin NPs with a mean diameter of 136 nm exhibited better efficacy on the inhibition of mycelial growth and spore germination of *F. oxysporum* as TDS, and their inhibitory effect at a dosage of 100 ppm was similar or even better than that of the commercial herbicide dazomet (Cho et al. 2013).

Mesoporous SiO_2 nanospheres with the mean particle diameter of 162 nm and mean pore size of 3.2 nm loaded with metalaxyl exhibited sustained release of fungicide and significantly delayed its release in soil, while compared to 76% of free metalaxyl, which was released in soil within a period of 30 days, fungicide release from the mesoporous framework was only 11.5% (Wanyika 2013). Validamycin-loaded nanosized calcium carbonate was found to improve germicidal efficacy against *R. solani* compared to conventional technical validamycin after about 7 days, and it extended the release time of the pesticide to 2 weeks (Qian et al. 2011).

8.3.2 Carbon-Based Nanofungicides and Nanobactericides

Several carbon nanomaterials (CNMs), e.g. single-walled (SWCNTs) or multi-walled carbon nanotubes (MWCNTs), graphene oxide (GO), reduced graphene oxide (rGO), fullerene (C_{60}) and activated carbon (AC), were tested on their activity against phytopathogens (Wang et al. 2014, 2017; Chen et al. 2014b; Sarlak et al. 2014; Sawangphruk et al. 2012). Wang et al. (2014) studied the antifungal activity of six different CNMs against *F. graminearum* and *F. poae* and found that it decreased in the following order, SWCNTs (128 nm) >>> MWCNTs (78.8 nm) > GO (68.06 nm) > > rGO (105.7) nm, while C_{60} (220.6 nm) and AC (190.1 nm) showed no significant antifungal activity. Because in the antifungal activities of carbon nanomaterials their direct contact with spores may play an important role, in the case of C_{60} and AC, lack of tight or direct contacts is responsible for their low antifungal activity. The CNMs could inhibit spore germination just by interfering with the process of water uptake before inducing plasmolysis. After the deposition of spores on CNMs, the CNMs may cause the blockage of the water channels of spores. Superior toxic effect on *Alternaria alternata* fungi was shown also by zineb and mancozeb encapsulated into hybrid materials prepared by polymerization of citric acid onto the surface of oxidized MWCNTs (Sarlak et al. 2014). The estimated IC_{50} values related to the inhibition of the mycelial growth of *F. oxysporum*, *A. niger* and *A. oryzae* by reduced GO nanosheets were estimated as 50, 100 and 100 μg/mL, respectively. The effective inhibition of mycelial growth by reduced GO having sharp edges is connected with its direct contact with the cell walls of fungi and a subsequent chemical reaction of the reactive oxygen-containing functionalities of small rGO nanosheets with the organic functional groups of chitin and other polysaccharides on cell walls of fungi (Sawangphruk et al. 2012).

Covalent functionalization of MWCNTs by lysine and arginine under microwave radiation resulted in improved antifungal activity of functionalized MWCNTs against *A. niger* and *F. culmorum* compared to pristine MWCNT reaching a 1.9- and 1.1-fold increase, respectively, for MWCNTs-lysine, and 2- and 1.7-fold increase, respectively, for MWCNTs-arginine (Zare-Zardini et al. 2013). Nitrogen-doped carbon nanohorns (NCNHs) with the size of 50–60 nm applied at a dose of 150 μg/mL inhibited *R. solani* after 72 h. It could be assumed that in the toxic effect against *R. solani*, primarily the interaction of NCNHs with the pathogens by mechanically wrapping could be considered, which may be one of the major toxicity actions of NCNHs against *R. solani*, and targeting of the endochitinase of *R. solani* by NCNHs results in deactivation of the enzyme (Dharni et al. 2016).

The application of 500 μg/mL GO was found to kill about 90% of *Pseudomonas syringae* and *Xanthomonas campestris* pv. *undulosa* and repress 80% macroconidia germination along with partial cell swelling and lysis in *F. graminearum* and *F. oxysporum*. It could be supposed that GO interwinds the bacteria and fungal spores with a wide range of aggregated GO sheets causing the local perturbation of their cell membrane with a subsequent decrease of the bacterial membrane potential and the leakage of electrolytes of fungal spores. Thus, the toxic effect of GO on phytopathogens is caused by its interaction with these pathogens by mechanical wrapping and local damaging the cell membrane, which finally results in cell lysis, and therefore GO could be successfully used also for resisting crop diseases (Chen et al. 2014a, b). Wang et al. (2017) developed GO-Fe_3O_4 nanocomposites that efficiently repressed the germination of sporangia of *Plasmopara viticola* and inhibited the development of downy mildew, showing also potent curative effects. The GO-Fe_3O_4 nanocomposites applied at concentration 50 μg/mL exhibited superb protective and fungicidal activities, and treatment of grapevine leaves in the field with a dose of 250 μg/mL could result in a notable decrease of the severity of downy mildew.

8.3.3 Metal-Based Nanofungicides and Nanobactericides

Copper belongs to elements that are essential for plants, and benign fungi occurring in the roots of plants could detoxify the excess of copper uptaken by plants (Vitanovic 2012; Anjum et al. 2015). Nanoscale Cu was also found to suppress the growth of bacterial pathogen *Xanthomonas axonopodis* pv. *punicae* causing bacterial blight of pomegranate at 0.2 ppm, i.e. >10,000-fold lower concentration than the usually applied Cu-oxychloride, resulting in cell wall degradation in nano-Cu treated bacterial cells that failed to colonize plant tissues and to produce water-soaked lesions (Mondal and Mani 2012). Ghasemian et al. (2012) studied the antifungal effect of CuNPs of the average particle size of 8 nm on filamentous fungi by agar dilution method and estimated following MIC values: ≤40 mg/L for *Penicillium chrysogenum*, ≤60 mg/L for *A. alternata*, ≤60 mg/L for *F. solani* and ≤80 mg/L for *A. flavus*, suggesting that fungal sensitivity to CuNPs varies depending on the fungal species. Also, Giannousi et al. (2013) tested three different Cu-based (Cu_2O, CuO and Cu/Cu_2O) NPs of similar sizes (11–14 nm) and nearly spherical shape in the field against *Phytophthora infestans* on tomato and found that all the tested Cu-based NPs were more effective in lower formulated product and active ingredient rate than the four registered copper-based agrochemicals Kocide 2000, Kocide Opti, Cuprofix Disperss and Ridomil Gold Plus, without causing any deleterious effect on plants.

Cu-CS NPs with particle sizes ranging from 180.0 to 487.9 nm and zeta potential of +88 mV applied at 0.1% concentration caused notable inhibition of the growth of phytopatogenic fungi *A. alternata* (89.5%), *Macrophomina phaseolina* (63.0%) and *R. solani* (60.1%) in vitro and also exhibited 87.4% inhibition of spore germination of *A. alternata*. The antifungal effectiveness of Cu-CS NPs is connected with their appropriate surface charge density (zeta potential of +88 mV) providing them greater binding affinity for negatively charged fungal membrane as well as with the production of toxic H_2O_2 at the reduction of Cu(II) to Cu(I) in fungi causing destruction of the cell viability (Saharan et al. 2013). Cu-CS NPs with hydrodynamic diameter 374 nm and zeta potential of +22.6 mV applied at concentration 0.12% caused 70% and 73% inhibition of mycelia growth and inhibition of spore germination in *A. solani* (70% and 61%, respectively) and *F. oxysporum* (73% and 83%, respectively). In pot experiments at the treatment with the same concentration of Cu-CS NPs, the observed percentage efficacy of disease control in tomato plants was 88% in early blight and 61% in *Fusarium* wilt. The higher antifungal activity of Cu-CS NPs in pot experiments as compared to Petri plate experiments could be connected with strong elicitor properties of CS in the plant defence mechanism and with the fact that during infection of plants by fungi, different levels of acids produced by fungi decreased the pH resulting in the protonation of CS NH_2 groups and subsequent release of Cu^{2+} ions from Cu-CS nanoformulation, and also highly reactive hydroxyl radicals were produced which caused serious damage of biomolecules (Saharan et al. 2015). Cu(II)-loaded CS nanohydrogels, in which the formation of a Cu(II)-CS complex significantly depends on pH (the decrease of pH results in the release of Cu(II)) and the hydrogels are a suitable substrate for CS hydrolytic enzymes showed a notable synergistic effect between CS and Cu in inhibiting *F. graminearum* growth (Brunel et al. 2013).

The ultrafine colloidal CuNPs (2–5 nm in diameter) prepared using PVA capping polymer and citrate dispersant were found to exhibit notable antifungal activity against *Corticium salmonicolor*, a fungus causing pink disease in citrus and coffee and rubber trees, and showed high killing ability at concentration of 7 ppm and 10 ppm, respectively. A single spraying of 10 ppm CuNPs completely killed *C. salmonicolor* fungi, and treating diseased rubber trees with ultrafine CuNPs resulted in significant reduction of the disease index after twice spraying (Cao et al. 2014). Copper bionanoparticles with spherical shape and the size ranging from 5 to 15 nm synthesized using leaf aqueous extract of *Datura innoxia* effectively inhibited *Xanthomonas oryzae* pv. *oryzae*, the causative organism of bacterial leaf blight of paddy (Kala et al. 2016).

CuNPs prepared using the cetyltrimethylammonium bromide exhibited antifungal activity against three different crop pathogenic fungi that decreased in the following order: *Fusarium equiseti* > *F. oxysporum* > *F. culmorum* (Bramhanwade et al. 2016). The significant antifungal activity of CuNPs coated by cetyltrimethylammonium bromide with particle size ranging from 3 to 10 nm against plant pathogenic fungi *Phoma destructiva*, *Curvularia lunata*, *A. alternata* and *F. oxysporum* was observed by Kanhed et al. (2014). The antifungal activity of CuNPs, which was found to be better than that of the commercially available fungicide bavistin against all the four plant pathogenic fungi, could be connected with their large surface area to volume ratio.

Mageshwari and Sathyamoorthy (2013) designed 3D flower-shaped CuO microspheres with the average diameter of about 1–2 μm, and it was observed that flower-shaped hierarchical microspheres are composed of interpenetrating 2D nanosheet subunits as building blocks, which were self-organized to form spherical assemblies, and the spacing among the nanosheets in the flower-like superstructure favours greater interaction of microbes with the NPs, thereby enhancing the antimicrobial activity. These flower-shaped CuO nanostructures showed antifungal activity against *Mucor*, *Penicillium notatum*, *A. flavus*, *A. niger*, *A. alternata*, *Rhizopus oryzae*, *Cladosporium carrionii* and *A. flavus*. Spherical CuO NPs with the mean diameter of 28 ± 4 nm biosynthesized using *E. crassipes* leaf extract as reducing and capping agents exhibited antifungal activity against plant pathogens that decreased in the following order: *F. culmorum* > *A. niger* > *F. oxysporum* > *A. flavus* > *A. fumigatus* (Vanathi et al. 2016).

Mishra and Singh (2015) in their review paper highlighted the potential applications of AgNPs in the agricultural sector, particularly for plant disease management, focused attention on major interactions of AgNPs with soil, soil biota and plants and analysed the toxicity-determining factors which could be associated with their usage in agriculture. Spraying of 500 kg of colloidal Ag solution with a concentration of 10 ppm on 3306 m^2 large area polluted by rose powdery mildew resulted in fading out (>95%) of the white rose powdery mildew after 2 days, and it did not recur for a week (Kim et al. 2008). AgNPs caused also detrimental effects not only on fungal hyphae but also on conidial germination of ambrosia fungus *Raffaelea* sp. that has been responsible for the mortality of a large number of oak trees in Korea (Kim et al. 2009). Also AgNPs were found to increase the antifungal activity of fluconazole against *Phoma glomerata*, while no significant enhancement of activity was observed against *Phoma herbarum* and *Fusarium semitectum* (Gajbhiye et al. 2009). AgNPs with particle size <5 nm in the commercial product Pyto-patch® exhibited strong inhibition of spore germination rate and mycelial growth of *C. gloeosporioides*, *B. cinerea* and *Sclerotinia sclerotiorum* in vitro; the germination rate of spore of *C. gloeosporioides* dipped in 5 ppm phyto-patch dilute was suppressed to 13.2%; and a dose of 10 ppm proved to inhibit mycelial growth for 2 weeks. While in the field test in untreated plot, the anthracnose development after 21 days reached 40%, treatment with 4 ppm phyto-patch reduced it to 7%, and application of Pyto-patch® spraying (10 ppm) every 7 days in heavy rainfall season was found to ensure the potent control of pepper anthracnose (6% infected fruits compared to 95% in untreated plot). On the other hand, even though during drying period the effectiveness of Phyto-patch® was slightly lower (the portion of diseased fruits was 24.2%), however, in the untreated plot all pepper fruits were completely destroyed within 3 days. These findings indicate that mulching textile coated with AgNPs represents a suitable preparation for the potent prevention of late blight of pepper and it could delay the occurrence of the disease for about 1 month (Il and Kim 2012). AgNPs significantly inhibited the colony formation of *Bipolaris sorokiniana* and *Magnaporthe grisea*, whereby the corresponding IC_{50} values estimated for *B. sorokiniana* were higher than for *M. grisea*. The application of AgNPs exhibited also considerable reduction of fungal diseases on perennial ryegrass (*Lolium perenne*) caused by these two phytopatogens, and for the most effective reduction of disease severity, treatment at 3 h before spore inoculation was necessary (Jo et al. 2009). Kim et al. (2012) investigated the antifungal activity of AgNPs against 11 different plant pathogenic fungi, which were cultivated on potato dextrose agar (PDA), malt extract agar and corn meal agar plates. The most significant inhibition of plant pathogenic fungi was observed on PDA: concentration of 100 ppm caused 100% inhibition of *B. cinerea*, *Cladosporium cucumerinum*, *Corynespora cassiicola*, *Cylindrocarpon destructans*, *F. oxysporum* f. sp. *cucumerinum*, *F. oxysporum*, *Fusarium* sp., *Glomerella cingulata*, *Monosporascus cannonballus*, *P. aphanidermatum* and *Pythium spinosum* and >90% inhibition of *A. alternata*, *Alternaria brassicicola*, *A. solani*, *Didymella bryoniae*, *F. oxysporum* f. sp. *lycopersici*, *F. solani and Stemphylium lycopersici*. AgNPs affected the metabolism and toxicity of moulds and when applied in a higher concentration decreased the mycotoxin production of *Aspergillus* sp. (81–96%), and the highest decrease of mycotoxin amount was noticed for ochratoxin A (*A. westerdijikiae*). In the presence of AgNPs in the culture medium, a decrease in the organic acid production from the 3rd day of incubation was estimated, and the production of organic acids was inhibited to a greater extent in *P. chrysogenum* than in *A. niger*. The most intensive suppression was estimated for oxalic acid and the lowest one for malic acid production. Moreover, treatment with AgNPs resulted in a change in the extracellular enzyme profile of *A. niger* and *P. chrysogenum* and an increase of the total enzymatic activity (Pietrzak et al. 2015). Circular AgNPs with the mean particle size of 30–90 nm prepared using cow milk applied at concentration 2 mM exhibited 87%, 86% and 84% inhibition of the growth of *Colletotrichum coccodes*, *Monilinia* sp. and *Pyricularia* sp. (Lee et al. 2013). Spherical AgNPs of the size 40–60 nm exhibited reduction in the growth of six different *R. solani* anastomosis groups infecting cotton plants in vitro using PDA and Czapek Dox agar (CDA), while generally, higher suppression of fungal radial growth was noticed at a concentration of 1.9 mmol/L (Elgorban et al. 2016a). A notable *F. culmorum*-induced reduction in wheat seedling blight was estimated following treatment with AgNPs, and a serious disintegration of the cell membranes of roots was observed as well. Increased quantum efficiency of energy trapping in the PSII reaction centre (F_v/F_m) with a simultaneous decrease in energy dissipation in the form of heat due to treatment with AgNPs resulted in the higher total dry weight of plants (Gorczyca et al. 2015). Incubation of *F. culmorum* (W.G. Smith) Sacc. (FC) spores with AgNPs resulted in a considerable reduction of mycelial growth, which did not depend significantly on the AgNPs concentration up to 2.5 ppm, and the number of spores formed by mycelia increased in the culture after contact with AgNPs relative to control samples, mainly on the nutrient-poor PDA medium (Kasprowicz et al. 2010). The application of 100 ppm AgNPs effectively inhibited the growth of fungal hyphae as well as conidial germination of *Colletotrichum* species in vitro compared to the control, while in field trials the application of AgNPs before disease outbreak on pepper plants resulted in the considerable inhibition of fungi (Lamsal et al. 2011a). On the other hand, the application of 100 ppm AgNPs in the field tests showed the highest inhibition rate both before and after the outbreak of powdery mildew disease on cucumbers and pumpkins, and this dose of AgNPs also exhibited maximum inhibition for the growth of fungal hyphae and conidial germination in in vivo tests (Lamsal et al. 2011b). Coating of wheat seeds with AgNPs did not reduce seed germinability, and even soil conditions did not affect seed protection provided by AgNPs against fungi, which was comparable to the effect of a conventional preplanting fungicide Carboxitiram, suggesting that also this nanocoating may be considered as potential preplanting fungicide (Karimi et al. 2012).

Biosynthesized spherical AgNPs with the size ranging from 5 to 30 nm exhibited considerable antifungal activity against white mould (*S. sclerotiorum*) and grey mould (*B. cinerea*) in strawberry (*Fragaria x ananassa*), treatment with 150 ppm of AgNPs being the most effective (Elgorban et al. 2016b). Bioactive bile salt sodium deoxycholate-capped AgNPs tested against *C. gloeosporioides* exhibited fivefold higher inhibitory effect than their bioactive capping agent without causing phytotoxicity to treated plants (Muthuramalingam et al. 2015). AgNPs biosynthesized using aqueous extract of *Artemisia absinthium* and applied at the dose of 10 μg/mL inhibited the mycelial growth of *Phytophthora parasitica*, *P. infestans*, *P. pabnivora*, *P. cinnamomi*, *P. tropicalis*, *P. capsici* and *P. katsurae* in vitro, being very efficient against *P. parasitica* and *P. capsici* with IC_{50} values 2.1–8.3 μg/mL and showing complete inhibition (100%) of mycelial growth, zoospore germination, germ tube elongation and zoospore production, and in greenhouse experiments AgNPs prevented *Phytophthora* infection and improved plant survival (Ali et al. 2015). Biosynthesized AgNPs prepared using *Descurainia sophia* applied at concentration 25 μg/mL inhibited the mycelium growth of *R. solani* (>86%), and the minimum inhibitory concentration and the minimum bactericidal concentration of these AgNPs against *Agrobacterium tumefaciens* (strain GV3850) and *A. rhizogenes* (strain 15,843) were estimated as 4 and 8 μg/mL, respectively (Khatami et al. 2016a). Treatment with 40 ppm of spherical AgNPs (mean particle size of 17 nm) biosynthesized using *Trifolium resupinatum* (Persian clover) seed exudates resulted in 94.1% and 84% inhibition of fungal growth of *R.* solani and *Neofusicoccum parvum*, respectively (Khatami et al. 2016b). AgNPs synthesized using *Acalypha indica* leaf extract as reducing agents applied at a dose of 15 mg/10 μL on fungi cultivated on PDA medium showed excellent inhibitory activity against six plant pathogens (*A. alternata*, *B. cinerea*, *C. lunata*, *M. phaseolina*, *R. solani* and *S. sclerotiorum*) (Krishnaraj et al. 2012). Biosynthesized spherical AgNPs with particle size ranging from 7 to 21 nm exhibited notable antifungal activity against plant *F. oxysporum* at the concentration of 8 μg/mL (Gopinath and Velusamy 2013). AgNPs biosynthesized using *Serratia* sp. showing spherical shape and particle size ranging from 10 to 20 nm applied at concentrations 2, 4 and 10 μg/mL caused complete inhibition of conidial germination of *B. sorokiniana*, while in the control the conidial germination was 100%, and these AgNPs also significantly reduced *B. sorokiniana* infection in wheat plants under greenhouse conditions (Mishra et al. 2014). Balashanmugam et al. (2016) reported that using *Cassia roxburghii* aqueous leaf extract stable AgNPs with mean particle size 35 nm and zeta potential of −18.3 mV could be synthesized which could be used as effective growth inhibitors in controlling various plant diseases caused by fungi such as *R. solani*, *F. oxysporum* and *Curvularia* sp. (Balashanmugam et al. 2016).

A nanosized Ag–irradiated fungal CS composite showed strong botryticidal activity (MIC = 125 μg/mL), and its application to the grey mould fungus *B. cinerea* Pers resulted in an alteration in the mycelial shape and moderate lysis in fungal hyphae, which lysed into small and elastic fragments at prolonged treatment. Coating of strawberries using solution containing this nanocomposite effectively eliminated grey mould infection signs even in 90% of the contaminated fruits after 7 days of storage, securing the fresh-like appearance of strawberries in the whole storage period (Moussa et al. 2013). Ho et al. (2015) prepared Ag core−CS shell nanoclusters via chemical reduction using 3,4-dihydroxyphenyl acetic-conjugated oligochitosan as a reducing and protecting agent by its surface adhesion to 3,4-dihydroxyphenyl acetamide moieties. The size of Ag core was 26 ± 9 nm and shell layer thickness was 18 ± 8 nm. These nanoclusters applied at the dose of 9 ppm showed 80% inhibition of *P. capsici* growth, and IC_{50} value estimated for the growth of *Phytophthora nicotianae* and *P. colocasiae* was about 6 ppm. It could be assumed that the CS-based shell layer could act as an active targeting site and results in increasing interaction of the cationic CS shell layer on the Ag core in the nanoclusters and phospholipid layer on bacterial membrane via electrostatic interaction, and the sustained release of Ag^+ ions from nanoclusters situated on the surface of the microbes could kill the fungi. Ag/CS Janus particles applied at the concentration of 0.02 mg/mL suppressed the growth and germination of *B. cinerea* in vitro and in vivo (Jia et al. 2015).

It was estimated that Tween 80 is a preferable stabilizer of AgNPs due to the beneficial synergistic effects of AgNPs and the surfactant related to antibacterial activity against phytopathogenic bacterium *Ralstonia solanacearum*, and Tween 80-stabilized AgNPs caused more severe damage in direct contact with cells, causing mechanistic injury to the cell membrane and strongly modifying and destructing the cellular proteins; also in pot experiments the Tween 80-stabilized AgNPs showed high control efficiency on tobacco bacterial wilt representing 96.7% at 7 days and 84.2% at 21 days, respectively (Chen et al. 2016). A stable nanosized silica hybrid silver complex, in which AgNPs (3–10 nm) representing core part were loaded onto the outer parts of SiO_2 NPs (5–20 nm), decreased the growth of *R. solani* by more than 90% at treatment with 6 μg/mL concentration (Kim et al. 2011). The antifungal efficiency of Ag-SiO_2 NPs synthesized by γ-irradiation, in which AgNPs of about 7 nm were attached to the surface of SiO_2 NPs of approximately 350 nm, applied against *B. cinerea* at doses of 50 and 100 ppm was 99.9% (Oh et al. 2006). Protonated $H_2Ti_3O_7$ nanotubes of ca. 11 nm in diameter and four layers with surface areas 300 m^2/g functionalized with AgNPs (5 nm) effectively inactivated *B. cinerea* isolated from tomato infection under visible light, and cell death was connected with plasmalemma invagination due to oxidative stress and serious morphology damage expanding the conidia (Rodriguez-Gonzalez et al. 2016). In an in vitro experiment, the DNA-directed AgNPs grown on graphene oxide (GO) applied at the dose of approximately 10 μg/mL killed all bacterial cells of Cu-tolerant *Xanthomonas vesicatoria*, *X euvesicatoria* and *X. gardneri* strains and Cu-sensitive *X. perforans* strains in suspensions containing approximately 10^3 CFU/mL within 15 min, and the treatment of tomato plants with this nanoformulation (75 or 100 μg/mL) prior to artificial inoculation resulted in significant reduction of disease severity compared to Cu-mancozeb and negative controls (Strayer et al. 2016). The application of such nanocomposite at 100 ppm on tomato transplants in a greenhouse experiment considerably reduced the severity of bacterial spot disease caused by *Xanthomonas perforans* compared to untreated plants, showing comparable efficiency to current grower standard treatment and no signs of phytotoxicity (Ocsoy et al. 2013).

The investigation of the antifungal effect of AuNPs applied at concentration 0.05–0.2 mg/L in PDA media against *Fusarium verticillioides*, *Penicillium citrinum* and *A. flavus* that was evaluated at 2, 4, 6 and 8 days after incubation showed that not even the concentration of 0.2 mg/L was able to completely inhibit fungal growth; however, in contrast to the untreated control, damaged hyphae and unusual bulges were observed in the fungi structure, suggesting that AuNPs affected the production of toxins by these pathogenic fungi (Savi et al. 2012). AuNPs synthesized using seed aqueous extract of *Abelmoschus esculentus* with nearly spherical shape and particle size ranging from 45 to 75 nm showed higher antifungal activity against *Puccinia graminis tritici* and *Candida albicans* than against *A. flavus* and *A. niger* (Jayaseelan et al. 2013).

Commercially available Zn NPs (264 nm) and ZnO NPs (19.3 nm) were found to inhibit spore germination and infectivity on tobacco leaves resulting from exposure to the fungi-like oomycete pathogen *Peronospora tabacina*, and treatment with these NPs at 8 and 10 mg/L markedly inhibited leaf infection, and considerable higher dependence of these inhibitory effects on the concentration was estimated than could be readily explained by the presence of dissolved Zn (Wagner et al. 2016). Rajiv et al. (2013) biosynthesized spherical (27 ± 5 nm) and hexagonal (84 ± 2 nm) ZnO NPs using different (50% and 25%) concentrations of *Parthenium hysterophorus* L. leaf extracts that exhibited size-dependent antifungal activity against plant fungal pathogens *A. flavus*, *A. niger*, *A. fumigatus*, *F. culmorum* and *F. oxysporum*, showing the highest effectiveness against *A. niger* and *A. flavus*, and the antifungal activity of smaller-sized ZnO NPs exceeded that of larger NPs. The antifungal effectiveness of spherical biogenic ZnO NPs with the mean particle size of 12 ± 3 nm prepared using *Lantana aculeata* leaf extract against *A. flavus* and *F. oxysporum* was reported also by Narendhran and Sivaraj (2016). The antifungal activity of ZnO NPs prepared using reproducible bacteria *Aeromonas hydrophila* as an eco-friendly reducing and capping agent against *A. flavus* was described by Jayaseelan et al. (2012). ZnO NPs were reported to be also suitable for the control of rice blast and brown spot diseases. Spraying of ZnO NPs with the concentrations of 0.2% and 0.5% 5 days before inoculation with a spore suspension of *P. grisea* was effective in controlling rice blast disease, while spraying of ZnO NPs 2 days before inoculation with a spore suspension of *Helminthosporium oryzae* gave the best effect in controlling rice brown spot disease (Kalboush et al. 2016). ZnO NPs applied at the concentration of 100 mM were found to completely inhibit the growth of *Penicillium citrinum* and significantly reduced the growth of *F. verticillioides* and *A. flavus,* and the conidia production of all fungi also was reduced. In treated fungi, hyphae morphological alterations showing hyphae damage as a result of ROS production were observed (Savi et al. 2013; Savi and Scussel 2014). ZnO NPs with sizes of 70 ± 15 nm applied at concentration 3 mmol/L notably inhibited the growth of *B. cinerea* and *Penicillium expansum*, *P. expansum* being more sensitive to the treatment, and it was found that the growth inhibition of *B. cinerea* is connected with the alteration of cellular functions caused by ZnO NPs, while in *P. expansum* the ZnO NPs prevented the development of conidiophores and conidia, causing eventually the death of fungal hyphae (He et al. 2011). In addition, ZnO NPs in the presence of visible light exhibited strong antifungal activity, and treatment with a suspension at the concentration of 5 mM ZnO NPs and incubation time of 24 h resulted in 58% photoinactivation of *B. cinerea* (Kairyte et al. 2013). The antifungal activity of ZnO NPs against *S. rolfsii* and *Pythium debaryanum* was reported by Sharma et al. (2011), and antifungal effectiveness depended on the size, morphology and contact of ZnO NPs with the fungal cell. Polyurethane membranes modified by ZnO NPs were found to exhibit important antifungal properties against *Aspergillus brasiliensis* (ATCC 16404 strain of *A.*) (Vlad et al. 2012). The antifungal activity of ZnO NPs with the size of 35–45 nm against *M. phaseolina* was reported by Shyla et al. (2014).

ZnO NPs were found to be twofold more effective against *Aspergilus niger* than ZnO microparticles (MIC values of 2.5 and 5 mg/L, respectively), and the MIC value of 1.25 mg/L estimated for the antifungal efficiency of ZnO NPs doped with 5% nano-Pd suggested that the antifungal activity of nanoscale ZnO could be improved by loading with nano-Pd (Gondal et al. 2012). The CdSe/ZnS quantum dots coated with 3-mercaptopropionic acid were found to be considerably taken up by the fungal hyphae of *F. oxysporum*, showing their potential for the development of novel control approaches of *F. oxysporum* and related pathogenic fungi following appropriate functionalization (Rispail et al. 2014).

A CS/TiO_2 nanocomposite at the ratio of 1:5 exhibited effective inhibition of the growth of rice bacterial pathogen *Xanthomonas oryzae* pv. *oryzae*, which exceeded that of the two individual components under both light and dark conditions (Li et al. 2016). A CS/TiO_2 hybrid film exhibited excellent antifungal activity against *Bipolaris maydis* under both visible-light irradiation and in a dark environment and showed superior antifungal efficacy of 100% even after 4 h under the irradiation of visible light. A large amount of positive charges on the structure of the hybrid film, which interacted with the negative charges from the cell as well as the detrimental effect of hydroxyl radicals generated by photocatalysis of TiO_2 contributed to the strong antifungal efficacy of the CS/TiO_2 hybrid film (Huang et al. 2013a). Pure and Ag-doped solid and hollow TiO_2 NPs exhibited antifungal activity against *F. solani* causing *Fusarium* wilt disease in potato, tomato, etc. and *Venturia inaequalis* causing apple scab disease, hollow NPs being the most active, and the activity was greater under visible-light exposure due to generation of harmful ROS during photocatalysis causing damage of cell wall with consecutive cell death.

Moreover, in the presence of Ag, stable Ag–S and disulphide bonds (R–S–S–R) in cellular proteins could be formed, which also results in cell damage. It was also observed that at a very low dose (0.015 mg/plate), the NPs successfully arrested the production of toxic naphthoquinone pigment for *F. solani*, which is related to the fungal pathogenicity, and the NPs were found to protect potatoes affected by *F. solani* from spoiling (Boxi et al. 2016). In greenhouse experiments, light-activated Zn-doped nanoscale TiO_2 formulations applied at doses 500–800 ppm considerably reduced bacterial spot severity in tomato transplants artificially infected with *Xanthomonas perforans* compared with untreated and Cu control. They exhibited similar protection as the grower standard, Cu + mancozeb, and also notably reduced disease incidence in three of four trials compared with untreated transplants and Cu control, whereby their effect was comparable or better than that of the grower standard (Paret et al. 2013). Visible-light-activated Pd-modified nitrogen-doped TiO_2 NPs strongly adsorbed onto the surface of *F. graminearum* macroconidium could contribute to the photocatalytic disinfection of these macroconidia causing cell wall/membrane damage by formed ROS (Zhang et al. 2013a).

8.3.4 Other Inorganic Nanofungicides

Maize plants treated with SiO_2 NPs (20–40 nm) showed a higher expression of phenolic compounds and a lower expression of stress-responsive enzymes against both tested fungi (*Aspergillus* spp. and *Fusarium* spp.), and treatment with 10 and

15 kg SiO_2 NPs/ha resulted in significantly higher resistance in maize than treatment with bulk SiO_2, and maize plants expressed more resistance to *Aspergillus* spp. than to *Fusarium* spp. (Suriyaprabha et al. 2014).

Sulphur NPs with the particle size of 35 nm effectively prevented the fungal growth of *F. solani* and *Venturia inaequalis*, and the fungicidal effect was connected mainly with the deposition of NPs on the cell wall and the subsequent damage of the cell wall (Rao and Paria 2013). Orthorhombic (spherical; ~10 nm) and monoclinic (cylindrical; ~50 nm) sulphur NPs significantly reduced the total lipid content of treated isolates of *A. niger*, caused notable downregulation of the expression of various desaturase enzymes (linoleoyl-CoA desaturase, stearoyl-CoA 9-desaturase and phosphatidylcholine desaturase) and noteworthy high accumulation of saturated fatty acids with depleted lipid layer, which could be one of the major reasons of sulphur NP-mediated fungistasis (Choudhury et al. 2012). Surface-modified sulphur NPs prepared using PEG 400 as a surface-stabilizing agent showed promising inhibitory effect on fungal growth and sporulation and significantly reduced phospholipid content in *A. niger* and *F. oxysporum* (Choudhury et al. 2011).

8.4 Nanoinsecticides

Insecticides are compounds that are able to kill insects in various stages of development/growth, i.e. can be applied against insect eggs, larvae or adult insects. Insects represent a class of invertebrates, and among these, there are some insect pests that destroy crops and infest stored grains. Thus, application of nanoscale pesticides can be helpful in the management of insect pests in agriculture without harming the nature (Jampílek and Kráľová 2015, 2017b). An overview regarding the prospects for the development of nanoencapsulated pesticides in sustainable agriculture was presented by Grillo et al. (2016). Agents with activity against insects can be classified according to origin as natural (pure compounds or mixtures) and synthetic or, according to their composition, as organic insecticides and inorganic compounds (metal-based substances, metalloids, clays). The last ones can be used as carriers, or they possess own intrinsic insect-killing effect.

8.4.1 Nanoinsecticides Based on Plant Extracts and Essential Oils

Insecticidal activity of colloidal suspensions of PCL NPs containing neem (*Azadirachta indica*) products as well as nanocapsule spray-dried powders was tested against *Plutella xylostella* by Forim et al. (2013). On day 9, the control neem oil and the colloidal suspension of neem-loaded NPs caused 100% larval mortality, while NPs in powder caused 91.7% larval mortality, and all neem treatments were found to be more efficient than the control insecticide deltamethrin 25 EC after day

5 of the experiments. The nanoformulation also exhibited the improved stability of neem products against ultraviolet radiation and increased their dispersion in the aqueous phase. Giongo et al. (2016) offered corn leaves treated with nanoformulations of neem in colloidal suspension or powder, containing PCL, poly(β-hydroxybutyrate) or poly(methyl methacrylate) in capsules or spheres to first instar larvae of fall armyworm during 10 days and observed that some nanoformulations caused mortality and sublethal effects up to 3 and up to 7 days after spraying; however the residual effect of commercial neem oil was not outperformed. Although all treatments showed phagodeterrence at day 1 after spraying, this was lost over time indicating limited or no release of active ingredient by NPs. Microcapsules of sugarcane bagasse lignin loaded with organic extracts of neem tested as potential bioinsecticides against *Spodoptera frugiperda* and *Diatraea saccharalis* were found to have increased thermal and photo stability compared to the control, and following their administration, for 100% mortality of insects, shorter time was needed than in the controls, indicating that neem extracts loaded into microcapsules not only retained their biopesticidal activity but also exhibited better resistance against the abiotic factor (Costa et al. 2017).

Comparison of the insecticidal activity of NPs loaded with neem products and enriched botanical extract was performed by da Costa et al. (2014). Nanoformulated neem products in the form of powder, soluble powder prepared with neem oil and neem oil emulsifiable concentrate tested against bean weevil *Zabrotes subfasciatus* showed that the treatment of the insect with 1000–4000 ppm neem oil in emulsifiable concentrate resulted in the highest mortality, while the greatest UV stability was observed with nanoformulated neem products in powder. Jamal et al. (2013) investigated the efficacy of nanoencapsulated formulation of essential oil from *Carum copticum* seeds on feeding behaviour of *Plutella xylostella* (Lep.: Plutellidae) larvae and observed that the increase of oil concentration resulted in a decrease of relative consumption rate, relative growth rate, efficacy of conversion of ingested food and efficacy of conversion of digested food, and 72 h after feeding, also a notable reduction of digestibility was estimated indicating that application of this nanoformulation could result in an increase in post-ingestive toxicity of the insect. *Carum copticum* essential oil-loaded myristic acid-CS nanogel was found to exhibit considerably higher toxicity against *Sitophilus granarius* and *Tribolium confusum* than pure oil even after 48 h, being ca. nine- and fourfold more toxic than the pure oil against *S. granarius* and *T. confusum*, respectively. Moreover, as far as the effectiveness of pure oil decreased in the early days of application, this nanoformulation lost its insecticidal effectiveness after 21 days post-application for *S. granarius* and 33 days in the case of *T. confusum* (Ziaee et al. 2014a).

Cuminum cyminum L. oil-loaded myristic acid-CS nanogels exhibited higher toxicity against beetle pests *S. granarius* L. and *T. confusum*, and after 12 days these nanoformulations lost about 60% of their activity when applied against *S. granarius* and 15% for *T. confusum*, while at the same period the complete loss of *C. cyminum* oil insecticidal activity was estimated (Ziaee et al. 2014b). Spherical nanocapsules of essential oil from *C. cyminum* L. with the particle size of about 30 nm in diameter exhibited significantly higher fumigant toxicity against 1–3-day-old adult insects of

Tribolium castaneum (LC_{50} = 16.25 ppm) than pure essential oil (LC_{50} = 32.12 ppm) after 7 days of exposure (Negahban et al. 2012).

PEG NPs containing geranium or bergamot essential oils (EOs), in which the ratio EO:PG was 10%, were characterized with mean diameter <235 nm and loading efficacy >75%, and good stability enhanced the EO contact toxicity and altered the nutritional physiology of both stored product pests *T. castaneum* and *Rhyzopertha dominica*. Due to slow and persistent release of the active ingredients, they considerably increased also residual contact toxicity (Gonzalez et al. 2014). PEG-coated NPs loaded with garlic essential oil with the average diameter <240 nm showing slow and persistent release of active components from the NPs preserved over 80% of their control efficacy against adult *T. castaneum* even after 5 months, while for the free garlic essential oil applied at similar concentration (640 mg/kg), it achieved only 11%. It could be noted that the abundance and percentage content of the major components of nanoencapsulated and free oil were found to be practically the same (Yang et al. 2009).

A nanoemulsion of purslane essential oil exhibited notable strong insecticidal activity against almond moth (*Ephestia cautella*) causing mostly a complete inhibition of moth's emergence, which is attributed to the sterilizing effect of purslane oil on the moths as well as its toxicity to the deposited eggs and adult emergence during storage intervals up to 125 days. The adverse effect of essential oils nanoformulations against larvae of *E. cautella* decreased in the following order: purslane oil > mustard oil > castor oil (Sabbour and Abd El-Aziz 2016a). Also in another experiment focused on the testing of the insecticidal activity of these three oils applied in a nanoform against the granary weevil *S. granarius* under laboratory and stored conditions, the nano-purslane was found to show the highest sterilizing effect, which was reflected in a significant reduction of the mean number of eggs/female compared to control. After 125 days of storage, the percentage of emerged weevils was 7% for treatment with nano-purslane, while at application of bulk purslane, it was 21% and for the untreated control even 98% (Sabbour and Abd El-Aziz 2016b).

For nanoemulsions of essential oils from *Ageratum conyzoides*, *Achillea fragrantissima* and *Tagetes minuta* plants showing significant ovicidal, adulticidal and residual activities against the cowpea beetle, *Callosobruchus maculatus*, which were tested as fumigants, estimated LC_{50} values 96 h after treatment ranged from 16.1 to 40.5 μL/L air and 4.5–243 μL/L air against eggs and adults, respectively, and the insecticidal activity of nanoformulations notably exceeded that of bulk oils (Nenaah et al. 2015). Rani et al. (2014) prepared formulations of α-pinene and linalool with SiO_2 NPs and evaluated their antifeedant activity against the tobacco cutworm (*Spodoptera litura* F.) and the castor semilooper (*Achaea janata* L.) in laboratory bioassays. The hydrodynamic parameters of nanoformulated α-pinene (APSI) and linalool were 46 nm and 48 nm, respectively, and the zeta potential of both nanoformulations was −39.7 mV. Both nanoformulations showed higher antifeedant activity than the corresponding essential oils, and 0.1% nanoformulations of α-pinene and linalool showed 100% feeding deterrence at a dose of 0.1 μL/cm^2, while the parent terpenes produced <50% activity even at 2 μL/cm^2. The antifeedant activity of α-pinene against both species was higher than that of linalool formulations. Compared to the effect of parent terpenes, the nanoformulation was found to be 25-fold more effective against *S. litura* and 10-fold more effective against *A. janata*, while SiO_2 NPs alone did not produce any antifeedant effect on tested insects even at higher concentrations (15 μL/cm^2). The nanoformulations prolonged the shelf life of the terpenes. The observed death of larvae 3 days after treatment suggested that larvae died from starvation. On the other hand, nanoformulations did not exhibit repellent activity, since larvae had reached the treated leaf surface and even attempted to feed at all doses tested.

The insecticidal properties of formulations based on *Ocimum gratissimum* and montmorillonite-Na$^+$ (MMT-Na) as well as cetyltrimethylammonium-modified MMT-Na (MMT-Na-CTMA) were tested against the maize weevil *Sitophilus zeamais* (Nguemtchouin et al. 2013). While 7 days following treatment, the mortality of *S. zeamais* treated with essential oil without adsorbent application was not estimated, it decreased from 100% to 87% for the essential oil adsorbed on unmodified clay and to 95% for the essential oil adsorbed on modified clay. The complete loss of insecticidal activity of the formulation prepared with unmodified clay was observed after 30 days, while the formulation with organo-modified clay retained 40% of its full insecticidal efficiency at the same time. The amount of formulation required to kill 50% of *S. zeamais* adults was estimated as 1.01 g and 0.69 g for MMT-Na-*O. gratissimum* and MMT-Na-CTMA-*O. gratissimum*, respectively, indicating higher toxicity of MMT-Na-CTMA, probably due to the incorporation of more compounds with insecticidal activity (i.e. terpenic components) in this formulation.

The entomocidal activity of powders and extracts of medicinal plants *Azadirachta indica*, *Zanthoxylum zanthoxyloides*, *Anacardium occidentale* and *Moringa oleifera* against *Sitophilus oryzae* (L), *Oryzaephilus mercator* (Faur) and *Rhyzopertha dominica* (Fabr.) was reported by Ileke and Ogungbite (2014). Findings focused on the effectiveness of plant extracts, essential oils, their isolated pure compounds and plant-based nanoformulations as well as their mode of action against storage insects with special reference to maize were summarized by Soujanya et al. (2016). Preparation methods related to the encapsulation of vegetable oils and applications of encapsulated vegetable oils as antimicrobials, insecticides, pesticides and pest repellents were summarized by Sagiri et al. (2016).

8.4.2 Synthetic Nanoinsecticides

Functional nano-dispensers of imidacloprid (IMI) encapsulated in PLGA with particle sizes 5–10 μm were found to cause equivalent mortality of Asian citrus psyllids (*Diaphorina citri*) as a current commercial formulation, however at a dosage 200-fold lower (Meyer et al. 2015). Kumar et al. (2014) performed field evaluation of IMI-loaded sodium ALG NPs with particle size ranging from 50 to 100 nm, 98.66% EE and 2.46% loading. Although the pesticide content in the nanoformulation was only 2.46%, its application in the form of spray on leaves of *Abelmoschus esculentus* was found to be effective up to the 15th day in reduction of leafhopper population and exhibited not only better insecticidal activity but also lower toxicity than pure pesticide. Guan et al. (2008) prepared photodegradable insecticide by direct encapsulation of IMI microcrystals with CS and sodium ALG through layer-by-layer self-assembly using sodium dodecyl sulphate (SDS)/Ag/TiO_2 as an effective photocatalyst. The IMI microcrystals had the mean length of 7 μm and the zeta potential of −37.5 mV. The IMI-loaded microparticles showed encapsulation efficiency 81.57 ± 0.96%, and the percentage of the drug-loading content was approx. 56.15 ± 0.96% after encapsulated for ten polyelectrolytes layers. The release rate of the IMI microcrystal decreased with an increase in the layer number of microcapsules, and the total release time for the corresponding microcapsules with 4, 10 and 20 layers was approximately 2-, 4- and 8-fold longer, respectively, than that of the uncoated pesticide.

Amphiphilic nano-polymers synthesized using different molecular weight PEGs (300, 600 and 1000) as a hydrophilic head and aliphatic diacids (glutaric acid, adipic acid, pimelic acid and suberic acid) as a hydrophobic moiety were used to prepare controlled release formulation for IMI. The micelle size of the polymers ranged from 127 to 354 nm, the loading capacity of the polymers ranged from 6.8% to 8.9% and the encapsulation efficiencies for different formulations were in the range from 75.0% to 97.9%. The value of half-life $t_{1/2}$ (i.e. time taken for 50% release) of IMI encapsulated in polymers ranged from 2.3 to 9.3 days, being higher for the formulation containing PEG 1000 than for polymers having PEG 300 and PEG 600 moiety, and $t_{1/2}$ was found to increase with the increasing molecular weight of PEG for diacids, namely, adipic acid and suberic acid. Thus, imidacloprid applications can be optimized to achieve insect control for the desired period using a suitable matrix of the polymer (Adak et al. 2012).

Memarizadeh et al. (2014) encapsulated IMI into ABA triblock linear-dendritic copolymers composed of polycitric acid (PCA) as A block and polyethylene glycol (PEG) as B block, the encapsulation process being performed by self-assembly of PCA-PEG-PCA in the presence of IMI in different solvents. The morphology of nano-IMI varied from fibre-like to globular and tubular, while its size varied from 10 nm to several mm, depending on the type of solvent, time and concentration. The loading capacity of the copolymers at pH 7 was estimated as 53% and, at pH 10, it was 80%. While IMI release at pH 7 slowly increased for 6 h and then remained constant, its release rate into phosphate-buffered saline solution with pH 10 increased up to 24 h, and higher percentage of pesticide was released than at pH 7. The insecticidal efficiency of nano-IMI and bulk insecticide diluted in water was investigated by leaf-dip bioassay tests on the *Glyphodes pyloalis*. The LC_{50} values estimated for the nano-IMI decreased over free IMI as exposure time increased, and after 4 and 5 days of exposure, they were five- and ninefold lower, respectively, than those observed for the bulk form. In the topical bioassay, the performance of nano-IMI prepared in ethanol was tested. Comparison of LC_{50} values observed at 24, 48, 72 and 96 h showed that at all periods of exposure, the LC_{50} values were considerably lower for the nanopesticide formulation than for free IMI. The increased penetration of the effective compound by means of citric acid molecules to the metathoracic tergum membrane cells of *G. pyloalis* larvae contributed to the higher effectiveness of nano-IMI prepared with ethanol. The higher loading capacity and the slower release rate of the pesticide from nano-IMI formulation at pH 10 corresponding to optimum pH of *G. pyloalis* gut compared to neutral pH suggest its selective and controllable action. Lower doses of nano-IMI compared to its bulk form required for pest control can also significantly reduce the environmental risk.

In contact toxicity bioassay using adult *Martianus dermestoides*, the 142-h LC_{50} values estimated with 50% nano-SDS/Ag/TiO_2-IMI were 9.86 mg/L compared to 13.45 mg/L observed with 95% IMI. The use of bentonite and/or activated carbon sorbents reduced the release rate of IMI and isoproturon in comparison with the technical product and with ALG formulation without modifying agents. The formulation with the highest percentage of activated carbon exhibited the highest decrease in the release rate, and the release rate was higher in imidacloprid systems than in those prepared with isoproturon (Garrido-Herrera et al. 2006).

Neonicotinoid acetamiprid-loaded nanocapsules prepared by polyelectrolyte complexation of ALG and CS showed controlled release in vitro, with maximum release at pH 10, the released amount decreasing with decreasing pH, and a controlled release pattern was observed also in soil, indicating that such nanoformulation could reduce the frequency of application of pesticides and reduce their side effects (Kumar et al. 2015b). Amphiphilic copolymers prepared from PEGs and various aliphatic and aromatic diacids, which self-assemble into nanomicellar aggregates, were used to prepare controlled release formulations of thiamethoxam, a systemic insecticide from the class of neonicotinoids exhibiting a broad spectrum of activity against many types of insects. The average micelle size of different formulations was approx. 138 nm, and the size of pyridalyl was <100 nm. The release of the insecticide from these nanoformulations was slower than from a commercial formulation with $t_{1/2}$ values ranging from 3.5 to 6 days, and the formulations showed non-Fickian transport (Sarkar et al. 2012).

The toxicity of the suspension of ALG nanocapsules containing pyridalyl against the larval stage of *Helicoverpa armigera* was tested using the leaf dip as well as the topical methods and compared with the toxicity of a technical material and a commercial formulation. The excellent insecticidal activity of the nanoformulation was confirmed by the estimated LC_{50} values of 40 and 80 μg/mL using the two above-mentioned methods. In the form of the nanoformulation, pyridalyl was found to be ca. two- and sixfold more effective against *H. armigera* as stomach poison than the technical product and the commercial formulation, respectively, while the LC_{50} values estimated by the topical method were 80, 150 and 250 μg/mL for nanoformulation, technical material and commercial formulation, respectively. The higher insecticidal effect of the nanoformulation could be connected with the better penetration of NPs through the epithelial lining of digestive tract and better penetration in capillaries to get into the systemic circulation, affecting the tertiary structure of protein, resulting finally in the malfunctioning and the death of the insect (Saini et al. 2014).

Organophosphate insecticide chlorpyrifos loaded CS-PLA-1,2-dipalmitoyl-sn-glycero-3-phosphoethanolamine copolymer NPs with the particle size of 100–300 nm exhibited controlled release by adjusting the ratio of copolymer to chlorpyrifos, showed an initial burst release and then a steadier release profile, the released amount depending on the amount of chlorpyrifos entrapped in the NPs, and the increased amount of the insecticide within NPs resulted in its decreased release (Zhang et al. 2013b). A nanohybrid prepared by intercalation of a chlorpyrifos inclusion complex with carboxymethyl-β-cyclodextrin into the interlayer of Zn-Al-layered double hydroxides showed distinct slow release, unlike to nanohybrid, in which sulphonated hydroxyethyl-β-cyclodextrin was used for chlorpyrifos inclusion complex and the kinetic process of pesticide release could be fitted well by the pseudo-second-order and the parabolic diffusion models (Liu et al. 2016). A nanocomposition prepared by encapsulation of organophosphate acephate with the particle size of 80–120 nm and irregular shape showed high efficacy against *S. litura*, *Lipaphis erysimi* (mustard aphid) and *Bemisia tabaci* (whitefly) both in vitro and in vivo and was found to be more effective than a commercial bulk formulation. Treatment with this nanocomposite at 300 ppm resulted in approximately 100% mortality of *S. litura* within 7 days, and at application of 240 ppm and 180 ppm almost 75% and 20% larvae, respectively, were killed. Higher concentrations also reduced the fecundity of larvae when they reached adulthood. Treatment with 300 ppm of nanocomposite caused about 100% mortality of the mites after 5 days of treatment. In the field study, foliar spray of a nanocomposition at 180 ppm, 240 ppm and 300 ppm ensured the superior control of *S. litura* and *Lipaphis erysimi* compared to the commercial one. Reduced acetylcholinesterase activity due to nanoacephate treatments indicated more binding of the active constituent of acephate with thiocholine, which could be a probable reason for breaking resistance in lepidopteron pest. Consequently, it could be expected that this nanoformulation might overcome the problem of reduced target site sensitivity, one of the major causes of resistance development in insects (Pradhan et al. 2013).

The toxic effect of urea-based insecticide novaluron NPs (50–200 nm) on Egyptian cotton leaf worm *Spodoptera littoralis* larvae was found to be similar to that of the commercial formulation (Elek et al. 2010). The effects of two nanotypes of pyrifluquinazon and a non-nanotype of this insecticide which modifies insect behaviour by rapidly stopping feeding so that insects starve to death on the mortality of the green peach aphid *Myzus persicae* were investigated using concentrations of 25, 50 and 100 ppm. The nanoformulations of pyrifluquinazon were prepared using a different molecular weight CS as coating material (CS 30,000 (0.1%) and CS 3000 (0.3%). The best controlled release feature was observed with the CS 3000, 0.3% nanotype pyrifluquinazon. Both CS-containing nanoformulations were effective against *M. persicae* at 14 days after treatment, and the reaction time slowed from 14 to 30 days after treatment in the aphids treated with CS 3000 (0.3%), while the best lethal efficiency of non-nanotype insecticide applied at 50 and 25 ppm was estimated at 2 days after treatment (Kang et al. 2012).

Liu et al. (2008) reported flash nanoprecipitation using a multi-inlet vortex mixer as the technology to produce bifenthrin NPs suspensions with sizes between 60 and 200 nm, the stability of which depended on the properties of the polymeric stabilizer. The most stable NPs with the narrowest size distribution were prepared using a block copolymer of polyacrylic acid and polybutylacrylate, but stable NPs were fabricated also with polyvinylpyrrolidone and polyvinyl alcohol. Bang et al. (2011) prepared CS-coated nanoliposomes containing etofenprox or α-cypermethrin using different types and concentrations of CS to regulate the mean size and the surface charge and found that as the CS concentration (0.1–0.5%, w/v) and the degree of deacetylation increased, surface charge also increased, and the release period of the entrapped insecticide could be prolonged by increasing the intrinsic surface charge or concentration of the coating material. By encapsulation of β-cyfluthrin in PEGs of different molecular weights (600, 1000, 1500 and 2000), insecticidal controlled release nanoformulations were prepared, and their effect on the mortality of *C. maculatus* (Coleoptera: Bruchidae) was tested. The approximate EC_{50} values of different test formulations against *C. maculatus* for 1, 3, 7, 14, 21 to 30 days in water after 24 h exposure of each day were estimated. At the 7th day, the formulations with PEG 600 and PEG 100 showed lower EC_{50} values than PEG 1500 and PEG 2000 due to faster release of the pesticide. The formulations prepared with PEG 1500 and PEG 2000 showed minimum EC_{50} on 14th day (2.20 and 1.58 mg/L, respectively) and mean EC_{50} value during 30 days (36.98 and 32.23 mg/L, respectively), whereby all prepared nanoformulations were more effective than a commercial preparation with the mean EC_{50} value of 124.29 mg/mL during 30 days (Loha et al. 2012).

Biocompatible oil-core silica-shell nanocapsules designed for sustained release of fipronil insecticide, in which release of insecticide can be tuned through control of the silica-shell thickness (i.e. 8–44 nm), showed insecticidal effect against economically important subterranean termites (Wibowo et al. 2014). Guo et al. (2015) fabricated enzyme-responsive emamectin benzoate microcapsules based on a copolymer matrix of SiO_2-epichlorohydrin-carboxymethyl cellulose showing excellent protection of the active ingredient against photo- and thermal degradation, notable cellulase stimuli-responsive properties as well as sustained insecticidal efficacy against *M. persicae*, and their genotoxicity was less than that of grade emamectin benzoate. Because mineral particles can scratch the exoskeletons of insects resulting in wounds, block their spiracles, reduce activity and cause strong dehydration of the insect resulting in death, they are used in the protection of plants against tiny insects.

8.4.3 Insecticides Based on Nanoscale Metals

The investigation of the impact of AgNPs on the life history parameters of two agricultural pest insect species, *Heliothis virescens* (tobacco budworm) and *Trichoplusia ni* (cabbage looper), and a beneficial predatory insect species, *Podisus maculiventris* (spined soldier bug), showed that AgNPs retarded the development, reduced the adult weight and fecundity and increased mortality in the predator, although they practically did not affect the developmental times, pupal weights and adult emergence. Thus, the adverse effects of AgNPs on the beneficial insect species require considering carefully the risk of their widespread application in insect pest management (Afrasiabi et al. 2016). The application of AgNPs with the particle size ranging from 42 to 98 nm prepared using *Sargassum muticum* extract resulted in significant changes in the protein profile of hemolymph, morphology of hemocytes and deteriorated midgut inclusions such as lumen, basement membrane, fat body and gastric caeca of *Ergolis merione*. In treated larvae, the hemocytes had thin or no outer membrane, and in the fat body, the lipid content was denatured and washed out, which led to opening its inclusion to the lumen of the midgut and attaining irregular shape and size (Moorthi et al. 2015). Yasur and Pathipati (2015) investigated the susceptibility of two lepidopteran pests of castor plant (*Ricinus communis* L.), asian armyworm, *S. litura* F. and castor semilooper, *A. janata* L. to polyvinylpyrrolidone (PVP)-coated AgNPs with particle size <100 nm and zeta potential of 22.3 ± 5.78 mV. They fed larvae with castor leaves treated with AgNPs or $AgNO_3$ and observed a decrease in larval and pupal body weights of both insects as well as changes in the antioxidative and detoxifying enzymes of the treated larva indicating that exposure of larvae to AgNPs led to induction of oxidative stress, which was countered by antioxidant enzymes. The LD_{50} and LD_{90} values of AgNPs synthesized using aqueous leaves extracts of *Euphorbia prostrata* having the rod shape and the size of 25–80 nm with the average size of 52 nm against *S. oryzae* L. were estimated as 45 mg/kg and 168 mg/kg, respectively, and they were found to be significantly lower than the corresponding values estimated for $AgNO_3$ (248 and 2675 mg/kg, respectively). Moreover, no fresh insect infestation was found in the AgNP-treated stored rice even after 2 months of treatment, indicating the superb potential of AgNPs as a stored grain and seed protecting agent if applied with proper safety measures (Zahir et al. 2012). Remarkable pesticidal activity on *S. oryzae* was shown also by AgNPs (15–25 nm) synthesized using *Avicennia marina* (Sankar and Abideen 2015). Rouhani et al. (2013) reported LC_{50} values related to insecticidal effect of AgNPs on the cowpea seed beetle, *C. maculatus* F. (Coleoptera: Bruchidae), as 2.06 g/kg (adults) and 1.00 g/kg (larvae), respectively. The experiments with a model insect *Drosophila melanogaster* showed that the activity of Cu-dependent enzymes, namely, tyrosinase and Cu-Zn superoxide dismutase, significantly decreased following the consumption of AgNPs, despite the constant level of Cu present in the tissue, which resulted in cuticular demelanization, because tyrosinase activity is essential for melanin biosynthesis (Armstrong et al. 2013). In the third instar larvae of *D. melanogaster* that were fed with a diet of standard cornmeal media mixed with AgNPs at the concentrations of 50 and 100 μg/mL for 24 and 48 h, the AgNPs induced heat-shock stress, oxidative stress, DNA damage and apoptosis (Ahamed et al. 2010).

AuNPs showing multiple irregular shape, crystalline nature and particle size in the range 20–50 nm prepared using latex of *Jatropha curcas* inhibited catalytic potential of trypsin due to the formation of trypsin–AuNPs complex because of covalent and electrostatic interactions of AuNPs with proteins and binding to –SH groups of aminoacids. This finding was supported also by investigations performed in vivo on serum of several vectors and agriculturally important pests (Patil et al. 2016). The citrate-capped AuNPs exhibited significant in vivo toxicity in the model insect *D. melanogaster* upon ingestion, which was reflected in a significant reduction of the life span and fertility, presence of DNA fragmentation as well as a significant overexpression of the stress proteins (Pompa et al. 2011).

As a safe alternative to insecticides in protection of rice grains against *S. oryzae* (Linnaeus), natural rock powder and ZnO NPs could be used (Hamza 2012). Shu et al. (2012) investigated the response of *S. litura* to zinc stress and found that the treatment with 50–500 mg Zn/kg resulted in notable induction of both metallothionein content and metallothionein gene expression in the midgut as well as changes in cell ultrastructure (mainly the presence of electron-dense granules in the cytoplasm of the midgut cells), showing significant positive correlation with Zn accumulation in the midgut, which could be considered as effective detoxification mechanisms in the common cutworm. Derbalah et al. (2014) tested the insecticidal activity of ZnO NPs and SiO_2 NPs against the pink bollworm *Pectinophora gossypiella*, which is one of the key pests of cotton in the world, and compared it with that of conventional insecticide pyriproxyfen. In these tests the effects of individual materials on some liver function enzymes, carbohydrate hydrolyzing enzymes, total protein and total lipids of the 4th instar larvae of the *P. gossypiella* pest were also investigated, and it was found that ZnO NPs were the most effective against the newly hatched larvae.

Biosynthesized NiNPs with cubical shape and the average particle size of 47 nm showed insecticidal activity against agricultural pest *Callosobruchus maculatus* resulting in 97% mortality (Elango et al. 2016).

8.4.4 Insecticides Based on Nanoscale Metalloids

Amorphous SiO_2 NPs (15–30 nm) were found to be highly effective against insect pest *S. oryzae* causing more than 90% mortality, indicating the effectiveness of SiO_2 NPs to control insect pests (Debnath et al. 2011). Spherical amorphous SiO_2 NPs with the size 70–80 nm were also found to be highly effective against stored grain pest *Corcyra cephalonica*, causing 100% mortality, suggesting their potential to control insect pests (Vani and Brindhaa 2013). The study focused on the cellular uptake of amorphous SiO_2 NPs (<30 nm) in the midgut of the third instar larvae of *D. melanogaster* that were exposed orally to 1−100 μg/mL of SiO_2 NPs for 12−36 h showed considerably increased expression of hsp70 and hsp22 along with caspase activation, membrane destabilization and mitochondrial membrane potential loss (Pandey et al. 2013). The experiment with adults of *R. dominica* F. and *T. confusum* Jacquelin du Val. that were exposed to SiO_2 NPs Aerosil® and Nanosav at the rate of 0.2 mg/cm^2 for 1 and 2 days on filter paper inside plastic Petri dishes confirmed the significant toxic effects of SiO_2 NPs on both insects, *R. dominica* being more susceptible. At low concentrations, Aerosil® was more effective than Nanosav, and the effectiveness of SiO_2 NPs in wheat grains was higher than in barley (Ziaee and Ganji 2016). Santo-Orihuela et al. (2016) tested bare SiO_2 NPs (14, 380 and 1430 nm) and amine-modified SiO_2 NPs (131 and 448 nm) on the viability of *S. frugiperda* cells (Sf9 cell line) and found that 14 nm NPs were the most effective. Exposure to 0.12 mg/mL SiO_2 NPs during 24 h resulted in the reduced viability of the cells by 60% compared to the control, and activity of cells was lowered also in the presence of other negatively charged NPs. On the other hand, positively charged NPs applied at concentrations 0.12 and 0.6 mg/mL were found to promote the proliferation of the cells, while the effect of higher concentrations (7.2 mg/mL) was comparable with that of the control. Silica NPs caused mortalities to carmine spider mite and two-spotted spider mite with mean lethal concentrations 317, 116 and 112, 83 ppm, 7 days after treatment, for *Tetranychus cinnabarinus* and *Tetranychus urticae* adult females and eggs, respectively, as well as the mortality of their predatory species *Stethorus punctillum* (97%), *Phytoseiulus persimilis* (35%) and *Orius insidiosus* (32%) (Hala and Elsamahy 2016). Soil and foliar treatments with SiO_2 NPs led to 37% and 44% feeding inhibition rate in oriental armyworm *Mythimna separata* (Walker) and elongation of the larval stage period to 31 days compared to 26 days observed in the control, and mortality percentages of larvae at SiO_2 NPs administration in the form of spray was 67%, while in the control it represented only 10% (Mousa et al. 2014). Rouhani et al. (2013) investigated the insecticidal effect of SiO_2 NPs on the cowpea seed beetle, *C. maculatus* F. (Coleoptera: Bruchidae), and estimated LC_{50} value of 0.68 g/kg on adults and 1.03 g/kg on larvae, respectively, and the high efficiency of SiO_2 NPs on adults was also reflected in 100% mortality. *Capsicum annuum* proteinase inhibitor immobilized on SiO_2-based nanospheres and rods showed bioactive peptide loading 62% at acidic pH and 56% of peptide release at pH 10, simulating gut milieu of the target pest *H. armigera*, and in vivo study showed that on the 8th day after feeding with this nanoformulation, about 40% reduction in insect body mass was estimated compared to control insects. This indicates the potential of peptide nanocarriers in delivering diverse biologically active complexes specific to gut pH of *H. armigera* (Khandelwal et al. 2015).

Spherical SiO_2 NPs synthesized by sol-gel method and surface functionalized in situ with 3-mercaptopropyltrimethoxysilane (MPTS) and hexamethyldisilazane (HMDS) with the size ranging from 15 to 20 nm for HMDS and from 29 to 37 nm for MPTS were found to exhibit insecticidal activity against the second instar larvae of *S. litura*, and their application at a dose 125 mg/cm^2 resulted in 58% (HMDS) and 64% (MPTS) mortality. Treatment with 0.25 mg/cm^2 of MPTS functionalized SiO_2 NPs killed all insect larvae, while the application of HMDS functionalized SiO_2 NPs resulted in 84% insect mortality at the same dose, and no survivors were estimated after application of both SiO_2 NPs at a dose of 0.5 mg/cm^2. The dead bodies of the insects were found to be remarkably dehydrated indicating that abrasion or, to some extent, also the absorption of lipids present in cuticle caused by SiO_2 NPs damaged the cuticular water barrier of *S. litura* resulting in the loss of water from the body and subsequent death due to desiccation (Debnath et al. 2012).

Amorphous nanosilica and nanoalumina were also found to be highly effective against mustard aphid *Lipaphis pseudobrassicae* (Debnath et al. 2010). Goswami et al. (2010) applied solid SiO_2, Al_2O_3, TiO_2 and ZnO NPs at doses 0.5, 1.0 and 2.0 g/kg against rice weevil *S. oryzae* and found that on the first day the treatment with 1 g/kg of hydrophilic SiO_2 NPs was the most effective. At application of 2 g/kg of SiO_2 NPs and Al_2O_3 NPs, the mortality on day 2 represented 90%, and after 7 days of exposure, 95% and 86% mortality was obtained with hydrophilic and hydrophobic SiO_2 NPs at 1 g/kg, while the treatment of rice with lipophilic SiO_2 NPs at 1 g/kg resulted in approximately 70% mortality, and Al_2O_3 NPs killed almost all the insects using a dose of 0.1 g/kg dose.

The experiments of Buteler et al. (2015) who tested the effect of three unique types of nanoalumina dust with particles <50 nm as an insecticide against 0–6-week-old adults of the rice weevil *S. oryzae* and the lesser grain borer *R. dominica*, two species that differ in their susceptibility to inert dusts, showed that insecticidal activity depended on particle size, particle morphology and surface area; however minimizing particle size and maximizing surface area were not the sole dominant factors influencing the efficacy. All dust types were more effective on *S. oryzae* than on *R. dominica*, and the dust synthesized using a modified glycine-nitrate combustion process consistently yielded greater mortality rates. In general, the superb efficacy of the dusts for both insect species was observed at low humidity, which decreased significantly at elevated humidity indicating that dusts can adsorb either water or cuticle waxes and thus atmospheric water reduces the effectiveness of all the tested dusts by competing with the cuticle hydrocarbons. The inert dusts absorb epicuticular hydrocarbons by capillary forces, and dusts with smaller particle size will cause greater insect mortality. Stadler et al. (2010a) tested insecticidal activity of nanostructured Al_2O_3 applied as dry dust against *S. oryzae* L. and *R. dominica* (F.), which are major insect pests in stored food supplies. Exposure of the insects to *T. aestivum* plants treated with nanostructured Al_2O_3 reduced survival in both species, whereby mortality in both species increased with increasing exposure interval and product concentration. While after 3 days of continuous exposure to 500 mg/kg, the mortality of *S. oryzae* and *R. dominica* adults represented 20% and 40%, respectively, at treatment with 250 mg/kg during 9 days, 80% of the adults of both species were dead. The LD_{50} values estimated after 9 days of exposure were 149 mg/kg (*R. dominica)* and 177 mg/kg (*S. oryzae*), respectively. It is suggested that nanoalumina kills arthropods by adsorbing epicuticular lipid layers through capillarity, causing excessive water loss through the cuticle (Stadler et al. 2010b), and those with a small particle size and high surface areas having a composition conducive to wetting of the specific hydrocarbons present on the surface of the insect could be suggested as the most effective dusts. In addition, nanostructured alumina particles were found to be more effective in killing *S. oryzae* than dry dust applications of Protect® diatomaceous earth, were equally toxic to *R. dominica* and caused also the reduction of progeny production, and *S. oryzae* showed higher susceptibility to inert dusts than *R. dominica* (Stadler et al. 2012). Huang et al. (2013b) investigated the effect of Al_2O_3 NPs on the rhythmic activities in the antennal lobe of *Drosophila* using patch clamps to record electrophysiological activities and found that 15 min after their application the average frequencies of spontaneous activities were significantly decreased compared with control groups indicating that these NPs might have adverse effects on the central nervous system in *Drosophila*.

8.4.5 Other Inorganic Nanoinsecticides

The estimated LC_{50} values at application of a 20% calcium carbonate suspension concentrate with particle sizes about 100 nm and a 95% bulk calcium carbonate powder (>1 μm) on infestations of peach aphids (*M. persicae*) were 2685 and 93,036 ppm, respectively, indicating that in controlling peach aphids, the 20% calcium carbonate suspension concentrate was the most effective (Liu et al. 2014). The study of the influence of temperature and humidity on the insecticidal effect of three diatomaceous earth formulations (Protect-It, PyriSec and DEA-P) against larger grain borer *Prostephanus truncatus* (Horn) (Coleoptera: Bostrychidae) adults in stored maize (*Z. mays* L.) at three temperatures (20, 25 and 30 °C) and 55% and 75% relative humidity levels showed that DEA-P was the most effective and caused complete mortality to the exposed insect, even at the lowest dose rate (75 ppm), completely suppressed progeny production, and its efficacy was continuously high in all temperatures and relative humidities examined (Athanassiou et al. 2007). The acaricidal effect of different diatomaceous earth formulations (SilicoSec, PyriSec, Insecto, Protect-It and DEA-P) applied at the dose 0.2 and 0.5 g/kg against *Tyrophagus putrescentiae* on stored wheat was investigated by Iatrou et al. (2010) by measuring the mortality of mite individuals after 5 days of exposure and checking for *T. putrescentiae* offspring on the treated wheat after 30 days. The application of 0.2 or 0.5 g/kg caused mortality of both adults and immatures >78%, and treatment with the dose 0.5 g/kg resulted in 100% mortality. The immature stages of the insect were less tolerant to diatomaceous earth formulations than the adults, and PyriSec was found to be the most effective against adults, whereby increasing of the dose led to considerable reduction of progeny production.

8.5 Risks of Nanopesticide Applications

In recent decades, advances in nanotechnology engineering have given rise to the rapid development of many novel applications in various industrial fields. Nanoscale materials exhibit unusual physical, chemical and biological properties, differing in important ways from the properties of bulk materials and single atom or molecule (National Nanotechnology Initiative 2008; Medina et al. 2007; Dolez 2015; Borm et al. 2006; Buzea et al. 2007; Fröhlich 2013). It is no wonder that nanotechnology has also found its use in agriculture and the food industry. Nanosystems/nanomaterials have been used for plant protection and nutrition in the form of nanopesticides or nanofertilizers or other plant growth-stimulating nanoscale materials (Jampílek and Kráľová 2015, 2017b, c; Masarovičová and Kráľová 2013; Masarovičová et al. 2014; Kookana et al. 2014; Bleeker et al. 2015). Various nanocomposites have been applied for protection of different foodstuffs, smart active or responsive packaging materials and edible coatings. Diverse nanosensors applicable for monitoring of food quality, safety and integrity can be found as well (Kookana et al. 2014; Bleeker et al. 2015; Jampílek and Kráľová 2015, 2018b). Thus, the application of nanotechnology for sustainable intensification of agricultural production, such as crop protection agrochemicals, but also agents facilitating the protection of plants against pesticides, enhancing plant growth, securing rise of global food production, guaranteeing enhanced food quality and minimizing the waste, can be considered as an excellent solution, but the most critical is stability and degradability all these nanomaterials (Prasad et al. 2014, 2017; Bleeker et al. 2015; Jampílek and Kráľová 2015, 2017b; Andronescu et al. 2016; Sangeetha et al. 2017a, b, c).

Due to their direct and intentional application in the environment, nanoagrochemicals may be regarded as particularly critical in terms of possible environmental impact, as they would represent the only intentional diffuse source of engineered NPs in the environment (Kah et al. 2013; Kah 2015). Although many nanomaterials showed benefits, nanosystems used in agriculture and the food industry, especially when they are exceedingly stable in the environment, can contaminate water resources and/or ground and return to the life cycle. On the other hand, they can contaminate food products by residues of packaging materials/edible coatings. Thus, NPs can be associated with some risks (toxicity) for human and environmental health. Possible routes of entry into the body include inhalation, absorption through the skin or digestive tract (Jampílek and Kráľová 2015, 2017a, b, c, 2018a; Khan 2013; Bleeker et al. 2015; Andronescu et al. 2016).

Different permeation through cell walls/membranes into cells is probably the most affected and the most valuable parameter in case of nanopesticides and their application for crop protection. In this context, especially particles with particle size <100 nm are critical, because they are able to practically unlimitedly permeate through biomembranes. In general, the ability of NPs "to permeate anywhere" is connected primarily with their particle size and shape (Hagens et al. 2007; Buzea et al. 2007; Keck and Müller 2013; Nehoff et al. 2014). These NPs may be more easily taken up by any organism, which could result in their longer persistence in environmental systems. The small size (an extrinsic property) of NPs influences these effects more significantly than a unique nanoscale property representing an intrinsic property (Buzea et al. 2007; De Jong and Borm 2008; Auffan et al. 2009; Kumar et al. 2012; Brayner et al. 2013; Janrao et al. 2014). Mainly adverse effects of NPs accumulated in the cell leading to intracellular changes such as disruption of organelle integrity, gene alterations, etc., or cytotoxic effects by generation of ROS as well as reactive nitrogen species resulting in the damage of plasma membrane, cell organelles and intracellular proteins are critical. Considering the potential toxicity of NPs and different nanomaterials on living organisms and also on human health, it is indispensable to minimize their entry into the environment (Ventola 2012; Berkner et al. 2016; Vestel et al. 2016).

Based on particle size definitions of NPs (European Commission 2011; National Nanotechnology Initiative 2008), a classification of NPs (Sioutas et al. 2005) according to biodegradability (ability of a compound to degrade in organism/environment) into four classes was suggested as follows: (*i*) size >100 nm and biodegradable, (*ii*) size >100 nm and non-biodegradable, (*iii*) size <100 nm and biodegradable and (*iv*) size <100 nm and non-biodegradable (Keck and Müller 2013). Logically, the last class of NPs is considered as the most dangerous for human health. Toxicological properties of NPs are affected by particle shape, size, surface area, surface charge and the adsorption properties of the material as well as by abiotic factors such as pH, ionic strength, water hardness and the presence of organic matter (Handy et al. 2008). While inside cells, NPs might directly provoke alterations of membranes and other cell structures and molecules as well as protective mechanisms, NPs can exhibit also indirect effects depending on their chemical and physical properties, e.g. physical restraints (clogging effects) or production of ROS (Navarro et al. 2008).

In the light of these facts, some multinational corporations, such as BASF, Bayer, Monsanto, DuPont and Syngenta, focused among others on pesticide production and started to invest in the development of nanopesticides, concealing this development before public, because prefix "nano" cannot be currently perceived so positively (Suppan 2013), which does not, however, mean that the first nanopesticides cannot be already applied (Gewin 2015). On the other hand, many nanoagrochemicals described in the scientific literature do not meet the cost-benefit requirements, and their production is not profitable (Aschberger et al. 2015).

It is also important to note that nanopesticides offer many advantages, such as increased efficacy, dose reduction, lower exposure to nontarget organisms or lower risk of resistance development. It can be stated that launched nanoagrochemical products mostly consist of "nano" formulations of already registered ingredients, and thus they are very similar to many agrochemical products currently available on the market (e.g. emulsions, suspensions) (Aschberger et al. 2015; Gewin 2015; Kah 2015).

Due to the above-mentioned facts, regulatory authorities will play a crucial role in the future development of other nanoagrochemicals. The facts that NPs are more efficient or that their application has some benefits for crop protection have been proven and are unexceptionable. Several international organizations coordinated seminars on nanotechnology applications for agriculture (e.g. FAO 2010, 2013; JRC-IPTS 2014). The activities of governments and regulatory authorities dealing with the development of legislation adapted to nanoagrochemicals vary considerably (FAO 2013; APVMA 2014). The extent to which nanoagrochemicals are developed will be strongly influenced by the regulatory system that controls their entry into the market. There are currently great geographic discrepancies that can influence applications in a given market (Watson et al. 2011). There are considerable issues relating to the definition of NPs and how the proposed criteria can be applied to nanopesticides (Kah et al. 2013; EC 2014; JRC-IPTS 2014).

When considering all the nanoproducts that appear in the agriculture and food sectors, there is a generally accepted consensus that there has been currently insufficient reliable data and the level of knowledge to allow a clear safety/risk assessment (FAO 2013; JRC-IPTS 2014). However, prohibiting the application of nanopesticides until they are proven entirely safe is unrealistic, as all pesticides are inherently toxic (at least to the target pest) and, thus, associated with some risk. When considering only nanoagrochemicals, a conventional approach to risk assessment – the hazard×exposure paradigm – would result in a number of pitfalls (Kookana et al. 2014). The exposure assessment is based on investigations into the environmental fate of a compound. There have been a limited number of studies investigating nanoagrochemicals (Kah et al. 2013; Kah and Hofmann 2014). It is also probable that endpoints of fate and dangers are not sufficiently determined by the use of protocols previously developed for other types of chemicals (Kah et al. 2014; Kookana et al. 2014). Thus, a fair assessment of nanopesticides should be focused on the evaluation of both the risks and benefits associated with their use relative to current solutions. While this may not be possible when considering all products discussed so far in literature, restricting the analysis to products that are likely to emerge in the next decade shows that a fair assessment may be possible (Kah et al. 2015).

The effects of agrochemical formulations on the environment and the effect of active ingredients have been evaluated within the EU under Directive 91/414. The new EU pesticide regulation (1107/2009) states that the impact of formulations should be taken into account, but it also comes with recommendations that it is reasonable to assume that "formulation does not affect the fate and behaviour of the active ingredient in the environment" (e.g. European Commission 2009). The authorization of pesticides has long been the subject of a rigorous and constantly protective regulatory risk assessment. Safety factors are commonly used to uncover uncertainties and provide a margin of safety. It is likely that the effects of formulations (nano or not) fall within this boundary. This is probably the reason why a representative of the European Crop Protection Association considered that "under the current procedure for traditional crop protection products, the safety of nanomaterials would also be properly assessed" (JRC-IPTS 2014), although the current scientific paradigm cannot be reasonable.

The use of the highly conservative risk assessment strategy mentioned above does not support the level of R&D investment needed to design risk-reducing formulations. Impacts of (nano) formulations on fate and effects of active substances have been repeatedly reported in scientific literature, but the relevant mechanisms remain poorly understood (Jampílek and Kráľová 2018c). Elucidation of these processes and the analysis of environmental impacts require the use of experimental protocols, analytical techniques and theories that differ from typical applications to agrochemicals. Kookana et al. (2014) discussed that combining and adapting approaches developed for pesticides and NPs could in many cases provide a reasonable assessment of the risks associated with nanopesticides. The same approach could be used successfully in assessing the impact of formulations that can exhibit colloidal behaviour in application (whether or not designated as "nano" according to criteria used in research, public or industrial) (Kah 2015).

8.6 Conclusion

It can be stated that plant protection plays an extremely important role in increasing the production of agricultural crops and in protecting them. Nanotechnology and nanoscale science afford unambiguously a great potential in innovative and improved solutions. Nanosized materials change their physical, chemical and biological properties in comparison with bulk materials, and some of them can really help to improve and innovate some pesticides for a more efficient combat against plant diseases, weeds and various pests. The requirements of the latest EU directive regarding a better evaluation of formulations should not be perceived as constraints, but as a tool that should prevent nanopesticides to become the next emerging category of contaminants of environment and human health risks associated with agriculture. To investigate nanopesticide risks, i.e. to minimize nanopesticide impacts on environment and human, cooperation among expert teams at all stages of the development and evaluation of nanopesticides (e.g. formulators, botanists, agricultural scientists and nano(eco)toxicologists) should originate and be intensified to result in the development of successful products, meeting the multiple constraints of the agrochemical sector, and this would bring an added value in relation to existing products.

Acknowledgements This study was supported by SANOFI-AVENTIS Pharma Slovakia, s.r.o.

References

Abigail MEA, Samuel SM, Chidambaram R (2016) Application of rice husk nanosorbents containing 2,4-dichlorophenoxyacetic acid herbicide to control weeds and reduce leaching from soil. J Taiwan Inst Chem Eng 63:318–326. https://doi.org/10.1016/j.jtice.2016.03.024

Adak T, Shakil NA, Walia S (2012) Development of controlled release formulations of imidacloprid employing novel nano-ranged amphiphilic polymers. Environ Sci Health B 47:217–225. https://doi.org/10.1080/03601234.2012.634365

Afrasiabi Z, Popham HJR, Stanley D, Suresh D, Finley K, Campbell J, Kannan R, Upendran A (2016) Dietary silver nanoparticles reduce fitness in a beneficial, but not pest, insect species. Arch Insect Biochem Physiol 93:190–201. https://doi.org/10.1002/arch.21351

Agrios GN (2005) Plant pathology, 5th edn. Elsevier Academic Press, San Diego.

Ahamed M, Posgai R, Gorey TJ, Nielsen M, Hussain SM, Rowe JJ (2010) Silver nanoparticles induced heat shock protein 70, oxidative stress and apoptosis in *Drosophila melanogaster*. Tox App Pharm 242:263–269. https://doi.org/10.1016/j.taap.2009.10.016

Ali M, Kim B, Elfield KDB, Norman D, Brennan M, Ali GS (2015) Inhibition of *Phytophthora parasitica* and *P-capsici* by silver nanoparticles synthesized using aqueous extract of *Artemisia absinthium*. Phytopathology 105:1183–1190. https://doi.org/10.1094/PHYTO-01-15-0006-R

Alromeed AA, Scrano L, Bufo SA, Undabeytia T (2015) Slow-release formulations of the herbicide MCPA by using clay-protein composites. Pest Manag Sci 71:1303–1310. https://doi.org/10.1002/ps.3929

Andronescu E, Brown JM, Oktar FN, Agathopoulos S, Chou J, Obata A (2016) Nanomaterials for medical applications: benefits and risks. J Nanomaterial 2016:8284319. https://doi.org/10.1155/2016/8284319

Anjum NA, Adam V, Kizek R, Duarte AC, Pereira E, Iqbal M, Lukatkin AN, Ahmad I (2015) Nanoscale copper in the soil–plant system–toxicity and underlying potential mechanisms. Environ Res 138:306–325. https://doi.org/10.1016/j.envres.2015.02.019

Anusuya S, Sathiyabama M (2014) Preparation of beta-D-glucan nanoparticles and its antifungal activity. Int J Biol Macromol 70:440–443. https://doi.org/10.1016/j.ijbiomac.2014.07.011

APVMA (2014) A draft report from the australian pesticides and veterinary medicines authority. Regulatory considerations for nanopesticides and veterinary nanomedicines. Available online at: http://apvma.gov.au/sites/default/files/docs/report-draft-regulatory-considerations-nanopesticides-veterinary-nanomedicines.pdf

Arasimowicz M, Floryszak-Wieczorek J (2007) Nitric oxide as a bioactive signaling molecule in plant stress responses. Plant Sci 172:876–887. https://doi.org/10.1016/j.plantsci.2007.02.005

Armstrong N, Ramamoorthy M, Lyon D, Jones K, Duttaroy A (2013) Mechanism of silver nanoparticles action on insect pigmentation reveals intervention of copper homeostasis. PLoS One 8:e53186. https://doi.org/10.1371/journal.pone.0053186

Aschberger K, Gottardo S, Amenta V, Arena M, Moniz FB, Bouwmeester H, Brandhoff P, Mech A, Quiros Pesudo L, Rauscher H, Schoonjans R, Vittoria Vettori M, Peters R (2015) Nanomaterials in food – current and future applications and regulatory aspects. J Phys Conf Ser 617:1–6.

Athanassiou CG, Kavallieratos NG, Peteinatos GG, Petrou SE, Boukouvala MC, Tomanovic Z (2007) Influence of temperature and humidity on insecticidal effect of three diatomaceous earth formulations against larger grain borer (Coleoptera: Bostrychidae). J Econ Entomol 100:599–603.

Auffan M, Rose J, Bottero JY, Lowry GV, Jolivet JP, Wiesner MR (2009) Towards a definition of inorganic nanoparticles from an environmental, health and safety perspective. Nat Nanotechnol 4:634–641. https://doi.org/10.1038/nnano.2009.242

Balashanmugam P, Balakumaran MD, Murugan R, Dhanapal K, Kalaichelvan PT (2016) Phytogenic synthesis of silver nanoparticles, optimization and evaluation of *in vitro* antifungal activity against human and plant pathogens. Microbiol Res 192:52–64. https://doi.org/10.1016/j.micres.2016.06.004

Balaure PC, Gudovan D, Gudovan I (2017) Nanopesticides – a new paradigm in crop protection. In: Grumezescu AM (ed) Nanotechnology in food industry, New pesticides and soil sensors, vol 10. Academic Press & Elsevier, London, pp 129–192. https://doi.org/10.1016/B978-0-12-804299-1.00005-9

Bang SH, Hwang IC, Yu YM, Kwon HR, Kim DH, Park HJ (2011) Influence of chitosan coating on the liposomal surface on physicochemical properties and the release profile of nanocarrier systems. J Microencapsul 28:595–604. https://doi.org/10.3109/02652048.2011.557748

Berkner S, Schwirn K, Voelker D (2016) Nanopharmaceuticals: tiny challenges for the environmental risk assessment of pharmaceuticals. Environ Toxicol Chem 35:780–787. https://doi.org/10.1002/etc.3039

Beyki M, Zhaveh S, Khalili ST, Rahmani-Cherati T, Abollahi A, Bayat M, Tabatabaei M, Mohsenifar A (2014) Encapsulation of *Mentha piperita* essential oils in chitosan-cinnamic acid nanogel with enhanced antimicrobial activity against *Aspergillus flavus*. Ind Crop Prod 54:310–319. https://doi.org/10.1016/j.indcrop.2014.01.033

Bhattacharyya A, Duraisamy P, Govindarajan M, Buhroo AA, Prasad R (2016) Nano-biofungicides: emerging trend in insect pest control. In: Prasad R (ed) Advances and applications through fungal nanobiotechnology. Springer International Publishing, Cham, pp 307–319.

Bheemaraya, Gogoi R, Aggarwal R, Kumar A, Rajesh Kumar, Rai SN, Hossain F, Bisht IS (2014) Effect of nano-hexaconazole on the phenotype and pathogenicity of *Rhizoctonia solani* f. Sp sasakii causing banded leaf and sheath blight in maize. J Pure Appl Microbiol 8:4579–4592.

Bleeker EAJ, Evertz S, Geertsma RE, Peijnenburg WJGM, Westra J, Wijnhoven SWP (2015) Assessing health and environmental risks of nanoparticles. National Institute for Public Health and the Environment, Bilthoven Available online at: http://www.rivm.nl/dsresource?objectid=9adbf65a-25dc-4f45-a580-dcf98bb3b5ac&type=org&disposition=inline

Borm PJA, Robbins D, Haubold S, Kuhlbusch T, Fissan H, Donaldson K, Schins R, Stone V, Kreyling W, Lademann J, Krutmann J, Warheit D, Oberdoster E (2006) The potential risks of nanomaterials: a review carried out for ECETOC. Part Fibre Toxicol 3:11. https://doi.org/10.1186/1743-8977-3-11

Boxi SS, Mukherjee K, Paria S (2016) Ag doped hollow TiO_2 nanoparticles as an effective green fungicide against *Fusarium solani* and *Venturia inaequalis* phytopathogens. Nanotechnology 27:085103. https://doi.org/10.1088/0957-4484/27/8/085103

Bramhanwade K, Shende S, Bonde S, Gade A, Rai M (2016) Fungicidal activity of Cu nanoparticles against *Fusarium* causing crop diseases. Environ Chem Lett 14:229–235. https://doi.org/10.1007/s10311-015-0543-1

Brayner R, Fiévet F, Coradin T (2013) Nanomaterials: a danger or a promise? A chemical and biological perspective. Springer, London.

Brunel F, El Gueddari NE, Moerschbacher BM (2013) Complexation of copper(II) with chitosan nanogels: toward control of microbial growth. Carbohydr Polym 92:1348–1356. https://doi.org/10.1016/j.carbpol.2012.10.025

Buteler M, Sofie SW, Weaver DK, Driscoll D, Muretta J, Stadler T (2015) Development of nanoalumina dust as insecticide against *Sitophilus oryzae* and *Rhyzopertha dominica*. Int J Pest Manag 61:80–89. https://doi.org/10.1080/09670874.2014.1001008

Buzea C, Pacheco I, Robbie K (2007) Nanomaterials and nanoparticles: sources and toxicity. Biointerphases 2:MR17–MR71. https://doi.org/10.1116/1.2815690

Cabrera A, Celis R, Hermosin MC (2016) Imazamox-clay complexes with chitosan- and iron(III)-modified smectites and their use in nanoformulations. Pest Manag Sci 72:1285–1294. https://doi.org/10.1002/ps.4106

Cao VD, Nguyen PP, Khuong VQ, Nguyen CK, Nguyen XC, Dang CH, Tran NQ (2014) Ultrafine copper nanoparticles exhibiting a powerful antifungal/killing activity against *Corticium salmonicolor*. Bull Kor Chem Soc 35:2645–2648. https://doi.org/10.5012/bkcs.2014.35.9.2645

Carris LM, Little CR, Stiles CM (2012) Introduction to fungi. The Plant Health Instructor. American Phytopathological Society. Available online at: http://www.apsnet.org/edcenter/intropp/pathogengroups/pages/introfungi.aspx

Cea M, Cartes P, Palma G, Mora ML (2010) Atrazine efficiency in an andisol as affected by clays and nanoclays in ethylcellulose controlled release formulations. J Soil Sci Plant Nutr 10:62–77. https://doi.org/10.4067/S0718-27912010000100007

Chaudhry Q, Castle L (2011) Food applications of nanotechnologies: an overview of opportunities and challenges for developing countries. Trends Food Sci Technol 22:595–603. https://doi.org/10.1016/j.tifs.2011.01.001

Chauhan N, Dilbaghi N, Gopal M, Kumar R, Kim KH, Kumar S (2017) Development of chitosan nanocapsules for the controlled release of hexaconazole. Int J Biol Macromol 97:616–624. https://doi.org/10.1016/j.ijbiomac.2016.12.059

Chen H, Seiber JN, Hotze M (2014a) ACS select on nanotechnology in food and agriculture: a perspective on implications and applications. J Agric Food Chem 62:1209–1212. https://doi.org/10.1021/jf5002588

Chen JN, Peng H, Wang XP, Shao F, Yuan ZD, Han HY (2014b) Graphene oxide exhibits broad-spectrum antimicrobial activity against bacterial phytopathogens and fungal conidia by intertwining and membrane perturbation. Nanoscale 6:1879–1889. https://doi.org/10.1039/C3NR04941H

Chen JN, Li SL, Luo JX, Wang RS, Ding W (2016) Enhancement of the antibacterial activity of silver nanoparticles against phytopathogenic bacterium *Ralstonia solanacearum* by stabilization. J Nanomater 2016:7135852. https://doi.org/10.1155/2016/7135852

Cho JS, Seo YC, Yim TB, Lee HY (2013) Growth and spore germination of *Fusarium oxysporum* f. Sp *raphani*. Int J Mol Sci 14:4283–4297. https://doi.org/10.3390/ijms14024283

Choudhury SR, Ghosh M, Mandal A, Chakravorty D, Pal M, Pradhan S, Goswami A (2011) Surface-modified sulfur nanoparticles: an effective antifungal agent against *Aspergillus niger* and *Fusarium oxysporum*. Appl Microbiol Biotechnol 90:733–743. https://doi.org/10.1007/s00253-011-3142-5

Choudhury SR, Ghosh M, Goswami A (2012) Inhibitory effects of sulfur nanoparticles on membrane lipids of *Aspergillus niger*: a novel route of fungistasis. Curr Microbiol 65:91–97. https://doi.org/10.1007/s00284-012-0130-7

Cindi MD, Shittu T, Sivakumar D, Bautista-Banos S (2015) Chitosan boehmite-alumina nanocomposite films and thyme oil vapour control brown rot in peaches (*Prunus persica* L.) during postharvest storage. Crop Protect 72:127–131. https://doi.org/10.1016/j.cropro.2015.03.011

Clemente Z, Grillo R, Jonsson M, Santos NZ, Feitosa LO, de Lima R, Fraceto LF (2013) Ecotoxicological evaluation of poly(epsilon-caprolactone) nanocapsules containing triazine herbicides. J Nanosci Nanotechnol 14:4911–4917.

Coles D, Frewer LJ (2013) Nanotechnology applied to European food production: a review of ethical and regulatory issues. Trends Food Sci Technol 34:32–43. https://doi.org/10.1016/j.tifs.2013.08.006

Costa ES, Perlatti B, Silva EM, Matos AP, Silva MFGF, Fernandes JB, Zuin VG, da Silva CMP, Forim MR (2017) Use of lignins from sugarcane bagasse for assembling microparticles loaded with *Azadirachta indica* extracts for use as neem-based organic insecticides. J Braz Chem Soc 28:126–135. https://doi.org/10.5935/0103-5053.20160155

Crop Protection Definitions (2017) STANDS4 LLC. Available online at: http://www.definitions.net/definition/crop%20protection

da Costa JT, Forim MR, Costa ES, de Souza JR, Mondego JM, Boica AL (2014) Effects of different formulations of neem oil-based products on control *Zabrotes subfasciatus* (Boheman, 1833) (Coleoptera: Bruchidae) on beans. J Stored Prod Res 56:49–53. https://doi.org/10.1016/j.jspr.2013.10.004

De Jong WH, Borm PJ (2008) Drug delivery and nanoparticles: applications and hazards. Int J Nanomedicine 3:133–149. https://doi.org/10.2147/ijn.s596

De Lucca AJ (2007) Harmful fungi in both agriculture and medicine. Rev Iberoam Micol 24:3–13.

de Oliveira JL, Campos EVR, da Silva CMG, Pasquoto T, de Lima R, Fraceto LF (2015) Solid lipid nanoparticles co-loaded with simazine and atrazine: preparation, characterization, and evaluation of herbicidal activity. J Agric Food Chem 63:422–432. https://doi.org/10.1021/jf5059045

Debnath N, Das S, Brahmachary RL, Chandra R, Sudan S, Goswami A (2010) Entomotoxicity assay of silica, zinc oxide, titanium dioxide, aluminium oxide nanoparticles on *Lipaphis pseudobrassicae*. AIP Conf Proc 1276:307–310. https://doi.org/10.1063/1.3504316

Debnath N, Das S, Seth D, Chandra R, Bhattacharya SC, Goswami A (2011) Entomotoxic effect of silica nanoparticles against *Sitophilus oryzae* (L.). J Pestic Sci 84:99–105. https://doi.org/10.1007/s10340-010-0332-3

Debnath N, Mitra S, Das S, Goswami A (2012) Synthesis of surface functionalized silica nanoparticles and their use as entomotoxic nanocides. Powder Technol 221:252–256. https://doi.org/10.1016/j.powtec.2012.01.009

Derbalah AS, Khidr AA, Moustafa HZ, Taman A (2014) Laboratory evaluation of some non-conventional pest control agents against the pink bollworm *Pectinophora gossypiella* (Saunders). Egypt J Biol Pest Control 24:363–368.

Dharni S, Sanchita, Unni SM, Kurungot S, Samad A, Sharma A, Patra DD (2016) *In vitro* and in silico antifungal efficacy of nitrogen- doped carbon nanohorn (NCNH) against *Rhizoctonia solani*. J Biomol Struct Dyn 34:152–162. https://doi.org/10.1080/07391102.2015.1018841

Dolez PI (2015) Nanoengineering: global approaches to health and safety issues. Elsevier, Amsterdam.

Du WC, Tan WJ, Peralta-Videa JR, Gardea-Torresdey JL, Ji R, Yin Y, Guo HG (2017) Interaction of metal oxide nanoparticles with higher terrestrial plants: physiological and biochemical aspects. Plant Physiol Biochem 110:210–225. https://doi.org/10.1016/j.plaphy.2016.04.024

EC (2014) JRC nanomaterials repository. Available online at: https://ec.europa.eu/jrc/en/scientific-tool/jrc-nanomaterials-repository

Elango G, Roopan SM, Dhamodaran KI, Elumalai K, Al-Dhabi NA, Arasu MV (2016) Spectroscopic investigation of biosynthesized nickel nanoparticles and its larvicidal, pesticidal activities. J Photochem Photobiol B 162:162–167. https://doi.org/10.1016/j.jphotobiol.2016.06.045

Elek N, Hoffman R, Raviv U, Resh R, Ishaaya I, Magdassi S (2010) Novaluron nanoparticles: formation and potential use in controlling agricultural insect pests. Colloids Surf A Physicochem Eng Asp 372:66–72. https://doi.org/10.1016/j.colsurfa.2010.09.034

Elgorban AM, El-Samawaty AEM, Yassin MA, Sayed SR, Adil SF, Elhindi KM, Bakri M, Khan M (2016a) Antifungal silver nanoparticles: synthesis, characterization and biological evaluation. Biotechnol Biotechnol Equip 30:56–62. https://doi.org/10.1080/13102818.2015.1106339

Elgorban AM, Aref SM, Seham SM, Elhindi KM, Bahkali AH, Sayed SR, Manal MA (2016b) Extracellular synthesis of silver nanoparticles using *Aspergillus versicolor* and evaluation of their activity on plant pathogenic fungi. Mycosphere 7:844–852. https://doi.org/10.5943/mycosphere/7/6/15

Essential Chemical Industry (2017) Crop protection chemicals. Centre for Industry Education Collaboration, Department of Chemistry, University of York, UK. Available online at: http://www.essentialchemicalindustry.org/materials-and-applications/crop-protection-chemicals.html

European Commission (2009) Regulation EC 1107/2009 concerning the placing of plant protection products on the market. Guidance on section 5 – fate and behaviour in the environment. Available online at: http://eur-lex.europa.eu/legal-content/EN/TXT/?uri=celex:32009R1107

European Commission (2011) Definition of a nanomaterial. Available online at: http://ec.europa.eu/environment/chemicals/nanotech/faq/definition_en.htm
FAO (2009) Global agriculture towards 2050. .High level expert forum – how to feed the world in 2050. Available online at: http://www.fao.org/fileadmin/templates/wsfs/docs/Issues_papers/HLEF2050_Global_Agriculture.pdf
FAO (2010) Food and Agriculture Organization of the United Nations and World Health Organization expert meeting on the application of nanotechnologies in the food and agriculture sectors. Potential food safety implications, in Meeting Report (Rome). Available online at: http://whqlibdoc.who.int/publications/2010/9789241563932_eng.pdf
FAO (2013) Food and Agriculture Organization of the United Nations and World Health Organization. State of the art on the initiatives and activities relevant to risk assessment and risk management of nanotechnologies in the food and agriculture sectors, in FAO/WHO Technical Paper. Available online at: http://www.fao.org/docrep/018/i3281e/i3281e.pdf
Fisher MC, Henk DA, Briggs CJ, Brownstein JS, Madoff LC, McCraw SL, Gurr SJ (2012) Emerging fungal threats to animal, plant and ecosystem health. Nature 484:186–194. https://doi.org/10.1038/nature10947
Forim MR, Costa ES, da Silva GFMF, Fernandes JB, Mondego JM, Boica AL (2013) Development of a new method to prepare nano−/microparticles loaded with extracts of *Azadirachta indica*, their characterization and use in controlling *Plutella xylostella*. J Agric Food Chem 61:9131–9139. https://doi.org/10.1021/jf403187y
Fraceto LF, Grillo R, de Medeiros GA, Scognamiglio V, Rea G, Bartolucci C (2016) Nanotechnology in agriculture: which innovation potential does it have? Front Environ Sci 4:20. https://doi.org/10.3389/fenvs.2016.00020
Fröhlich E (2013) Cellular targets and mechanisms in the cytotoxic action of non-biodegradable engineered nanoparticles. Curr Drug Metab 14:976–988. https://doi.org/10.2174/1389200211314090004
Gajbhiye M, Kesharwani J, Ingle A, Gade A, Rai M (2009) Fungus-mediated synthesis of silver nanoparticles and their activity against pathogenic fungi in combination with fluconazole. Nanomedicine Nanotechnol Biol Med 5:382–386. https://doi.org/10.1016/j.nano.2009.06.005
Garcia M, Forbe T, Gonzalez E (2010) Potential applications of nanotechnology in the agro-food sector. Food Sci Technol 30:573–581. https://doi.org/10.1590/S0101-20612010000300002
Garrido-Herrera FJ, Gonzlez-Pradas E, Fernndez-Prez M (2006) Controlled release of isoproturon, imidacloprid, and cyromazine from alginate-bentonite-activated carbon formulations. J Agric Food Chem 54:10053–10060. https://doi.org/10.1021/jf062084m
Gauthier GM, Keller NP (2013) Crossover fungal pathogens: the biology and pathogenesis of fungi capable of crossing kingdoms to infect plants and humans. Fungal Genet Biol 61:146–157. https://doi.org/10.1016/j.fgb.2013.08.016
Gewin W (2015) Everything you need to know about nanopesticides. Modern Farmer Article. Available online at: http://modernfarmer.com/2015/01/everything-need-know-nanopesticides
Ghasemian E, Naghoni A, Tabaraie B, Tabaraie T (2012) *In vitro* susceptibility of filamentous fungi to copper nanoparticles assessed by rapid XTT colorimetry and agar dilution method. J Mycol Med 22:322–328. https://doi.org/10.1016/j.mycmed.2012.09.006
Ghormade V, Deshpande MV, Paknikar KM (2011) Perspectives for nano-biotechnology enabled protection and nutrition of plants. Biotechnol Adv 29:792–803. https://doi.org/10.1016/j.biotechadv.2011.06.0
Giannousi K, Avramidis I, Dendrinou-Samara C (2013) Synthesis, characterization and evaluation of copper based nanoparticles as agrochemicals against *Phytophthora infestans*. RSC Adv 3:21743–21752. https://doi.org/10.1039/c3ra42118j
Giongo AMM, Vendramim JD, Forim MR (2016) Evaluation of neem-based nanoformulations as alternative to control fall armyworm. Cienc Agrotec 40:26–36. https://doi.org/10.1590/S1413-70542016000100002
Gondal MA, Alzahrani AJ, Randhawa MA, Siddiqui MN (2012) Morphology and antifungal effect of nano-ZnO and nano-Pd-doped nano-ZnO against *Aspergillus* and *Candida*. J Environ Sci

Health A Tox Hazard Subst Environ Eng 47:1413–1418. https://doi.org/10.1080/10934529.2012.672384

Gonzalez JOW, Gutierrez MM, Ferrero AA, Band BF (2014) Essential oils nanoformulations for stored-product pest control – characterization and biological properties. Chemosphere 100:130–138. https://doi.org/10.1016/j.chemosphere.2013.11.056

Gopinath V, Velusamy P (2013) Extracellular biosynthesis of silver nanoparticles using *Bacillus* sp GP-23 and evaluation of their antifungal activity towards *Fusarium oxysporum*. Spectrochim Acta A Mol Biomol Spectrosc 106:170–174. https://doi.org/10.1016/j.saa.2012.12.087

Gorczyca A, Pociecha E, Kasprowicz M, Niemiec M (2015) Effect of nanosilver in wheat seedlings and *Fusarium culmorum* culture systems. Eur J Plant Pathol 142:251–261. https://doi.org/10.1007/s10658-015-0608-9

Goswami A, Roy I, Sengupta S, Debnath N (2010) Novel applications of solid and liquid formulations of nanoparticles against insect pests and pathogens. Thin Solid Films 519:1252–1257. https://doi.org/10.1016/j.tsf.2010.08.079

Grillo R, dos Santos NZP, Maruyama CR, Rosa AH, de Lima R, Fraceto LF (2012) Poly(epsilon-caprolactone) nanocapsules as carrier systems for herbicides: physico-chemical characterization and genotoxicity evaluation. J Hazard Mater 231:1–9. https://doi.org/10.1016/j.jhazmat.2012.06.019

Grillo R, Pereira AE, Nishisaka CS, de Lima R, Oehlke K, Greiner R, Fraceto LF (2014) Chitosan/tripolyphosphate nanoparticles loaded with paraquat herbicide: an environmentally safer alternative for weed control. J Hazard Mater 278:163–171. https://doi.org/10.1016/j.jhazmat.2014.05.079

Grillo R, Abhilash PC, Fraceto LF (2016) Nanotechnology applied to bio-encapsulation of pesticides. J Nanosci Nanotechnol 16:1231–1234. https://doi.org/10.1166/jnn.2016.12332

Guan HN, Chi DF, Yu JM, Li XC (2008) A novel photodegradable insecticide: preparation, characterization and properties evaluation of nano-imidacloprid. Pestic Biochem Physiol 92:83–91. https://doi.org/10.1016/j.pestbp.2008.06.008

Guo YZ, Yang Q, Yan W, Li B, Qian K, Li T, Xiao W, He L (2014) Controlled release of acetochlor from poly (butyl methacrylate-diacetone acrylamide) based formulation prepared by nanoemulsion polymerisation method and evaluation of the efficacy. Int J Environ Anal Chem 94:1001–1012. https://doi.org/10.1080/03067319.2014.930844

Guo MC, Zhang WB, Ding GL, Guo D, Zhu JL, Wang BT, Punyapitak D, Cao YS (2015) Preparation and characterization of enzyme-responsive emamectin benzoate microcapsules based on a copolymer matrix of silica-epichlorohydrin-carboxymethylcellulose. RSC Adv 5:93170–93179

Hagens WI, Oomen AG, de Jong WH, Cassee FR, Sips AJAM (2007) What do we (need to) know about the kinetic properties of nanoparticles in the body? Regul Toxicol Pharmacol 49:217–229. https://doi.org/10.1016/j.yrtph.2007.07.006

Hala HA, Elsamahy MFM (2016) Relative toxicity of silica nanoparticles to two tetranychids and three associated predators. Egypt J Biol Pest Control 26:283–286.

Hamza AM (2012) Efficacy and safety of non-traditional methods as alternatives for control of *Sitophilus oryzae* (L.) (Coleoptera: Curculionidae) in rice grains. Egypt J Biol Pest Control 22:103–108.

Handy RD, Owen R, Valsami-Jones E (2008) The ecotoxicology of nanoparticles and nanomaterials: current status, knowledge gaps, challenges, and future needs. Ecotoxicology 17:315–325. https://doi.org/10.1007/s10646-008-0206-0

He LL, Liu Y, Mustapha A, Lin MS (2011) Antifungal activity of zinc oxide nanoparticles against *Botrytis cinerea* and *Penicillium expansum*. Microbiol Res 166:207–215. https://doi.org/10.1016/j.micres.2010.03.003

Hiraga S, Sasaki K, Ito H, Ohashi Y, Matsui H (2001) A large family of class III plant peroxidases. Plant Cell Physiol 42:462–468. https://doi.org/10.1093/pcp/pce061

Ho VA, Le PT, Nguyen TP, Nguyen CK, Nguyen VT, Tran NQ (2015) Silver core-shell nanoclusters exhibiting strong growth inhibition of plant-pathogenic fungi. J Nanomater 2015:241614. https://doi.org/10.1155/2015/241614

Huang QZ, Jiao ZJ, Li M, Qiu DF, Liu KC, Shi HZ (2013a) Preparation, characterization, antifungal activity, and mechanism of chitosan/TiO_2 hybrid film against *Bipolaris maydis*. J Appl Polym Sci 128:2623–2629. https://doi.org/10.1002/APP.38322

Huang NY, Yan Y, Xu Y, Jin Y, Lei J, Zou X, Ran D, Zhang H, Luan S, Gu H (2013b) Alumina nanoparticles alter rhythmic activities of local interneurons in the antennal lobe of *Drosophila*. Nanotoxicology 7:212–220. https://doi.org/10.3109/17435390.2011.648668

Hussein MZ, Yahaya AH, Zainal Z, Kian LH (2005) Nanocomposite-based controlled release formulation of an herbicide, 2,4-dichlorophenoxyacetate encapsulated in zinc-aluminium-layered double hydroxide. Sci Technol Adv Mater 6:956–962.

Iatrou SA, Kavallieratos NG, Palyvos NE, Buchelos CT, Tomanovic S (2010) Acaricidal effect of different diatomaceous earth formulations against *Tyrophagus putrescentiae* (Astigmata: Acaridae) on stored wheat. J Econ Entomol 103:190–196.

Il KS, Kim I (2012) Synthesis of Pyto-patch as silver nanoparticle product and antimicrobial activity. Protect Hort Plant Fac 21:140–146.

Ileke KD, Ogungbite OC (2014) Entomocidal activity of powders and extracts of four medicinal plants against *Sitophilus oryzae* (L), *Oryzaephilus mercator* (Faur) and *Rhyzopertha dominica* (Fabr.). Jordan J Biol Sci 7:57–62.

Ing LY, Zin NM, Sarwar A, Katas H (2012) Antifungal activity of chitosan nanoparticles and correlation with their physical properties. Int J Biomater 2012:632698. https://doi.org/10.1155/2012/632698

Ismail M, Prasad R, Ibrahim AIM, Ahmed ISA (2017) Modern prospects of nanotechnology in plant pathology. In: Prasad R, Kumar M, Kumar V (eds) Nanotechnology. Springer Nature Singapore Pte Ltd, Singapore, pp 305–317.

Jamal M, Moharramipour S, Zandi M, Negahban M (2013) Efficacy of nanoencapsulated formulation of essential oil from *Carum copticum* seeds on feeding behavior of *Plutella xylostella* (Lep.: Plutellidae). J Entomol Soc Iran 33:23–31.

Jampílek J (2016) Potential of agricultural fungicides for antifungal drug discovery. Expert Opin Drug Dis 11:1–9. https://doi.org/10.1517/17460441.2016.1110142

Jampílek J, Kráľová K (2015) Application of nanotechnology in agriculture and food industry, its prospects and risks. Ecol Chem Eng S 22:321–361. https://doi.org/10.1515/eces-2015-0018

Jampílek J, Kráľová K (2017a) Nano-antimicrobials: activity, benefits and weaknesses. In: Ficai A, Grumezescu AM (eds) Nanostructures in therapeutic medicine, Nanostructures for antimicrobial therapy, vol 2. Elsevier, Amsterdam., Chapter 2, pp 23–54. https://doi.org/10.1016/B978-0-323-46152-8.00002-0

Jampílek J, Kráľová K (2017b) Nanopesticides: preparation, targeting and controlled release. In: Grumezescu AM (ed) Nanotechnology in food industry, New pesticides and soil sensors, vol 10. Academic Press & Elsevier, London, pp 81–127. https://doi.org/10.1016/B978-0-12-804299-1.00004-7

Jampílek J, Kráľová K (2017c) Nanomaterials for delivery of nutrients and growth-promoting compounds to plants. In: Prasad R, Kumar M, Kumar V (eds) Nanotechnology: an agricultural paradigm. Springer, Berlin/Heidelberg., Chapter 9, pp 177–226. https://doi.org/10.1007/978-981-10-4573-8_9

Jampílek J, Kráľová K (2018a) Application of nanobioformulations for controlled release and targeted biodistribution of drugs. In: Sharma AK, Keservani RK, Kesharwani RK (eds) Nanobiomaterials: applications in drug delivery. Apple Academic Press & CRC Press, Oakville Part II, Chapter 5, 131–208.

Jampílek J, Kráľová K (2018b) Nanomaterials applicable in food protection. In: Rai RV, Bai JA (eds) Nanotechnology applications in food industry. CRC Press/Taylor & Francis Group, Boca Raton, Chapter 5, 75–96.

Jampílek J, Kráľová K (2018c) Impact of nanoparticles on living organisms and human health. In: Nalwa HS (ed) Encyclopedia of nanoscience and nanotechnology. American Scientific Publishers, Valencia. ISBN 1-58883-212-0

Jampílek J, Opatřilová R, Coufalová L, Černíková A, Dohnal J (2013) Utilization of alaptide as transdermal penetration modifier in pharmaceutical compositions for human and veterinary applications containing anti-inflammatory drugs and/or antimicrobial chemotherapeutics. WO2013020527 A1

Jampílek J, Opatřilová R, Řezáčová A, Oktábec Z, Dohnal J (2014) Alaptide: methods of effecting its solubility, membrane permeation and pharmaceutical compositions for human and/or veterinary applications. WO2014019556 A1

Jampílek J, Záruba K, Oravec M, Kuneš M, Babula P, Ulbrich P, Brezaniová I, Opatřilová R, Tříska J, Suchý P (2015) Preparation of silica nanoparticles loaded with nootropics and their *in vivo* permeation through blood–brain barrier. Biomed Res Int 2015:812673. https://doi.org/10.1155/2015/812673

Janrao K, Gadhave MV, Banerjee SK, Gaikwad DD (2014) Asian J Biomed Pharm Sci 4:1–7.

Jayaseelan C, Rahuman AA, Kirthi AV, Marimuthu S, Santhoshkumar T, Bagavan A, Gaurav K, Karthik L, Rao KV (2012) Novel microbial route to synthesize ZnO nanoparticles using *Aeromonas hydrophila* and their activity against pathogenic bacteria and fungi. Spectrochim Acta A Mol Biomol Spectrosc 90:78–84. https://doi.org/10.1016/j.saa.2012.01.006

Jayaseelan C, Ramkumar R, Rahuman AA, Perumal P (2013) Green synthesis of gold nanoparticles using seed aqueous extract of *Abelmoschus esculentus* and its antifungal activity. Ind Crop Prod 45:423–429. https://doi.org/10.1016/j.indcrop.2012.12.019

Jia RX, Jiang HJ, Jin MY, Wang XY, Huang JY (2015) Silver/chitosan-based Janus particles: synthesis, characterization, and assessment of antimicrobial activity *in vivo* and *vitro*. Food Res Int 78:433–441. https://doi.org/10.1016/j.foodres.2015.08.035

Jiang LC, Basri M, Omar D, Rahman MBA, Salleh AB, Rahman RNZRA, Selamat A (2012) Green nano-emulsion intervention for water-soluble glyphosate isopropylamine (IPA) formulations in controlling *Eleusine indica* (*E. indica*). Pest Biochem Physiol 102:19–29. https://doi.org/10.1016/j.pestbp.2011.10.004

Jo YK, Kim BH, Jung G (2009) Antifungal activity of silver ions and nanoparticles on phytopathogenic fungi. Plant Dis 93:1037–1043. https://doi.org/10.1094/PDIS-93-10-1037

JRC-IPTS (2014) Nanotechnology for the agricultural sector: from research to the field. In: Parsi C, Vigani M, Rodriguez-Cerezo E (eds) Proceedings of a workshop organized by Joint Research Centre – Institute for Prospective Technological. Available online at: https://ec.europa.eu/jrc/sites/default/files/ipts_jrc_89736_(online)__final.pdf

Kah M (2015) Nanopesticides and nanofertilizers: emerging contaminants or opportunities for risk mitigation? Front Chem 3:64. https://doi.org/10.3389/fchem.2015.00064

Kah M, Hofmann T (2014) Nanopesticides research: current trends and future priorities. Environ Int 63:224–235. https://doi.org/10.1016/j.envint.2013.11.015

Kah M, Beulke S, Tiede K, Hofmann T (2013) Nanopesticides: state of knowledge, environmental fate and exposure modelling. Crit Rev Environ Sci Technol 43:1823–1867. https://doi.org/10.1080/10643389.2012.671750

Kah M, Machinski P, Koerner P, Tiede K, Grillo R, Fraceto LF, Hofmann T (2014) Analysing the fate of nanopesticides in soil and the applicability of regulatory protocols using a polymer-based nanoformulation of atrazine. Environ Sci Pollut Res 21:11699–11707. https://doi.org/10.1007/s11356-014-2523-6

Kairyte K, Kadys A, Luksiene Z (2013) Antibacterial and antifungal activity of photoactivated ZnO nanoparticles in suspension. J Photochem Photobiol B 128:78–84. https://doi.org/10.1016/j.jphotobiol.2013.07.017

Kala A, Soosairaj S, Mathiyazhagan S, Raja P (2016) Green synthesis of copper bionanoparticles to control the bacterial leaf blight disease of rice. Curr Sci 110:2011–2014. https://doi.org/10.18520/cs/v110/i10/2005-2011

Kalboush ZA, Hassan AA, Gabr WE (2016) Control of rice blast and brown spot diseases by synthesized zinc oxide nanoparticles. Egypt J Biol Pest Control 26:713–720.

Kang MA, Seo MJ, Hwang IC, Jang C, Park HJ, Yu YM, Youn YN (2012) Insecticidal activity and feeding behavior of the green peach aphid, *Myzus persicae*, after treatment with nano types of pyrifluquinazon. J Asia Pac Entomol 15:533–541. https://doi.org/10.1016/j.aspen.2012.05.015

Kanhed P, Birla S, Gaikwad S, Gade A, Seabra AB, Rubilar O, Duran N, Rai M (2014) *In vitro* antifungal efficacy of copper nanoparticles against selected crop pathogenic fungi. Mater Lett 115:13–17. https://doi.org/10.1016/j.matlet.2013.10.011

Kanimozhi V, Chinnamuthu CR (2012) Engineering core/hallow shell nanomaterials to load herbicide active ingredient for controlled release. Res J Nanosci Nanotechnol 2:58–69. https://doi.org/10.3923/rjnn.2012.58.69

Karimi N, Minaei S, Almassi M, Shahverdi AR (2012) Application of silver nano-particles for protection of seeds in different soils. Afr J Agric Res 7:1863–1869. https://doi.org/10.5897/AJAR11.1150

Kasprowicz MJ, Koziol M, Gorczyca A (2010) The effect of silver nanoparticles on phytopathogenic spores of *Fusarium culmorum*. Can J Microbiol 56:247–253. https://doi.org/10.1139/w10-012

Keck CM, Müller RH (2013) Nanotoxicological classification system (NCS) – a guide for the risk-benefit assessment of nanoparticulate drug delivery systems. Eur J Pharm Biopharm 84:445–448. https://doi.org/10.1016/j.ejpb.2013.01.001

Khalili ST, Mohsenifar A, Beyki M, Zhaveh S, Rahmani-Cherati T, Abdollahi A, Bayat M, Tabatabaei M (2015) Encapsulation of thyme essential oils in chitosan-benzoic acid nanogel with enhanced antimicrobial activity against *Aspergillus flavus*. LWT Food Sci Technol 60:502–508. https://doi.org/10.1016/j.lwt.2014.07.054

Khan FH (2013) Chemical hazards of nanoparticles to human and environment (a review). Orient J Chem 29:4. https://doi.org/10.13005/ojc/290415

Khandelwal N, Doke DS, Khandare JJ, Jawale PV, Biradar AV, Giri AP (2015) Bio-physical evaluation and *in vivo* delivery of plant proteinase inhibitor immobilized on silica nanospheres. Colloids Surf B Biointerfaces 130:84–92. https://doi.org/10.1016/j.colsurfb.2015.03.060

Khatami M, Mehnipor R, Poor MHS, Jouzani GS (2016a) Facile biosynthesis of silver nanoparticles using *Descurainia sophia* and evaluation of their antibacterial and antifungal properties. J Clust Sci 27:1601–1612. https://doi.org/10.1007/s10876-016-1028-5

Khatami M, Nejad MS, Salari S, Almani PGN (2016b) Plant-mediated green synthesis of silver nanoparticles using *Trifolium resupinatum* seed exudate and their antifungal efficacy on *Neofusicoccum parvum* and *Rhizoctonia solani*. IET Nanobiotechnol 10:237–243. https://doi.org/10.1049/iet-nbt.2015.0078

Kheiri A, Jorf SAM, Malihipour A, Saremi H, Nikkhah M (2016) Application of chitosan and chitosan nanoparticles for the control of *Fusarium* head blight of wheat (*Fusarium graminearum*) *in vitro* and greenhouse. Int J Biol Macromol 93:1261–1272. https://doi.org/10.1016/j.ijbiomac.2016.09.072

Khot LR, Sankaran S, Maja JM, Ehsani R, Schuster EW (2012) Applications of nanomaterials in agricultural production and crop protection: a review. Crop Prot 35:64–70. https://doi.org/10.1016/j.cropro.2012.01.007

Kim HS, Kang HS, Chu GJ, Byun HS (2008) Antifungal effectiveness of nanosilver colloid against rose powdery mildew in greenhouses. Solid State Phenom 135:15–18. https://doi.org/10.4028/www.scientific.net/SSP.135.15

Kim SW, Kim KS, Lamsal K, Kim YJ, Kim SB, Jung M, Sim SJ, Kim HS, Chang SJ, Kim JK, Lee YS (2009) An *in vitro* study of the antifungal effect of silver nanoparticles on oak wilt pathogen *Raffaelea* sp. J Microbiol Biotechnol 19:760–764. https://doi.org/10.4014/jmb.0812.649

Kim HJ, Park HJ, Choi SH (2011) Antimicrobial action effect and stability of nanosized silica hybrid Ag complex. J Nanosci Nanotechnol 11:5781–5787. https://doi.org/10.1166/jnn.2011.4492

Kim SW, Jung JH, Lamasal K, Kim YS, Min JS, Lee YS (2012) Antifungal effects of silver nanoparticles (AgNPs) against various plant pathogenic fungi. Mycobiology 40:53–58. https://doi.org/10.5941/MYCO.2012.40.1.053

Koli P, Singh BB, Shakil NA, Kumar J, Kamil D (2015) Development of controlled release nanoformulations of carbendazim employing amphiphilic polymers and their bioefficacy evaluation against *Rhizoctonia solani*. J Environ Sci Health B 50:674–681. https://doi.org/10.1080/03601234.2015.1038961

Kookana RS, Boxall ABA, Reeves PT, Ashauer R, Beulke S, Chaudhry Q, Cornelis G, Fernandes TF, Gan J, Kah M, Lynch I, Ranville J, Sinclair C, Spurgeon D, Tiede K, Van den Brink PJ (2014) Nanopesticides: guiding principles for regulatory evaluation of environmental risks. J Agric Food Chem 62:4227–4240. https://doi.org/10.1021/jf500232f

Krishnaraj C, Ramachandran R, Mohan K, Kalaichelvan PT (2012) Optimization for rapid synthesis of silver nanoparticles and its effect on phytopathogenic fungi. Spectrochim Acta Part A Mol Biomol Spectrosc 93:95–99. https://doi.org/10.1016/j.saa.2012.03.002

Kumar V, Kumari A, Guleria P, Yadav SK (2012) Evaluating the toxicity of selected types of nanochemicals. Rev Environ Contam Toxicol 215:39–121. https://doi.org/10.1007/978-1-4614-1463-6_2

Kumar S, Bhanjana G, Sharma A, Sidhu MC, Dilbaghi N (2014) Synthesis, characterization and on field evaluation of pesticide loaded sodium alginate nanoparticles. Carbohydr Polym 101:1061–1067. https://doi.org/10.1016/j.carbpol.2013.10.025

Kumar R, Nair KK, Alam MI, Gogoi R, Singh PK, Srivastava C, Gopal M, Goswami A (2015a) Development and quality control of nanohexaconazole as an effective fungicide and its biosafety studies on soil nitifiers. J Nanosci Nanotechnol 15:1350–1356. https://doi.org/10.1166/jnn.2015.9088

Kumar S, Chauhan N, Gopal M, Kumar R, Dilbaghi N (2015b) Development and evaluation of alginate-chitosan nanocapsules for controlled release of acetamiprid. Int J Biol Macromol 81:631–637. https://doi.org/10.1016/j.ijbiomac.2015.08.062

Kumar N, Kumar R, Shakil NA, Das TK (2016) Nanoformulations of pretilachlor herbicide: preparation, characterization and activity. J Sci Ind Res 75:676–680

Kumar S, Bhanjama G, Sharma A, Dilbaghi N, Sidhu MC, Kim KH (2017) Development of nanoformulation approaches for the control of weeds. Sci Total Environ 586:1272–1278. https://doi.org/10.1016/j.scitotenv.2017.02.138

Lamsal K, Kim SW, Jung JH, Kim YS, Kim KS, Lee YS (2011a) Application of silver nanoparticles for the control of *Colletotrichum* species *in vitro* and pepper anthracnose disease in field. Mycobiology 39:194–199. https://doi.org/10.5941/MYCO.2011.39.3.194

Lamsal K, Kim SW, Jung JH, Kim YS, Kim KS, Lee YS (2011b) Inhibition effects of silver nanoparticles against powdery mildews on cucumber and pumpkin. Mycobiology 39:26–32. https://doi.org/10.4489/MYCO.2011.39.1.026

Lee KJ, Park SH, Govarthanan M, Hwang PH, Seo YS, Cho M, Lee WH, Lee JY, Kamala-Kannan S, Oh BT (2013) Synthesis of silver nanoparticles using cow milk and their antifungal activity against phytopathogens. Mater Lett 105:128–131. https://doi.org/10.1016/j.matlet.2013.04.076

Li J, Jiang M, Wu H, Li Y (2009) Addition of modified bentonites in polymer gel formulation of 2,4-D for its controlled release in water and soil. J Agric Food Chem 57:2868–2874. https://doi.org/10.1021/jf803744w

Li B, Zhang Y, Yang YZ, Qiu W, Wang XX, Liu BP, Wang Y, Sun G (2016) Synthesis, characterization, and antibacterial activity of chitosan/TiO_2 nanocomposite against *Xanthomonas oryzae* pv. oryzae. Carbohydr Polym 152:825–831. https://doi.org/10.1016/j.carbpol.2016.07.070

Lim CJ, Basri M, Omar D, Rahman MBA, Salleh AB, Rahman RNZRA (2012) Physicochemical characterization and formation of glyphosate-laden nano-emulsion for herbicide formulation. Ind Crop Prod 36:607–613. https://doi.org/10.1016/j.indcrop.2011.11.005

Lim CJ, Basri M, Omar D, Rahman MBA, Salleh AB, Rahman RNZRA (2013) Green nanoemulsion-laden glyphosate isopropylamine formulation in suppressing creeping foxglove

(*A. gangetica*), slender button weed (*D. ocimifolia*) and buffalo grass (*P. conjugatum*). Pest Manag Sci 69:104–111. https://doi.org/10.1002/ps.3371

Liu Y, Tong Z, Prud'homme RK (2008) Stabilized polymeric nanoparticles for controlled and efficient release of bifenthrin. Pest Manag Sci 64:808–812. https://doi.org/10.1002/ps.1566

Liu TC, Wang HC, Hua KH, Hsu JC (2014) The efficiency of calcium carbonate particles with different grain sizes for controlling peach aphid (*Myzus persicae*). Formos Entomol 34:49–54.

Liu JX, Ren JH, Zhao Q, Shi TJ, Liu ZF, Luo Z, Zhang XG (2016) Preparation and characterization of chlorpyrifos/cyclodextrin complex intercalation into ZnAl-layered double hydroxide. Acta Phys -Chim Sin 32:558–564. https://doi.org/10.3866/PKU.WHXB201511271

Loha KM, Shakil NA, Kumar J, Singh M, Srivastava C (2012) Bio-efficacy evaluation of nanoformulations of β-cyfluthrin against *Callosobruchus maculatus* (*Coleoptera*: *Bruchidae*). J Environ Sci Health B 47:687–691. https://doi.org/10.1080/03601234.2012.669254

Ma CX, White JC, Dhankher OP, Xing BS (2015) Metal-based nanotoxicity and detoxification pathways in higher plants. Environ Sci Technol 49:7109–7122. https://doi.org/10.1021/acs.est.5b00685

Mageshwari K, Sathyamoorthy R (2013) Flower-shaped CuO nanostructures: synthesis, characterization and antimicrobial activity. J Mater Sci Technol 29:909–914. https://doi.org/10.1016/j.jmst.2013.04.020

Manikandan A, Sathiyabama M (2016) Preparation of chitosan nanoparticles and its effect on detached rice leaves infected with *Pyricularia grisea*. Int J Biol Macromol 84:58–61. https://doi.org/10.1016/j.ijbiomac.2015.11.083

Maqueda C, Villaverde J, Sopena F, Undabeytia T, Morillo E (2009) Effects of soil characteristics on metribuzin dissipation using clay−gel-based formulations. J Agric Food Chem 57:3273–3278. https://doi.org/10.1021/jf803819q

Maruyama CR, Guilger M, Pascoli M, Bileshy-Jose N, Abhilash PC, Fraceto LF, de Lima R (2016) Nanoparticles based on chitosan as carriers for the combined herbicides imazapic and imazapyr. Sci Rep 6:19768. https://doi.org/10.1038/srep19768

Masarovičová E, Kráľová K (2013) Metal nanoparticles and plants. Ecol Chem Eng S 20:9–22. https://doi.org/10.2478/eces-2013-0001

Masarovičová E, Kráľová K, Zinjarde SS (2014) Metal nanoparticles in plants. Formation and action. In: Pessarakli M (ed) Handbook of plant and crop physiology, 3rd edn. CRC Press, Boca Raton, pp 683–731. https://doi.org/10.1201/b16675-39

Medina C, Santos-Martinez MJ, Radomski A, Corrigan O, Radomski MW (2007) Nanoparticles: pharmacological and toxicological significance. Br J Pharmacol 150:552–558. https://doi.org/10.1038/sj.bjp.0707130

Memarizadeh N, Ghadamyari M, Adeli M, Talebi K (2014) Preparation, characterization and efficiency of nanoencapsulated imidacloprid under laboratory conditions. Ecotoxicol Environ Saf 107:77–83. https://doi.org/10.1016/j.ecoenv.2014.05.009

Meyer WL, Gurman P, Stelinski LL, Elman NM (2015) Functional nano-dispensers (FNDs) for delivery of insecticides against phytopathogen vectors. Green Chem 17:4173–4177. https://doi.org/10.1039/C5GC00717H

Mishra S, Singh HB (2015) Biosynthesized silver nanoparticles as a nanoweapon against phytopathogens: exploring their scope and potential in agriculture. Appl Microbiol Biotechnol 99:1097–1107. https://doi.org/10.1007/s00253-014-6296-0

Mishra S, Singh BR, Singh A, Keswani C, Naqvi AH, Singh HB (2014) Biofabricated silver nanoparticles act as a strong fungicide against *Bipolaris sorokiniana* causing spot blotch disease in wheat. PLoS One 9:e97881. https://doi.org/10.1371/journal.pone.0097881

Mittler R (2002) Oxidative stress, antioxidants and stress tolerance. Trends Plant Sci 7:405–410. https://doi.org/10.1016/S1360-1385(02)02312-9

Mohammadi A, Hashemi M, Hosseini SM (2015) Nanoencapsulation of *Zataria multiflora* essential oil preparation and characterization with enhanced antifungal activity for controlling *Botrytis cinerea*, the causal agent of gray mould disease. Innov Food Sci Emerg Technol 28:73–80.

Mondal KK, Mani C (2012) Investigation of the antibacterial properties of nanocopper against *Xanthomonas axonopodis* pv. punicae, the incitant of pomegranate bacterial blight. Ann Microbiol 62:889–893. https://doi.org/10.1007/s13213-011-0382-7

Mondal P, Kumar R, Gogoi R (2017) Azomethine based nano-chemicals: development, *in vitro* and *in vivo* fungicidal evaluation against *Sclerotium rolfsii*, *Rhizoctonia bataticola* and *Rhizoctonia solani*. Bioorg Chem 70:153. https://doi.org/10.1016/j.bioorg.2016.12.006

Moorthi PV, Balasubramanian C, Mohan S (2015) An improved insecticidal activity of silver nanoparticle synthesized by using *Sargassum muticum*. Appl Biochem Biotechnol 175:135–140. https://doi.org/10.1007/s12010-014-1264-9

Mousa KM, El Sharkawy MM, Khodeir IA, El-Dakhakhni TN, Youssef AE (2014) Growth perturbation, abnormalities and mortality of oriental armyworm *Mythimna separata* (Walker) (Lepidoptera: Noctuidae) caused by silica nanoparticles and *Bacillus thuringiensis* toxin. Egypt J Biol Pest Control 24:347–351.

Moussa SH, Tayel AA, Alsohim AS, Abdallah RR (2013) Botryticidal activity of nanosized silver-chitosan composite and its application for the control of gray mold in strawberry. J Food Sci 78:M1589–M1594. https://doi.org/10.1111/1750-3841.12247

Mukhopadhyay SS (2014) Nanotechnology in agriculture: prospects and constraints. Nanotechnol Sci Appl 7:63–71. https://doi.org/10.2147/NSA.S39409

Muthuramalingam TR, Shanmugam C, Gunasekaran D, Duraisamy N, Nagappan R, Krishnan K (2015) Bioactive bile salt-capped silver nanoparticles activity against destructive plant pathogenic fungi through *in vitro* system. RSC Adv 5:71174–71182. https://doi.org/10.1039/c5ra13306h

Nadiminti PP, Rookes JE, Dong YD, Sayer C, Boyd BJ, Cahill DM (2016) Nanostructured liquid crystalline particle assisted delivery of 2,4-dichlorophenoxyacetic acid to weeds, crops and model plants. Crop Prot 82:17–29. https://doi.org/10.1016/j.cropro.2015.12.018

Nair R, Varghese SH, Nair BG, Maekawa T, Yoshida Y, Sakthi Kumar D (2010) Nanoparticulate material delivery to plants. Plant Sci 179:154–163. https://doi.org/10.1016/j.plantsci.2010.04.012

Namasivayam SKR, Aruna A, Gokila (2014) Evaluation of silver nanoparticles-chitosan encapsulated synthetic herbicide paraquat (AgNp-CS-PQ) preparation for the controlled release and improved herbicidal activity against *Eichhornia crassipes*. Res J Biotechnol 9:19–27. https://doi.org/10.1371/journal.pone.0132971

Narendhran S, Sivaraj R (2016) Biogenic ZnO nanoparticles synthesized using *L. aculeata* leaf extract and their antifungal activity against plant fungal pathogens. Bull Mater Sci 39:1–5. https://doi.org/10.1007/s12034-015-1136-0

Nasseri M, Golmohammadzadeh S, Arouiee H, Jaafari MR, Neamati H (2016) Antifungal activity of *Zataria multiflora* essential oil-loaded solid lipid nanoparticles *in vitro* condition. Iran J Basic Med Sci 19:1231–1237. https://doi.org/10.22038/ijbms.2016.7824

National Nanotechnology Initiative (2008) Nanotechnology: big things from a tiny world. Arlington. Available online at: https://www.nano.gov/sites/default/files/pub_resource/nanotechnology_bigthingsfromatinyworld-print.pdf

Navarro E, Baun A, Behra R, Hartmann NB, Filser J, Miao AJ, Quigg A, Santschi PH, Sigg L (2008) Environmental behavior and ecotoxicity of engineered nanoparticles to algae, plants, and fungi. Ecotoxicology 17:372–386. https://doi.org/10.1007/s10646-008-0214-0

Negahban M, Moharramipour S, Zandi M, Hashemi SA, Ziayee F (2012) Nano-insecticidal activity of essential oil from *Cuminum cyminum on Tribolium castaneum*. In: Navarro S, Banks HJ, Jayas DS, Bell CH, Noyes RT, Ferizli AG, Emekci M, Isikber AA, Alagusundaram K (eds) Proceedings of 9th international conference on controlled atmosphere and fumigation in stored products, Antalya, pp 63–68. Available online at: http://ftic.co.il/2012AntalyaPDF/SESSION%2001%20PAPER%2009.pdf

Nehoff H, Parayath NN, Domanovitch L, Taurin S, Greish K (2014) Nanomedicine for drug targeting: strategies beyond the enhanced permeability and retention effect. Int J Nanomedicine 9:2539–2555. https://doi.org/10.2147/IJN.S47129

Nel A, Xia T, Madler L, Li N (2006) Toxic potential of materials at nanolevels. Science 311:622–627. https://doi.org/10.1126/science.1114397

Nenaah GE, Ibrahim SIA, Al-Assiuty BA (2015) Chemical composition, insecticidal activity and persistence of three *Asteraceae* essential oils and their nanoemulsions against *Callosobruchus maculatus* (F.). J Stored Prod Res 61:9–16. https://doi.org/10.1016/j.jspr.2014.12.007

Nguemtchouin MGM, Ngassoum MB, Chalier P, Kamga R, Ngamo LST, Cretin M (2013) *Ocimum gratissimum* essential oil and modified montmorillonite clay, a means of controlling insect pests in stored products. J Stored Prod Res 52:57–62. https://doi.org/10.1016/j.jspr.2012.09.006

Nishisaka C, Grillo R, Sanches G, Fraceto L, Lima R (2014) Analysis of the effects of pesticides and nanopesticides on the environment. BMC Proc 8(Suppl 4):P100. https://doi.org/10.1186/1753-6561-8-S4-P100

Nuruzzaman M, Rahman MM, Liu Y, Naidu R (2016) Nanoencapsulation, nano-guard for pesticides: a new window for safe application. J Agric Food Chem 64(7):1447–1483. https://doi.org/10.1021/acs.jafc.5b05214

Ocsoy I, Paret ML, Ocsoy MA, Kunwar S, Chen T, You MX, Tan WH (2013) Nanotechnology in plant disease management: DNA-directed silver nanoparticles on graphene oxide as an antibacterial against *Xanthomonas perforans*. ACS Nano 7:8972–8980. https://doi.org/10.1021/nn4034794

Oh SD, Lee S, Choi SH, Lee IS, Lee YM, Chun JH, Park HJ (2006) Synthesis of Ag and Ag-SiO_2 nanoparticles by gamma-irradiation and their antibacterial and antifungal efficiency against *Salmonella enterica* serovar *Typhimurium* and *Botrytis cinerea*. Colloids Surf A Physicochem Eng Asp 275:228–233.

Oliveira HC, Stolf-Moreira R, Martinez CBR, Grillo R, de Jesus MB, Fraceto LF (2015) Nanoencapsulation enhances the post-emergence herbicidal activity of atrazine against mustard plants. PLoS One 10:e0132971. https://doi.org/10.1371/journal.pone.0132971

Pandey A, Chandra S, Chauhan LKS, Narayan G, Chowdhuri DK (2013) Cellular internalization and stress response of ingested amorphous silica nanoparticles in the midgut of *Drosophila melanogaster*. Biochim Biophys Acta Gen Subjects 1830:2256–2266. https://doi.org/10.1016/j.bbagen.2012.10.001

Paret ML, Vallad GE, Averett DR, Jones JB, Olson SM (2013) Photocatalysis: effect of light-activated nanoscale formulations of TiO_2 on *Xanthomonas perforans* and control of bacterial spot of tomato. Phytopathology 103:228–236. https://doi.org/10.1094/PHYTO-08-12-0183-R

Parisi C, Vigani M, Rodríguez-Cerezo E (2015) Agricultural nanotechnologies: what are the current possibilities? NanoToday 10:124–127. https://doi.org/10.1016/j.nantod.2014.09.009

Patil CD, Borase HP, Suryawanshi RK, Patil SV (2016) Trypsin inactivation by latex fabricated gold nanoparticles: a new strategy towards insect control. Enzym Microb Technol 92:18–25. https://doi.org/10.1016/j.enzmictec.2016.06.005

Pereira AE, Grillo R, Mello NF, Rosa AH, Fraceto LF (2014) Application of poly(epsilon-caprolactone) nanoparticles containing atrazine herbicide as an alternative technique to control weeds and reduce damage to the environment. J Hazard Mater 268:207–215. https://doi.org/10.1016/j.jhazmat.2014.01.025

Pérez-de-Luque A, Hermosín MC (2013) Nanotechnology and its use in agriculture. In: Bagchi D, Bagchi M, Moriyama H, Shahidi F (eds) Bio-nanotechnology: a revolution in food, biomedical and health sciences. Wiley-Blackwell, West Sussex, pp 299–405.

Pietrzak K, Twaruzek M, Czyzowska A, Kosicki R, Gutarowska B (2015) Influence of silver nanoparticles on metabolism and toxicity of moulds. Acta Biochim Pol 62:851–857. https://doi.org/10.18388/abp.2015_1146

Pompa PP, Vecchio G, Galeone A, Brunetti V, Sabella S, Maiorano G, Falqui A, Bertoni G, Cingolani R (2011) *In vivo* toxicity assessment of gold nanoparticles in *Drosophila melanogaster*. Nano Res 4:405–413. https://doi.org/10.1007/s12274-011-0095-z

Population Institute (2017) FAO says food production must rise by 70%. Available online at: https://www.populationinstitute.org/resources/populationonline/issue/1/8

Pradhan S, Roy I, Lodh G, Patra P, Choudhury SR, Samanta A, Goswami A (2013) Entomotoxicity and biosafety assessment of PEGylated acephate nanoparticles: a biologically safe alternative to neurotoxic pesticides. J Environ Sci Health B 48:559–569. https://doi.org/10.1080/03601234.2013.774891

Prado AGS, Moura AO, Nunes AR (2011) Nanosized silica modified with carboxylic acid as support for controlled release of herbicides. J Agric Food Chem 59:8847–8852. https://doi.org/10.1021/jf202509g

Prasad R, Kumar V, Prasad KS (2014) Nanotechnology in sustainable agriculture: present concerns and future aspects. Afr J Biotechnol 13:705–713. https://doi.org/10.5897/AJBX2013.13554

Prasad R, Bhattacharyya A, Nguyen QD (2017) Nanotechnology in sustainable agriculture: recent developments, challenges, and perspectives. Front Microbiol 8:1014. https://doi.org/10.3389/fmicb.2017.01014

Qian K, Shi TY, Tang T, Zhang SL, Liu XL, Cao YS (2011) Preparation and characterization of nano-sized calcium carbonate as controlled release pesticide carrier for validamycin against *Rhizoctonia solani*. Microchim Acta 173:51–57. https://doi.org/10.1007/s00604-010-0523-x

Rajiv P, Rajeshwari S, Venckatesh R (2013) Bio-fabrication of zinc oxide nanoparticles using leaf extract of *Parthenium hysterophorus* L. and its size-dependent antifungal activity against plant fungal pathogens. Spectrochim Acta A Mol Biomol Spectrosc 112:384–387. https://doi.org/10.1016/j.saa.2013.04.072

Raliya R, Tarafdar JC, Gulecha K, Choudhary K, Ram R, Mal P, Saran RP (2013) Scope of nanoscience and nanotechnology in agriculture. J Appl Biol Biotechnol 1:41–44. https://doi.org/10.7324/JABB.2013.1307

Rani PU, Madhusudhanamurthy J, Sreedhar B (2014) Dynamic adsorption of α-pinene and linalool on silica nanoparticles for enhanced antifeedant activity against agricultural pests. J Pestic Sci 87:191–200. https://doi.org/10.1007/s10340-013-0538-2

Rao KJ, Paria S (2013) Use of sulfur nanoparticles as a green pesticide on *Fusarium solani* and *Venturia inaequalis* phytopathogens. RSC Adv 3:10471–10478. https://doi.org/10.1039/c3ra40500a

Rashidi L, Khosravi-Darani K (2011) The applications of nanotechnology in food industry. Crit Rev Food Sci Nutr 51:723–730. https://doi.org/10.1080/10408391003785417

Rispail N, De Matteis L, Santos R, Miguel AS, Custardoy L, Testillano PS, Risueno MC, Pérez-de-Luque A, Maycock C, Fevereiro P, Oliva A, Fernández-Pacheco R, Ibarra MR, de la Fuente JM, Marquina C, Rubiales D, Prats E (2014) Quantum dot and superparamagnetic nanoparticle interaction with pathogenic fungi: internalization and toxicity profile. ACS Appl Mater Interfaces 6:9100–9110. https://doi.org/10.1021/am501029g

Rizwan M, Ali S, Qayyum MF, Ok YS, Adrees M, Ibrahim M, Zia-ur-Rehmand M, Farid M, Abbas F (2017) Effect of metal and metal oxide nanoparticles on growth and physiology of globally important food crops: a critical review. J Hazard Mater 322:2–16. https://doi.org/10.1016/j.jhazmat.2016.05.061

Rodriguez-Gonzalez V, Dominguez-Espindola RB, Casas-Flores S, Patron-Soberano OA, Camposeco-Solis R, Lee SW (2016) Antifungal nanocomposites inspired by titanate nanotubes for complete inactivation of *Botrytis cinerea* isolated from tomato infection. ACS Appl Mater Interfaces 8:31625–31637. https://doi.org/10.1021/acsami.6b10060

Rouhani M, Samih MA, Kalantari S (2013) Insecticidal effect of silica and silver nanoparticles on the cowpea seed beetle, *Callosobruchus maculatus* F. (Col.: Bruchidae). J Entomol Res 4:297–305.

Sabbour MM, Abd El- Aziz SE (2016a) Roll of three essential oils and their Nano against *Ephestia cautella* (Lepidoptera-Pyralidae) under laboratory and store conditions. Int J PharmTech Res 9:194–200.

Sabbour MM, Abd El- Aziz SE (2016b) Efficacy of three essential oils and their nano-particles against *Sitophilus granarius* under laboratory and store conditions. J Entomol Res 40:229–234. https://doi.org/10.5958/0974-4576.2016.00042.6

Sagiri SS, Anis A, Pal K (2016) Review on encapsulation of vegetable oils: strategies, preparation methods, and applications. Polym-Plast Technol 55:291–311. https://doi.org/10.1080/03602559.2015.1050521

Saharan V, Mehrotra A, Khatik R, Rawal P, Sharma SS, Pal A (2013) Synthesis of chitosan based nanoparticles and their *in vitro* evaluation against phytopathogenic fungi. Int J Biol Macromol 62:677–683. https://doi.org/10.1016/j.ijbiomac.2013.10.012

Saharan V, Sharma G, Yadav M, Choudhary KM, Sharma SS, Pal A, Raliya R, Biswas P (2015) Synthesis and *in vitro* antifungal efficacy of Cu–chitosan nanoparticles against pathogenic fungi of tomato. Int J Biol Macromol 75:346–353. https://doi.org/10.1016/j.ijbiomac.2015.01.027

Saini P, Gopal M, Kumar R, Srivastava C (2014) Development of pyridalyl nanocapsule suspension for efficient management of tomato fruit and shoot borer (*Helicoverpa armigera*). J Environ Sci Health B 49:344–351. https://doi.org/10.1080/03601234.2014.882168

Sandhya, Kumar S, Kumar D, Dilbaghi N (2017) Preparation, characterization, and bio-efficacy evaluation of controlled release carbendazim-loaded polymeric nanoparticles. Environ Sci Pollut Res 2:926–937. https://doi.org/10.1007/s11356-016-7774-y

Sangeetha J, Thangadurai D, Hospet R, Purushotham P, Karekalammanavar G, Mundaragi AC, David M, Shinge MR, Thimmappa SC, Prasad R, Harish ER (2017a) Agricultural nanotechnology: concepts, benefits, and risks. In: Prasad R, Kumar M, Kumar V (eds) Nanotechnology. Springer Nature Singapore Pte Ltd, Singapore, pp 1–17.

Sangeetha J, Thangadurai D, Hospet R, Purushotham P, Manowade KR, Mujeeb MA, Mundaragi AC, Jogaiah S, David M, Thimmappa SC, Prasad R, Harish ER (2017b) Production of bionanomaterials from agricultural wastes. In: Prasad R, Kumar M, Kumar V (eds) Nanotechnology. Springer Nature Singapore Pte Ltd, Singapore, pp 33–58.

Sangeetha J, Thangadurai D, Hospet R, Harish ER, Purushotham P, Mujeeb MA, Shrinivas J, David M, Mundaragi AC, Thimmappa AC, Arakera SB, Prasad R (2017c) Nanoagrotechnology for soil quality, crop performance and environmental management. In: Prasad R, Kumar M, Kumar V (eds) Nanotechnology. Springer Nature Singapore Pte Ltd, Singapore, pp 73–97.

Sankar MV, Abideen S (2015) Pesticidal effect of green synthesized silver and lead nanoparticles using *Avicennia marina* against grain storage pest *Sitophilus oryzae*. Int J Nanomat Biostruct 5:32–39.

Santo-Orihuela PL, Foglia ML, Targovnik AM, Miranda MV, Desimone MF (2016) Nanotoxicological effects of SiO_2 nanoparticles on *Spodoptera frugiperda* Sf9 cells. Curr Pharm Biotechnol 17:465–470.

Sarijo SH, Hussein MZ, Yahaya AH, Zainal Z, Yarmo MA (2010a) Synthesis of phenoxyherbicides-intercalated layered double hydroxide nanohybrids and their controlled release property. Curr Nanosci 6:199–205. https://doi.org/10.2174/157341310790945614

Sarijo SH, Hussein MZ, Yahaya AHJ, Zainal Z (2010b) Effect of incoming and outgoing exchangeable anions on the release kinetics of phenoxyherbicides nanohybrids. J Hazard Mater 182:563–569. https://doi.org/10.1016/j.jhazmat.2010.06.070

Sarkar DJ, Kumar J, Shakil NA, Walia S (2012) Release kinetics of controlled release formulations of thiamethoxam employing nano-ranged amphiphilic PEG and diacid based block polymers in soil. J Environ Sci Health A Tox Hazard Subst Environ Eng 47:1701–1712. https://doi.org/10.1080/10934529.2012.687294

Sarlak N, Taherifar A, Salehi F (2014) Synthesis of nanopesticides by encapsulating pesticide nanoparticles using functionalized carbon nanotubes and application of new nanocomposite for plant disease treatment. J Agric Food Chem 62:4833–4838. https://doi.org/10.1021/jf404720d

Sathiyabama M, Manikandan A (2016) Chitosan nanoparticle induced defense responses in fingermillet plants against blast disease caused by *Pyricularia grisea* (Cke.) Sacc. Carbohydr Polym 154:241–246. https://doi.org/10.1016/j.carbpol.2016.06.089

Savi GD, Scussel VM (2014) Inorganic compounds at regular and nanoparticle size and their antitoxigenic fungi activity. In: Book of abstracts of the 11th international working conference on stored product protection, Chiang Mai, pp 589–598.

Savi GD, da Silva PMM, Possato JC, Barichello T, Castagnaro D, Scussel VM (2012) Biological activity of gold nanoparticles towards filamentous pathogenic fungi. J Nano Res 20:11–20. https://doi.org/10.4028/www.scientific.net/JNanoR.20.11

Savi GD, Bortoluzzi AJ, Scussel VM (2013) Antifungal properties of zinc compounds against toxigenic fungi and mycotoxin. Int J Food Sci Technol 48:1834–1840. https://doi.org/10.1111/ijfs.12158

Sawangphruk M, Srimuk P, Chiochan P, Sangsri T, Siwayaprahm P (2012) Synthesis and antifungal activity of reduced graphene oxide nanosheets. Carbon 50:5156–5161. https://doi.org/10.1016/j.carbon.2012.06.056

Sekhon BS (2014) Nanotechnology in agri-food production: an overview. Nanotechnol Sci Appl 7:31–53. https://doi.org/10.2147/NSA.S39406

Sharma D, Sharma S, Kaith BS, Rajput J, Kaur M (2011) Synthesis of ZnO nanoparticles using surfactant free in-air and microwave method. Appl Surf Sci 257:9661–9672. https://doi.org/10.1016/j.apsusc.2011.06.094

Sharon A, Shlezinger N (2013) Fungi infecting plants and animals: killers, non-killers, and cell death. PLoS Pathog 9:e1003517. https://doi.org/10.1371/journal.ppat.1003517

Shu Y, Zhang G, Wang J (2012) Response of the common cutworm *Spodoptera litura* to zinc stress: Zn accumulation, metallothionein and cell ultrastructure of the midgut. Sci Total Environ 438:210–217. https://doi.org/10.1016/j.scitotenv.2012.06.065

Shyla KK, Natarajan N, Nakkeeran S (2014) Antifungal activity of zinc oxide, silver and titanium dioxide nanoparticles against *Macrophomina phaseolina*. J Mycol Plant Pathol 44:268–273.

Siddiqi KS, Husen A (2017) Plant response to engineered metal oxide nanoparticles. Nanoscale Res Lett 12:92. https://doi.org/10.1186/s11671-017-1861-y

Silva MD, Cocenza DS, de Melo NFS, Grillo R, Rosa AH, Fraceto LF (2010) Alginate nanoparticles as a controlled release system for clomazone herbicide. Quim Nova 33:1868–1873. https://doi.org/10.1590/S0100-40422010000900009

Silva MD, Cocenza DS, Grillo R, de Melo NFS, Tonello PS, de Oliveira LC, Cassimiro DL, Rosa AH, Fraceto LF (2011) Paraquat-loaded alginate/chitosan nanoparticles: preparation, characterization and soil sorption studies. J Hazard Mater 190:366–374. https://doi.org/10.1016/j.jhazmat.2011.03.057

Sioutas C, Delfino RJ, Singh M (2005) Exposure assessment for atmospheric ultrafine particles (UFPs) and implications in epidemiologic research. Environ Health Perspect 113:947–955. https://doi.org/10.1289/ehp.7939

Soujanya PL, Sekhar JC, Kumar P, Sunil N, Prasad CV, Mallavadhani UV (2016) Potentiality of botanical agents for the management of post harvest insects of maize: a review. J Food Sci Technol Mysore 53:2169–2184. https://doi.org/10.1007/s13197-015-2161-0

Stadler T, Buteler M, Weaver DK (2010a) Novel use of nanostructured alumina as an insecticide. Pest Manag Sci 66:577–579. https://doi.org/10.1002/ps.1915

Stadler T, Buteler M, Weaver DK (2010b) Nanoinsecticidas: Nuevas perspectivas para el control de plagas [Nanoinsecticides: new perspectives for pest management]. Rev Soc Entomol Argent 69:149–156.

Stadler T, Buteler M, Weaver DK, Sofie S (2012) *Sitophilus oryzae* (L.) and *Rhyzopertha dominica* (F.) at varying ambient humidity levels. J Stored Prod Res 48:81–90. https://doi.org/10.1016/j.jspr.2011.09.004

Strayer A, Ocsoy I, Tan W, Jones JB, Paret ML (2016) Low concentrations of a silver-based nanocomposite to manage bacterial spot of tomato in the greenhouse. Plant Dis 100:1460–1465. https://doi.org/10.1094/PDIS-05-15-0580-RE

Sun Z, Shi C, Wang X, Fang Q, Huang J (2017) Synthesis, characterization, and antimicrobial activities of sulfonated chitosan. Carbohydr Polym 155:321–328. https://doi.org/10.1016/j.carbpol.2016.08.069

Suppan S (2013). Nanomaterials in soil: our future food chain? institute for agriculture and trade policy. Available online at: http://www.iatp.org/documents/nanomaterials-in-soil-our-future-food-chain

Suriyaprabha R, Karunakaran G, Kavitha K, Yuvakkumar R, Rajendran V, Kammam N (2014) Application of silica nanoparticles in maize to enhance fungal resistance. IET Nanobiotechnol 8:133–137. https://doi.org/10.1049/iet-nbt.2013.0004

Tan DY, Yuan P, Annabi-Bergaya F, Dong FQ, Liu D, He HP (2015) A comparative study of tubular halloysite and platy kaolinite as carriers for the loading and release of the herbicide amitrole. Appl Clay Sci 114:190–196. https://doi.org/10.1016/j.clay.2015.05.024

Vaculíková E, Černíková A, Plachá D, Pisarčík M, Dedková K, Peikertová P, Devínsky F, Jampílek J (2016a) Cimetidine nanoparticles for permeability enhancement. J Nanosci Nanotechnol 16:7840–7843. https://doi.org/10.1166/jnn.2016.12562

Vaculíková E, Černíková A, Plachá D, Pisarčík M, Peikertová P, Dedková K, Devínsky F, Jampílek J (2016b) Preparation of hydrochlorothiazide nanoparticles for solubility enhancement. Molecules 21:1005. https://doi.org/10.3390/molecules21081005

Vanathi P, Rajiv P, Sivaraj R (2016) Synthesis and characterization of *Eichhornia*-mediated copper oxide nanoparticles and assessing their antifungal activity against plant pathogens. Bull Mater Sci 39:1165–1170. https://doi.org/10.1007/s12034-016-1276-x

Vani C, Brindhaa U (2013) Silica nanoparticles as nanocides against *Corcyra cephalonica* (S.), the stored grain pest. Int J Pharm Bio Sci 4(B):1108–1118.

Ventola CL (2012) The nanomedicine revolution: part 3: regulatory and safety challenges. Pharm Therapeutics 37:631–639.

Vestel J, Caldwell DJ, Constantine L, D'Aco VJ, Davidson T, Dolan DG, Millard SP, Murray-Smith R, Parke NJ, Ryan JJ, Straub JO, Wilson P (2016) Use of acute and chronic ecotoxicity data in environmental risk assessment of pharmaceuticals. Environ Toxicol Chem 35:1201–1212. https://doi.org/10.1002/etc.3260

Vitanovic E (2012) Use of Cu fungicides in vineyards and olive groves. In: Dhanasekaran D, Thajuddin N, Panneerselvam A (eds) Fungicides for plant and animal diseases. InTech, Rijeka, pp 279–298. https://doi.org/10.5772/26953

Vlad S, Tanase C, Macocinschi D, Ciobanu C, Balaes T, Filip D, Gostin IN, Gradinaru LM (2012) Antifungal behaviour of polyurethane membranes with zinc oxide nanoparticles. Digest J Nanomater Biostruct 7:51–58.

Wagner G, Korenkov V, Judy JD, Bertsch PM (2016) Nanoparticles composed of Zn and ZnO inhibit peronospora tabacina spore germination *in vitro* and *P. tabacina* infectivity on tobacco leaves. Nano 2016:6. https://doi.org/10.3390/nano6030050

Wang XP, Liu XQ, Chen JN, Han HY, Yuan ZD (2014) Evaluation and mechanism of antifungal effects of carbon nanomaterials in controlling plant fungal pathogen. Carbon 68:798–806. https://doi.org/10.1016/j.carbon.2013.11.072

Wang X, Cai A, Wen X, Jing D, Qi H, Yuan H (2017) Graphene oxide-Fe_3O_4 nanocomposites as high-performance antifungal agents against *Plasmopara viticola*. Sci China Mater 60:258–268. https://doi.org/10.1007/s40843-016-9005-9

Wanyika H (2013) Sustained release of fungicide metalaxyl by mesoporous silica nanospheres. J Nanopart Res 15:UNSP 1831. https://doi.org/10.1007/s11051-013-1831-y

Watson SB, Gergely A, Janus ER (2011) Where is "Agronanotechnology" heading in the United States and european union? Nat Resour Environ 26:8–12.

Wen YZ, Zhang LJ, Chen ZW, Sheng XL, Qiu JG, Xu DM (2016) Co-exposure of silver nanoparticles and chiral herbicide imazethapyr to *Arabidopsis thaliana*: enantioselective effects. Chemosphere 145:207–214. https://doi.org/10.1016/j.chemosphere.2015.11.035

Wennersten R, Fidler J, Spitsyna A (2008) Nanotechnology: a new technological revolution in the 21st century. In: Misra KB (ed) Handbook of performability engineering. Springer, London, pp 943–952.

Wibowo D, Zhao CX, Peters BC, Middelberg AP (2014) Sustained release of fipronil insecticide *in vitro* and *in vivo* from biocompatible silica nanocapsules. J Agric Food Chem 62:12504–12511. https://doi.org/10.1021/jf504455x

Wilpiszewska K, Spychaj T, Pazdzioch W (2016) Carboxymethyl starch/montmorillonite composite microparticles: properties and controlled release of isoproturon. Carbohydr Polym 136:101–106. https://doi.org/10.1016/j.carbpol.2015.09.021

Xing K, Shen XQ, Zhu X, Ju XY, Miao XM, Tian J, Feng ZZ, Peng X, Jiang JH, Qin S (2016) Synthesis and *in vitro* antifungal efficacy of oleoyl-chitosan nanoparticles against plant pathogenic fungi. Int J Biol Macromol 82:830–836. https://doi.org/10.1016/j.ijbiomac.2015.09.074

Xu L, Cao LD, Li FM, Wang XJ, Huang QL (2014) Utilization of chitosan-lactide copolymer nanoparticles as controlled release pesticide carrier for pyraclostrobin against *Colletotrichum gossypii* Southw. J Dispers Sci Technol 35:544–550. https://doi.org/10.1080/01932691.2013.800455

Yang FL, Li XG, Zhu F, Lei CL (2009) Structural characterization of nanoparticles loaded with garlic essential oil and their insecticidal activity against *Tribolium castaneum* (Herbst) (Coleoptera: Tenebrionidae). J Agric Food Chem 57:10156–10162. https://doi.org/10.1021/jf9023118

Yasur J, Pathipati UR (2015) Lepidopteran insect susceptibility to silver nanoparticles and measurement of changes in their growth, development and physiology. Chemosphere 124:92–102. https://doi.org/10.1016/j.chemosphere.2014.11.029

Yearla SR, Padmasree K (2016) Exploitation of subabul stem lignin as a matrix in controlled release agrochemical nanoformulations: a case study with herbicide diuron. Environ Sci Pollut Res 23:18085–18098. https://doi.org/10.1007/s11356-016-6983-8

Zahid N, Alderson PG, Ali A, Maqbool M, Manickam S (2013) *In vitro* control of *Colletotrichum gloeosporioides* by using chitosan loaded nanoemulsions. Acta Hortic 1012:769–774. https://doi.org/10.17660/ActaHortic.2013.1012.104

Zahir AA, Bagavan A, Kamaraj C, Elango G, Abdul Rahuman A (2012) Efficacy of plant-mediated synthesized silver nanoparticles against *Sitophilus oryzae*. J Biopest 5:95–102.

Zare-Zardini H, Amiri A, Shanbedi M, Memarpoor-Yazdi M, Asoodeh A (2013) Studying of antifungal activity of functionalized multiwalled carbon nanotubes by microwave-assisted technique. Surf Interace Anal 45:751–755. https://doi.org/10.1002/sia.5152

Zhang JT, Liu Y, Li Q, Zhang XP, Shang JK (2013a) Antifungal activity and mechanism of palladium-modified nitrogen-doped titanium oxide photocatalyst on agricultural pathogenic fungi *Fusarium graminearum*. ACS Appl Mater Interfaces 5:10953–10959. https://doi.org/10.1021/am4031196

Zhang JK, Li M, Fan TF, Xu Q, Wu Y, Chen CY, Huang QL (2013b) Construction of novel amphiphilic chitosan copolymer nanoparticles for chlorpyrifos delivery. J Polym Res 20:107. https://doi.org/10.1007/s10965-013-0107-7

Zhou Y, Jiang SJ, Jiao YJ, Wang H (2017) Synergistic effects of nanochitin on inhibition of tobacco root rot disease. Int J Biol Macromol 99:205–212. https://doi.org/10.1016/j.ijbiomac.2017.02.069

Ziaee M, Ganji Z (2016) Insecticidal efficacy of silica nanoparticles against *Rhyzopertha dominica* F. and *Tribolium confusum* Jacquelin du Val. J Plant Prot Res 56:250–256. https://doi.org/10.1515/jppr-2016-0037

Ziaee M, Moharramipour S, Mohsenifar A (2014a) Toxicity of *Carum copticum* essential oil-loaded nanogel against *Sitophilus granarius* and *Tribolium confusum*. J Appl Entomol 138:763–771. https://doi.org/10.1111/jen.12133

Ziaee M, Moharramipour S, Mohsenifar A (2014b) MA-chitosan nanogel loaded with *Cuminum cyminum* essential oil for efficient management of two stored product beetle pests. J Pest Sci 87:691–699. https://doi.org/10.1007/s10340-014-0590-6

Chapter 9
Applications of Silver Nanoparticles in Plant Protection

Nomita Gupta, Chandrama Prakash Upadhyaya, Amar Singh, Kamel A. Abd-Elsalam, and Ram Prasad

9.1 Introduction

The word "nano" is the one billionth of a meter or 10^{-9}. The term nanotechnology was coined by Professor Norio Taniguchi of Tokyo University of Science in 1974 who illustrated the precise manufacturing of materials at the nanoscale level (Taniguchi 1974). In green nanotechnology, microorganisms and plants are used for the synthesis of nanoparticles (NPs). It is well known that many microorganisms are capable of aggregating inorganic material within or outside the cell to form NPs. However, an enormous number of microbial species are capable of producing metal NPs, and the mechanism of NPs biosynthesis is very important. Microbial synthesis of NPs is an approach that interconnects nanotechnology and microbial biotechnology. Biosynthesis of many metals nanoparticles like gold, silver, gold-silver alloy, selenium, tellurium, platinum, palladium, silica, titania, zirconia, quantum dots, magnetite, and uraninite by bacteria, actinomycetes, fungi, yeasts, algae, and viruses

N. Gupta
Amity Institute of Microbial Technology, Amity University, Noida, India

C. P. Upadhyaya
Department of Biotechnology, DR Harisingh Gour Central University, Sagar, Madhya Pradesh, India

A. Singh
Lal Bahadur Shastri Memorial College, Jamshedpur (Kolhan University, Chaibasa), Jamshedpur, India

K. A. Abd-Elsalam
Plant Pathology Research Institute, Agricultural Research Center (ARC), Giza, Egypt

R. Prasad (✉)
School of Environmental Science and Engineering, Sun Yat-Sen University, Guangzhou, China

Amity Institute of Microbial Technology, Amity University, Noida, India

K. A. Abd-Elsalam, R. Prasad (eds.), *Nanobiotechnology Applications in Plant Protection*, Nanotechnology in the Life Sciences,
https://doi.org/10.1007/978-3-319-91161-8_9

has been reported (Narayanan and Sakthivel 2010; Prasad et al. 2016). Silver nanoparticles (AgNPs) have become one of the most commonly used nanomaterials in consumer products, and for several decades, silver (Ag^+) has been studied as an antimicrobial agent against various harmful microorganisms (Prasad 2014).

Due to plant disease agricultural production is reduced worldwide every year; therefore, millions of moneys have been invested in efforts to control the plant diseases. Various natural and artificial methods of control for protection of plants from these diseases have been applied. Among methods for disease control, use of pesticides is the most prevalent. In recent years, environmental hazards caused by excessive use of pesticides have been widely discussed; therefore, researchers in the agricultural field are searching for alternative measures against pesticides. Nanotechnological applicability in crop disease protection offers a great promise in the management of insects and pathogens. AgNPs are very effective against phytopathogens with low toxicity and lead to broad range of applicability in pesticidal activity. It is efficiently used for site-targeted delivery of important agrochemical products and for diagnosis purpose tools in case of prior detection of plant diseases (Chowdappa and Shivakumar 2013). AgNPs are the most studied and utilized NPs in the field of agricultural research to improve the efficiency, yield, and sustainability of agricultural crops. It has long been known to have strong pesticidal, antifungal, antiviral, and bactericidal effects (Chen and Schluesener 2008). Due to its broad spectrum of antimicrobial activities, AgNPs have the prospect to increase food quality, global food production, plant protection, detection and regulation of plant diseases, monitoring of plant growth, and pest control for "sustainable agricultural development" (Kim et al. 2012; Khan and Rizvi 2014; Prasad et al. 2017a, 2017b). AgNPs are highly stable and very well dispersive in aqueous solution. It is being used as foliar spray to inhibit the growth of fungi, molds, rot, and several other plant diseases (Singh et al. 2015). Moreover, AgNPs are also used as an excellent plant-growth stimulator. It provides novel tool for the management of diseases, rapid disease detection, and minimizing nutrient losses in fertilization through an optimized nutrient management (Pérez-de-Luque and Rubiales 2009). As an alternative to chemically manufactured pesticides, use of AgNPs as antimicrobial agents has become more common as technological advances make their production more economical. One of the important appliances of AgNPs is in the management of plant diseases. Silver displays multiple modes of inhibitory action against microorganisms; therefore, it may be used with relative safety for control of various plant pathogens, compared to synthetic fungicides (Aziz et al. 2016; Prasad et al. 2014, 2017a, 2017b).

Artificial chemical antimicrobials are widely used in modern agriculture to control plant diseases. Environmental hazards caused by excessive use of pesticides pose health problems as modern society is becoming more health-conscious. Therefore, agricultural scientists are searching for alternative eco-friendly and less capital-intensive approaches to control plant diseases. As an alternative to chemically manufactured pesticides, use of AgNPs as antimicrobial agents has become more common as technological advances make their production more economical. So, the focus of this chapter is to study the possibilities of using the synthesized AgNPs in plant protection (Fig. 9.1).

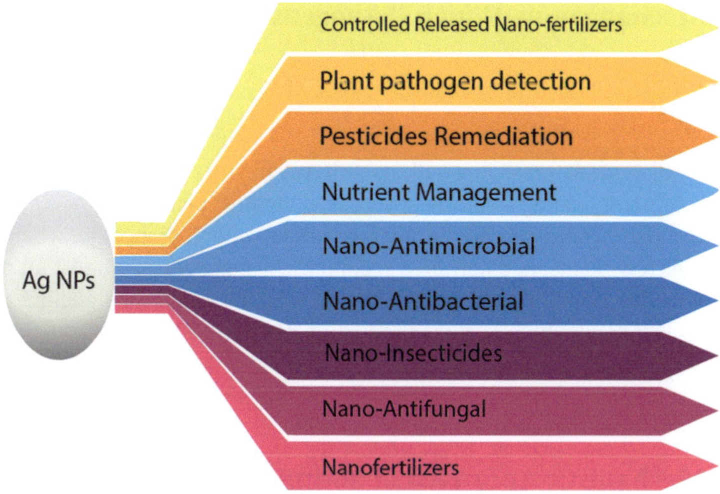

Fig. 9.1 Silver nanoparticle application in plant protections

9.2 Chemical and Biological Silver Nanoparticles (AgNPs)

The physical and chemical methods are numerous in number for the synthesis of AgNPs, and many of these methods are expensive or use toxic substances which are major factors that make them "not so favored" methods of synthesis. Various types of physical and chemical methods that are employed in the production of nanoparticles are top-down method, bottom-up method, reduction of silver metal salt, electroreduction of $AgNO_3$ in aqueous solution in the presence of polyethylene glycol, son decomposition, photoreduction in reverse micelle, and many more. The quest for such a method has led to the need for biomimetic production of silver nanoparticles whereby biological methods (using plants and microbes) are used to synthesize AgNPs. Biologically, various types of microbes and plants species are used for the biosynthesis of nanoparticles. In utmost cases, the chemical synthesis methods lead to some chemically toxic materials being absorbed on the surface and can hinder their practice in medical applications (Parashar et al. 2009; Swamy and Prasad 2012; Prasad and Swamy 2013; Prasad et al. 2016). Biological synthesis of AgNPs is a bottom-up method that typically involves reduction/oxidation reactions. The microbial enzymes or the plant phytochemicals with antioxidant or reducing properties act on the respective compounds and give the desired nanoparticles. The three major components involved in the biological synthesis of nanoparticles: solvent medium, the eco-friendly reducing agent, and a nontoxic stabilizing agent. The nanoparticles so produced have protein cap over it imparted by biological host. This capping helps in easy entry in pathogenic fungi. Presence of peak at 1654 in FTIR spectrum indicates the presence of amide bond in the sample indicating the presence of protein capping which is responsible for stabilizing the synthesized nanoparticles (Aziz et al. 2015).

9.3 Mechanism by Which Pathogens Cause Diseases

Pathogens cause infection via various mechanisms. Some common known mechanisms are:

1. Cell wall degradation: some pathogens have enzymes which can degrade the cell walls of plant leading to the easy access to the host plant.
2. Toxins: these chemicals are further categorized into host specific, which are specific for few plants, and non-host specific, which are active against all plants.
3. Effector proteins: these proteins interfere with the chemical signaling pathways of the host plant which results in reduction of phytochemical production (Winbo 2011).

9.4 Mechanistic Approach of antimicrobial activity of Silver Nanoparticles and Controls of the Growth of Pathogens

There are numerous mechanisms by which AgNPs control the growth of pathogens. The exact mechanism of AgNPs by which it causes antimicrobial effect is not evidently known and is one of the debated topics. Numerous theories on which action of AgNPs based by which it cause antimicrobial effects. Silver nanoparticles possess the ability to anchor the bacterial cell wall and penetrate it, causing structural changes in the cell membrane like cell membrane permeability and cell death. There are formation of "pits" and accumulation of the nanoparticles on the cell surface. The formation of free radicals by the AgNPs may be considered as another mechanism by which the cells die. The electron spin resonance spectroscopy studies suggest that there is a formation of free radicals when AgNPs contact the bacteria and these free radicals make membrane porous which ultimately leads to cell death (Danilcauk et al. 2006; Kim et al. 2007). It has also been proposed that there may be a release of silver ions by the nanoparticles (Feng et al. 2008) and these ions have inbuilt property to interact with the thiol (-SH) groups of many crucial enzymes and inactivate them (Matsumura et al. 2003). The bacterial cells come in contact with silver ions, which inhibit many functions and damage the cells which result in the generation of reactive oxygen species (ROS) that may also be produced possibly through the inhibition of a respiratory enzyme by silver ions and attack the cell itself. The action of these nanoparticles on the cell can cause the reaction to take place and subsequently lead to cell death. Another important fact is that the DNA contains sulfur and phosphorus as their major components; the nanoparticles can act on these components and destroy the DNA which ultimately lead to cell death (Hatchett and Henry 1996). The interaction of the silver nanoparticles with the sulfur and phosphorus of the DNA can lead to problems in the DNA replication of the bacteria and thus terminates their growth. It has also been found that the AgNPs can modulate signal transduction in bacteria. It is a well-known fact that phosphorylation of protein substrates in bacteria persuade bacterial signal

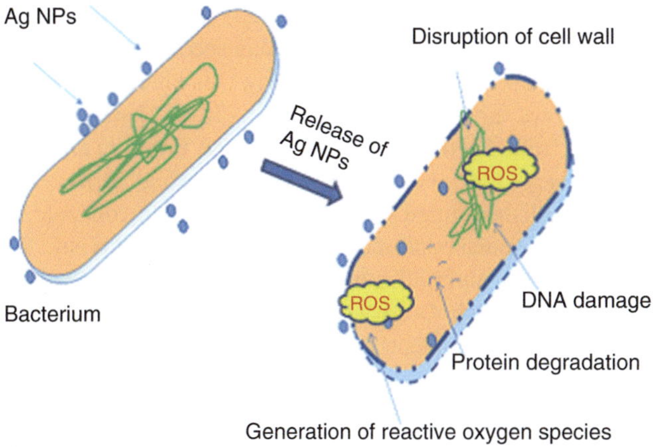

Fig. 9.2 Mechanistic approach of the antibacterial action indicating ROS generation induced by AgNPs. (Reprinted from Aziz et al. 2015)

transduction. Dephosphorylation is only reported in the tyrosine residues of gram-negative bacteria. The phosphor tyrosine profile of bacterial peptides is altered by the nanoparticles. It was found that the nanoparticles dephosphorylate the peptide substrates on tyrosine residues, which leads to signal transduction inhibition and thus the stoppage of growth. It is however necessary to understand that further research is required on the topic to thoroughly establish the claims (Shrivastava et al. 2007) (Fig. 9.2).

9.5 Effect of Silver Nanoparticles on Phytopathogens

Phytopathogens, viz., bacteria, fungi, viruses and nematodes, are key limiting factors in the production of food material. Several methods are used to control pathogens but not a perfect method to control of the disease. Hence, a great prospect exists for the manipulation of nanotechnology for the management of plant pathogens. Silver is considered the most capable nanomaterials with fungicidal, bactericidal, and viricidal properties owing to its wide-ranging effectiveness, low toxicity, ease of use, charge capacity, high surface-to-volume ratios, crystallographic structure, and adaptability to several substrates (Nangmenyi and Economy 2009). AgNPs act as robust antimicrobial agent due to strong inhibitory effects against various microorganisms (Clement and Jarrett 1994). Nanosilver exhibits high level of toxicity to the microorganisms and lower toxicity to the mammalian cells. It was observed that the microbe-killing effects of AgNPs were size dependent (Raza et al. 2017). The AgNPs/PVP were tested for fungicidal activity against different yeasts and molds such as *Candida albicans*, *C. krusei*, *C. tropicalis*, *C. glabrata*, and *Aspergillus brasiliensis*. The hybrid materials showed strong antifungal effects against the tested microbes (Bryaskova et al. 2011). Traditional microbiological plating, scanning electron microscopy, and Raman spectroscopy were used to study antifungal activities of AgNPs and to characterize the changes in morphology and cellular compositions of fungal hyphae. Aziz et al. (2016) observed the effect of biogenic nanoparticles from AgNPs on pathogenic fungi, *Candida albicans*, *Fusarium oxysporum*, and *Aspergillus flavus*, and these antimicrobial attributes were comparable to those of established fungicides (amphotericin B, fluconazole, and ketoconazole). Importantly, these nanoparticles show significant synergistic characteristics when combined with the antibiotics and fungicides to offer substantially greater resistance to microbial growth. Ocsoy et al. (2013) developed nanocomposite DNA-directed AgNPs grown on graphene oxide (Ag@dsDNA@GO). These composites effectively decrease *Xanthomonas perforans* cell viability in culture and on plants. At the very low concentration (16 ppm), composites show excellent antibacterial ability with significant benefits in improved stability and higher antibacterial activity. Also, in most cases, inhibition increased as the concentration of AgNPs increased. This could be due to the high density at which the solution was able to saturate and cohere to fungal hyphae and to deactivate plant pathogenic fungi (Kim et al. 2012). Synthesis of nanoparticles chemically requires chemical substances that are toxic in nature. Even after purification, there is a chance of chemical contamination which leads to unsafe use of nanoparticles. Chemical and physical methods need expensive chemicals and instruments, which leads to hike in production cost. However, biologically synthesized nanoparticles don't undergo any toxic and expensive procedure. Laboratory test reveals that biologically synthesized nanoparticles are safest to use and these nanoparticles possess protein caps which allow easy excess to the pathogen cell membrane.

Nano Silver is one of the known strong bacteriostats and possesses broad-spectrum antimicrobial activity. It has been reported that well-dispersed nanosilver colloid is more adhesive to bacteria and fungi leading to enhanced antimicrobial activity (Kim et al. 2008). In nature many agricultural crops and forestry are attacked by many microorganisms resulting in loss of agricultural product and death of tree species. AgNPs came up as a new hope and control disease mechanism by damaging fungal hyphae, interference with nutrient absorption, and enhanced inhibition of fungal growth and germination. The mechanism involved may be the influence of silver ions and nanoparticles on spore formation and disease progression in plant pathogenic fungi. Hence, AgNPs prove to have high potential to be used as nanopesticides for controlling phytopathogens (Alghuthaymi et al. 2015).

9.5.1 Nano-antibacterial

Silver nanoparticles in agricultural soil affect several bacterial communities which are beneficial/harmful for plant and environment (Panyala et al. 2008). AgNPs act as strong antimicrobial agent due to strong inhibitory effects against various bacterial species (Clement and Jarrett 1994; Joshi et al. 2018). Kamran et al. (2011) reported that the nanosilver and nano-TiO_2 with a good potential may be used for removing the bacterial contaminants in the tobacco plant. AgNP exposure causes toxicity to bacteria, and treatment can prevent replication and protein synthesis (Chaloupka et al. 2010). Notably, the most common application problem involves the agglomeration and diffusion of these nanoparticles, which reduce antibacterial activity. These studies revealed that used various organic (Jo et al. 2009) and inorganic substances (Lamsal et al. 2011) as well as powerful carriers (Ouda 2014) to stabilize AgNPs. These substances can strongly influence the antibacterial activity and reduce the biological toxicity of nanoparticles. Also a synergistic antimicrobial effect is achieved when AgNPs are hybrid with other metal nanoparticles or oxides acting as a shell or a core to form bimetallic nanoparticles (Chou and Chen 2007). Chen et al. (2016) study revealed that bacteriostatic and bactericidal activity of the pure and surfactant-stabilized AgNPs (SDS-Tween 80-CTAB-PVP) capped silver nanoparticles. *Ralstonia solanacearum* (phytopathogenic fungi), which causes severe bacterial wilt in tobacco, is used to investigate the bacteriostatic and bactericidal activity of pure and surfactant-stabilized AgNPs. The surfactants affected the antibacterial activity of AgNPs toward *R. solanacearum* to different extents.

9.5.1.1 AgNP Antibacterial Mechanism

The accumulation of AgNPs in the cellular membrane led to an increase of its permeability and eventually to the death of bacterial cells. Also, they attempted to understand their mechanism of action. Presently, there are five main explanations that have been proposed to describe the antibacterial activity (Lemire et al. 2013):

1. Release of toxic ions that bind to sulfur-containing proteins – this accumulation avoids the proper functioning of proteins in the membrane and interfere in cell permeability (Sondi and Salopek-Sondi 2004).
2. They may be genotoxic – toxic ions can DNA destruction which leads to death of cell.
3. Interruption of electron transport chain, protein oxidation system, and collapse of membrane potential.
4. Generation of reactive oxygen species (ROS)-mediated cellular damage and different metal-catalyzed oxidation reactions might underlie specific types of DNA, protein, and membrane damage (Banerjee et al. 2010; Zeng et al. 2007; Aziz et al. 2015).
5. Interruption with uptake of nutrients (Pal et al. 2007).

These mechanisms might not operate separately which suggests that more than one mechanism occur simultaneously. These multiple targets of action might access NPs to fight effectively against different plant pathogens.

9.5.2 Nano-antiviral

For AgNP role against plant viruses, there are few reports documented. For instance, the effective control of bean yellow mosaic virus (BYMV), genus *Potyvirus*, family *Potyviridae*, would be of high interest for many African countries, which can suffer significant yield reductions in fava bean crops upon viral infection leading to considerable economic losses (Radwan et al. 2008). It has been suggested that AgNPs inhibit viral nucleic acid replication, while their antiviral activity depends on the particle size, as well as on the distribution of interacting ligand/receptor molecules (Lü et al. 2009; Papp et al. 2010). Elbeshehy et al. (2015) studied the effect of biosynthesized AgNPs on leaves of fava bean infected with BYMV which showed severe symptoms, including yellow mosaic, mottling, crinkling, size reduction, and deformation, symptoms that were absent from the non-infected leaves.

9.5.3 Nano-antifungal

The antifungal effect of AgNPs has established only insignificant attention and with very few publications (Roe et al. 2008; Kim et al. 2008). There are few studies available dealing precisely with their mechanism of action against clinical isolates and American type culture collection strains of *Candida* spp. and *Trichophyton mentagrophytes* (Li et al. 2012; Panáček et al. 2006; Min et al. 2009). The use of AgNPs as antimicrobial agents becomes more widespread as technological advances make their production more economical. Control of phytopathogens is one of the probable applications in which silver can be utilized in the management of plant diseases. Since silver displays a collection of modes of inhibitory action to plant pathogens (Park et al. 2006), it might be used as controlling agent for various plant pathogens in a moderately safer way as compared to synthetic fungicides. Ag-SiO NPs have a strong antifungal effect against *Botrytis cinerea* (Oh et al. 2006). The combined effect of fluconazole and AgNPs for their antifungal activity was evaluated by Gajbhiye et al. (2009) against *Phoma glomerata*, *Phoma herbarum*, *F. semitectum*, *Trichoderma* sp., and *Candida albicans* by disc diffusion technique. Ag_2S nanocrystals on amorphous silica particles show antifungal activity against *Aspergillus niger* (Fateixa et al. 2009). The phytogenic AgNPs were tested against three different plant pathogenic fungi such as *Rhizoctonia solani*, *Fusarium oxysporum*, and *Curvularia* sp. and showed antifungal activity against *R. solani* followed by *F. oxysporum* and *Curvularia* sp. (Balashanmugam et al. 2016). Amphotericin B showed moderate antifungal activity against the three plants pathogenic fungi (Fig. 9.3).

9.6 Nano-insecticides

Advance investigation highlighted the extensive application of AgNPs for insecticidal application to kill the mosquitoes and fleas. AgNPs possessed excellent anti-lice and mosquito larvicidal activity, having dynamic application in community

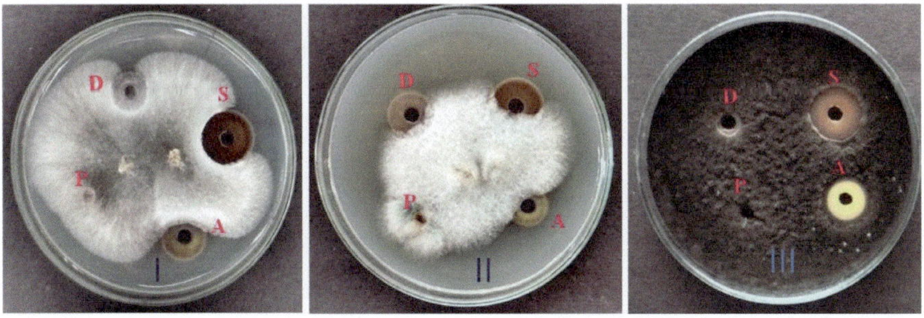

Fig. 9.3 Antifungal activity of phytosynthesized AgNPs against different plant pathogens on the third day (I) *Rhizoctonia solani*, (II) *Fusarium oxysporum*, (III) *Curvularia* sp. D-silver nitrate, S-phytosynthesized AgNPs, A-amphotericin-B, P-*C. roxburghii* aqueous leaf extract. (Reprinted from Balashanmugam et al. 2016)

health improvement. There are tremendous researches investigating the efficacies of biogenic synthesized AgNPs as mosquito larvicidal agent against different species of mosquitoes, i.e., *Culex quinquefasciatus*, *Heteroscodra maculata*, *Rhipicephalus microplus*, and *Anopheles subpictus*, and suggesting that it can be used as an ideal eco-friendly approach for their control (Marimuthu et al. 2011; Suman et al. 2013; Mondal et al. 2014). Jayaseelan et al. (2011) documented on the pediculicidal and larvicidal activity of synthesized AgNPs (from leaf extract of *Tinospora cordifolia*) against the head louse *Pediculus humanus* and larvae of *Anopheles subpictus* and *Culex quinquefasciatus* and showed maximum mortality. Rouhani et al. (2012) evaluated the insecticidal activity of AgNPs against the *Aphis nerii*. Soni and Prakash (2015) have described the larvicidal and pupicidal properties of biologically produced AgNPs (from fungal strain of *Aspergillus niger*) against the mosquito larvae of *Aedes aegypti*, *Culex quinquefasciatus*, and *Anopheles stephensi.*

9.7 Controlled Released Nanofertilizers

Silver nanoparticles are very stable and biodegradable and it also displays slow release of agrochemicals. So, it can be used for formation of nanocapsules for slow and optimized delivery of agrochemicals, pesticides, and fertilizers in agricultural practices (Chowdappa and Shivakumar 2013). Nanoencapsulated agrochemicals are designed to possess the desired properties including effective optimum concentration, time-controlled release, enhanced activity on target site, and least toxic effects (Tsuji 2001). It helps in slow release of agrochemical in controlled way to the particular host through dissolution, biodegradation, diffusion, and osmotic pressure with specific pH. Nanotagged agrochemicals reduce the damage to nontarget plant tissues and reduce risk of nonspecific chemical contamination in the surrounding environment (González-Melendi et al. 2008; Rai and Ingle 2012). Combinations of inorganic fertilizer mainly supply three nutrients, nitrogen (N), potassium (K), and phosphorus (P), to various crops at different growing conditions. This brings out the idea of developing encapsulated fertilizers, in which NPK fertilizers are entrapped within nanosilver to boost nutrient management.

9.8 Positive Effect on Plant Growth

Some nanoparticles have a beneficial effect on some plant species manifested by enhancing seed germination, enhancing crop yield, or suppressing plant disease (Servin et al. 2015; Arruda et al. 2015). As we know that NPs have both positive and negative effects on plant growth and development. Recently, Krishnaraj et al. (2012) studied the effect of biogenic AgNPs on hydroponically grown *Bacopa monnieri* growth metabolism, showed a significant effect on seed germination, and induced the synthesis of protein and carbohydrate and decreased the total phenol contents and catalase and peroxidase activities. Also, Savithramma et al. (2012) revealed that biologically synthesized AgNPs enhanced seed germination and seedling growth of trees *Boswellia ovalifoliolata*. AgNPs increased plants morphological (shoot and root length, leaf area) and biochemical attributes (chlorophyll, carbohydrate and protein contents, antioxidant enzymes) of *Brassica juncea*, common bean, and corn (Salama 2012; Sharma et al. 2012). However, Gruyer et al. (2013) reported AgNPs have both positive and negative effect on root elongation depending on the plant species. They reported that root length was increased in barley but was inhibited in lettuce. Also, Yin et al. (2012) studied the effects of AgNPs on germination of 11 wetland plants species (*Lolium multiflorum*, *Panicum virgatum*, *Carex lurida*, *C. scoparia*, *C. vulpinoidea*, *C. crinita*, *Eupatorium fistulosum*, *Phytolacca americana*, *Scirpus cyperinus*, *Lobelia cardinalis*, *Juncus effusus*) and found AgNPs enhanced the germination rate of 1 species (*E. fistulosum*). AgNPs induce root growth by blocking ethylene signaling in *Crocus sativus* (Rezvani et al. 2012).

Silver nanoparticles may have both a positive and a negative impact on plants, depending on size, concentration, chemical composition, zeta potential, stability, and the shape of nanoparticles (Mirzajani et al. 2013; Tripathi et al. 2015, 2017; Costa and Sharma 2016). Several studies have depicted a negative impact of nanoparticles on plants in the form of decrease in plant growth, productivity, and pigments (Tripathi et al. 2017). On the other hand, robust and smart engineered nanoparticles are also explored for the betterment of agricultural crop production, as growth stimulators, nanopesticides, nanofertilizers, soil-improving agents, or sensors for monitoring different agricultural parameters in the field (Fraceto et al. 2016; Prasad et al. 2016, 2017a). Due to the increased interest in the area, most of the research depicting the influence of industrial nanoparticles on plants has been performed in recent years. Recent studies have exposed that when AgNP was combined with different treatment/compounds, it may have a different impact on plants (Berahmand et al. 2012; Belava et al. 2017), due to the influence of other phenomena/compound on AgNPs. AgNP treatment in combination with magnetic field was observed to improve quantitative yields in *Zea mays* (Berahmand et al. 2012). In the wheat-pathogen phytosystem, an increase of lipid peroxidation was observed, when compared with NP or pathogen alone (Belava et al. 2017). The biosynthesized nanoparticles induced the protein and carbohydrate synthesis and decreased the total phenol contents, which can be considered as a positive effect; it may be due to the presence of altered size (2–50 nm) of nanoparticle or the different chemical property of biogenic NPs. The size of AgNPs of 200–800 nm was observed to enhance the plant growth (Jasim et al. 2016) whereas 35–40 nm observed to positively influence the root and shoot growth of different plant (Pallavi et al. 2016), which may be due to the inability of the penetration of large nanoparticles (Mirzajani et al. 2013). AgNPs (of comparatively small size, i.e., <30 nm) when applied in high concentration were observed to inhibit the root and shoot growth in different plants studied (Dimkpa et al. 2013; Vinković et al. 2017). Treatment of *Lupinus termis* seedlings with 100 ppm bio-AgNPs might improve the growth profile, while exposure of seedlings to high concentrations (300 and 500 ppm) resulted in a highly significant reduction in all growth parameters and growth indices (Al-Huqail et al. 2018) (Fig. 9.4 and Table 9.1).

9.9 Pesticide Remediation

As an alternative, AgNPs can be applied for the degradation of pesticides to overcome these problems. In modern nanotechnological research, applicability of AgNPs in pesticide mineralization is well reported. Nair and Pradeep (2003) confirmed the halocarbon mineralization and catalytic destruction by means of silver and gold nanoparticles. Manimegalai et al. (2011) reported the applicability of AgNPs for the removal of pesticides "chlorpyrifos" and "malathion" from water. These nanoparticles have been shown to completely remove the pesticides as it actively anchored the pesticide to its inert surfaces.

9.10 Plant Pathogen Detection

It is worth mentioning here that newly developed smart nanomaterials with special nanoscale characteristics offer tremendous breakthrough in plant pathogen detection and diagnosis technology (Khiyami et al. 2014). Striping voltammetry as an electrochemical technique can be applied to detect the metal nanoparticles directly making the assay simple to perform. Gold and silver nanoparticles can be used in these methods including different inorganic nanocrystals (ZnS, PbS and CdS) for analytic detection (Upadhyayula 2012). Schwenkbier et al. (2015) developed a helicase-dependent isothermal amplification in combination with on-chip hybridization for the detection of *Phytophthora* species. This approach allows efficient amplification of the yeast GTP-binding protein (Ypt1) target gene region at one constant temperature in a miniaturized heating device. The assay's specificity was

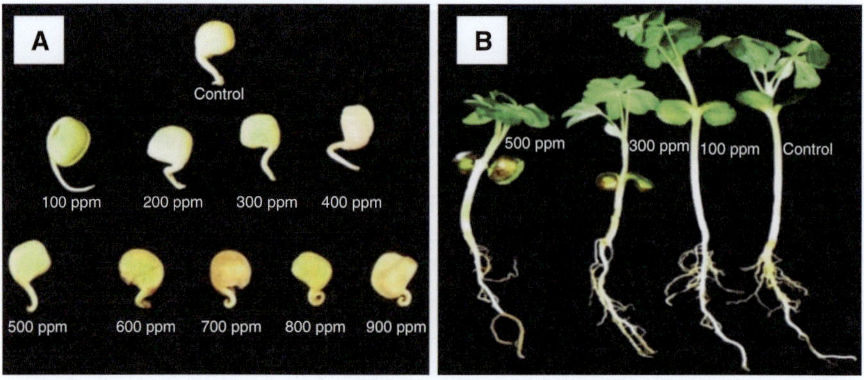

Fig. 9.4 (**a**) Effect of different concentrations of CSL-AgNPs (0–900 ppm) on *Lupinus termis* L. seed germination. (**b**) Effect of different concentrations of CSL-AgNPs (0, 100, 300 and 500 ppm) on growth parameters of *Lupinus termis* L. (Reprinted from Al-Huqail et al. 2018)

determined by on-chip DNA hybridization and subsequent AgNP deposition. The silver deposits serve as stable endpoint signals that enable the visual as well as the electrical readout. These advancements point to the direction of a near future on-site application of the combined techniques for a reliable detection of several kinds of plant pathogens.

9.11 Conclusion

Silver has been constantly superb antimicrobial (antibacterial and antifungal) and has been used for the purpose for ages. The unique physicochemical properties of AgNPs only increase the efficacy of silver. Chemical and physical methods of AgNP synthesis were being followed by several periods, but they are expensive, and the use of several toxic chemicals for their synthesis makes the biological synthesis the more desired possibility. Though microbial and plant extract sources can be used for AgNP synthesis, the easy availability, the nontoxic nature, the various options available, and the advantage of quicker synthesis make plant extracts the best and an excellent choice for biogenic AgNP synthesis. The uses of AgNPs are varied and many, but the most exploited and desired aspect is their antimicrobial and anti-inflammatory activities. The disadvantage of AgNPs is that they can induce toxicity at various degrees. It is recommended that higher concentrations of AgNPs are toxic and can cause innumerable health problems. It also revealed that the nanoparticles of silver can induce various ecological problems and disturb the ecosystem if released into the environment. Hence, this chapter concludes the application of silver nanoparticles in plant disease management, nanofertilizers, nanopesticides, pesticide remediation, and plant pathogen detections; with that there would be mechanisms devised to nullify any toxicity caused by nanosilver to humans and the environment so that the unique properties of this substance can be put to great use for human betterment without any controversies.

Table 9.1 Positive effect of AgNP on plant species

Size (diameter in nm)	Concentration	Exposure methodology	Plant studied	Impact	References
20	40 gha^{-1}	Field, through irrigation water, (nanoparticle applied with 10 mT magnetic field)	*Zea mays*	Combination of silver nanoparticles and magnetic field led to improved quantitative yields of fodder maize	Berahmand et al. (2012)
20 (polyvinylpyrrolidine-coated, PVP-NP) 6 (gum Arabic coated, GA-NP)	1, 10, 40 mg/L (toxic study performed with 40 mg/L in pure culture experiment)	Petri plates (treatment on seeds)	Eleven species of common wetland plants	PVP-NP significantly increases leave length in *Scirpus cyperinus* and *Carex lurida* whereas decreases in *Lolium multiflorum*. GA-NP shows a significant decrease in leave length except *Phytolacca americana* Root growth was observed to be positively affected by PVP-NP in *Phytolacca americana*, *Panicum virgatum*, and *Carex lurida*, whereas six other species have been observed to have negative effect of PVP-NP	Yin et al. (2012)
200–800	1 mg/L	Growth medium with Agar + pots with soil (treatment on germinated seeds)	*Trigonella foenum-graecum*	Enhancement in plant growth and diosgenin synthesis was observed	Jasim et al. (2016)

(continued)

Table 9.1 (continued)

Size (diameter in nm)	Concentration	Exposure methodology	Plant studied	Impact	References
35–40	50, 75 mg/L	Pots (foliar treatment on grown plant)	*Triticum aestivum*, *Vigna sinensis*, *Brassica juncea*	Relatively unaffected (wheat) The optimum growth promotion and increased root nodulation were observed at 50 ppm treatment (cowpea) Improved shoot parameters were recorded at 75 ppm (*Brassica*)	Pallavi et al. (2016)
12.9 ± 9.1 (90%) nanoparticles in ultrapure water	0.01, 0.05, 0.1, 0.5, 1 mg/L	Pots with soil (treatment on seedling)	*Capsicum annuum*	Concentration-dependent decrease in plant growth Concentration-dependent increase in cytokinin concentration	Vinković et al. (2017)
<100	1.5 mg/L	Hydroponic and pots (treatment on seeds)	*Triticum aestivum* (wheat-*Pseudocercosporella herpotrichoides* phytosystem)	In Myronivska 808 the lipid peroxidation was observed to be significantly high where nanoparticle was present with pathogen	Belava et al. (2017)

References

Alghuthaymi MA, Almoammar H, Rai M, Said-Galiev E, Abd-Elsalam KA (2015) Myconanoparticles: synthesis and their role in phytopathogens management. Biotechnol Biotechnol Equip 29:221–236

Al-Huqail AA, Hatata MM, AL-Huqail AA, Ibrahim MM (2018) Preparation, characterization of silver phyto nanoparticles and their impact on growth potential of *Lupinus termis* L. seedlings. Saudi J Biol Sci 25:313–319

Arruda SCC, Silva ALD, Galazzi RM, Azevedo RA, Arruda MAZ (2015) Nanoparticles applied to plant science: a review. Talanta 131:693–705

Aziz N, Faraz M, Pandey R, Sakir M, Fatma T, Varma A, Barman I, Prasad R (2015) Facile algae-derived route to biogenic silver nanoparticles: synthesis, antibacterial and photocatalytic properties. Langmuir 31:11605–11612. https://doi.org/10.1021/acs.langmuir.5b03081

Aziz N, Pandey R, Barman I, Prasad R (2016) Leveraging the attributes of *Mucor hiemalis*-derived silver nanoparticles for a synergistic broad-spectrum antimicrobial platform. Front Microbiol 7:1984. https://doi.org/10.3389/fmicb.2016.01984

Balashanmugam P, Balakumaran MD, Murugan R, Dhanapal K, Kalaichelvan PT (2016) Phytogenic synthesis of silver nanoparticles, optimization and evaluation of *in vitro* antifungal activity against human and plant pathogens. Microbiol Res 192:52–64

Banerjee M, Mallick S, Paul A, Chattopadhyay A, Ghosh SS (2010) Heightened reactive oxygen species generation in the antimicrobial activity of a three component iodinated chitosan-silver nanoparticle composite. Langmuir 26:5901–5908

Belava VN, Panyuta OO, Yakovleva GM, Pysmenna YM, Volkogon MV (2017) The effect of silver and copper nanoparticles on the Wheat- *Pseudocercosporella herpotrichoides* pathosystem. Nanoscale Res Lett 12(250):250. https://doi.org/10.1186/s11671-017-2028-6

Berahmand AA, Ghafariyan-Panahi A, Sahabi H, Feizi H, Rezvani-Moghaddam P, Shahtahmassebi N et al (2012) Effects silver nanoparticles and magnetic field on growth of fodder maize (*Zea mays* L.). Biol Trace Elem Res 149:419–424

Bryaskova R, Pencheva D, Nikolov S, Kantardjiev T (2011) Synthesis and comparative study on the antimicrobial activity of hybrid materials based on silver nanoparticles (AgNps) stabilized by polyvinylpyrrolidone (PVP). J Chem Biol 4(4):185–191

Chaloupka K, Malam Y, Seifalian AM (2010) Nanosilver as a new generation of nanoproduct in biomedical applications. Trends Biotechnol 28:580–588

Chen J, Li S, Luo J, Wang R, Ding W (2016) Enhancement of the antibacterial activity of silver nanoparticles against phytopathogenic bacterium *Ralstonia solanacearum* by stabilization. J Nanomater 2016:7135852, 15 pages. https://doi.org/10.1155/2016/7135852

Chen X, Schluesener HJ (2008) Nanosilver: a nanoproduct in medical application. Toxicol Lett 176:1–12

Chou KS, Chen CC (2007) Fabrication and characterization of silver core and porous silica shell nanocomposite particles. Microporous Mater 98:208–213

Chowdappa P, Shivakumar G (2013) Nanotechnology in crop protection: status and cope. Pest Manag Hortic Ecosys 19:131–151

Clement JL, Jarrett PS (1994) Antibacterial silver. Met Based Drugs 1:467–482

Costa MVJD, Sharma PK (2016) Effect of copper oxide nanoparticles on growth, morphology, photosynthesis, and antioxidant response in *Oryza sativa*. Photosynthetica 54:110–119

Danilcauk M, Lund A, Saldo J, Yamada H, Michalik J (2006) Conduction electron spin resonance of small silver particles. Spectrochimaca Acta Part A 63:189–191

Dimkpa CO, McLean JE, Martineau N, Britt DW, Haverkamp R, Anderson AJ (2013) Silver nanoparticles disrupt wheat (*Triticum aestivum* L.) growth in a sand matrix. Environ Sci Technol 47:1082–1090

Elbeshehy EF, Elazzazy AM, Aggelis G (2015) Silver nanoparticles synthesis mediated by new isolates of Bacillus spp., nanoparticle characterization and their activity against Bean Yellow Mosaic Virus and human pathogens. Front Microbiol 6:453

Fateixa S, Neves MC, Almeida A, Oliveira J, Trindade T (2009) Anti-fungal activity of SiO_2/Ag_2S nanocomposites against *Aspergillus niger*. Colloids Surfaces B 74:304–308

Feng QL, Wu J, Chen GQ, Cui FZ, Kim TN, Kim JO (2008) A mechanistic study of the antibacterial effect of silver ions on *Escherichia coli* and *Staphylococcus aureus*. J Biomed Mater Res 52:662–668

Fraceto LF, Grillo R, de Medeiros GA, Scognamiglio V, Rea G, Bartolucci C (2016) Nanotechnology in agriculture: which innovation potential does it have? Front Environ Sci 4(20). https://doi.org/10.3389/fenvs.2016.00020

Gajbhiye M, Kesharwani J, Ingle A, Gade A, Rai M (2009) Fungus mediated synthesis of silver nanoparticles and their activity against pathogenic fungi in combination with fluconazole. Nanomedicine 5:382–386

González-Melendi P, Fernández-Pacheco R, Coronado MJ, Corredor E, Testillano PS, Risueño MC, Marquina C, Ibarra MR, Rubiales D, Pérez-de-Luque A (2008) Nanoparticles as smart treatment-delivery systems in plants: assessment of different techniques of microscopy for their visualization in plant tissues. Ann Bot 101:187–195

Gruyer N, Dorais M, Bastien C, Dassylva N, Triffault-Bouchet G (2013) Interaction between sliver nanoparticles and plant growth. In: International symposium on new technologies for environment control, energy-saving and crop production in greenhouse and plant factory– greensys, Jeju, Korea, 6–11 Oct 2013

Hatchett DW, Henry S (1996) Electrochemistry of sulfur adlayers on the low-index faces of silver. J Phys Chem 100:9854–9859

Jasim B, Thomas R, Mathew J, Radhakrishnan EK (2016) Plant growth and diosgenin enhancement effect of silver nanoparticles in Fenugreek (*Trigonella foenumgraecum* L.). Saudi Pharm J 25:443–447

Jayaseelan C, Rahaman AA, Rajkumar G, Vishnu Kirthi A, Santhoshkumar T, Marimuthu S, Bagavan A, Kamaraj C, Zahir AA, Elango G (2011) Synthesis of pediculocidal and larvicidal silver nanoparticles by leaf extract from heart leaf moonseed plant, *Tinospora cordifolia* Miers. Parasitol Res 109:185–194

Jo YK, Kim BH, Jung G (2009) Antifungal activity of silver ions and nanoparticles on phytopathogenic fungi. Plant Dis 93(10):1037–1043

Joshi N, Jain N, Pathak A, Singh J, Prasad R, Upadhyaya CP (2018) Biosynthesis of silver nanoparticles using *Carissa carandas* berries and its potential antibacterial activities. J Sol-Gel Sci Techn https://doi.org/10.1007/s10971-018-4666-2

Kamran S, Forogh M, Mahtab E, Mohammad A (2011) In vitro antibacterial activity of nanomaterials for using in tobacco plants tissue culture. World Acade Sci Eng Technol 79:372–373

Khan MR, Rizvi TF (2014) Nanotechnology: scope and application in plant disease management. Plant Pathol J 13:214–231

Khiyami MA, Almoammar H, Awad YM, Alghuthaym MA et al (2014) Plant pathogen nanodiagnostic techniques: forthcoming changes? Biotechnol Biotechnol Equip 28(5):775–785

Kim H, Kang H, Chu G, Byun H (2008) Antifungal effectiveness of nanosilver colloid against rose powdery mildew in greenhouses. Solid State Phenomenon 135:15–18

Kim JS, Kuk E, Yu K, Kim JH, Park SJ, Lee HJ, Kim SH, Park YK, Park YH, Wang C-Y, Kim YK, Lee YS, Jeong DH, Cho MH (2007) Antimicrobial effects of silver nanoparticles. Nanomedicine 3:95–101

Kim SW, Jung JH, Lamsal K, Kim YS, Min JS, Lee YS (2012) Antifungal effects of silver nanoparticles (AgNPs) against various plant pathogenic fungi. Mycobiology 40(1):53–58

Krishnaraj C, Jagan EG, Ramachandran R, Abirami SM, Mohan N, Kalaichelvan PT (2012) Effect of biologically synthesized silver nanoparticles on *Bacopa monnieri* (Linn.) Wettst. Plant growth metabolism. Process Biochem 47(4):51–658

Lamsal K, Kim SW, Jung JH, Kim YS, Kim KS, Lee YS (2011) Inhibition effects of silver nanoparticles against powdery mildews on cucumber and pumpkin. Mycobiology 39(1):26–32

Lemire JA, Harrison JJ, Turner RJ (2013) Antimicrobial activity of metals: mechanisms, molecular targets and applications. Nat Rev Microbiol 11(6):371–384

Li G, He D, Qian Y, Guan B, Gao S, Cui Y, Yokoyama K, Wang L (2012) Fungus-mediated green synthesis of silver nanoparticles using *Aspergillus terreus*. Int J Mol Sci 3:466–476

Lü JM, Wang X, Marin-Muller C, Wang H, Lin PH, Yao Q, Chen C (2009) Current advances in research and clinical applications of PLGA based Nanotechnology. Expert Rev Mol Diagn 9:325–341

Manimegalai G, Kumar SS, Sharma C (2011) Pesticide mineralization in water using silver nanoparticles. Int J Chem Sci 9:1463–1471

Marimuthu S, Rahuman AA, Rajkumar G, Santhoshkumar T, Kirthi AV, Jayaseelan C, Bagavan A, Zahir AA, Elango G, Kamaraj C (2011) Evaluation of green synthesized silver nanoparticles against parasites. Parasitol Res 108:1541–1549

Matsumura Y, Yoshikata K, Kunisaki S, Tsuchido T (2003) Mode of bacterial action of silver zeolite and its comparison with that of silver nitrate. Appl Environ Microbiol 69:4278–4281

Min JS, Kim KS, Kim SW, Jung JH, Lamsal K, Kim SB, Jung M, Lee YS (2009) Effects of colloidal silver nanoparticles on sclerotium-forming phytopathogenic fungi. Plant Pathol J 25:376–380

Mirzajani F, Askari H, Hamzelou S, Farzaneh M, Ghassempour A (2013) Effect of silver nanoparticles on *Oryza sativa* L. and its rhizosphere bacteria. Ecotoxicol Environ Saf 88:48–54

Mondal NK, Chowdhury A, Dey U et al (2014) Green synthesis of silver nanoparticles and its application for mosquito control. Asian Pac J Trop Dis 4:S204–S210

Nair S, Pradeep T (2003) Halocarbon mineralization and catalytic destruction by metal nanoparticles. Curr Sci 84:12

Nangmenyi G, Economy J (2009) Nonmetallic particles for oligodynamic microbial disinfection. In: Street A, Sustich R, Duncan J, Savage N (eds) Nanotechnol application for clean water. William Andrew, Norwich, NY, pp 3–15

Narayanan KB, Sakthivel N (2010) Biological synthesis of metal nanoparticles by microbes. Adv Colloid Interf Sci 156:1–13

Ocsoy I, Paret ML, Ocsoy MA, Kunwar S, Chen T, You M, Tan W (2013) Nanotechnology in plant disease management: DNA-directed silver nanoparticles on graphene oxide as an antibacterial against *Xanthomonas perforans*. ACS Nano 7(10):8972–8980

Oh SD, Lee S, Choi SH, Lee IS, Lee YM, Chun JH, Park HJ (2006) Synthesis of Ag and Ag-SiO_{2abil} nanoparticles by y-irradiation and their antibacterial and antifungal efficiency against *Salmonellaenteric* serovar *Typhimurium* and *Botrytis cinerea*. Colloids Surf A 275:228–233

Ouda SM (2014) Antifungal activity of silver and copper nanoparticles on two plant pathogens, *Alternaria alternata* and *Botrytis cinerea*. Res J Microbiol 9(1):34–42

Pal S, Tak UK, Song JM (2007) Does the antibacterial activity of silver nanoparticles depend on the shape of the nanoparticle? A study of the gram-negative bacterium *Escherichia coli*. App Environ Microbiol 73:1712–1720

Pallavi MCM, Srivastava R, Arora S, Sharma AK (2016) Impact assessment of silver nanoparticles on plant growth and soil bacterial diversity. Biotech 6(254):254. https://doi.org/10.1007/s13205-016-0567-7

Panáček A, Kolář M, Večeřová R, Prucek R, Soukupová J, Kryštof V, Park HJ, Kim SH, Kim HJ, Choi SH (2006) A new composition of nanosized silica-silver for control of various plant diseases. Plant Pathol J 22:295–302

Panyala NR, Pena Mendez EM, Havel J (2008) Silver or silver nanoparticle: a hazardous treat to the environment and human health? J Appl Med 6:117–129

Papp I, Sieben C, Ludwig K, Roskamp M, Böttcher C, Schlecht S et al (2010) Inhibition of influenza virus infection by multivalent sialic-acid-functionalized gold nanoparticles. Small 6:2900–2906

Parashar UK, Saxena SP, Srivastava A (2009) Bioinspired synthesis of silver nanoparticles. Dig J Nanomat Biostruct 4:159–166

Park H-J, Kim SH, Kim HJ, Choi S-H (2006) A new composition of nanosized silica-silver for control of various plant diseases. Plant Pathol J 22(3):295–302

Pérez-de-Luque A, Rubiales D (2009) Nanotechnology for parasitic plant control. Pest Manag Sci 65:540–545

Prasad R, Swamy VS (2013) Antibacterial activity of silver nanoparticles synthesized by bark extract of *Syzygium cumini*. J Nanopart 2013:1. https://doi.org/10.1155/2013/431218

Prasad R (2014) Synthesis of silver nanoparticles in photosynthetic plants. J Nanopart 963961., https://doi.org/10.1155/2014/963961:1

Prasad R, Bhattacharyya A, Nguyen QD (2017a) Nanotechnology in sustainable agriculture: recent developments, challenges and perspectives. Front Microbiol 8:1014. https://doi.org/10.3389/fmicb.2017.01014

Prasad R, Gupta N, Kumar M, Kumar V, Wang S, Abd-Elsalam KA (2017b) Nanomaterials act as plant defense mechanism. In: Prasad R, Kumar V, Kumar M (eds) Nanotechnology. Springer, Singapore, pp 253–269

Prasad R, Kumar V, Prasad KS (2014) Nanotechnology in sustainable agriculture: present concerns and future aspects. Afr J Biotechnol 13(6):705–713

Prasad R, Pandey R, Barman I (2016) Engineering tailored nanoparticles with microbes: quo vadis. WIREs Nanomed Nanobiotechnol 8:316–330. https://doi.org/10.1002/wnan.1363

Radwan DEM, Fayez KA, Mahmoud SY, Hamad A (2008) Protective action of salicylic acid against bean yellow mosaic virus infection in *Vicia faba* leaves. J Plant Physiol 165:845–857

Rai M, Ingle A (2012) Role of nanotechnology in agriculture with special reference to management of insect pests. Appl Microbiol Biotechnol 94:287–293

Raza MA, Kanwal Z, Rauf A, Sabri AN, Riaz S, Shahzad Naseem S (2017) Size- and shape-dependent antibacterial studies of silver nanoparticles synthesized by wet chemical routes. Nano 6(74):1–15

Rezvani N, Sorooshzadeh A, Farhadi N (2012) Effect of nano-silver on growth of saffron in flooding stress. World Acad Sci Eng Technol 1:517–522

Roe D, Karandikar B, Bonn-Savage N, Gibbins B, Roullet JB (2008) Antimicrobial surface functionalization of plastic catheters by silver nanoparticles. J Antimicrob Chemother 61:869–887

Rouhani M, Samih MA, Kalantri S (2012) Insecticidal effect of silica and silver nanoparticles on the cowpea seed beetle, *Callosobruchus maculatus* F (Col: Bruchidae). J Entomol Res 4:297–305

Salama HMH (2012) Effects of silver nanoparticles in some crop plants, common bean (*Phaseolus vulgaris* L.) and corn (*Zea mays* L.). Int Res J Biotech 3(10):190–197

Savithramma N, Ankanna S, Bhumi G (2012) Effect of nanoparticles on seed germination and seedling growth of *Boswellia ovalifoliolata* an endemic and endangered medicinal tree taxon. Nano Vision 2:61–68

Schwenkbier L, Pollok S, König S, Urban M, Werres S, Dana Cialla-May D, Karina Weber K, Popp J (2015) Towards on-site testing of *Phytophthora* species. Anal Methods 7:211–217

Servin A, Elmer W, Mukherjee A, De La Torre-Roche R, Hamdi H, White JC, Bindraban P, Dimkpa C (2015) A review of the use of engineered nanomaterials to suppress plant disease and enhance crop yield. J Nanopart Res 17:1–21

Sharma P, Bhatt D, Zaidi MG, Saradhi PP, Khanna PK, Arora S (2012) Silver nanoparticle mediated enhancement in growth and antioxidant status of *Brassica juncea*. Appl Biochem Biotechnol 167:2225–2233

Shrivastava S, Bera T, Roy A, Singh G, Ramachandrarao P, Dash D (2007) Characterization of enhanced antibacterial effects of novel silver nanoparticles. Nanotechnology 18:1–9

Singh S, Singh BK, Yadav SM, Gupta AK (2015) Applications of nanotechnology in agricultural and their role in disease management. Res J Nanosci Nanotechnol 5:1–5

Sondi I, Salopek-Sondi B (2004) Silver nanoparticles as antimicrobial agent: a case study on *E. coli* as a model for gram-negative bacteria. J Colloid Interface Sci 275:177–182

Soni N, Prakash S (2015) Antimicrobial and mosquitocidal activity of microbial synthesized silver nanoparticles. Parasitol Res 114:1023–1030

Swamy VS, Prasad R (2012) Green synthesis of silver nanoparticles from the leaf extract of Santalum album and its antimicrobial activity. J Optoelectron Biom Mater 4(3):53–59

Suman TY, Elumali D, Kaleena PK (2013) GCMS analysis of bioactive components and synthesis of silver nanoparticle using *Ammannia baccifera* aerial extract and its larvicidal activity against malaria and fiariasis vectors. Ind Crop Prod 47:239–245

Taniguchi N (1974) On the basic concept of 'nano-technology'. Proceedings of the international conference on production engineering Tokyo, Part II; Tokyo: Japan Soc Precision Engineering. pp. 18–23

Tripathi DK, Singh S, Singh S, Srivastava PK, Singh VP, Singh S et al (2017) Nitric oxide alleviates silver nanoparticles (AgNps)-induced phytotoxicity in *Pisum sativum* seedlings. Plant Physiol Biochem 110:167–177

Tripathi DK, Singh VP, Prasad SM, Chauhan DK, Dubey NK (2015) Silicon nanoparticles (SiNp) alleviate chromium (VI) phytotoxicity in *Pisum sativum* (L.) seedlings. Plant Physiol Biochem 96:189–198

Tsuji K (2001) Microencapsulation of pesticides and their improved handling safety. J Microencapsul 18:137–147

Upadhyayula VKK (2012) Functionalized gold nanoparticle supported sensory mechanisms applied in detection of chemical and biological threat agents: a review. Anal Chim Acta 715:1–18

Vinković T, Novák O, Strnad M, Goessler W, Jurašin DD, Paradiković N, Vrček IV (2017) Cytokinin response in pepper plants (*Capsicum annuum* L.) exposed to silver nanoparticles. Environ Res 156:10–18

Winbo Ma (2011) How do plants fight disease? Breakthrough research by UC Riverside plant pathologist offers a clue. http://newsroom.ucr.edu/2587

Yin L, Colman BP, McGill BM, Wright JP, Bernhardt ES (2012) Effects of silver nanoparticle exposure on germination and early growth of eleven wetland plants. PLoS One 7:e47674. https://doi.org/10.1371/journal.pone.0047674

Zeng F, Hou C, Wu SZ, Liu XX, Tong Z, Yu SN (2007) Silver nanoparticles directly formed on natural macroporous matrix and their anti-microbial activities. Nanotechnology 18:1–8

Chapter 10
Positive Impacts of Nanoparticles in Plant Resistance against Different Stimuli

Tahsin Shoala

10.1 Introduction

Application of Nanotechnology is studying the new physical properties of different materials in the nanosize. Nanomaterials in the nanosize have changed the vision of the world towards physical structure and size of the same material. Nanomaterials have altered the direction of the biological research towards nanoscience and became one of the most important and applicable sciences in different biological aspects. Application of nanomaterials in biological research enhanced the human health and longevity by curing dangerous disease like cancer; increased the sensitivity and specificity of detection and diagnosis methods, vaccinations and drug delivery; decreased pre- and postharvest diseases; and decreased the pollution in the environment (Prasad et al. 2014, 2016, 2017a). Different materials in the normal size could be toxic so that converting those materials to nanosize will change the physical properties and increase the level of toxicity. Number of metallic nanoparticles could be toxic in the normal size, and the toxicity increased in the nanosize because of the easiest entrance inside the cell, so that using metallic nanoparticles in detection and identification of certain diseases will increase the positive side of those nanomaterials, like using nanosilver and nano-gold in the DNA extraction and PCR techniques. Converting natural and safe products to nanosize could play dual role in the nanoscience by enhancing the nanoactivity of natural products against different pathogens, increasing shelf life fruits and vegetables in room temperature, decreasing the side effect and toxicity of using pesticides, curing different human diseases and decreasing the negative or bad side of using unsafe nanomaterials. Application of nanotechnology could take positive or negative impact according to your first choice of your material. So biological research committee should assign

T. Shoala (✉)
College of Biotechnology, Misr University for Science and Technology, Giza, Egypt
e-mail: Tahsen.shoala@must.edu.eg

K. A. Abd-Elsalam, R. Prasad (eds.), *Nanobiotechnology Applications in Plant Protection*, Nanotechnology in the Life Sciences,
https://doi.org/10.1007/978-3-319-91161-8_10

specific rules, obligations and licences regarding converting specific material to nanosize. Positive impact of nanotechnology increased the chances of changing the biological research and application towards nanomaterials.

Nanotechnology became one of the main technologies that could be used to solve many issues in our life. Application of nanotechnology changed the overviews for using many materials and altered the features for specific elements to make them useful for human being. Nowadays, nanotechnology could be used to solve many problems and also has many applications in different areas. Application of nanotechnology in renewable energy is the magical key to produce clean energy and sort out energy problems worldwide. Also nanotechnology could help the world by sorting out water purity by using nanomaterials to decrease water contaminant. By 2050, the whole world will suffer from the available fresh water because of increase population and increase contaminated water. Application of nanotechnology could improve human health and longevity by curing many diseases like cancer, vaccination and infectious diseases. Additionally, nanomaterials could help the universe by decreasing pollution in the environment.

Nanotechnologies face many problems in the application sides. Although application of nanotechnology could be useful at some points, using nanomaterials could be dangerous. So all the nanotechnologist should stop applying and start thinking about toxicity, safety, new combinations, mechanism, mood of action, accumulation and biology of using nanomaterials. Many studies should be applied to discover the bad and ugly face of nanomaterials. So new science could be invented to study bad side of nanotechnology and could be named as (nanotoxicity) and focus in the study of toxic side of different martials in the nano form.

Researchers worldwide started already studying how to manage materials in the nanosize to understand the danger side of nanoparticles before application. Even if using nanomaterials in the diagnosis techniques might be useful, but we should be aware of how to get rid of it as hazard material.

10.2 Plant-Nanomaterial Interaction

Plant-nanomaterial interactions have taken many steps forward during the last few years. Different nanoparticles have variable mode of action according to the type and sizes of nanomaterials. Nanoparticles could have direct effect on the plant surfaces or may induce different metabolic pathways which lead to either resistance or susceptibility against different stimuli. Nanoparticles may have either positive or negative effects on plants which lead to either resistance or susceptibility of plants against different stimuli (Fig. 10.1).

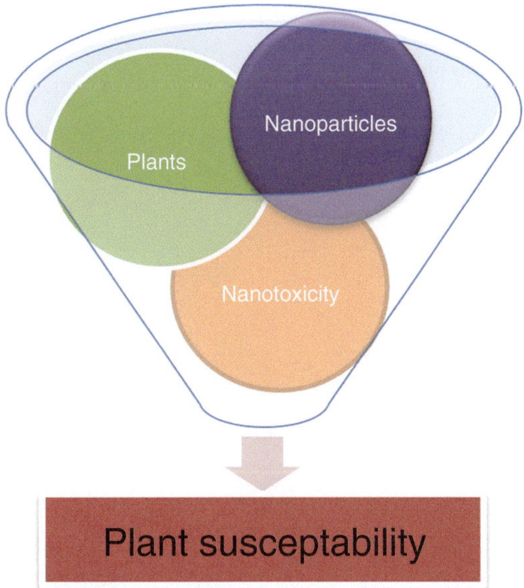

Fig. 10.1 Nanoparticle-plant interaction towards plant susceptibility

10.2.1 The Role of Nanoparticles in Plant Resistance against Different Pathogens

Recently, applicable nanotechnology has been increased towards the development of innovative antimicrobials for controlling the phytopathogenic pathogens which affect agricultural crops. Nanomaterials could be applied directly to suppress pathogen infection which led indirectly to an increase in the plant growth and crop production (Prasad et al. 2014, 2017a, b). Remarkably, several of the nanoparticles could be effective against different microorganisms, and also they are required as micronutrients for plants. Therefore, nanomaterials could play dual role in plants by enhancing the growth level in both roots and shoots plus targeting the microorganisms which resulted in increasing plant resistance against plant pathogens. So, nanoparticles could be applied as nanofertilizers and nanopesticides at once (Fig. 10.2).

Specially, substantial progress in nanomaterials synthesis, for instance, polymeric, carbon-based and metallic, has fascinated researchers' attraction in the direction of applications in controlling plant pathogens. Additionally, toxicity has been taken in consideration to decrease the negative impacts of nanotechnology in the environment by redesigning and modification the sizes, shapes and surfaces of the nanoparticles towards the nanomicrobial activities and decreases the toxic effect on the environment (Neal 2008).

Numerous research studies have been done to protect crops from plant pathogens, e.g. development of polymeric and lipid-based edible film sheets (Cagri et al. 2004). Many metal ions, e.g. Ag^{+}, Zn^{2+} and Cu^{2+}, have been applied to deactivate

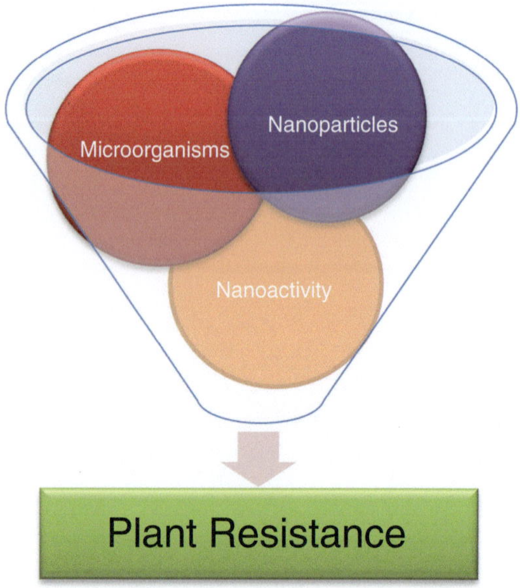

Fig. 10.2 Nanoparticle-microorganism interaction towards plant resistance

bacterial growth (Kim et al. 1998). Many metallic nanoparticles have been synthesized and used, e.g. Ag, Cu, CuO, ZnO and TiO_2 carbon nanotubes (CNTs) as well as graphene oxide (GO) to manage bacterial plant pathogens (Yoon et al. 2007; Mallick et al. 2012; Karlsson et al. 2008; Hu et al. 2010). Amongst diverse forms of nanoparticles, Ag NPs have been used as effectively against different types of pathogens, e.g. fungi, bacteria and viruses (Sondi and Salopek-Sondi 2004; Elechiguerra et al. 2005; Kim et al. 2009; Zodrow et al. 2009; Prasad and Swamy 2013; Swamy and Prasad 2012; Aziz et al. 2014, 2015, 2016). So, the mechanisms of nanosilver activities against microbes could be explained by binding of Ag^+ ions to cysteine-containing proteins on the plasma membrane, initiating interruption in both physiological and biochemical process that compromise membrane integrity. Accordingly, critical enzymes systems will be deactivated followed by cell death as a result of Ag^+ penetration (Ocsoy et al. 2013). Treated perennial ryegrass (*Lolium perenne*) with Ag NPs reduced the pathogenic fungal infection (*Neotyphodium lolii*) to 50% in colony formation. Also, in field trials, Ag NPs suppressed and reduced the activity of *Colletotrichum* spp. (anthracnose pathogen) by penetrating the cell membrane of the fungus. Lamsal et al. (2011) described that Ag NPs in the nanometre sizes (~8 nm) reduced the fungal infection by inducing the resistance mechanisms which work effectively towards the plant resistance (Jo et al., 2009; Lamsal et al. 2011). A ZnO nanoparticle is considered as another effective nanomaterial against plant pathogenic fungi. ZnO NPs have many advantages compared to Ag NPs, lower toxicity, source of soil fertility plus nanofungal activity. ZnO NPs showed 26%

nanofungal activity against *Fusarium graminearum* in a mung bean broth agar medium compared to bulk oxide and control (Dimkpa et al. 2013a, b). Researcher studies reported that ZnO NPs inhibited significantly the fungal growth of inhibited *Botrytis cinerea* (63–80%) and *Penicillium expansum* (61–91%) in a plating assay and reported the systemic disruption which caused hyphal deformity and fungal death in both fungi (He et al. 2011). Another research study proved that biosynthesized ZnO NPs in the concentration (25 μg/mL) supressed the growth of pathogenic bacteria (*Pseudomonas aeruginosa*) and fungus *Aspergillus flavus* (Jayaseelan et al. 2012). ZnO NPs and MgO NPs at low concentration (100 mg/L) suppressed significantly the germination rate of the fungal spores *Alternaria alternata*, *Fusarium oxysporum*, *Rhizopus stolonifer* and *Mucor plumbeus* (Wani and Shah 2012). Huang et al. (2005) have revealed the significant nanomicrobial activity of MgO against bacteria as a result of possible interactions between MgO NPs and negative surface membranes of bacterial spores (Huang et al. 2005). Metallic nanoparticles TiO_2 have photocatalytic and antimicrobial properties and also could be used as agricultural and soil amendments. Previous research studies revealed the nanoactivity reduction of TiO_2 NPs against cucumber pathogenic bacterial *Pseudomonas syringae* pv. *lachrymans* and *P. cubensis* by (69–91%), respectively, moreover increasing the percentage of photosynthetic activity to 30%. Also, TiO_2 NPs showed an increase in the photoactivity which resulted in increasing the resistance of Rosa 'Noare' and tomato against the causal agent of bacterial spot disease (*Xanthomonas* sp.) compared to conventional treatment (Cui et al. 2009; Paret et al. 2013a, b). Copper in the nanosize showed significant results against different microorganisms. Treated tomato (*Lycopersicon esculentum*) plants with Cu NPs showed an increase in the resistant levels by 73.5% against fungal pathogen *Phytophthora infestans* compared to (57.8%) in the control (Giannousi et al. 2013). On the other hand, chemically synthesized Cu NPs showed great nanofungal activities against plant pathogenic fungi *Phoma destructiva*, *Alternaria alternata*, *Curvularia lunata* and *Fusarium oxysporum* (Kanhed et al. 2014).

Chitosan is considered as a biomaterial, biocompatible, biodegradable and nontoxic material which could be applied effectively in different fields. Chitosan has been applied in agriculture either in enhancing plant growth or plant pathogen management (Katiyar et al. 2015; Xing et al. 2015; Prasad et al. 2017b). Chitosan in the nanosize has different physical and chemical properties which increase the potential application of chitosan NPs compared to normal size. Chitosan NPs have an effective role in plant growth and also as an inducer of the antioxidants and defence systems in the plants (Sathiyabama et al. 2016; Chandra et al. 2015). Transcriptome analysis of plants treated with chitosan NPs showed that an increase in the expression levels of defence-related genes resulted in enhancing the defence level and innate immunity of plants against different stimuli (Chandra et al. 2015). Previous research studies reported the effective role of combination between copper and chitosan NPs as an antifungal and plant growth regulator (Saharan et al. 2015) (Fig. 10.2).

10.2.2 *Positive Impacts of Nanoparticles on Plant Growth and Development*

Nanoparticle-plant interaction may cause several morphological and physiological changes according to physical properties of nanoparticles. Effective role of nanomaterials is determined according to the applied dose, chemical composition, size, surface area and reactivity. Previous studies showed that positive and negative impacts of nanoparticles towards plant growth and development (Khodakovskaya et al. 2012) (Fig. 10.2).

10.2.2.1 Zinc Oxide Nanoparticles

Several studies suggested that the effective role of ZnO NPs towards plant growth and development including ZnO nanorod-embedded fungus *Piriformospora indica* enhances the germination and biomass of *Brassica oleracea* var. *botrytis* plant (Singhal et al. 2017). Previous research studies reported that low concentrations of ZnO NPs revealed positive influence on seed germination in wheat and onion; conversely, higher ZnO NPs decreased seed germination. The impact of NPs on seed germination varies from one plant to another and relies on the NP concentrations (Ramesh et al. 2014; Raskar and Laware, 2014). Previous research studies reported that treatment of cucumber, alfalfa and tomato seeds with different concentrations of ZnO NPs enhanced the germination level in cucumber only (de la Rosa et al. 2013). Murashige and Skoog (MS) media provided with ZnO NPs stimulated shooting, regeneration and somatic embryogenesis of plantlets and encouraged proline synthesis, superoxide dismutase, catalase and and peroxidase activities in this manner improving tolerance to biotic stress (Helaly et al. 2014).

10.2.2.2 Silicon Dioxide Nanoparticles

SiO_2 NPs play an important role in seed germination and plant development. SiO2 nanoparticles in low concentrations enhanced tomato and maize seed germination by increasing availability of nutrient materials and better pH (Siddiqui and Al-Whaibi 2014; Suriyaprabha et al. 2012). Exogenous application of SiO_2 on Changbai larch (*Larix olgensis*) seedlings enhanced growth and quality and also encouraged chlorophyll synthesis (Bao-shan et al. 2004). Furthermore, exogenous application of SiO_2 NPs on tomato and squash seeds enhanced seed germination under abiotic stress, induced antioxidant under salinity stress and also enhanced leaf fresh and dry weight, chlorophyll content and proline accumulation (Haghighi et al. 2012; Siddiqui et al. 2014).

10.2.2.3 Titanium Dioxide Nanoparticles

Titanium dioxide nanoparticles are playing an important role in plant resistance in response to abiotic stress and plant development. TiO_2NPs have been reported as an enhancer of canola seed germination and seedling growth. Furthermore, TiO_2NPs increased wheat growth and yield under drought stress. Also, titanium dioxide nanoparticles control enzyme activities involved in nitrogen metabolism, e.g. nitrate reductase, glutamate dehydrogenase, glutamine synthase and glutamic-pyruvic transaminase, which help the plants in different physiological processes like nitrate absorption, converting inorganic nitrogen to organic nitrogen to form protein and chlorophyll (Mahmoodzadeh et al. 2013; Jaberzadeh et al. 2013; Mishra et al. 2014).

10.2.2.4 Carbon Nanotubes

Carbon nanotubes have unique characters, e.g. chemical, thermal and mechanical, that acquired an important position between other nanomaterials. As a result of the distinctive features of CNT NPs, they have the capability to enter cell wall and membrane. CNT NPs have been applied extremely in human and animals compared to plant cells and metabolism. The single-walled CNTs (SWCNTs) perform as nanocarriers for DNA and dye molecule delivery into plant cells (Srinivasan and Saraswathi 2010). Conversely, in several studies researchers conveyed that multi-walled CNTs (MWCNTs) influence distinctively the seed germination and plant growth and also act as delivery system of DNA and chemicals to plant cells. MWCTs induce the uptake efficiency of water and essential nutrients, e.g. Ca and Fe, which could improve the seed germination and plant development and growth (Villagarcia et al. 2012; Tiwari et al. 2014). Research studies reported that oxidized MWCNTs enhanced significantly root cell elongation and stimulated dehydrogenase activity (Wang et al. 2012) Tripathi and Sarkar (2014) assured the presence of water-soluble CNTs inside wheat plants by using scanning electron and fluorescence microscope (Fig. 10.2).

10.2.3 Negative Impacts of Nanomaterials on Plants

Toxicity of nanoparticles could have negative impacts on physiological process in plants which could be recognized by different factors like reduction of seed germination percentage, suppression of root and shoot elongation, decrease leaf number and plant death (Lee et al. 2010). Several nanotoxicity research studies on plants showed the decrease of seed germination percentage, suppression of shoot length and downregulation of gene expression on wheat (*Triticum aestivum*), maize, barley, rye grass and soybean when exposed to multiwalled carbon nanotubes (MWCNTs), single-walled carbon nanotubes, ZnO NPs, Ag NPs and Fe NPs

(Dimkpa et al. 2012; El-Temsah & Joner 2012; Riahi-Madvar et al. 2012; Yan et al. 2013; Ghosh et al. 2015). *Bacillus thuringiensis* (Bt)-transgenic cotton showed decrease in the growth as a result of SiO2 NP treatment (Le Van et al. 2014). CuO NPs reduced significantly the fresh weights and root length of *Arabidopsis* seedlings and also affect the germination rate and biomass of rice seeds (Shaw and Hossain, 2013). TiO_2 NPs positively enhanced the total contents of chlorophyll and catalase (CAT) and suppressed ascorbate peroxidase (APX) level in leaves (Servin et al. 2013). Nanoparticles could affect the plant at different growth stages, for example, seedling leaf size of cucumber decreased as a result of treatment with 200 mg/L of CeO_2 NPs and CuO NPs compared to control. However the mature leaves had no differences compared to the control plants (Fig. 10.1).

Nanoparticles could affect and disrupt the reactive oxygen species at different growth stages which may lead to plant death. Reactive oxygen species (ROS) enzymes catalase (CAT) and ascorbate peroxidase (APX) activities in the cucumber roots were decreased 15 days after treatment with CeO_2 NPs and CuO NPs compared to control. Additionally, H_2O_2 concentration level increased ten times in the treated plants with CeO_2 NPs compared to control and ended up with plant death (Hong et al. 2016; Majumdar et al. 2014; Zhao et al. 2012). ROS accumulation can direct the cell to oxidative damage in the nuclear, chloroplast and mitochondrial DNA (Imlay and Linn 1988). Also, Yasur and Rani (2013) reported that increase ROS reproduction and associated antioxidant defence enzymes, e.g. superoxide dismutase enzyme (SOD) and peroxidase (POD), in castor bean seeds (*Ricinus communis*) in response to Ag NP treatment resulted in an increase in phenolic acids (Yasur and Rani 2013). ROS and oxidative stress by products could affect other proteins and disrupt physiological process and many pathways which cause an increase in the plant susceptibility followed by plant death (Fig. 10.1).

Research study occurred in wheat seedlings treated with 10 mg/L Ag NPs to study the phytotoxicity and genotoxicity of Ag NPs, and the results showed an alteration of all different types of cell metabolism proteins (Vannini et al. 2014). *Oryza sativa* plants treated with Ag NPs showed that downregulation and upregulation of 28 responsive proteins involved in various important physiological and biochemical processes in the plants, e.g. oxidative stress tolerance, Ca^{2+} signalling, cell wall and RNA/DNA/protein direct damage, transcription and protein degradation, lead to cell division and apoptosis (Mirzajani et al. 2014). Also, proteomic research study occurred to study the nanotoxicity of Al_2O_3, ZnO and Ag NPs in soybean seedlings, and the obtained results showed significant change in 16 common proteins associated with protein degradation and photosystem (Hossain et al. 2016).

Metallic carbon nanoparticles may cause direct damage to the genetic material DNA which led to disruption of proteins encoding ended up with inactivation of encoded proteins. Furthermore, high concentrations of CeO_2 NPs caused DNA damage on *Glycine max*, and bismuth oxide NPs caused total chromosomal abnormalities and increased mitotic index on the roots of *Allium cepa*. Also, another research study proved the genotoxicity, chromosome and micronuclei damages and DNA fragmentation of TiO_2 NPs on several species of plants, e.g. *Nicotiana taba-*

cum and *A. cepa* (López-Moreno et al. 2010; Shen et al. 2010; Khodakovskaya et al. 2013; Dimkpa et al. 2013a, b; Liman, 2013; Demir et al. 2014; Ghosh et al. 2010).

10.3 Conclusion and Future Prospects

Nanotechnology became the magic and master key for any problem that might harm plants, animals and humans. Nanotechnology could have positive and negative impacts on plants and other organisms according to our view, but in fact what we can see is very little compared to undiscovered beauty or ugly facts. Application of nanotechnology in plants could play dual role in towards enhancing plant resistance by decreasing number of harming microorganisms and enhancing plant growth and development. On the other hand, nanomaterials could induce specific genes and proteins which play towards the microorganism by disrupting defence system in the plant which may result in plant susceptibility followed by cell death. Metallic nanoparticles might be effective against several microorganisms but at the same time could be accumulated in the plant cell which is the main food for both human and animals. Nanotoxicity studies should be applied in all nanoparticles to verify that they are safe for human, animals and environment. Natural products in the nanosize might be good solution to avoid the danger of metallic nanoparticles towards human and will be safe to use in the environment.

References

Aziz N, Faraz M, Pandey R, Sakir M, Fatma T, Varma A, Barman I, Prasad R (2015) Facile algae-derived route to biogenic silver nanoparticles: synthesis, antibacterial and photocatalytic properties. Langmuir 31:11605–11612. https://doi.org/10.1021/acs.langmuir.5b03081

Aziz N, Pandey R, Barman I, Prasad R (2016) Leveraging the attributes of *Mucor hiemalis*-derived silver nanoparticles for a synergistic broad-spectrum antimicrobial platform. Front Microbiol 7:1984. https://doi.org/10.3389/fmicb.2016.01984

Aziz N, Fatma T, Varma A, Prasad R (2014) Biogenic synthesis of silver nanoparticles using Scenedesmus abundans and evaluation of their antibacterial activity. J Nanopart Article ID 689419., https://doi.org/10.1155/2014/689419:1

Bao-shan L, Shao-qi D, Chun-hui L, Li-jun F, Shu-chun Q, Min Y (2004) Effect of TMS (nanostructured silicon dioxide) on growth of Changbai larch seedlings. J Forest Res 15:138–140

Cagri A, Ustunol Z, Ryser ET (2004) Antimicrobial edible films and coatings. J Food Prot 67:83–848

Chandra S et al (2015) Chitosan nanoparticles: a positive modulator of innate immune responses in plants. Sci Rep 5:15195

Cui H, Zhang P, Gu W, Jiang J (2009) Application of anatasa TiO_2 sol derived from peroxotitannic acid in crop diseases control and growth regulation. NSTI-Nanotech 2:286–289

de la Rosa G, Lopez-Moreno ML, de Haro D, Botez CE, Peralta-Videa JR, Gardea-Torresdey JL (2013) Effects of ZnO nanoparticles in alfalfa, tomato, and cucumber at the germination stage: root development and X-ray absorption spectroscopy studies. Pure Appl Chem 85(12):2161–2174

Dimkpa CO, McLean JE, Latta DE, Manangón E, Britt DW, Johnson WP, Boyanov MI, Anderson AJ (2012) Cuo and ZnO nanoparticles: phytotoxicity, metal speciation, and induction of oxidative stress in sand-grown wheat. J Nanopart Res 14:1–15

Dimkpa CO, McLean JE, Britt DW, Anderson AJ (2013a) Antifungal activity of ZnO nanoparticles and their interactive effect with a biocontrol bacterium on growth antagonism of the plant pathogen *Fusarium graminearum*. Biometals 26(6):913–924

Dimkpa CO, McLean JE, Martineau N, Britt DW, Haverkamp R, Anderson AJ (2013b) Silver nanoparticles disrupt wheat (*Triticum aestivum* L.) growth in a sand matrix. Environ Sci Technol 47:1082–1090

Demir E, Kaya N, Kaya B (2014) Genotoxic effects of zinc oxide and titanium dioxide nanoparticles on root meristem cells of *Allium cepa* by comet assay. Turk J Biol 38: 31–39

Elechiguerra J, Burt J, Morones J, Camacho-Bragado A, Gao X, Lara H, Yacaman M (2005) Interaction of silver nanoparticles with HIV-1. J Nanobiotechnology 3:1–10

El-Temsah YS, Joner EJ (2012) Impact of Fe and Ag nanoparticles on seed germination and differences in bioavailability during exposure in aqueous suspension and soil. Environ Toxicol 27:42–49

Ghosh M, Bandyopadhyay M, Mukherjee A (2010) Genotoxicity of titanium dioxide (TiO2) nanoparticles at two trophic levels: plant and human lymphocytes. Chemosphere 81:1253–1262

Ghosh M, Bhadra S, Adegoke A, Bandyopadhyay M, Mukherjee A (2015) MWCNT uptake in *Allium cepa* root cells induces cytotoxic and genotoxic responses and results in DNA hypermethylation. Mutat Res Fundam Mol Mech Mutagen 774:49–58

Giannousi K, Avramidis I, Dendrinou-Samara C (2013) Synthesis, characterization and evaluation of copper based nanoparticles as agrochemicals against *Phytophthora infestans*. RSC Adv 3(44):21743–21752

Haghighi M, Afifipour Z, Mozafarian M (2012) The effect of N-Si on tomato seed germination under salinity levels. J Biol Environ Sci 6:87–90

Helaly MN, El-Metwally MA, El-Hoseiny H, Omar SA, El-Sheery NI (2014) Effect of nanoparticles on biological contamination of in vitro cultures and organogenic regeneration of banana. Aust J Crop Sci 8:612–624

Hong J, Wang L, Sun Y, Zhao L, Niu G, Tan W, Rico CM, Peralta-Videa JR, Gardea-Torresdey JL (2016) Foliar applied nanoscale and microscale CeO_2 and CuO alter cucumber (*Cucumis sativus*) fruit quality. Sci Total Environ 563–564:904–911

Hu W, Peng C, Luo W, Lv M, Li X, Li D, Huang Q, Fan C (2010) Graphene-based antibacterial paper. ACS Nano 4:4317–4323

Huang L, Li DQ, Lin YJ, Wei M, Evans DG, Duan X (2005) Controllable preparation of nano-MgO and investigation of its bactericidal properties. J Inorg Biochem 99(5):986–993

He L, Liu Y, Mustapha A, Lin M (2011) Antifungal activity of zinc oxide nanoparticles against *Botrytis cinerea* and *Penicillium expansum*. Microbiol Res 166:207–215

Hossain Z, Mustafa G, Sakata K, Komatsu S (2016) Insights into the proteomic response of soybean towards Al_2O_3, ZnO, and Ag nanoparticles stress. J Hazard Mater 304:291–305

Imlay JA, Linn S (1988) DNA damage and oxygen radical toxicity. Sci (Washington) 240:1302–1309

Jaberzadeh A, Moaveni P, Moghadam HRT, Zahedi H (2013) Influence of bulk and nanoparticles titanium foliar application on some agronomic traits, seed gluten and starch contents of wheat subjected to water deficit stress. Not Bot Horti Agrobo 41:201–207

Jayaseelan C et al (2012) Novel microbial route to synthesize ZnO nanoparticles using Aeromonas hydrophila and their activity against pathogenic bacteria and fungi. Spectrochim Acta A 90:78–84

Jo YK, Kim BH, Jung G (2009) Antifungal activity of silver ions and nanoparticles on phytopathogenic fungi. Plant Dis 93(10):1037–1043

Kanhed P et al (2014) In vitro antifungal efficacy of copper nanoparticles against selected crop pathogenic fungi. Mater Lett 115:13–17

Karlsson HL, Cronholm P, Gustafsson J, Möller L (2008) Copper oxide nanoparticles are highly toxic:a comparison between metal oxide nanoparticles and carbon nanotubes. Chem Res Toxicol 21:1726–1732

Katiyar D, Hemantaranjan A, Singh B (2015) Chitosan as a promising natural compound to enhance potential physiological responses in plant: a review. Indian J Plant Physiol 20:1–9

Khodakovskaya MV, de Silva K, Biris AS, Dervishi E, Villagarcia H (2012) Carbon nanotubes induce growth enhancement of tobacco cells. ACS Nano 6(3):2128–2135

Khodakovskaya MV, Kim BS, Kim JN, Alimohammadi M, Dervishi E, Mustafa T, Cernigla CE (2013) Carbon nanotubes as plant growth regulators: effects on tomato growth, reproductive system, and soil microbial community. Small 9:115–123

Kim KJ, Sung W, Suh B, Moon SK, Choi JS, Kim J, Lee D (2009) Antifungal activity and mode of action of silver Nano-particles on *Candida Albicans*. Biometals 22:235–242

Kim TN, Feng QL, Kim JO, Wu J, Wang H, Chen GC, Cui FZ (1998) Antimicrobial effects of metal ions (ag^{+}, Cu^{2+}, Zn^{2+}) in hydroxyapatite. J Mater Sci Mater Med 9:129–134

Lamsal K, Kim SW, Jung JH, Kim YS, Kim KS, Lee YS (2011) Application of silver nanoparticles for the control of *Colletotrichum species* in vitro and pepper anthracnose disease in field. Mycobiology 39(3):194–199

Lee CW, Mahendra S, Zodrow K, Li D, Tsai YC, Braam J, Alvarez PJ (2010) Developmental phytotoxicity of metal oxide nanoparticles to *Arabidopsis thaliana*. Environ Toxicol Chem 29:669–675

Le Van N, Rui Y, Gui X, Li X, Liu S, Han Y (2014) Uptake, transport, distribution and bio-effects of SiO2 nanoparticles in Bt-transgenic cotton. J Nanobiotechnol 12:50

Liman R (2013) Genotoxic effects of bismuth (III) oxide nanoparticles by allium and comet assay. Chemosphere 93:269–273

López-Moreno ML, de la Rosa G, Hernández-Viezcas JÁ, Castillo- Michel H, Botez CE, Peralta-Videa JR, Gardea-Torresdey JL (2010) Evidence of the differential biotransformation and genotoxicity of ZnO and CeO2 nanoparticles on soybean (*Glycine max*) plants. Environ Sci Technol 44:7315–7320

Mahmoodzadeh H, Nabavi M, Kashefi H (2013) Effect of nanoscale titanium dioxide particles on the germination and growth of canola (*Brassica napus*). J Ornamental Hortic Plants 3:25–32

Majumdar S, Peralta-Videa JR, Bandyopadhyay S, Castillo-Michel H, Hernandez-Viezcas JA, Sahi S, Gardea-Torresdey JL (2014) Exposure of cerium oxide nanoparticles to kidney bean shows disturbance in the plant defense mechanisms. J Hazard Mater 278:279–287

Mallick S, Sharma S, Banerjee M, Ghosh SS, Chattopadhyay A, Paul A (2012) Iodine-stabilized cu nanoparticle chitosan composite for antibacterial applications. ACS Appl Mater Interfaces 4:1313–1323

Mirzajani F, Askari H, Hamzelou S, Schober Y, Römpp A, Ghassempour A, Spengler B (2014) Proteomics study of silver nanoparticles toxicity on *Oryza sativa* L. Ecotoxicol Environ Safe 108:335–339

Mishra V, Mishra RK, Dikshit A, Pandey AC (2014) Interactions of nanoparticles with plants: an emerging prospective in the agriculture industry. In: Ahmad P, Rasool S (eds) Emerging technologies and management of crop stress tolerance: biological techniques, vol 1, pp 159–180

Neal A (2008) What can be inferred from bacterium-nanoparticle interactions about the potential consequences of environmental exposure to nanoparticles? Ecotoxicology 17:362–371

Ocsoy I, Paret ML, Ocsoy MA, Kunwar S, Chen T, You M, Tan W (2013) Nanotechnology in plant disease management: DNA-directed silver nanoparticles on graphene oxide as an antibacterial against *Xanthomonas perforans*. ACS Nano 7(10):8972–8980

Paret M, Palmateer A, Knox G (2013a) Evaluation of a light activated nanoparticle formulation of TiO_2/Zn for managementof bacterial leaf spot on Rosa 'Noare. Hortscience 48(2):189–192

Paret ML, Vallad GE, Averett DR, Jones JB, Olson SM (2013b) Photocatalysis: effect of light-activated nanoscale formulations of TiO_2 on *Xanthomonas perforans* and control of bacterial spot of tomato. Phytopathology 103(3):228–236

Prasad R, Kumar V, Prasad KS (2014) Nanotechnology in sustainable agriculture: present concerns and future aspects. Afr J Biotechnol 13(6):705–713

Prasad R, Pandey R, Barman I (2016) Engineering tailored nanoparticles with microbes: quo vadis. WIREs Nanomed Nanobiotechnol 8:316–330. https://doi.org/10.1002/wnan.1363

Prasad R, Bhattacharyya A, Nguyen QD (2017a) Nanotechnology in sustainable agriculture: recent developments, challenges, and perspectives. Front Microbiol 8:1014. https://doi.org/10.3389/fmicb.2017.01014

Prasad R, Gupta N, Kumar M, Kumar V, Wang S, Abd-Elsalam KA (2017b) Nanomaterials act as plant defense mechanism. In: Prasad R, Kumar V, Kumar M (eds) Nanotechnology. Springer, Singapore, pp 253–269

Prasad R, Swamy VS (2013) Antibacterial activity of silver nanoparticles synthesized by bark extract of *Syzygium cumini*. J Nanopart 2013:1–6. http://dx.doi.org/10.1155/2013/431218

Ramesh M, Palanisamy K, Babu K, Sharma NK (2014) Effects of bulk & nano-titanium dioxide and zinc oxide on physio-morphological changes in *Triticum aestivum* Linn. J Glob Biosci 3:415–422

Raskar SV, Laware SL (2014) Effect of zinc oxide nanoparticles on cytology and seed germination in onion. Int J Curr Microbiol App Sci 3:467–473

Riahi-Madvar A, Rezaee F, Jalali V (2012) Effects of alumina nanoparticles on morphological properties and antioxidant system of *Triticum aestivum*. Iran J Plant Physiol 3:595–603

Saharan V et al (2015) Synthesis and in vitro antifungal efficacy of cu–chitosan nanoparticles against pathogenic fungi of tomato. Int J Biol Macromol 75:346–353

Sathiyabama M, Bernstein N, Anusuya S (2016) Chitosan elicitation for increased curcumin production and stimulation of defence response in turmeric (*Curcuma longa* L.). Ind Crop Prod 89:87–94

Servin AD, Morales MI, Castillo-Michel H, Hernandez-Viezcas JA, Munoz B, Zhao L, Nunez JE, Peralta-Videa JR, Gardea-Torresdey JL (2013) Synchrotron verification of TiO_2 accumulation in cucumber fruit: a possible pathway of TiO_2 nanoparticle transfer from soil into the food chain. Environ Sci Technol 47:11592–11598

Shaw AK, Hossain Z (2013) Impact of nano-CuO stress on rice (*Oryza sativa* L.) seedlings. Chemosphere 93:906–915

Shen CX, Zhang QF, Li J, Bi FC, Yao N (2010) Induction of programmed cell death in Arabidopsis and rice by single-wall carbon nanotubes. Am J Bot 97:1602–1609

Singhal U, Khanuja M, Prasad R, Varma A (2017) Impact of synergistic association of ZnO-nanorods and symbiotic fungus *Piriformospora indica* DSM 11827 on *Brassica oleracea* var. botrytis (broccoli). Front Microbiol 8:1909. https://doi.org/10.3389/fmicb.2017.01909

Siddiqui MH, Al-Whaibi MH (2014) Role of nano-SiO2 in germination of tomato (*Lycopersicum esculentum* seeds mill.). Saudi Biol Sci 21:13–17

Siddiqui MH, Al-Whaibi MH, Faisal M, Al Sahli AA (2014) Nano-silicon dioxide mitigates the adverse effects of salt stress on Cucurbita pepo L. Environ Toxicol Chem 33(11):2429–2437

Sondi I, Salopek-Sondi B (2004)Silver nanoparticles as antimicrobial agent: a case study on *E. coli* a model for gram-negative Bacteria. J Colloid Interface Sci 275:177–182

Srinivasan C, Saraswathi R (2010) Nano-agriculture-carbon nanotubes enhance tomato seed germination and plant growth. Curr Sci 99:273–275

Suriyaprabha R, Karunakaran G, Yuvakkumar R, Rajendran V, Kannan N (2012) Silica nanoparticles for increased silica availability in maize (*Zea mays* L) seeds under hydroponic conditions. Curr Nanosci 8:902–908

Swamy VS, Prasad R (2012) Green synthesis of silver nanoparticles from the leaf extract of *Santalum album* and its antimicrobial activity. J Optoelectronic and Biomedical Materials 4(3):53–59

Tiwari DK, Dasgupta-Schubert N, Villaseñor-Cendejas LM, Villegas J, Carreto-Montoya L, Borjas-García SE (2014) Interfacing carbon nanotubes (CNT) with plants: enhancement of growth, water and ionic nutrient uptake in maize (*Zea Mays*) and implications for nanoagriculture. Appl Nanosci 4:577–591

Tripathi S, Sarkar S (2014) Influence of water soluble carbon dots on the growth of wheat plant. Appl Nanosci

Vannini C, Domingo G, Onelli E, De Mattia F, Bruni I, Marsoni M, Bracale M (2014) Phytotoxic and genotoxic effects of silver nanoparticles exposure on germinating wheat seedlings. J Plant Physiol 171:1142–1148

Villagarcia H, Dervishi E, Silva K, Biris AS, Khodakovskaya MV (2012) Surface chemistry of carbon nanotubes impacts the growth and expression of water channel protein in tomato plants. Small 8:2328–2334

Wang X, Han H, Liu X, Gu X, Chen K, Lu D (2012) Multi-walled carbon nanotubes can enhance root elongation of wheat (*Triticum aestivum*) plants. J Nanopart Res 14(6):1–10

Wani AH, Shah MA (2012) A unique and profound effect of MgO and ZnO nanoparticles on some plant pathogenic fungi. J App Pharm Sci 2(3):40–44

Xing K, Zhu X, Peng X, Qin S (2015) Chitosan antimicrobial and eliciting properties for pest control in agriculture: a review. Agronomy Sustain Develop 35:569–588

Yan S, Zhao L, Li H, Zhang Q, Tan J, Huang M, He S, Li L (2013) Single walled carbon nanotubes selectively influence maize root tissue. J Hazard Mater 246–247:110–118

Yasur J, Rani PU (2013) Environmental effects of nanosilver: impact on castor seed germination, seedling growth, and plant physiology. Environ Sci Pollut Res 20:8636–8648

Yoon KY, Hoon Byeon J, Park JH, Hwang J (2007) Susceptibility constants of *Escherichia coli* and *Bacillus Subtilis* to silver and copper nanoparticles. Sci Total Environ 373:572–575

Zhao L, Peng B, Hernandez-Viezcas JA, Rico C, Sun Y, Peralta-Videa JR, Tang X, Niu G, Jin L, Varela-Ramirez A et al (2012) Stress response and tolerance of Zea mays to CeO2 nanoparticles: cross talk among H_2O_2, heat shock protein, and lipid peroxidation. ACS Nano 6:9615–9622

Zodrow K, Brunet L, Mahendra S, Li D, Zhang A, Li Q, Alvarez PJJ (2009) Polysulfone ultrafiltration membranes impregnated with silver nanoparticles show improved biofouling resistance and virus removal. Water Res 43:715–723

Chapter 11
Nanoantimicrobials Mechanism of Action

Manal Mostafa, Amal-Asran, Hassan Almoammar, and Kamel A. Abd-Elsalam

11.1 Introduction

Nanotechnology, the procedure to produce, control, and release nanomaterials, represents a zone holding huge guarantee for the agricultural scenario (Baruah and Dutta 2009; Navrotsky 2000; Kuzma 2007). Nanoparticles (NPs) having at least one dimension in the order of 100 nm or less (Auffan et al. 2009), in light of the fact that it is at this scale, the properties of materials vary as for their physical, substance, and organic properties from those at a higher scale. Nanostructuring increases the value of customary materials by improving their mechanical quality, superconductivity, and capacity to join and effectively convey dynamic substances into biological systems, at low expenses and with restricted agroecosystem effect (García-Rincóna et al. 2010). From an agricultural viewpoint, nanotechnology can possibly turn into a helpful method for plant pathologists in the diagnosis and treatment of plant diseases by the utilization of nano-based kits, pathogen detection plan by the application of nanosensors, improved ability of plants for micronutrients absorption, maximized plant yield by nanoporous zeolites, and plant insect management by

M. Mostafa
CIHEAM IAMB – Mediterranean Agronomic Institute of Bari, Valenzano (BA), Italy

Microbiology Department, Faculty of Agriculture, Cairo University, Giza, Egypt

Plant Pathology Research Institute, Agricultural Research Center (ARC), Giza, Egypt

Amal-Asran · K. A. Abd-Elsalam (✉)
Plant Pathology Research Institute, Agricultural Research Center (ARC), Giza, Egypt

H. Almoammar
National Centre for Biotechnology, King Abdulaziz City for Science and Technology (KACST), Riyadh, Saudi Arabia

K. A. Abd-Elsalam, R. Prasad (eds.), *Nanobiotechnology Applications in Plant Protection*, Nanotechnology in the Life Sciences,
https://doi.org/10.1007/978-3-319-91161-8_11

using nanocapsules (Chaudhry et al. 2008; Sharon et al. 2010; Rai and Ingle 2012; Prasad 2014; Prasad et al. 2014, 2017a; Ismail et al. 2017). Researchers are occupied with preparing various types of natural, inorganic, and crossover nanoparticles having physical, optical, and organic properties (Salata 2004; Rai and Ingle 2012; Prasad et al. 2016). Nanobiotechnology works at a similar level with viral infection or disease-infecting particle and in this manner holds the potential for primordial identification and suppression. It additionally holds out the likelihood that smart sensors and conveyance frameworks will enable the agricultural industry to fight viruses and other plant pathogens (Prasad et al. 2014, 2017a). Then again, nanobio-innovation can enhance our comprehension of the nanobiology of different crops and consequently can possibly improve yields or nutritional values and also create enhanced systems for monitoring agroecosystem and improving the capacity of plants to preserve micronutrients or pesticides (Fakruddin et al. 2012; Tarafdar et al. 2013).

In order to completely and accurately define the mechanisms of toxicity that nanomaterials exhibit in a cellular environment, researchers at the forefront of this field will need to take a highly interdisciplinary approach utilizing both chemical and biological techniques to substantiate their claims. The current chapter will summarize recent progress toward an understanding of the antimicrobial mechanisms of nanostructures, with a focus on studies providing evidence for oxidative stress induction, membrane disruption, and genotoxicity. With a specific end goal to totally and precisely define the mechanisms of toxicity that nanomaterials display in a cellular environment, scientists at the forefront of this field should adopt an exceedingly interdisciplinary strategy using both concoction and natural procedures to substantiate their cases. The current chapter will summarize recent progress toward an understanding of the antimicrobial mechanisms of nanostructures, with a focus on studies providing evidence for oxidative stress induction, membrane disruption, and genotoxicity.

11.2 Resistance to Conventional Antimicrobial

Antimicrobial resistance is a regularly developing issue, yet what is an antimicrobial resistance? The expression "antimicrobial resistance" is not only a potential danger, it is a severe health problem that is quickly spreading around the world. Over the past decades, people depended on regular antibiotic and antifungal eventually this led to improve the genotype of microorganisms to become more resistant, thus made researcher become more interested to provide new solutions using nanostructures, heavy metals are known to be toxic to various pathogens. In nature, microbial resistance from most dangerous substantial metals is because of their compound detoxification and because of vitality subordinate particle efflux from the cell by layer proteins that capacity either as ATPase or as chemiosmotic cation or proton antiporters. Modification in dissolvability additionally assumes a part in microbial resistance (Liu et al. 2011b).

11.3 Nanostructures as an Antimicrobial

Morphology and surface properties of colloidal nanoparticles are imperative. So that smaller nanoparticles than larger nanoparticles have further antimicrobial activity (Chwalibog et al. 2010), polymer-based copper nanocomposites have been examined for antifungal efficacy against plant pathogenic fungi; also silver nanoparticles are used in controlling spore-producing fungal plant pathogens and also showed the highest inhibition rate for both before and after the outbreak of disease on cucumbers and pumpkins and maximum inhibition for the growth of fungal hyphae and conidial germination in vivo assay (Kim et al. 2009; Cioffi et al. 2004). Also, a portion of the nanoparticles that have entered into the arena of controlling plant diseases are nanocarbon, silica, and aluminosilicates (Prasad et al. 2014). Ag NPs have an inhibitory effect on fungal colony and on spores of *F. oxysporum*. Silver nanoparticles might be directly devoted to penetrate cell layer membranes to kill spores (Abkhoo and Panjehkeh 2017; Morones et al. 2005). As of late, chitosan has turned out to be viable against microorganisms, plant pathogens, and viral pathogens (Xing et al. 2015). Furthermore zinc oxide (ZnO) and magnesium oxide (MgO) nanoparticles are effective antibacterial and anti-odor agents (Shah and Towkeer 2010; Bhuyan et al. 2015), and platinum nanoparticles TiO_2 (Goswami et al. 2010) were achieved using the metal ion-reducing bacterium *Shewanella algae* growth (Konishi et al. 2007). Likewise it is notable that graphene oxide (GO) and its composites possess antimicrobial properties and have been used as antibacterial and antifungal agents (Santos et al. 2012; De Faria et al. 2014), and alumina NPs have been evaluated on bacteria such as *Escherichia coli*, *Pseudomonas aeruginosa*, and *Bacillus subtilis* and algae *Scenedesmus* sp. and *Chlorella* sp., and the toxic mechanism has been suggested as the interaction of the nanoparticles with the cell surface which cause leakage in the membrane (Sadiq et al. 2011). In addition, the likelihood to conjugate the surface of the NSs or to consolidate them with different materials (e.g., polymers, characteristic and manufactured fibers, clay) enables to achieve nanoantimicrobials with tunable properties as far as productive bioactivity against the focus of microorganisms and constrained, assuming any, toxicity toward human cells.

11.4 Applied Nanostructures and Their Mechanisms

The biocidal mode of action of the biofungicides is diverse relying upon the type of microorganisms utilized, viz., rhizosphere fitness, parasitism, and antibiosis, activating metabolic changes, promoting plant growth, and so forth (Saraf et al. 2014; Shrivastava et al. 2014). At the point when nanomaterials bind electrostatically to the bacterial cell wall and membranes, prompting modification of film potential, film depolarization, and loss of integrity, thus, result in roughness of transport, hampered breath, interruption of energy transduction and/or cell lysis, and eventually cell death (Pelgrift and Friedman 2013). Subsequently, the fundamental mechanisms that have been proposed to clarify the antimicrobial activity of inorganic and metal nanoparticles were ROS, which prompt oxidative stress and liberate superoxide, free radicals, and particles that can respond with the peptide linkages in the cell wall of microscopic organisms and in this manner upset them (Makhluf et al. 2005). To explain In the mitochondria of cells, ATP is synthesized by reduction of molecular oxygen to water through a sequence of coupled proton and electron transfer reactions. During this process, a small percentage of the oxygen is not reduced completely, resulting in the formation of superoxide anion radicals, and subsequently other oxygen-containing radicals. Thus, ROS need aid impacts from claiming cell division oxidative metabolism, much of which occurs in the mitochondria. Biologically relevant ROS include superoxide anion radicals, hydroxyl radicals, singlet oxygen, and hydrogen peroxide (H_2O_2) (Yin et al. 2012; Prasad et al. 2017b). The burst of ROS causes, through extreme oxidative stress, harm to all the cell's macromolecules, prompting lipid peroxidation, adjustment of proteins, interruption of enzymes, and RNA and DNA destruction. At high concentrations the ROS lead to cell death and at low concentrations cause serious DNA damage and mutations (Wang et al. 2010; Matˇeejka and Tokarsk´y 2014). The nanoparticle is unable to cross the nuclear membrane and thus accumulates in the cytoplasm, where they can gain access to the nucleus during mitosis when the nuclear membrane breaks down (Singh et al. 2009). The direct interaction of nanoparticles with the DNA and DNA-related protein may lead to physical damage in the genetic material. Interference with the structure or function of the DNA repair enzymes in the nucleus may be another reason for DNA damage. The nanoparticle cannot cross the nuclear membrane and accordingly aggregates in the cytoplasm, where they can access the core amid mitosis when the nuclear membrane separates (Singh et al. 2009). The direct interaction of nanoparticles with the DNA and DNA-related protein may lead to physical destruction of the nucleic acids. Interference with the structure or function of the DNA repair enzymes in the nucleus might be an extra reason for DNA damage (Huang et al. 2015). And the alkalin effect also there is ionic mimicry mechanism that based on the donor atom selectivity and/or speciation of metals: metal ions in general bind to some atoms of donor ligands, such as O, N and S, through strong and selective, depending on whether metal ions or metal complexes are involved. In this way, some metals can stimulate the damage of Fe-S clusters, for instance, from bacterial dehydratases that are particularly susceptible to site-specific inactivation by toxic metals interactions, exterior metal ions, or their complexes that can replace original metals present in biomolecules leading to cellular dysfunction. Metals can also replace non-catalytic metal-binding sites inhibiting enzyme activity (Lemire et al. 2013; Grass et al. 2011; Ruparelia et al. 2008; Xu and Imlay 2012).

11.4.1 Silver NPs

Silver NPs are being utilized as part of various advancements and consolidated into a wide exhibit of customer items that exploit their attractive optical, conductive, antifungal, and antibacterial properties (Aziz et al. 2016; Joshi et al. 2018). The fundamental utilization of Ag NPs is, be that as it may, as antimicrobial agents (Cioffi and Rai 2012; Prasad et al. 2012). Silver-based nanocomposites have been utilized broadly as antimicrobial agents in various areas including therapeutic, pharmaceutical, material, food safety, ecological, and agrarian applications (Kim et al. 2007). With the antimicrobial activity of AgNPs, it is encouraging to note that they are predominantly utilized for plant disease control (Jo et al. 2009; Kim et al. 2011). Although, AgNPs have been proved effective against over 650 microorganisms including bacteria (both Gram-positive and negative), fungi and viruses; however, the exact mechanism of silver action on microbes is still not known, but the possible mechanism of action of metallic silver, silver ions, and silver NPs have been suggested according to the injuries and changes, induced in microbial cells (Malarkodi et al. 2013). Several mechanisms maybe involved in the antimicrobial activity of Ag NP's most of them damage the microbial's cell structure integrity and result in leakage of intracellular compounds, and eventually cell death (Durán et al. 2016). Concerning the activity of $Ag+$, the subsequent actions are caused: (1) binding to negatively charged proteins and nucleic acids (particularly with functional groups like imidazole, indole, hydroxyl, phosphate, thiol) causing changes in structure. As an example, it is known that $Ag +$ ions bind to cysteine-containing proteins on plasma membranes, causing both physiological and biochemical destructions that compromise membrane integrity. Subsequent penetration of Ag into the cytoplasm causes the inactivation of critical enzyme systems and condenses DNA which then reacts with the thiol group proteins and triggers cell death (Ocsoy et al. 2013; Aziz et al. 2014, 2015, 2016). The antifungal activity of AgNPs takes place due to the high attraction between the nanoparticles and the functional chemical groups existing in the cell membrane of fungi and other microorganisms (Jo et al. 2009). It has been suggested that AgNPs with a positive surface charge are more easily internalized through the cell membrane than particles negatively charged or neutral (Jo et al. 2015). This is caused by 1) a procedure called electromagnetism that occurs between the positively charged AgNPs and negatively charged bacterial cell membranes, due to the presence of phosphate and carboxyl groups that 2) alter the functions of the ribosome, causing an inhibition of protein synthesis and locking mechanisms of transcription and translation (Abbaszadegan et al. 2015). DNA loses its ability to duplicate when the fungal culture was treated with AgC and $Ag+$, which may lead to a deactivated expression of ribosomal subunit proteins and to the synthesis of disabled enzymes and cellular proteins, important for the adenosine triphosphate production (Yamanaka et al. 2005; Sang et al. 2012) and which also prevent protein expression related to ATP fabrication (Kim et al. 2009). The biogenic silver NPs bind with protein of the outer cell wall of some pathogens including bacterial, fungal, or viral bodies that disrupt the lipoproteins of the microbial cell wall. Finally the cell division was stopped and cell leads to death (Kuppusamy et al. 2016; Prasad et al. 2016) and (3) deterioration of the external cell membrane, due to the fact that the synthesis of the component proteins is affected. When the membrane is not totally functional, the protein precursors of the envelope proteins and periplasmic constituents accumulate in the cytoplasm, such as when the membrane potential is dissipated by small molecules (Lok et al. 2006). Due to the abundance of sulfur-containing structural proteins and enzymes on the bacterial cell membrane, silver NPs can interact with them and in turn reduce cell functionality and viability. Furthermore, they interact with phosphorus-containing compounds like DNA. Nanoparticles less than 20 nm in diameter may attach to the cell membrane, make pores on the cell wall, leading to more permeability, and release cytoplasmic content outside the cells, destroying enzymes and attacking the respiratory chain and cell division, which cause the death of bacteria (Morones et al. 2005). Additionally, AgNPs are recognized to (4) generate ROS and liberated radicals, which lead both mitochondrial dysfunction and DNA harm. In addition, a study improved by Rai et al. (2009) indicated that the AgNPs affect the performance of reactive oxygen species (ROS) and lead to a process of lipid peroxidation of the cell membrane, changes in cell cycle, and damage to the DNA of the microorganisms. (5) intercalate between DNA bases, silver ions (particularly Ag^+) released from silver nanoparticles can interact with phosphorus moieties in DNA, resulting in inactivation of DNA replication, or can react with sulfur-containing proteins, leading to the inhibition of enzyme functions (Gupta and silver 1998; Matsumura et al. 2003). Silver NPs might prevent many oxidative enzymes, including alcohol dehydrogenase, and prohibit the uptake of succinate by the membrane vesicles. They cause oxidative DNA harm and interfere with DNA replication processes (Petrus et al. 2011; Prasad et al. 2017b). In short, the antimicrobial mode of action for AgNPs is associated with four well-defined mechanisms: (1) adherence of AgNPs onto the surface of cell wall and membrane, (2) AgNPs penetration inside the cell and damaging of intracellular structures (mitochondria, vacuoles, ribosomes) and biomolecules (protein, lipids, and DNA), (3) AgNPs induced cellular toxicity and oxidative stress caused by formation of reactive oxygen species (ROS) and free radicals, and (4) modulation of signal transduction pathways.

11.4.2 TiO_2 NPs

Titanium dioxide nanoparticles (TiO_2 NPs) can be utilized in different agricultural applications. The photocatalytic properties of TiO_2 NPs play a major role in the management of different pathogens. There have been many new reports on the inhibitions of microorganisms in the presence of pure TiO_2 nanoparticles with several crystalline phases under UV irradiation (Lin et al. 2014). When TiO_2 is irradiated with near-UV light, this semiconductor exhibits robust photocatalytic chemical reaction, which is a sophisticated chemical reaction method for the removal of trace contaminants and microorganism pathogens (Hossain et al. 2014). The photocatalytic antimicrobial activity of TiO_2 is attributed to lipid peroxidation that cause a rise within the membrane thinness, discontinuity of the cell structure, and production of ROS, as well as peroxide (H_2O_2), superoxide radical ($O^{2\bullet}$), and free radicals ($^{\bullet}OH$), upon exposure to near-UV and UVA radiation (Choi et al. 2007; Niazi and Gu 2009; Huh and Kwon 2011; Carré et al. 2014).

The generated ROS causes harm to the molecular structure of the cell, including DNA, lipid, and protein damage (Fu et al. 2014). Mathur et al. (2015) reported the effects of TiO_2 NPs on intracellular levels of two major types of ROS – hydroxyl radical (•OH) and superoxide radical $(O)^{-2}$. These free radicals can interact with macromolecules, such as lipids, proteins, enzymes, and nucleic acid molecules in bacteria, viruses, and different microorganisms, which can destroy cell structures through a series of chain reactions (Yu et al. 2002; Sonawane et al. 2003; Zhao et al. 2000; Prasad et al. 2016). Formation of oxidative stress (hydroxyl and superoxide radicals) was higher in bacterial cells presented to NPs and UVA light. Eventually, a significant increase in membrane permeability was noted in cells exposed to NPs and UVA light in comparison to that in dark and visible light conditions. Therefore, opinions exist that the primary mechanism of action of titania NPs is based on creation of ROS, which induce oxidative stress. TiO_2 NPs also have bactericidal effects in the absence of irradiation, suggesting that they use other antimicrobial mechanisms unrelated to photocatalytic ROS production (Choi et al. 2007). Inactivation of microorganisms depends upon several factors, e.g., concentration of TiO_2, type of microorganism, intensity and wavelength of light, degree of hydroxylation, pH, temperature, availability of oxygen, and ROS retention time (Markowska-Szczupak et al. 2011; Hossain et al. 2014).

The antimicrobial activity of titanium nanoparticles have been broadly examined throughout the years. TiO_2 is economical, nontoxic, and insoluble food additive. Titanium dioxide NP is photocatalytic; their poisonous quality is initiated by obvious light, close UV, or UV (Pelgrift and Friedman 2013). The antialgal activity of titania nanoparticles against microalgae species have likewise been led for which a focus subordinate reduction in chlorophyll content was noticed (Sadiq et al. 2011). Later reports have demonstrated its efficiency against different viral species and pathogens (Brady-Est´evez et al. 2008; Allahverdiyev et al. 2013). The effect of TiO_2 NPs on the symbiotic behaviour of symbiotic arbuscular mycorrhizal fungi (AMF) colonising rice at 0, 25, 50 and 100 mg plant-1 to the rhizosphere of mycorrhizal rice plants maintained in pots. TiO_2 NPs had an inhibitory affected AMF in plant roots (Priyanka et al. 2017). The TiO_2 nanoparticles could protect the wood against white- and brown-rot fungi (De Filpo et al. 2013). TiO_2 nanoparticles had viable antifungal properties at the concentration of 5.14 and 5.35 g/mL for fluconazole-susceptible and fluconazole-resistant strains of *Candida albicans* biofilms contrasted with fluconazole medication, respectively (Haghighi et al. 2012).

11.4.3 ZnO NPs

Zinc nanoparticles have been utilized as nanofertilizers on numerous plant species and indicated positive outcomes in ideal concentration; however ZnO NPs as fungicide against various plant pathogens is less studied. ZnO NPs are extremely successful antimicrobial agents and are effective against both, bacteria, fungi, toxicogenic fungi in addition to the thermophilic and barophilic spores (Sondi and Salopek-Sondi

2004; Raghupathi et al. 2011; Sierra-Fernandez et al. 2017). ZnO could also be utilized as antimicrobial agents against microorganisms that could be causes of foodborne pathogens and plant diseases. The exact mechanism of this activity is not fully understood yet. Be that as it may, one hypothesis proposes the development of a solid oxidant, hydrogen peroxide (H_2O_2). The superoxides and hydroxyl radicals cannot penetrate into the membrane because of their negative charges (Xie et al. 2011). Accordingly, these species are found on the external surface of bacteria, and by differentiation, H_2O_2 particles can go through the bacterial cell wall, in this way prompting wounds and destroying and lastly activating cell death (Zhang et al. 2007; Sawai et al. 1996). Some other possible mechanisms include cell membrane disruption, generation of ROS on the NP surface, the influx of zinc particle in the cell, membrane dysfunction, or internalization of NPs, which could help in its antimicrobial activity (Li et al. 2012a). ZnO NPs caused liquefaction of cytoplasmic substances, making the cytoplasm less electron-dense and creating an eminent separation of the fungal cell wall (Arciniegas-Grijalba et al. 2017). A few investigations recommended that ZnO NPs may cause support changes of the microbial cell membrane, causing cytoplasm leakage and in the long run the demise of bacterial cells (Sawai and Yoshikawa 2004; Brayner et al. 2006). ZnO nanoparticles showed noteworthy antibacterial activity and exhibited a deadly impact against *C. jejuni*, even at low concentrations. ZnO nanoparticles prompted noteworthy morphological changes and assessable membrane leakage (Xie et al. 2011). Different researchers show that the event of ROS is the principal mode of action responsible for the killing efficiency of ZnO NPs (Bhuyan et al. 2015; Arciniegas-Grijalba et al. 2017). The formation of ROS, for example, hydrogen peroxide H_2O_2, hydroxyl radical $^*OH$, and superoxide O^{*2-}, is the aftereffect of ZnO initiation by UV or unique light (photocatalysis).

The SEM micrographs of *Botrytis cinerea* mycelium treated with photoactivated ZnO NPs plainly show different degenerative modifications in the conidial heads and hyphal morphology. Twisted conidial heads and withered hyphal wall were usually seen in treated mycelia compared with control. Fungistatic mode of action of ZnO activity on fungi was seen by He et al. (2011). Authors discovered control of conidial advancement by the distortion of conidiophores in *Penicillium expansum*, while the fungal mat of *B. cinerea* was disfigured after treatment with ZnO NPs (He et al. 2011). Jayaseelan et al. (2012) showed that ZnO NPs suppress the development of pathogenic microbes *P. aeruginosa* and *Aspergillus flavus*. ZnO NPs rate the germination of fungal spores of *Alternaria alternata*, *Fusarium oxysporum*, *Rhizopus stolonifer*, and *Mucor plumbeus* (Wani and Shah 2012). SEM micrographs got by Kairyte et al. (2013) showed distorted conidial heads and withered hyphal wall in *B. cinerea* mycelia after treatment by photoactivated ZnO NPs (Fig. 11.1).

Distortion of the hyphal cell structure may be routed to over-the-top collection of nucleic acids and sugars, since ZnO NPs can influence cell physiology and trigger higher generation of nucleic acids. In addition, the increase in the production of nucleic acids can be considered as a stress response of fungal hyphae, and increase production of starches might be a consequence of cell self-protection from the activity of ZnO NPs (Perez Espitia et al. 2012). ZnO nanoparticles demonstrated less phytotoxicity on plants in contrast with AgNP and can be valid as a nanopesticide.

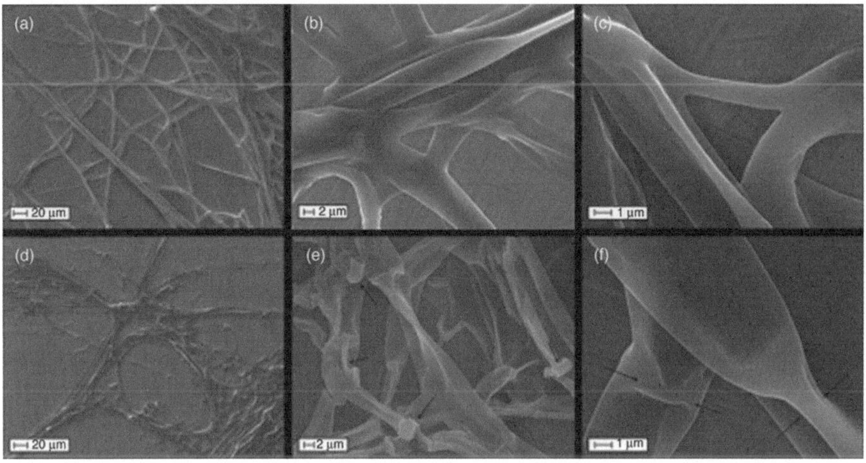

Fig. 11.1 Scanning electron microscopy of *B. cinerea* mycelium after treatment with 5 × 10^{-3} M photoactivated ZnO NPs (24 h incubation; 34.56 J/cm 2 illumination dose) (**d–f**) in comparison with control (**a–c**), non-treated ones. (Reprinted from Kairyte et al. 2013)

There have been various investigations that determined that controlling the measure of ZnO NPs is basic for antimicrobial-related applications. As needs be, the state of the zinc oxide nanostructures can influence penetration of cell membranes; spherical nanoparticles cannot enter as effectively as rod structures (Sirelkhatim et al. 2015). Be that as it may, the most essential part in antimicrobial movement is played by molecule size and concentration of ZnO NPs (Sirelkhatim et al. 2015). In this manner the high antimicrobial activity of ZnO NPs is normally identified with a substantial surface area and a high concentration of particles. This can be clarified by methods of entry into the bacterial cell wall; a smaller-sized particle can easily penetrate bacterial membranes and injure or kill the cells. The fungal wall, in controlling cell permeability, is the part of the cell that interacts with the external environment, and thus with the ZnO NPs present in the fungal culture of interest in this work. This part of the fungal cell is made primarily out of polysaccharides and proteins. In particular, there are β-1,3-D-glucan and β-1,6-D-glucan macroproteins, chitin, proteins, and lipids, and among the polysaccharides, chitin, glucan, and mannan or galactomannan (Pontón 2008) prevail. The last factors that are accepted to affect the component of antimicrobial activity are the changes that can happen on the surface of ZnO NPs, in this way making it possibly responsive. Another approach to improve a better antimicrobial agent is to functionalize the surface of zinc oxide nanostructures. In such manner, if the surface area of the zinc oxide nanostructures is reformed, it could advance the generation of ROS and the release of ZnO, consequently expanding antimicrobial activity. Other than the previously mentioned applications, the mix of photoactive nanomaterials and microorganisms can then again be utilized as a part of different new fields later on. For instance, the good bacteria which fight off the pathogens can be designed with nanomaterials to upgrade the general antimicrobial impact.

11.4.4 MgO NPs

Magnesium oxide nanoparticles (MgO NPs) are extremely fixed, biocompatible, and exceptionally efficient as an antibacterial agent. The formation of reactive oxygen species (ROS) mechanism has been proposed to explain the antimicrobial mechanism of MgO nanoparticales, the interaction of nanoparticles with bacteria, subsequently leads to damaging the bacterial cell (Tang and Bin-Feng 2014). It has been described that the development of the surface area of MgO particles prompts a rise of the O_2^- concentrations in solution and thus results in a more effective damage to the cell wall of the bacteria (Sawai et al. 2000; Yamamoto et al. 2000). It was proposed that the cell death was caused by the electrostatic contact between the bacterial cell surface and MgO nanoparticles. MgO nanoparticles showed high bactericidal effect against microbes because of the association of particles and bacteria (Peter et al. 2002; Makhluf et al. 2005). It was discovered that MgO nanoparticles could take up halogen gases because of the defective nature of their surface and its positive charge, which resulted in a strong interaction with bacteria, which are negatively charged (Stoimenov et al. 2002). Conversely, non-ROS-interceded bacterial toxicity was additionally found in MgO nanoparticles, proposing that oxidative stress would not be the mode of action for cell death (Leung et al. 2014). The alkaline effect has been considered as another primary factor in the antibacterial action of MgO nanoparticles (Sawai et al. 2001; Yamamoto et al. 2000). The possible antibacterial mode of action was the preservation of water humidity on the MgO nanoparticle surfaces, which could frame a thin water layer around the nanoparticles. The local pH of this thin water layer formed around the nanoparticles might be much higher than its equilibrium value in the solution. At the point when the nanoparticles are in contact with the bacteria, the high pH in this thin surface layer of water could harm the cell membrane, resulting in cell death (Sawai et al. 1997).

These antibacterial mechanisms of MgO NPs are dissimilar to the membrane lipid peroxidation caused by oxidative stress, based on the following three points:

1. When the bacterial cell membrane is broken and surface pores are visibly clear, MgO NPs are not detected in the cell. Moreover, no extreme Mg ions are visible in energy-dispersive X-ray spectroscopy spectra. Hence, the inhibitory impact of MgO harms the cell membrane.
2. Only one kind of MgO NP can identify little measures of ROS; the other two cannot.
3. Lipopolysaccharide (LPS) and phosphatidylethanolamine (PE) in the cell wall are not significantly changed by MgO NP treatment, which shows that MgO does not cause lipid peroxidation in the cell membrane (Leung et al. 2014; Wang et al. 2017a).

The antifungal impact of MgO NPs on a few fungal pathogens like *Alternaria alternata*, *Fusarium oxysporum*, *Rhizopus stolonifer*, and *Mucor plumbeus* detailed the most noteworthy impact utilizing the 30 and 50 nm nanoparticle size (Wani and Shah 2012). MgO nanoparticles have a good antibacterial effect against three essential food-borne pathogens. The interaction of nanoparticles with bacterial cells causes cell membrane outflow, stimulates oxidative stress, and at last prompts cell death (He et al. 2016). MgO nanoparticles induced systemic resistance in tomato against bacterial wilt disease. The quick fabrication of O2 or phenoxyl radicals in tomato roots treated with MgO NPs may assume a related part in the plant resistance response of tomatoes against *Ralstonia solanacearum* (Imada et al. 2016). Several mechanisms for the action of MgO nanoparticles on bacteria are the following: (1) MgO nanoparticles continuously produce a specific level of H_2O_2 while in suspension, which induces oxidative stress in cells; (2) physical interaction between nanoparticles and bacterial cell surface interrupts bacterial membrane integrity and causes membrane leakage; (3) higher concentrations of nanoparticles prompt serious membrane damage, cell content release, irreversible oxidization of biomolecules (e.g., DNA, proteins, and lipids), and eventually cell death (He et al. 2016).

11.4.5 Magnetic Nanoparticles (MNPs)

Nanotechnology application in plant disease management in the early stage. Magnetic nanomaterials could be used for site-targeted delivery of systemic nano-agrochemical plant protection, for improving plant disease resistance, increasing effective nutrient consumption, and improvement of plant growth (Nair and Kumar 2013). Fe_2O_3 NP-coated seeds have indicated enhanced seed germination and root and shoot lengths of *Solanum lycopersicum*. Based on the improved Fe2+/Fe total ratio found in iron extracted from dry biomass of the plant, we affirm the take-up of Fe_2O_3 NPs and their internalization and/or biomineralization in the plant body (Shankramma et al. 2016). Furthermore, superparamagnetic iron oxide could be able to destroy macromolecules, including DNA, lipids, and proteins, through the Fenton reaction, leading to bacterial death. Iron increases the generation of ROS through oxidative stress and promotes the electron transport chain to produce superoxides (O^{-2}), which destroy the iron clusters (Leuba et al. 2013). Consequently, more divalent iron participates in the oxidation in Fenton reaction, leading to the generation of more hydroxyl radicals (·OH) and stimulating the death of residual bacteria through the catabolism of the carbon source and the formation of nicotinamide adenine dinucleotide. The intake of superparamagnetic iron may also increase simultaneously because of functionalized polycarboxylate. The uncovering of magnetic fields to attract nanoparticles can change the organic movement and raise the bacteriostaticity of these nanocomposites in bacterial medium. Momentary nanomagnets may fill in as a helpful model framework to apply electromagnetic interactions of nanoparticles in agrobiology system (Sadjad et al. 2017). Fe_3O_4/ZnO/AgBr nanocomposites with diverse weight ratios of Fe_3O_4 to ZnO/AgBr were prepared using a facile microwaved-assisted technique, and their antifungal activities were investigated against *Fusarium graminearum* and *Fusarium oxysporum*, the causal agents of wheat head blight and lentil-vascular wilt diseases, respectively. Magnetic nanocomposites completely deactivated *F. oxysporum* after 60 min, which is a shorter time duration than for *F. graminearum* (Hoseinzadeh et al. 2016). The prepared magnetic nanocomposites could be used as an effective nanofungicide in plant pathology applications.

11.4.6 Ni NPs

The antimicrobial activity of nickel nanoparticles (Ni NPs) depends on the formation of ROS and release of nickel ions Ni(II). Enhanced leakage of proteins from Ni NPs treated bacterial and fungal cell membranes into culture medium is due to generation of free radicals from NiGs surface that induced membrane damage and leaked membrane and cellular contents (Choi and Hu 2008; Pandian et al. 2016). Stimulation of ROS synthesis leads to the generation of highly responsive radicals that destroy the cells by damaging cell membranes, proteins, DNA, and intracellular system (Kim et al. 2011; Jyoti et al. 2007). The Ni NPs can affect the quantity of lactate dehydrogenase, an important cytoplasmic enzyme (Pandian et al. 2016). Generally ROS generation has been suggested as a mode of action that clarifies the phytotoxicity effect of metal oxide NPs in the microbial cell. In the mitochondria of cells, ATP is produced by reduction of molecular oxygen to water through a sequence of attached proton and electron transfer reactions. Amid this procedure, the low level of oxygen is not reduced totally, bringing about the arrangement of superoxide anion radicals and therefore other oxygen-containing radicals. Hence, ROS are by-products of cellular oxidative metabolism, much of which happens in the mitochondria. Naturally released significant ROS include superoxide anion radicals, hydroxyl radicals, singlet oxygen, and hydrogen peroxide (H_2O_2) (Yin et al. 2012). The antifungal efficacy of Ni NPs against *Fusarium* wilt of tomato and lettuce was studied under in vitro and in vivo assay (Fig. 11.2). The Ni NPs suppressed the fungal growth of *F. oxysporum* f. sp. *lactucae* and *F. oxysporum* f. sp. *lycopersici* by 60.23 and 59.77%, respectively, at 100 ppm concentration compared with control (Ahmed et al. 2016). Nickel and cobalt ferrite nanoparticles ($NiFe_2O_4$ and $CoFe_2O_4$) are effectively confirmed for antifungal activity against three fungal plant pathogens: *Fusarium oxysporum*, *Colletotrichum gloeosporioides*, and *Dematophora necatrix*. Furthermore, it is also investigated that these ferrite nanoparticles decrease disease incidence of *Fusarium* wilt in pepper (Sharma et al. 2017).

11.4.7 Carbon-Based NPs

Carbon-based nanomaterials, for instance, carbon nanotubes (CNTs), single-walled carbon nanotubes (SWCNTs), multiwalled carbon nanotubes (MWCNTs), graphene oxide (GO) nanoparticles, and fullerenes, presented prospective antimicrobial activities (Wang et al. 2014; Dizaj et al. 2015). A specific investigation evaluated the microbial effects of carbon nanotubes and fullerenes on some pesticide uptake

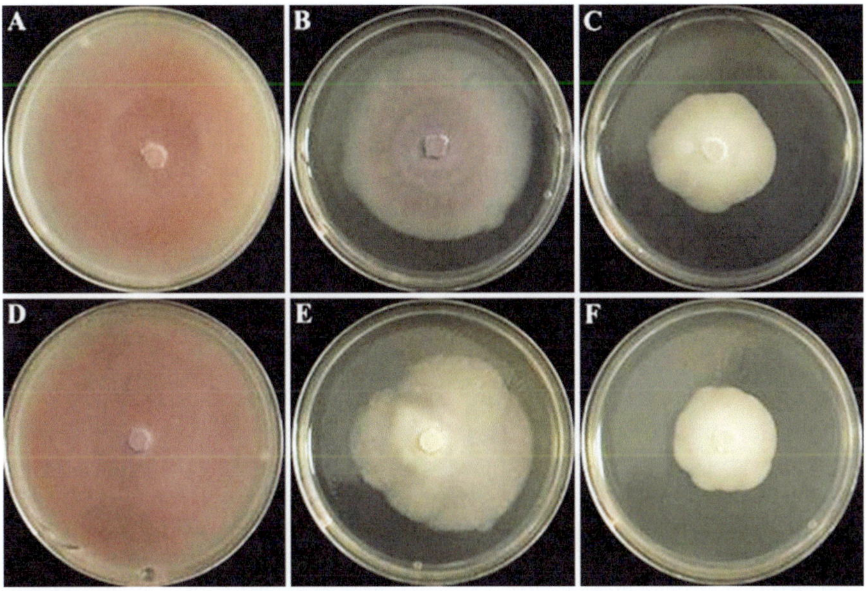

Fig. 11.2 Zone inhibition of *Fusarium* wilt pathogens. First row, *Fusarium oxysporum* f. sp. *lycopersici* (**a**, control; **b**, 50 ppm nickel nanoparticles; **c**, 100 ppm nickel nanoparticles); second row, *Fusarium oxysporum* f. sp. *lactucae* (**d**, control; **e**, 50 ppm nickel nanoparticles; **f**, 100 ppm nickel nanoparticles). (Reprinted from Ahmed et al. 2016)

by agricultural plants (De La Torre-Roche et al. 2013). The CNTs' antimicrobial mode of action is not totally clear. Former reports on CNTs divided the antimicrobial mechanism into two major kinds (Kang et al. 2007; Li et al. 2014). The first one is physical interaction, which includes membrane leakage or cell growth inhibition caused by interactions of the cell or cell membrane with the GNPs. The second category contains chemical reactive ions giving rise to the formation of reactive oxygen species (ROS). One of the most interesting, and the simplest, mechanism is the mechanical damage of bacterium cell envelope by some carbon forms. With the cell membrane damage resulting from direct contact with SWCNT, the membrane damage leads to leakage of intracellular materials (e.g., cytoplasm, ribosomes, and nucleic acids), which will eventually lead to cellular death (Kang et al. 2007; Jastrzębska et al. 2012). This mode of action will be affected significantly by some characteristics of the carbon nanomaterial, such as size, contact time, concentration, functionalization, and others. Another important factor affecting the antimicrobial efficacy of CNTs is emanated from their electronic structure.

Different investigational assays discovered that physical damage of pathogens resulted from their interaction with graphenes by two potential mechanisms: by extreme insertion and breaking of the cell membrane and by damaging extraction of phospholipids from lipid membranes (Zhou and Gao 2014). The oxidative stress mechanism has been suggested as a major cytotoxicity mechanism of graphene (Roda et al. 2014). Reactive oxygen species (ROS) are formed by GO, which would affect microorganisms sustainability. These ROS contain hydrogen peroxide, superoxide anion radicals, singlet oxygen, hydroxyl radicals, and nitric oxide. To help ensure against the dangerous impacts of ROS, oxygen-consuming life-forms and facultative anaerobic microorganisms deliver defensive cell reinforcement catalysts such as catalase, superoxide dismutase, and glutathione peroxidase. Catalases are proteins that catalyse the conversion of Hydrogen Peroxide (H_2O_2) to water and molecular oxygen, thereby protecting cells from the toxic effects of hydrogen peroxide Catalases are produced by all microorganisms utilized as part of this examination with the exception of *S. faecalis* which is microaerophilic (Roda et al. 2014). The antibacterial property of graphene does not come from ROS-initiated harm but rather through electron transfer communication from microbial membrane to graphene (Li et al. 2014). Other, more changeable mode of action for antimicrobial activity has been shown for graphene-based structures too. But previously mentioned disturbing the integrity of the cell wall (mechanical harm of the cell), these materials can (1) wrap around the microorganisms isolating them from the agro-system, (2) produce hurtful reactive oxygen species (ROS), (3) remove phospholipid atoms of the microorganisms by the presence of the lipophilic graphene, and (4) lower the metabolic activity of the bacterial cells (Akhavan et al. 2011; Liu et al. 2011a; Gurunathan et al. 2012; Sawangphruk et al. 2012; Tu et al. 2013; Chen et al. 2014).

There are not many examples of the antifungal mechanism of the graphene family materials in literature. According to Sawangphruk et al. (2012), inhibition of the mycelia growth results perhaps from direct interaction of rGO nanosheets with the cell wall (*Aspergillus niger*, *Aspergillus oryzae*, and *Fusarium oxysporum*). Researchers propose that there is a chemical reaction between oxygen (from rGO) and polysaccharides (e.g., chitin) in the wall. The six CNTs involved, SWCNTs, MWCNTs, graphene oxide (GO), reduced graphene oxide (rGO), fullerene (C60), and activated carbon (AC), were studied against two fungal pathogens: *Fusarium graminearum* and *Fusarium poae* (Wang et al. 2014). The SWCNTs had the maximum antifungal efficacy tracked by MWCNTs, GO, and rGO. C60 and AC showed no important antifungal activity. The antifungal properties of MWCNTs with diverse surface groups against *F. graminearum* were examined by Wang et al. (2017b). As per their discoveries, spore germination was strangely suppressed by surface-modified MWCNTs, with germination rate being 18%, threefold lower than for pristine MWCNTs.

The antifungal mode of action was assumed to target the spores in three steps: (1) depositing on the surface of the spores, (2) inhibiting water uptake, and (3) inducing plasmolysis. Different reports asserted no antifungal efficacy of GO toward *Candida albicans* and *Candida tropical* (Li et al. 2013b; Cui et al. 2014). The antimicrobial mode of action for fullerenes is still under open discussion. Precisely, fullerenes and their subordinates have exhibited effective antibacterial action against a wide range of microorganisms when presented to light (Chen et al. 2016). Fullerenes can integrate light and along these lines create reactive oxygen species (Kleandrova et al. 2015). Other possible mechanisms have additionally been accounted for, including impact on respiratory chain, disturbance of the cell membrane structure

(Cataldo and Da Ros 2008), interaction with membrane lipids, and intercalation into them (Cataldo and Da Ros 2008; Dizaj et al. 2015). The antimicrobial property of fullerene is likewise influenced by the size and surface area of it (Dizaj et al. 2015) and the form of functional group used (Li et al. 2012a). CNTs are unable to substitute/compete with the currently used antimicrobial materials (e.g., polymers, Cu NPs, and Ag-NPs) for many reasons, for instance, their toxicity profile for human cells has not been well addressed yet (Al-Jumaili et al. 2017). Currently, most antimicrobial carbon nanomaterials are still lack research/development.

11.4.8 Cu NPs

Furthermore to the control of growth of yeasts and molds,copper nanoparticles have also found to be effective against Gram-positive and Gram-negative bacteria, the activity of copper oxide nanoparticles (100–150 nm) coated to fabric showed 100 % reduction of *E. coli*, *S. aureus*, and *Aspergillus niger* after 48 h of incubation (Schrand et al. 2010; Theron et al. 2008; Usha et al. 2010). Copper nanoparticles gained position as innovative antimicrobial agents due to their high antimicrobial activities against widespread microorganisms including multidrug-resistant organisms. Similarly, copper is economical and simply obtainable, therefore synthesis of copper nanoparticles is cheap. One more advantage of copper nanoparticles is that they oxidize and form copper oxide nanoparticles, which can simply mix with polymers or macromolecules to produce nanocomposites, and are relatively stable in terms of both chemical and physical properties (Cioffi et al. 2005; Usman et al. 2012). Cu nanoparticles (Cu NPs) can penetrate the cell directly through the pores present in cell membrane due to their small size, or they enter through ion channels and transporter proteins present in the plasma membrane. Nanoparticles which are introduced into the cell can have direct contact with oxidative organelles such as mitochondria. Furthermore, redox-active proteins can stimulate reactive oxygen species (ROS) production in cells, and ions (Cu^{2+}) produced by nanoparticles can induce ROS by several chemical reactions. Also, Cu^{2+} ions have the ability to form chelates with biomolecules or remove the metal ions in specific metalloproteins, which may result in functional protein inhibitions. Cu^{2+} released by copper oxide nanoparticles increases their local concentration and disrupts cellular metal cation homeostasis resulting in cell toxicity (Chang et al. 2012). Linoleic acid capped copper nanoparticles after penetrating into the bacteria deactivate their enzymes, generating hydrogen peroxide resulting from ROS, which leads to bacterial cell death (Das et al. 2010). Schrand et al. (2010) hypothesized that copper nanoparticles act as actual antibacterial agent against the wide range of bacterial pathogens due to interactions with -SH groups, leading to protein denaturation. ROS may bind with DNA molecules and interrupt the helical structure by cross-linking within and between the nucleic acid strands and affect gene expression (Fig. 11.3). Copper ions inside bacterial cells also disrupt biochemical progressions (Kim et al. 2000; Stohs and Bagchi 1995). Still there is an absence of definite data regarding the mechanism

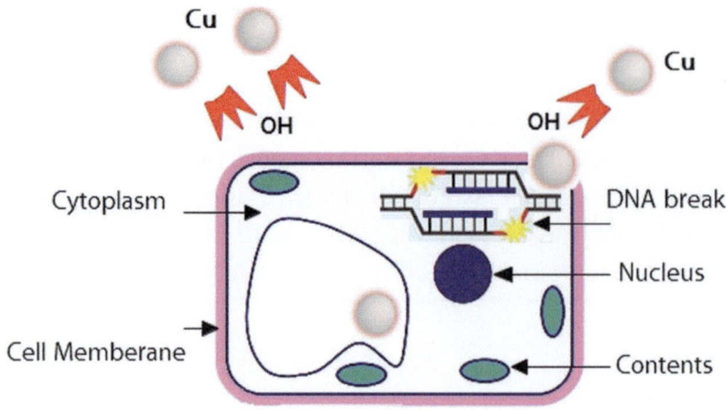

Fig. 11.3 Fungal cell illustration showing the different mechanisms associated with the toxicity of micro- or nano-copper against phytopathogenic fungi. Copper can form both free radicals disrupting the cell membrane and release ions able to produce genotoxic effects with proteins and DNA molecules

of action; however till now, completely different activity pathways are advised for the two copper oxides (I and II), with the involvement of ROS primarily within the case of CuO NPs and therefore the specific binding of Cu(I) to macromolecule surfaces for Cu_2O NPs. However, the exact mechanism behind bactericidal effect of copper nanoparticles is not known and needs to be further studied on the broader range of bacteria strains.

11.4.9 Al NPs

The marketable use of aluminum oxide nanoparticles Al_2O_3 extremely increased during the last decade, and, by no means, it enhances the risk of environmental pollution. Additionally, the toxic effects of Al_2O_3 NPs are also described on some model organisms for ecotoxicity assays, such as *Bacillus subtilis*, *Escherichia coli*, and *Pseudomonas fluorescens* (Jiang et al. 2009: Sadiq et al. 2009). There are chronic toxicity studies on Al_2O_3 nanoparticle exposure that cause neurotoxic effects on locomotion behaviors by prompting more ROS generation and interruption of ROS defense mode of actions in nematode *Caenorhabditis elegans* (Li et al. 2012b). The variation in cytotoxicity between micron-sized and nanosized alumina nanoparticles toward *Scenedesmus* sp. and *Chlorella* sp. was investigated. The antialgal inhibitory effect of the nanoparticle was studied against both the species, and an evident reduction in the chlorophyll content was also investigated in the cells treated with nanoparticles (Sadiq et al. 2011). Alumina nanoparticles showed a mild growth-inhibitory effect, only with very high concentration, which might be due to surface charge interactions between the particles and cells. Free-radical scavenging properties of the particles prevented cell wall disruption and drastic antimicrobial action (Sadiq et al. 2009). Alumina nanoparticles (Al_2O_3 NPs) showed a minor development inhibitory effect, so to speak, with extremely high focus, which may be because of surface charge naturally occurring between the particles and cells. Free-radical scrounging properties of the particles have foreseen cell wall disturbing influence and extraordinary antimicrobial action (Ruparelі et al. 2008; Sadiq et al. 2009). SEM images of nanoalumina-interacted cells of *E. coli* indicate the changes in cell shape and agglomerated particles on the cell wall. Moreover, TEM micrographs show disruption and disorganization of cell membrane and cell wall (Ansari et al. 2013). The cell membrane was widely injured and, most probably, the intracellular content has leaked out. Al_2O_3 carry a positive charge on its surface, electrostatic interaction between bacteria and NPs results in the adhesion of them on the bacterial surface and expressed antimicrobial activity. Al_2O_3 NPs not only adhered at the surface of cell membrane, but also penetrated inside the bacterial cells, cause formation of irregular-shaped pits and perforation on their surfaces and may also interact with the cellular macromolecules causing adverse effect including cell death. However more investigations as regards to the connection of alumina nanoparticles with cells should be done before its broad use in restorative and horticultural and crop application.

11.4.10 Au NPs

The antimicrobial property of Au NPs has been confirmed in various microbes, including Gram-positive bacteria, Gram-negative bacteria, and some pathogenic fungi. The green-synthesized gold nanoparticles (Au NPs) (45–75 nm) act as an active antifungal agent against wheat stem rust caused by *Puccinia graminis tritici*, and other fungal pathogens including *Aspergillus flavus*, *Aspergillus niger*, and *Candida albicans* using standardized well diffusion technique and hence have a great prospect in the preparation of fungicides used against different plant diseases (Jayaseelan et al. 2013). This action is credited to unique properties of Au NPs in illumination centering, solid cationic attractions to the negatively charged plasma layer of organisms, or conjugation with antimicrobial agents and antibodies.

In another way, it can bind to the DNA of microorganisms and repress DNA transcription (Rai et al. 2010). Gold NPs, 4,6-diaminopyrimidine thiol as a simple of bacterial tRNA base, has potential capability to prevent the subunit of ribosome for tRNA which influence on its capacity that affect protein synthesis (Carbon and David 1968; Sayed et al. 2006; Zhao et al. 2010). Au NPs can change membrane potential and suppress ATP synthase activities to decrease the ATP level, indicating a general decrease in metabolism, and also improve chemotaxis in the early-stage reaction (Cui et al. 2012). Those properties prompt cell layer disturbance, ROS collection, hindering DNA translation, and subsequent cell demise (Huh and Kwon 2011). In the first mechanism, Au nanoparticles generate holes in the cell wall, resulting in leakage of the cell contents, formation of biofilm, and finally cell death (Chwalibog et al. 2010). A second mechanism had been suggested that the strong electrostatic attractions among Au NPs and the cell wall surface of the pathogens introduce adhesin-mediated interaction between the pathogenic cells and the substrate surfaces (Yu et al. 2016).

11.4.11 Bimetallic NPs

Bimetallic nanoparticles made out of two kinds of metal components and metallic nanoparticles can be sorted as bimetallic or trimetallic relying upon the quantity of segment metallic fixings, for example, metal oxide NPs (Cu, Mg, Zn, and Ag) (Roopan et al. 2014). A synergistic antimicrobial effect is achieved when silver nanoparticles are hybrid with other metal nanoparticles or oxides acting as a shell or a core to form bimetallic nanoparticles (Chou and Chen 2007). The superparamagnetic bimetallic Ag/Co polymeric nanocomposite was evaluated to exhibit bactericidal activity during treatment of bacteria-contaminated aqueous solutions (Alonso et al. 2011). The Fe-Ag NPs showed high antimicrobial activity against *E. coli* (Gram-negative bacteria). Cu-chitosan and Zn-chitosan nanocomposites (NCs) were prepared by reduction of metal precursors in the presence of chitosan in sc CO2 medium and deposition of organosol on chitosan, respectively. Inorganic bimetallic blends (BBs) in light of understood fungicide nanoscale $Cu(OH)_2$ were acquired with the basic properties of salt hydrolysis. The BBs and Cu-chitosan demonstrated the most astounding antifungal adequacy against both *R. solani* anastomosis gatherings. The in vivo assessment of Cu-chitosan NC and *Trichoderma* hybrid with BBs indicated plant growth promotions and synergistic inhibitory impact against *R. solani.*

This exploration could prompt the likelihood of applying Cu-chitosan NCs, BBs and *Trichoderma* as nanobiofungicides at the field level. The most astounding group of BBs and NCs influenced DNA molarity and resulted in significant degradation. Copper particles discharged may likewise cooperate with DNA atoms and intercalate into nucleic corrosive strands. Cu nanoparticles degrade DNA in a single oxygen-mediated fashion even in the absence of any external agents like hydrogen peroxide or ascorbate. Low-molecular weight chitosan can enter cell dividers and interface with cell DNA of growths and microorganisms which therefore hinders mRNA interpretation and protein production (Abd-Elsalam et al. 2017). The antimicrobial mechanism of bimetallic might the ROS generation and cell wall damaged. Therefore, the combination of this metal oxide can be enhanced antimicrobial action (Liu et al. 2012; Vidic et al. 2013).

11.4.12 Chitosan NPs

Chitosan is a nontoxic, biodegradable biopolymer showing antimicrobial and plant immunity-eliciting properties. Nanochitosan has been shown to be useful in many different areas, specifically in agriculture, plant pathology, food, and biomedicine (Cota-Arriola et al. 2013; El Hadrami et al. 2010; Saharan et al. 2015; Abd-Elsalam et al. 2017). Chitosan stimulates various plant responses, including induction of disease and abiotic stress resistance, enhancement of plant growth and yield and shelf life of flowers and fruits, and activation of secondary metabolite production (Pichyangkura and Chadchawan 2015). Chitosan antimicrobial activities, mechanism, and induction of plant defense responses were reviewed and discussed (Xing et al. 2015). Chitosan demonstrated antimicrobial activities against different plant pathogens including parasites, microorganisms, and fungi and goes about as an elicitor of plant barrier systems. With a wide range of antimicrobial effects, chitosan has been used to reduce or keep the spread of pathogens (Mansilla et al. 2013) or to upgrade plant intrinsic resistance (Fondevilla and Rubiales 2012).

The correct mode of action for antimicrobial activity of chitin, chitosan, and their derivatives is as yet unclear, notwithstanding extraordinary systems that have been proposed. The positively charged amino gatherings of the glucosamine units connect with negatively charged particles on pathogen surfaces, which is named as electrostatic collaborations, and can destroy the cell structure, cause direct cell surface modifications, and increase membrane permeability and in this manner cause the demise of microscopic organisms (Helander et al. 2001; Rabea et al. 2009; Chung et al. 2004; Liu et al. 2004; Zakrzewska et al. 2005; Je and Kim 2006; Chung and Chen 2008).

The growth inhibition of *F. oxysporum* as a response to chitosan was accompanied by marked cellular changes, which included hyphal swelling, increased vacuolation, retraction and alteration of the plasma membrane, cytoplasm aggregation, and irregular cell wall deposition (Benhamou 1992). In electron micrographs, the outer membrane of chitosan-treated *E. coli* was disrupted and covered by an additional toothlike layer. In micrographs of chitosan-treated *S. aureus*, the membrane of dividing cells was disrupted in the constricting region with the loss of bacterial cell substances (Liu et al. 2004). Furthermore the efflux of potassium particles was recognized as an early reaction of the cell to the nearness of some cationic mixes. A quick efflux of potassium dependent on the chitosan fixation was investigated. Furthermore, there was an important inhibitory effect of chitosan on H^+-ATPase activity in the plasma membrane of *Rhizopus stolonifer*. The decrease in the H^+-ATPase's activity could provoke the accumulation of protons inside the cell, which would result in the suppression of the chemiosmotic-driven transport that allows the H^+/K^+ exchange (García-Rincóna et al. 2010).

A parallel confirmation of method of activity of chitosan has been shown in view of the interactions with DNA or RNA. Chitosan with low molecular weight can penetrate cell wall and interact with cellular DNA of fungi and bacteria which consequently prevents RNA and protein synthesis (Sudarshan et al. 1992; Goy et al. 2009), destroys intracellular components from colloidal state to flocculation and degeneration, disrupts the normal physiological metabolic activity of bacteria, or directly interferes with genetic materials (Come et al. 2003; Issam et al. 2005), and then stops the reproduction of bacteria, resulting in the death of microorganisms eventually. It is presumable that chitosan could bind with DNA and inhibit synthesis of messenger RNA (mRNA) through penetration toward the nuclei of the microorganisms and interfere with the synthesis of mRNA and proteins (Sudarshan et al. 1992; Rabea et al. 2009).

A report used fluorescence visualization to determine that oligochitosan can penetrate the cell membrane of *Phytophthora capsici* and *A. niger* and that, as it is positively charged, chitosan can bind to intracellular targets, such as DNA and RNA, which are negatively charged (Li et al. 2008). Infiltrated chitosan oligomers (molecular weight = 8000 and 5000) were suggested to block the transcription from DNA to inhibit the growth of bacteria (Liu et al. 2001) and then disrupt the related protein synthesis. The phosphate group might be an extracellular target contributing to its interaction with the positively charged chitosan, ultimately resulting in damage of vital bacterial activity. There are also phosphate groups in the primary structure of nucleic acid (DNA/RNA). It is possible that the amino groups of chitosan that possess positive charges would attract the negatively charged phosphate groups of DNA/RNA. The brightness of bands weakened gradually as the concentration of chitosan nanoparticles increased, showing the aggravation of chitosan-DNA/RNA interactions. The possible reason might be that negative charges of DNA/RNA had been counteracted by chitosan so that they could not move in the electric field accordingly. The gel retardation experiment pointed out that DNA and RNA might be the intracellular targets of chitosan (Xing et al. 2009).

11.4.13 Elicitation of Plant Defense Responses by Chitosan

Many reports presented that chitosan is not only an antimicrobial agent but also an active elicitor of plant systemic induced resistance to pathogens (El Hadrami et al. 2010; Xing et al. 2015). Most investigations with respect to the utilization of chitosan on rural items center on diseases caused by parasites at preharvest and postharvest stages. The use of adjusted chitosan derivatives and the blend of chitosan with different substances have as of late been looked into somewhere else (Das et al. 2015). Management of plant diseases especially postharvest diseases by chitosan or essential oil treatments seems to happen through two various modes of action: a direct germicide effect on plant pathogens and an indirect effect by inducing defense mechanisms in plant tissue (Zhang et al. 2011; Shao et al. 2013). Plant resistance toward pathogens occurs through hypersensitive responses that result in cell death at the penetration site, structural alterations, accumulation of reactive oxygen species (ROS), synthesis of secondary metabolites and defense molecules, and activation of pathogenesis-related (PR) proteins (Van-Loon and Van-Strien 1999). Chitosan can increase pathogenesis-related (PR) quality capacity through different modes, which incorporates enactment of cell surface or layer receptors, and inner impacts on the plant DNA compliance, which can, thus, influence gene translation (Hadwiger 1999). Chitosan application has been mentioned to increase phenylalanine ammonia lyase (PAL) activity in treated fruit tissue. PAL elicitation via chitosan was established with table grapes in the vineyard sprayed with or without *C. laurentii* and covered with chitosan at postharvest, and then stored at 0 °C (Meng and Tian 2009; Meng et al. 2010). Chitinase and β-1,3-glucanase are two PR proteins that participate in defense against pathogens, because they can partially degrade the fungal cell wall (Van-Loon and Van-Strien 1999). Increases within the activities of chitinase and β-1,3-glucanase are evidenced as a result of chitosan application in Valencia oranges (Canale Rappussi et al. 2009). Chitosan remedy may start compliant resistance of fruit through regulation of ROS levels, inhibitor enzymes, and also the ascorbate-glutathione cycle. Changes in the content of ROS, such as H_2O_2 and O^{2-}, are the earliest events that correlate plant resistance to pathogens (Wang et al. 2014), since ROS are involved in the development of disease resistance in fruit (Torres et al. 2003). This may be because of direct effects, as chitosan itself has inhibitor activity and scavenges hydroxyl group radicals (Yen et al. 2008), or to indirect effects, like chitosan causation of the plant inhibitor system. Chitosan treatment has been reported to influence inhibitor catalyst activities within the tissues of fruits and vegetables. Compared to untreated strawberries, those treated with chitosan maintained higher levels of many defense-related enzymes, such as catalase, glutathione peroxidase, guaiacol peroxidase, polyphenol oxidase, superoxide dismutase, dehydroascorbate reductase, and monodehydroascorbate reductase (Wang and Gao 2012). Chitosan significantly improved the production of polyphenol oxidase activity in rice seedlings following infestation with two rice phytobacteria (*Xanthomonas oryzae* pv. *oryzae* and *X. oryzae* pv. *oryzicola*) (Li et al. 2013a). Secondary metabolites are not directly involved in growth or reproduction, but they are often involved with plant defense. Elicitation is a method widely used for improving secondary metabolite yields (Xing et al. 2015). The antimicrobial efficacy of chitosan is impacted by a variety of things that represent the kind of chitosan, the degree of chitosan polymerization, relative molecular weight, solvent, pH, its charges, and solubility (Tavaria et al. 2013). The five fundamental modes of action of chitosan are electrostatic interactions, plasma membrane harm mechanism, chitosan-DNA/RNA interactions, metal chelation potential of chitosan, and deposition onto the microbial cell membrane. The mechanisms of how chitosan acted on plant immune system have now not been elucidated virtually. It is assumed that the mode of motion of chitosan is probably greatly complicated than assumed above; similarly researches need to clarify the precise mechanism.

11.4.14 Chitosan Nanocomposites

However, much work has to be done regarding the mode of action of chitosan-based nanomaterials against plant pathogens (El Hadrami et al. 2010; Abd-Elsalam et al. 2017). It seems that seed treatment with chitosan NCs ends in extra induction of plant defense mechanisms as previously demonstrated by using Saharan et al. (2013, 2015). Chitosan nanoparticles express more affinity towards pathogen's outer membrane and thus easily enter into the pathogens' cell (Van et al. 2013). Interactions between undoubtedly charged nanochitosan molecules and the polyanionic structures of microbial cellular membranes result in destabilization of cell membrane. This induces the leakage of intracellular contents and subsequently causes death of pathogens. Disruption of protein synthesis and membrane destabilizations is in all likelihood primary and secondary modes of antimicrobial activity of chitosan (Marquez et al. 2013). Furthermore, nanochitosan mechanism entails the penetration of low-molecular weight chitosan into the mobile, binding to DNA and subsequently inhibiting RNA and protein synthesis. Chitosan has also been shown to prompt several protection methods in plant tissues and inhibit the production of pollutants and microbial growth. Nanochitosan upregulates the plant defense mechanisms which involve enzymes such as phenylalanine ammonium lyase, polyphenol oxidase, tyrosine ammonium lyase, and antioxidative enzymes SOD, CAT, and POD (superoxide dismutase, catalase, and peroxidase) (Ma et al. 2014; Katiyar et al. 2015). It also induces the hypersensitivity-related reactions in different plant species to keep away or put off the invading pathogen from the cell (Chandra et al. 2015). Chitosan has been used to control seed-borne fungi of flowers as an elicitor as opposed to a fast-acting toxic agent because it has been stated that chitosan can activate plant defenses in infected flora (Kaur et al. 2012). Cu-chitosan nanoparticles enhanced enzyme activities involved in plant defense by chitosan participants in the reactive oxygen species (ROS) scavenging system (Saharan et al. 2013; Saharan et al. 2015). The plant microenvironment becomes acidic as a result of mycotic contamination which leads to the breakup of nanostructure and discharge of Cu particles (Brunel et al. 2013). The released Cu ions produce reactive hydroxyl radicals to prevent fungal pathogens (Borkow and Gabbay 2005). Cu-chitosan nanonetwork was evident through a higher Cu accumulation in porous areas which supported the ionic and chelating interaction mechanism to inhibit enzymes and toxins used by fungal pathogens throughout pathogenesis (Vahabi et al. 2011). Cu nanoparticles and Cu-chitosan and Zn-chitosan NCs have nearly the same mode of action as Cu-chitosan nanocomposites, for instance, the production of ROS, and membrane disruption (Xie et al. 2011; Ingle et al. 2013). Similarly, zinc is an important micronutrient for plant growth and is absorbed by plants through diffusion and specific transporters in the form of divalent ions. Another important mechanism includes penetration of the chitosan oligomer into the cells of microorganisms which inhibits the growth of microbial cells by stopping the transcription of DNA into mRNA (Hernández-Lauzardo et al. 2011). Cu-chitosan nanocomposites could penetrate cell walls of fungi and bind to DNA or mRNA. Disruption of fungal metabolism and duplication should in the long run lead to pathogen demise. The common nanometals used as antimicrobial agents collectively with their mechanisms of action are summarized in Table 11.1.

11.4.15 Nanoemulsions Mechanism

The nanoemulsion (NE) droplets with antimicrobial agents fuse with lipid containing organism thereby destroying them by numerous modes of action. The fusion between the nanodroplets is driven by the electrostatic attraction between the droplet charge and the charge on pathogens. When certain amounts of droplets fuse with the pathogens, the active ingredient from the nanodroplets is released to the lipid membrane causing lysis and death of pathogens. They observed that the antimicrobial activity depended on the target microorganism and nanoemulsions with smaller

Table 11.1 Mechanisms and applications nanoantimicrobial

Nanomaterials	Plant protection application	Mode of action
Ag NPs	Antifungal, antibacterial, antiviral	By increasing membrane permeability, inhibiting DNA replication, promoting free radical production, and inhibiting signal transduction
TiO NPs	Antifungal	Generation of hydroxyl radicals leading to initial oxidative attack on microbial cell membrane damage to microbial membrane inhibits important biological processes such as semipermeability, respiration, and oxidative phosphorylation reaction
Zn NPs	Antifungal	ROS generation on the surface of the particles, zinc ion release, membrane dysfunction, and nanoparticles internalization into cell
MgO NPs	Antifungal, antibacterial, *agent*	Damaging the cell membrane and then causing the leakage of intracellular contents and death of the bacterial cells. Formation of reactive oxygen species (ROS) mechanism
Magnetic NPs	Antifungal, antibacterial, *agent*	Damage macromolecules, including DNA, lipids, and proteins, through the Fenton reaction, leading to bacterial death. Iron increases the formation of ROS through oxidative stress and stimulates the electron transport chain to produce superoxides
Cu NPs	Antifungal, antibacterial	Generate toxic effects by generating ROS, which disturbs amino acid biosynthesis and DNA
Au Nps		Strong electrostatic attractions to negatively charged bilayer of cell membrane, ROS production
Nanochitosan	Antifungal, antibacterial	Binding of chitosan to negatively charged bacterial cell surface increases microbial cell wall permeability, chelation of trace metals inhibiting enzyme activities, and microbial growth
Nanocomposites	Antifungal, antibacterial, antiviral *agent*	(1) absorption of metal ions in cells and disruption of DNA replication, (2) oxidative damage via the generation of reactive oxygen species (ROS) on the surfaces of the NPs, and (3) free metal ion toxicity arising from dissolution and accumulation of the metals in the bacterial membrane that cause changes of theirpermeability and dissipation of the proton motive force
CNTs	Antibacterial, antifungal agent	Disruption of membrane integrity by electrostatic forces Reactive oxygen species generation harms biologically or prompt DNA destruction
Graphene-related families		Damaging to the cell membrane destructive of phospholipids from lipid membranes. Oxidative stress through ROS generation. Separating microorganisms from their microenvironment
Fullerenes	Antibacterial pesticides	Destruction of cell membrane integrity Inhibition of energy metabolism chain Upon light illumination, fullerenes generally yield high rate of ROS
Nanoemulsion	Antimicrobial,	Disruption of cell membrane
Bimetallic NPs	Antibacterial, antifungal agent	ROS generation and cell wall damage Interact with cellular DNA of fungi and bacteria and inhibit mRNA transcription and protein synthesis

Fig. 11.4 Action mechanism of NE against spores. (Reprinted from Kaur 2016)

diameters showing better antimicrobial activity due to the fast delivery via the cellular membrane of the target pathogens (Donsì et al. 2011). This fusion among the emulsion and the anionic rate of the pathogen could bring about the antimicrobials' lysis and death. Due to this unique, nonspecific mode of action, there are no possibilities for development of resistant microbial strains (Karthikeyan et al. 2011; Moghimi et al. 2016). Strong electrostatic attraction could improve the fusion, and then nanoemulsions with positive charge exhibited higher antimicrobial activity (Hamouda and Baker 2000). Strong electrostatic attraction ought to improve the fusion, after which nanoemulsions with positive charge exhibited better antimicrobial activity (Hamouda and Baker 2000). Synergistic effect between one-of-a-kind antimicrobial agents is constantly taken into consideration to improve the antimicrobial activity of nanoemulsions. The basic theory behind these studies was that NE particles were thermodynamically driven. The anionic charge on pathogen and the electrostatic appeal between the cationic charges of the emulsion complements their mixing capability. The fusion of an adequate number of nanoparticles with pathogens assists in the release of some of the energy trapped within the emulsion (http://nano.med.umich.edu/platforms/Antimicrobial-Nanoemulsion.html). It is this trapped electricity and the actively worried components that weaken the pathogen lipid membrane leading to destruction of cells and their final dying. However, NEs, to be more effective in the case of spores, extra germination enhancers are required to be integrated with emulsion. The moment germination begins, the germination spores end up vulnerable to antimicrobial activity of the NE (Fig. 11.4). One peculiar aspect of NE is that concentrations exert selective toxicity on microbes.

11.4.16 Photocatalyst Mechanism

Other than the previously mentioned mechanisms, there are other unique antimicrobial mechanisms of nanomaterials that are available in the literature. One such unique mechanism is photocatalysis-mediated antimicrobial activity. Photocatalysis has been shown to be capable of killing an wide range of microorganisms together with bacteria, fungi, algae, and viruses (Paspaltsis et al. 2006; Foster et al. 2011). One such nanomaterial showing antimicrobial properties during photoactivation was nano-TiO_2, a semiconductor (Mueller and Nowack 2008). Moreover, ZnO is also a semiconductor that, upon absorption of photons, transported its electrons between the valence and conduction band. Not all the nano-based material had photo-mediated antimicrobial capacity. Just semiconducting nano metal oxides like TiO_2 and ZnO nanomaterials are found to have this sort of photocatalytic antimicrobial impact. As the component manages the arrangement of ROS and hinderd microorganisms through photocatalytic impact, broad investigations must be completed to decrease the cytotoxicity among higher living beings because of the ROS delivered by these nano products.

11.5 How to Investigate the Mode of Actions?

While composing and examining nanomaterials for their antimicrobial abilities, data is not just required to focus on nanoparticles but also on measure disseminations and shape and level of collection of the particles, making representation techniques vital to material characterization. Moreover, with a specific end goal to evaluate the antimicrobial efficacy of nanosized materials, representation of the association among microorganisms and the material is required, and the result of such connections on the feasibility of the microbial cell must be known. The nanocidal abilities of a mass material, which itself has no investigated antimicrobial effects, depend on supported oxidation systems and consequently on little amounts of the silver particle that are discharged into the fluid condition. The systems hostile to microbial efficacy should likewise be anticipated. The nanomaterial may straightforwardly harm the target cell membrane (of both prokaryotic and eukaryotic cells), or the impact might be because of compound activity or enhanced membrane permeability, bringing about spillage of cell substance or interruption of DNA replication. These effects can often be directly seen using the latest high-resolution microscopy methods. The stability and enduring efficacy of antimicrobial activity also depends largely on the properties of the nanomaterials once combined with the bulk matrix, particularly final particle size, shape, and availability (Marambio-Jones and Hoek 2010). Consequently, direct visualization of the particles incorporated in situ within the bulk matrix is also required, although this is also perhaps one of the most difficult to achieve without the introduction of significant sampling artifacts. A suite of high-resolution microscopy techniques are now readily available, and most bulk sample types, including fully hydrated samples, can now be visualized by a range of methods. However, given the wide-ranging nature of materials in which nanoparticles are now being incorporated, including polymers, powders, aerosols, and zeolites, no single technique will be able to provide all the required information. Rather, as pointed out by Samberg et al. (2011) in their investigation of the antibacterial impacts of silver nanoparticles, an examination of techniques is required, keeping in mind the end goal is to accomplish characterization of any antimicrobial activities.

A review of the literature indicates that the majority of studies published to date on the antimicrobial nature of nanomaterials rely on several methods to visualize such materials. Electron microscopy techniques such as transmission electron microscopy (TEM) are frequently coupled with a scanning probe microscopy

(SPM) technique such as atomic force microscopy (AFM). Without a doubt the current has intense far-reaching accessibility and adaptable methods, for example, AFM has altered characterization of nanosized materials over the most recent two decades, and it can be influentially contended that the accessibility of AFM and related strategies are eventually in charge of the quick development of nanotechnology investigation, by and large and particularly interface and colloidal science (Butt et al. 2007). Several imaging and molecular techniques will perform to evaluate the molecular mechanisms that underlie the microorganisms response to the nanomaterials (NMs).

11.5.1 Electron Microscopy Techniques

The field of electron microscopy covers a wide assortment of systems that can be used for imaging both nanomaterials and the mass material in which the nanomaterials are consolidated. An assortment of both auxiliary and concoction data can be inferred, despite the fact that EM overwhelmingly gives subjective basic data about the examples inspected. Three fundamental electron microscopy methods are utilized for material characterization, transmission electron microscopy (TEM), scanning transmission electron microscopy (STEM), and scanning electron microscopy (SEM). Together these three imaging techniques and their derivatives are capable of offering a numerous range of records on a selected pattern, ranging from sub-nanometer resolution as in the case of TEM to structural information on bulk materials many centimeters in measurement which can be achieved through scanning electron microscopes with specially designed chambers. The disadvantages of EM techniques include the fact that EM can only deliver statistics in dimensions and lacks distinct statistics on 3-D morphology of samples without specialized software. However, the principle negatively looks at the end result from the destructive nature of the sample-guided techniques required for EM and the tough imaging situations essential for a postive microscopy.

11.5.2 Transmission Electron Microscopy Measurements

The interaction between the microorganisms and the different nanomaterials can be illustrated using bright-field TEM imaging of the bacteria treated with various NMs. Irrespective of the type of bacteria used, it was noticed that most of the nanoparticles were found attached to the surface of the bacterial cell wall, implying their higher affinity toward the cells. Although TEM imaging provides a direct measure of nanoparticle interactions with the microorganisms, the potential for imaging artifacts cannot be eliminated.

11.5.3 Atomic Force Microscopy (AFM) Measurements

AFM can be used to photograph fully hydrated samples and might correctly photo each nanoparticles and microorganisms in situ on most surfaces, including both difficult and gentle surfaces. Unlike electron microscopy strategies or STM strategies, the pattern may be either a conductor or insulator and no staining is needed with a purpose to obtain evaluation. AFM may also offer future insights into nano-antimicrobial substances and the interest of such materials. One of the limitations of AFM is that chemical specificity is lacking and capabilities are recognized based on size and form. Another limitation is that while molecular scale surface topographic details are well resolved, AFM has a limited ability to image features that do not provide sufficient topographical contrast such as peptides in a membrane. In tandem with developments in electron microscopy and scanning probe microscopy methods, optical microscopy techniques continue to evolve rapidly, and techniques such as confocal microscopy provide unprecedented insights into microbial interactions with nanomaterials.

11.5.4 Confocal Microscopy

The primary concept of confocal laser scanning microscopy (CLSM) and multiphoton strategies are integral techniques for visible characterization of nanomaterial-based antimicrobial materials. Particularly, CLSM facilitated the exploration of microbial habitats and allowed the commentary of host-associated microorganisms in situ with an unprecedented accuracy (Cardinale 2014). External modifications in mobile membrane integrity can be monitored to unravel the mechanisms of antimicrobial interest and the modifications attributable to contact with nanomaterials.

11.6 Reactive Oxygen Species Generation

ROS are chemically reactive molecules together with peroxides that incorporate oxygen. ROS are equally reactive because of the presence of unpaired valence shell electrons. ROS form as a herbal derivative of the everyday metabolism of oxygen and have vital roles in mobile signaling, homeostasis, and furthermore apoptosis. However, during times of environmental stress, in the present case in the form of nanoparticles, ROS levels are known to increase drastically which might result in significant damage to cell structures. This cumulates into the event known as oxidative stress.

ROS production can be monitored using various analytical methods such as XTT assay that yields a colorimetric signal when reduced by superoxides. ROS are best known to implicate toxicity to several prokaryotic and eukaryotic systems upon interaction with metal/metal oxide nanoparticles. ROS in the form of either superoxide radical (O^{2-}), hydrogen peroxide (H_2O_2), or hydroxyl radical (OH) causes oxidative stress, thereby causing damage to DNA, cell membranes, and cellular proteins, and finally leading to cell death. The presence of ROS was observed using an XTT assay, which yields a colorimetric signal when decreased by superoxides. ROS quantification flow cytometric assay was used to evaluate the production of free intracellular radicals as reported (Raimondi et al. 2008; Prasad et al. 2017b).

11.7 Omics Methodologies

The suffix "omics" stands for "as a whole" and omics technology consists of genomics, transcriptomics, proteomics, and metabolomics credited to modern breakthroughs in genome sequencing, bioinformatics, and analytic equipment including liquid and gas chromatography and mass spectrometry, in conjunction with high-throughput technology (Fig. 11.5). Omics technologies have provided crucial insights into processes related to microbial physiology, virulence, and stress and mechanisms of action (MOA) of nanoantimicrobial materials (Tang 2015; Fröhlich 2017). These methods differ from the microscopic observation of phenotypes in the way that they can provide primarily mechanistic information and may identify the pathway of toxicity for microbial pathogens. One advantage might be the

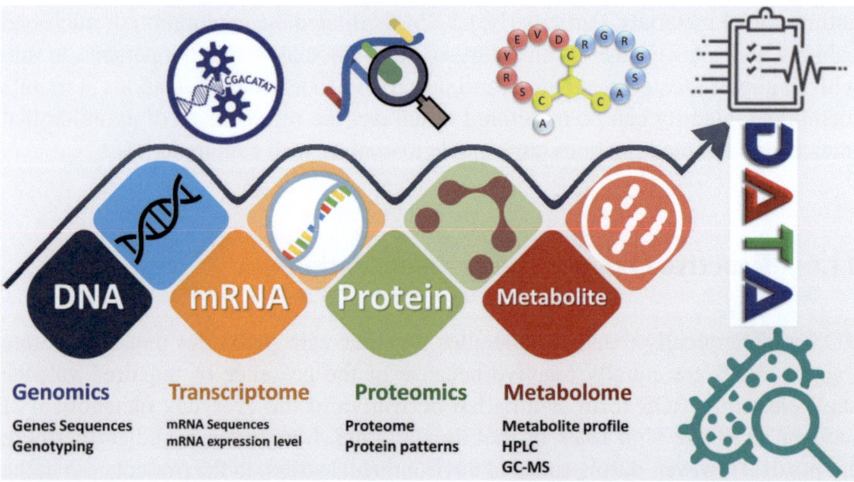

Fig. 11.5 Omics techniques used for the study of nanotoxicology include genomics, epigenomics (miRNomics and DNA modifications), transcriptomics, proteomics, and metabolomics. Genomics investigates genes and their functions through the use of recombinant DNA, DNA sequencing, and bioinformatics to analyze the function and structure of the genome. Proteomic and transcriptomic tools: target identification can be performed by evaluating the differential expression of genes in microbial strains treated or untreated with NMs; while metabolomic assay: the metabolic profile of a microbe treated with NMs can be compared with the profile of untreated strains

identification of new targets and markers for NP toxicity. Another benefit of the omics strategies might be their lower interference with NPs (Fröhlich 2017). DNA- and RNA-based research, genomics record analyses, transcriptomics, metagenomics, metabolomics, next-generation sequencing (NGS) technologies, and proteomics processes have proved to be precious techniques to examine plant-pathogen interactions and their associations. Various "omics" methods are a promising approach to recognize the advantages and the pathogenic effect of microbes in crop development. The plant native immunity has always been an important aspect of research and leads to some interesting information like the adaptation of unique immune mechanisms of plants against pathogens (Imam et al. 2016). Proteomics and metabolomics, pooled with systems biology, are outstanding tools to screen the results and toxicity mechanisms elicited by NPs. However, the metabolomics approach remains terribly difficult and still comparatively new in the field of nanotoxicology.

11.7.1 Genomics Assays

The mode of action of a nanomaterial compound on DNA integrity can be evaluated through DNA-binding evaluation, in which a pattern of purified plasmid DNA is blended with different concentrations of the examined compound. Pathways that might describe contrary effects on DNA are regulation of DNA destruction and repair of nucleic acid metabolism. Epigenetic fluctuations are involved in the transformation and mutation of cells and, thus, may assist as indicator for genotoxicity. With the advent of next genertions sequence technologies which results in the completion of genome sequencing and re-sequencing of the over-whelming numbers of plant and their pathogens generating huge amount of data, we are witnessing an era of genomics and post-genomics with a challenge to translate these plethora of information for the crop improvements with broader disease resistance spectra (Knief 2014). In the post-genomic era, the translational genomics presented a better solution in crop improvement against pathogenic bacteria, fungi, and viruses and prepare these crops in current thwarting climatic conditions (Knief 2014). DNA sequencing and bioinformatics investigate characteristic and shape of the genome. The purpose is to identify a selected sensitivity of people to a given toxin as opposed to the screening for toxicity of compounds or NPs. The epigenome may be altered via toxicants and, consequently, is useful for toxicity screening.

11.7.2 Transcriptomics Analysis

To evaluate the genetic-based response mechanisms, the global transcriptomic response of microorganisms upon exposure to nanomaterials can be assessed using whole-genome microarray analysis and compared to treatments with NMs or Milli-Q water control. Presently, DNA microarrays are the technology of choice for large-scale studies of gene expression. Microarray technology was developed using the information available from the genome projects and is based on the hybridization of cDNA (complementary DNA produced from mRNA) to oligonucleotide probes incorporated into a slide. Each probe has a sequence of a specific gene from the microorganism (Nambiar et al. 2010). Transcriptomics basically cognize DNA sequencing using NGS approaches and also quantitatively measure the expression of mRNA, and their variations occurring under diverse stress situations. The use of mRNA sequencing evaluation and microarray approach to generate transcriptome level facts is one of the important methodologies employed for studying plant-microbe interactions (Budak and Akpinar 2015). Furthermore, the development in bioinformatics methods, both as hardware and software program for the evaluation of statistics, allows us to enhance our information on ever-increasing management and mining of such big datasets. RNA-seq, a recently evolved approach for transcriptome profiling by way of deep-sequencing technologies, offers a precise dimension of the extent of gene transcripts and their isoforms than different strategies.

11.7.3 Proteomics Analysis

Proteomics is the systematic assessment of all proteins expressed through one particular mobile, tissue, organic fluid, or organism in a given time period. It is used to identify and quantify proteins targeted in selected biological parameters and also can be implemented to determine post-translational modifications, in addition to cellular origin and location of movement (Yates et al. 2009). Different methods that are being applied to promote our understanding of proteins are gel-based techniques like two-dimensional gel electrophoresis (2DE) and fluorescence two-dimensional difference gel electrophoresis (2DDIGE) and gel-free techniques like isotope-coded affinity tags (ICAT), isobaric tags for relative and absolute quantitation (iTRAQ), multidimensional protein identification technology (MudPIT), and the commonly used primary tool mass spectrophotometry (MS) and MALDI-TOF. High sensitivity is the advantage of MS.

11.7.4 Metabolomics Assays

The final "omics" technology to be provided in this review is metabolomics, which is defined as the examination of the global profile of metabolites found in a biological system under sure conditions and time (Nambiar et al. 2010). The approaches followed to generate metabolic signatures are usually nuclear magnetic resonance (NMR), high-performance liquid chromatography (HPLC), the combination of chromatography with MS that helps to detect more number of complex compounds. Widely used methods are combinations of gas chromatography (GC)-MS and liquid chromatography (LC)-MS (Zhang et al. 2012). Metabolomics techniques had been significantly accepted to discover the responses of microorganisms to numerous environmental stressors inclusive of heavy metals, temperature, and natural compounds (Lankadurai et al. 2013). Metabolomics may represent a promising approach to explain the mechanism of diverse nanomaterials. Datasets from diverse studies like genomics, transcriptomics, proteomics and metabolomics need to be combined using bioinformatics and statistical tools that will help to identify and integrate key biological processes.

11.8 Conclusion and Future Perspectives

Understanding the inactivation mechanisms is the key to the increasing use of nanoparticles (NPs) and enhances the practicability of their application aganist numerous plant pathogens under extraordinary environments. The potential mechanism of toxicity has been attributed to numerous possible techniques, and the dissolution or release of ions from the nanoparticles elicits either inflammatory response, mitochondrial dysfunction, disruption of mobile membrane integrity, oxidative strain, protein or DNA binding and harm, or reactive oxygen species (ROS) technology, affecting the proteins and phosphate lipids and in the long run inflicting cellular death. Particular emphasize become given to antimicrobial mechanisms with attention on formation of reactive oxygen species (ROS) consisting of hydrogen peroxide (H_2O_2), OH^- (hydroxyl radicals), and O_2^{-2} (peroxide). ROS has been a chief factor for several mechanisms consisting of cell wall harm due to NPs-localized interaction and more advantageous membrane permeability. Use of highly sophisticated techniques such as high resolution microscopic (AFM, FE-SEM, TEM, and XRD), spectroscopy (DLS, ESR spectroscopy, fluorescence spectroscopy, inductively coupled plasma optical emission spectroscopy, UV-vis), and molecular and biochemical techniques have provided deep mechanistic insights approximately in the mode of action of antimicrobial activity of AgNPs (Kim et al. 2007; Rai and Ingle 2012). The emerging discipline of nanotoxicogenomics (Waters et al. 2009) which attempts to correlate global gene expression profiles of cells or tissues exposed to NPs with organic/toxicological responses to the usage of cDNA microarray technology may offer beneficial information in this regard. Genotoxicity is a harmful impact which affects DNA integrity through the movement of harm-inducing markers (genotoxins) together with chemicals and radiation. There are many markers which can act as genotoxins, and they may be categorized as bodily (e.g., UV, X-rays), chemical (e.g., benzopyrene, ethidium bromide), and organic (e.g., virus, transposons). These agents can also be described according to their action mechanism such as oxidants (e.g, hydrogen peroxide), alkylating (e.g, methyl methanesulfonate), inductors of DNA breaking (e.g, ionizing radiation), and aneugenic agents which affect chromosome division (e.g, taxanes) (Parry and Parry 2012). As DNA damage may also both provoke and sell carcinogenesis or effect fertility, the genotoxicology technological know-how has become a crucial area governing regulatory health risk evaluation. Mechanistic investigations of these interactions may even enable a correct evaluation of fate and results within the crop or cropping system as a way to deal with issues over threat and meals safety. Great efforts have also been made to recognize the nanoantimicrobial mechanisms; researchers are still looking to recognize the mechanism of antimicrobial activity of nanostructures, and consequently they may be investigating the morphological changes in microbes as a result of nanoantimicrobials. No-clear cut conclusion for nanoantimicrobial mechanism through; damage through membrane disruption, DNA transformation, ROS production or other mechanisms, we still want for in addition research to find the exact mechanism.

Acknowledgment This research was supported by the Science and Technology Development Fund (STDF), Joint Egypt (STDF)-South Africa (NRF) Scientific Cooperation, Grant ID 27837 to Kamel Abd-Elsalam.

References

Abbaszadegan A, Ghahramani Y, Gholami A, Hemmateenejad B, Dorostkar S, Nabavizadeh M, Sharghi H (2015) The effect of charge at the surface of silver nanoparticles on antimicrobial activity against gram-positive and gram-negative bacteria: a preliminary study. J Nanomater ID 720654

Abd-Elsalam KA, Vasil'kov AY, Said-Galiev EE, Rubina MS, Khokhlov AR, Naumkin AV, Shtykova EV, Alghuthaymi MA (2017) Bimetallic and chitosan nanocomposites hybrid with trichoderma: novel antifungal agent against cotton soil-borne fungi. Euro J Plant Pathol. https://doi.org/10.1007/s10658-017-1349-8

Abkhoo J, Panjehkeh N (2017) Evaluation of antifungal activity of silver nanoparticles on *Fusarium oxysporum*. Int J Inf Secur 4(2):e41126. https://doi.org/10.5812/iji.41126

Ahmed IIS, Yadav DR, Lee YS (2016) Applications of nickel nanoparticles for control of fusarium wilt on lettuce and tomato. Int J Innov Res Sci Eng Technol 5:7378–7385

Akhavan O, Ghaderi E, Esfandiar A (2011) Wrapping bacteria by graphene nanosheets for isolation from environment, reactivation by sonication, and inactivation by near-infrared irradiation. J Phys Chem B 115:6279–6288

Al-Jumaili A, Alancherry S, Bazaka K, Jacob MV (2017) Review on the antimicrobial properties of carbon nanostructures. Materials 10:1066. https://doi.org/10.3390/ma10091066

Allahverdiyev AM, Abamor ES, Bagirova M, Baydar SY, Ates SC, Kaya F, Kaya C (2013) Investigation of antileishmanial activities of Tio2@Ag nanoparticles on biological properties of *L. tropica* and *L. infantum* parasites, in vitro. Experimental Parasitol 135:55–63

Alonso A, Vigués N, Muñoz-Berbel X, Macanás J, Muñoz M, Mas J, Muraviev DN (2011) Environmentally-safe bimetallic Ag@Co magnetic nanocomposites with antimicrobial activity. Chem Commun (Camb) 47:10464–10466

Ansari MA, Khan HM, Khan AA, Pal R, Cameotra SS (2013) Antibacterial potential of Al2O3 nanoparticles against multidrug resistance strains of *Staphylococcus aureus* isolated from skin exudates. J Nanopart Res 15:1970

Arciniegas-Grijalba PA, Patin˜o-Portela MC, Mosquera-Sa´nchez LP, Guerrero-Vargas JA, Rodrı´guez-Pa´ez JE (2017) ZnO nanoparticles (ZnO-NPs) and their antifungal activity against coffee fungus *Erythricium salmonicolor*. Appl Nanosci 7:225–241

Auffan M, Rose J, Bottero JY, Lowry GV, Jolivet JP, Wiesner MR (2009) Towards a definition of inorganic nanoparticles from an environmental, health and safety perspective. Nat Nanotechnol 4:634–664

Aziz N, Fatma T, Varma A, Prasad R (2014) Biogenic synthesis of silver nanoparticles using *Scenedesmus abundans* and evaluation of their antibacterial activity. J Nanopart:689419. https://doi.org/10.1155/2014/689419

Aziz N, Faraz M, Pandey R, Sakir M, Fatma T, Varma A, Barman I, Prasad R (2015) Facile algae-derived route to biogenic silver nanoparticles: synthesis, antibacterial and photocatalytic properties. Langmuir 31:11605–11612. https://doi.org/10.1021/acs.langmuir.5b03081

Aziz N, Pandey R, Barman I, Prasad R (2016) Leveraging the attributes of *Mucor hiemalis*-derived silver nanoparticles for a synergistic broad-spectrum antimicrobial platform. Front Microbiol 7:1984. https://doi.org/10.3389/fmicb.2016.01984

Baruah S, Dutta J (2009) Nanotechnology applications in pollution sensing and degradation in agriculture: a review. Environ Chem Lett 7:161–204

Benhamou N (1992) Ultrastructural and cytochemical aspects of chitosan on *Fusarium oxysporum* f. sp. *radices-lycopersici*, agent of tomato crown and root rot. Phytopathology 82:1185–1193

Bhuyan T, Mishra K, Khanuja M, Prasad R, Varma A (2015) Biosynthesis of zinc oxide nanoparticles from *Azadirachta indica* for antibacterial and photocatalytic applications. Mater Sci Semicond Process 32:55–61

Borkow G, Gabbay J (2005) Copper as a biocidal tool. Curr Med Chem 12:2163–2175

Brady-Est´evez AS, Kang S, Elimelech M (2008) A single walled-carbon-nanotube filter for removal of viral and bacterial pathogens. Small 4:481–484

Brayner R, Ferrari-Iliou R, Brivois N, Djediat S, Benedetti MF, Fievet M (2006) Toxicological impact studies based on *Escherichia coli* bacteria in ultrafine ZnO nanoparticles colloidal medium. Nano Lett 6:866–870

Brunel F, Gueddari NE, Moerschbacher BM (2013) Complexation of copper (II) with chitosan nanogels: toward control of microbial growth. Carbohydr Polym 92:1348–1356

Budak H, Akpinar BA (2015) Plant miRNAs: biogenesis, organization and origins. Funct Integr Genomics 15:523–531

Butt HJ, Berger R, Bonaccurso E, Chen Y, Wang J (2007) Impact of atomic force microscopy on interface and colloid science. Adv Colloid Interf Sci 133:91–104

Canale Rappussi MC, Pascholati SF, Aparecida Benato E, Cia P (2009) Chitosan reduces infection by *Guignardia citricarpa* in postharvest "Valencia" oranges. Braz Arch BioTechnol 52:513–521

Carbon J, David H (1968) Thiobases in *Escherichia coli* transfer RNA-2- thiocytosine and 5-methylaminomethyl-2-thiouracil. Science 161(3846):1146–1147

Cardinale M (2014) Scanning a microhabitat: plant-microbe interactions revealed by confocal laser microscopy. Front Microbiol 5:94. https://doi.org/10.3389/fmicb.2014.00094

Carré G, Hamon E, Ennahar S, Estner M, Lett MC, Horvatovich P, Gies JP, Kellerb V, Kellerb N, Andrea P (2014) TiO2 Photocatalysis damages lipids and proteins in *Escherichia coli*. Appl Environ Microbiol 80:2573–2581

Cataldo F, Da Ros T (2008) Medicinal chemistry and pharmacological potential of fullerenes and carbon nanotubes. Springer Science & Business Media, Dordrecht

Chandra S, Chakarborty N, Dasgupta A, Sarkar J, Panda K, Acharya K (2015) Chitosan nanoparticle: a positive modulator of innate immune responses in plants. Sci Rep 5:1–13

Chang YN, Zhang M, Xia L, Zhang J, Xing G (2012) The toxic effects and mechanisms of CuO and ZnO nanoparticles. Materials 5:2850–2871

Chaudhry Q, Scotter M, Blackburn J, Ross B, Boxall A, Castle L, Watkins R (2008) Applications and implications of nanotechnologies for the food sector. Food Addit Contam 25:241–258

Chen JP, Peng H, Wang X, Shao F, Yuan Z, Han H (2014) Graphene oxide exhibits broad-spectrum antimicrobial activity against bacterial phytopathogens and fungal conidia by intertwining and membrane perturbation. Nanoscale 6:1879–1889

Chen Q, Ma Z, Liu G, Wei H, Xie X (2016) Antibacterial activity of cationic cyclen-functionalized fullerene derivatives: membrane stress. Dig J Nanomater Biostruct (DJNB) 11:753–761

Choi O, Hu Z (2008) Size dependent and reactive oxygen species related nanosilver toxicity to nitrifying bacteria. Environ Sci Technol 42:4583–4588

Choi JY, Kim KH, Choy KC, Oh KT, Kim KN (2007) Photocatalytic antibacterial effect of TiO 2 fim formed on Ti and TiAg exposed to *Lactobacillus acidophilus*. J Biomed Mater Res Part B 80:353–359

Chou KS, Chen CC (2007) Fabrication and characterization of silver core and porous silica shell nanocomposite particles. Microporous and Mesoporous Mater 98:208–213

Chung YC, Chen CY (2008) Antibacterial characteristics and activity of acid-soluble chitosan. Bioresour Technol 99:2806–2814

Chung YC, Su YP, Chen CC, Jia G, Wang HL, Wu JC, Lin JG (2004) Relationship between antibacterial activity of chitosan and surface characteristics of cell wall. Acta Pharmacol Sin 25:932–936

Chwalibog A, Sawosz E, Hotowy A, Szeliga J, Mitura S, Mitura K et al (2010) Visualization of interaction between inorganic nanoparticles and bacteria or fungi. Int J Nanomed 5:1085–1094

Cioffi N, Rai M (2012) NanoAntimicrobials: progress and prospects springer. Springer-Verlag, Berlin, Heidelberg

Cioffi N, Torsi L, Ditaranto N, Sabbatini L, Zambonin PG, Tantillo G, Ghibelli LD, Alessio M, Bleve-Zacheo T, Traversa E (2004) Antifungal activity of polymer-based copper nanocomposite coatings. Appl Phys Lett 85:2417–2419

Cioffi N, Torsi L, Ditaranto N, Tantillo G, Ghibelli L, Sabbatini L (2005) Copper nanoparticle/polymer composites with antifungal and bacteriostatic properties. Chem Mater 17:5255–5262

Come V, Deschamps A, Mertial A (2003) Bioactive packaging materials from edible chitosan polymer-antimicrobial activity assessment on dairy-related contaminants. J Food Sci 68:2788–2792

Cota-Arriola O, Cortez-Rocha MO, Burgos-Hernández A, Ezquerra-Brauer JM, Plascencia-Jatomea M (2013) Controlled release matrices and micro/nanoparticles of chitosan with antimicrobial potential: development of new strategies for microbial control in agriculture. J Sci Food Agric 93:1525–1536

Cui Y, Zhao Y, Tian Y, Liu W, Ma W, Jiang X (2012) The molecular mechanism of action of bactericidal gold nanoparticles on *Escherichia coli*. Biomaterials 33:2327–2333

Cui J, Yang Y, Zheng M, Liu Y, Xiao Y, Lei B, Chen W (2014) Facile fabrication of graphene oxide loaded with silver nanoparticles as antifungal materials. Mater Res Express 1:045007

Das R, Gang S, Nath SS, Bhattacharjee R (2010) Linoleic acid capped copper nanoparticles for antibacterial activity. Bionanosci J 4:82–86

Das SN, Madhuprakash J, Sarma PVSRN, Purushotham P, Suma K, Manjeet K, Podile AR (2015) Biotechnological approaches for field applications of chitooligosaccharides (COS) to induce innate immunity in plants. Crit Rev Biotechnol 35:29–43

De Faria AF, De Moraes ACM, Alves OL (2014) Toxicity of nanomaterials to microorganisms: mechanisms, methods, and new perspectives. In: Duran N, Guterres SS, Alves OL (eds) Nano medicine and Nanotoxicology. Springer, New York, pp 363–405

De Filpo G, Palermo AM, Rachiele F, Nicoletta FP (2013) Preventing fungal growth in wood by titanium dioxide nanoparticles. Int Biodeterior Biodegradation 85:217–222

De La Torre-Roche R, Hawthorne J, Deng YQ, Xing BS, Cai WJ, Newman LA, Wang Q, Ma XM, Hamdi H, White JC (2013) Multiwalled carbon nanotubes and C-60 fullerenes differentially impact the accumulation of weathered pesticides in four agricultural plants. Environ Sci Technol 47:12539–12547

Dizaj SM, Mennati A, Jafari S, Khezri K, Adibkia K (2015) Antimicrobial activity of carbon-based nanoparticles. Adv Pharm Bull 5:19–23

Donsì F, Annunziata M, Sessa M, Ferrari G (2011) Nanoencapsulation of essential oils to enhance their antimicrobial activity in foods. LWT Food Sci Technol 44:1908–1914

Durán N, Durán M, Bispo M, Jesus d, Seabra AB, Fávaro WJ, Nakazato G (2016) Silver nanoparticles: a new view on mechanistic aspects on antimicrobial activity. Nanomedicine 12:789–799

El Hadrami A, Adam LR, El Hadrami I, Daayf F (2010) Chitosan in plant protection. Mar Drugs 8:968–987

Fakruddin MD, Hossain Z, Afroz H (2012) Prospects and applications of nanobiotechnology: a medical perspective. J Nanobiotechnol 10:31

Fondevilla S, Rubiales D (2012) Powdery mildew control in pea. A review. Agron Sustain Dev 32:401–409

Foster HA, Ditta IB, Varghese S, Steele A (2011) Photocatalytic disinfection using titanium dioxide: spectrum and mechanism of antimicrobial activity. Appl Microbiol Biotechnol 90:1847–1868

Fröhlich E (2017) Role of omics techniques in the toxicity testing of nanoparticles. J Nanobiotechnol 15:84. https://doi.org/10.1186/s12951-017-0320-3

Fu PP, Xia Q, Hwang HM, Ray PC, Yu H (2014) Mechanisms of nanotoxicity: generation of reactive oxygen species. J Food Drug Anal 22:64–75

García-Rincóna J, Vega-Pérezb J, Guerra-Sánchezb MG, Hernández-Lauzardo AN, Peña-Díazc A, Velázquez-Del Vallea MG (2010) Effect of chitosan on growth and plasma membrane properties of Rhizopus stolonifer (Ehrenb.:Fr.) Vuill. Pestic Biochem Phys 97:275–278

Goswami A, Roy I, Sengupta S, Debnath N (2010) Novel applications of solid and liquid formulations of nanoparticles against insect pests and pathogens. Thin Solid Films 519:1252–1257

Goy RC, Britto D, Assis OBG (2009) A review of the antimicrobial activity of chitosan. Polímeros 19:241–247

Grass G, Rensing C, Solioz M (2011) Metallic copper as an antimicrobial surface. Appl Environ Microbiol 77:1541–1548

Gupta A, Silver S (1998) Silver as a biocide: will resistance become a problem? Nat Biotechnol 16:888–890

Gurunathan S, Han JW, Dayem AA, Eppakayala V, Kim JH (2012) Oxidative stress-mediated antibacterial activity of graphene oxide and reduced graphene oxide in *Pseudomonas aeruginosa*. Int J Nanomedicine 7:5901–5914

Hadwiger LA (1999) Host-parasite interactions: elicitation of defense responses in plant with chitosan. Experientia Suppl 87:185–200

Haghighi F, Roudbar Mohammadi S, Mohammadi P, Eskandari M, Hosseinkhani S (2012) The evaluation of *Candida albicans* bio¿lms formation on silicone catheter, PVC and glass coated with titanium dioxide nanoparticles by XTT method and ATPase assay. Bratisl Lek Listy 113:711–715

Hamouda T, Baker J (2000) Antimicrobial mechanism of action of surfactant lipid preparations in enteric gram-negative bacilli. J Appl Microbiol 89:397–403

He L, Liu Y, Mustapha A, Lin M (2011) Antifungal activity of zinc oxide nanoparticles against *Botrytis cinerea* and *Penicillium expansum*. Microbiol Res 166:207–215

He Y, Ingudam S, Reed S, Gehring A, Strobaugh TPN, Irwin P (2016) Study on the mechanism of antibacterial action of magnesium oxide nanoparticles against foodborne pathogens. J Nanobiotechnol 14:54. https://doi.org/10.1186/s12951-016-0202-0

Helander IM, Nurmiaho-Lassila EL, Ahvenainen R, Rhoades J, Roller S (2001) Chitosan disrupts the barrier properties of the outer membrane of gram-negative bacteria. Int J Food Microbiol 71:235–244

Hernández-Lauzardo A, Velázquez M, Guerra-Sánchez M (2011) Current status of action mode and effect of chitosan against phytopathogens fungi. Afr J Microbiol Res 5:4243–4247

Hoseinzadeh A, Habibi-Yangjeh A, Davari M (2016) Antifungal activity of magnetically separable Fe3O4/ZnO/AgBr nanocomposites prepared by a facile microwave-assisted method. Prog Nat Sci: Mater Int J 26:334–340

Hossain F, Perales-Perez OJ, Hwang S, Román F (2014) Antimicrobial nanomaterials as water disinfectant: applications, limitations and future perspectives. Sci Total Environ 466–467:1047–1059

Huang S, Wang L, Liu L, Hou Y, Li L (2015) Nanotechnology in agriculture, livestock, and aquaculture in China. A review. Agron Sustain Dev 35:369–400

Huh AJ, Kwon YJ (2011) "Nanoantibiotics": a new paradigm for treating infectious diseases using nanomaterials in the antibiotics resistant era. J Control Release 156:128–145

Imada K, Sakai S, Kajihara H, Tanaka S, Ito S (2016) Magnesium oxide nanoparticles induce systemic resistance in tomato against bacterial wilt disease. Plant Pathol 65:551–560

Imam J, Singh PK, Shukla P (2016) Plant microbe interactions in post genomic era: perspectives and applications. Front Microbiol 7:1488. https://doi.org/10.3389/fmicb.2016.01488

Ingle AP, Duran N, Rai M (2013) Bioactivity, mechanism of action, and cytotoxicity of copper-based nanoparticles: a review. Appl Microbiol Biotechnol 98:1001–1009

Ismail M, Prasad R, Ibrahim AIM, Ahmed ISA (2017) Modern prospects of nanotechnology in plant pathology. In: Prasad R, Kumar M, Kumar V (eds) Nanotechnology. Springer Nature Singapore Pte Ltd, Singapore, pp 305–317

Issam ST, Adele MG, Adele CP, Stephane G, Veronique C (2005) Chitosan polymer as bioactive coating and film against *Aspergillus niger* contamination. J Food Sci 70:100–104

Jastrzębska AM, Kurtycz P, Olszyna AR (2012) Recent advances in graphene family materials toxicity investigations. J Nanopart Res 14(12):1–21

Jayaseelan C, Rahuman AA, Kirthi AV, Marimuthu S, Santhoshkumar T, Bagavan A, Gaurav K, Karthik L, Rao KV (2012) Novel microbial route to synthesize ZnO nanoparticles using *Aeromonas hydrophila* and their activity against pathogenic bacteria and fungi. Spectrochim Acta A Mol Biomol Spectrosc 90:78–84

Jayaseelan C, Ramkumar R, Abdul Rahuman A, Perumal P (2013) Green synthesis of gold nanoparticles using seed aqueous extract of *Abelmoschus esculentus* and its antifungal activity. Ind Crop Prod 45:423–429

Je JY, Kim SK (2006) Chitosan derivatives killed bacteria by disrupting the outer and inner membrane. J Agric Food Chem 54:6629–6633

Jiang W, Mashayekhi H, Xing B (2009) Bacterial toxicity comparison between nano- and micro-scaled oxide particles. Environ Pollut 157:1619–1625

Jo Y, Kim BH, Jun G (2009) Antifungal activity of silver ions and nanoparticles on phytopathogenic fungi. Plant Dis 93:1037–1043

Jo DH, Kim JH, Lee TG, Kim JH (2015) Review article: size, surface charge, and shape determine therapeutic effects of nanoparticles on brain and retinal diseases. Nanomedicine 11:1603–1611

Joshi N, Jain N, Pathak A, Singh J, Prasad R, Upadhyaya CP (2018) Biosynthesis of silver nanoparticles using *Carissa carandas* berries and its potential antibacterial activities. J Sol-Gel Sci Techn. https://doi.org/10.1007/s10971-018-4666-2

Jyoti S, Satendra S, Sushma S, Anjana T, Shashi S (2007) Antistressor activity of *Ocimum sanctum* (Tulsi) against experimentally induced oxidative stress in rabbits. Methods Find Exp Clin Pharmacol 29:411–416

Kairyte K, Kadys A, Luksiene Z (2013) Antibacterial and antifungal activity of photoactivated ZnO nanoparticles in suspension. Photochem J Photobiol B 128:78–84

Kang S, Pinault M, Pfefferle LD, Elimelech M (2007) Single-walled carbon nanotubes exhibit strong antimicrobial activity. Langmuir 23:8670–8673

Karthikeyan R, Amaechi BT, Rawls HR, Lee VA (2011) Antimicrobial activity of nanoemulsion on cariogenic *Streptococcus mutans*. Arch Oral Biol 56:437–445

Katiyar D, Hemantarajan A, Sing B (2015) Chitosan as a promising natural compound to enhance potential physiological responses in plant: a review. Ind J Plant Physiol 20:1–9

Kaur K (2016) Nanoemulsions as delivery vehicles for food and pharmaceuticals. In: Grumezescu AM (ed) Emulsions. Nanotechnology in the Agri-Food Industry, vol. 3. https://doi.org/10.1016/B978-0-12-804306-6.00018-0

Kaur P, Thakur R, Choudhary A (2012) An in vitro study of the antifungal activity of silver/chitosan nanoformulations against important seed borne pathogens. Int J Sci Technol Res 1:83–86

Kim J, Cho H, Ryu S, Choi M (2000) Effects of metal ions on the activity of protein tyrosine phosphatase VHR: highly potent and reversible oxidative inactivation by Cu2+ ion. Arch Biochem Biophys 382:72–80

Kim JS, Kuk E, Yu KN, Kim JH, Park SJ, Lee J, Kim SH, Park YK, Park YH, Hwang CY et al (2007) Antimicrobial effects of silver nanoparticles. Nanomed Nanotechnol Biol Med 3(95):101–157

Kim SW, Kim KS, Lamsal K, Kim YJ, Kim SB, Jung M, Sim SJ, Kim HS, Chang SJ, Kim JK, Lee YS (2009) An in vitro study of the antifungal effect of silver nanoparticles on oak wilt pathogen Raffaelea sp. J Microbiol Biotechnol 19:760–764

Kim SH, Lee HS, Ryu DS, Choi SJ, Lee DS (2011) Antibacterial activity of silver-nanoparticles against *Staphylococcus aureus* and *Escherichia coli*. Kor Microbiol Biotechnol 39:77–85

Kleandrova VV, Luan F, Speck-Planche A, Cordeiro M (2015) Review of structures containing fullerene-C60 for delivery of antibacterial agents. Multitasking model for computational assessment of safety profiles. Curr Bioinforma 10:565–578

Knief C (2014) Analysis of plant microbe interactions in the era of next generation sequencing technologies. Front Plant Sci 5:216. https://doi.org/10.3389/fpls.2014.00216

Konishi Y, Ohno K, Saitoh N, Nomura T, Nagamine S, Hishida H, Takahashi Y, Uruga T (2007) Bioreductive deposition of platinum nanoparticles on the bacterium *Shewanella algae*. J Biotechnol 128:648–653

Kuppusamy P, Yusoff MM, Maniam GP, Govindan N (2016) Biosynthesized metallic nanoparticles using for pharmacological applications. Saudi Pharmaceutical J 24:473–484

Kuzma J (2007) Moving forward responsibly oversight for the nanotechnology-biology interface. J Nanopart Res 9:165–182

Lankadurai BP, Nagato EG, Simpson MJ (2013) Environmental metabolomics: an emerging approach to study organism responses to environmental stressors. Environ Rev 21:180–205

Lemire JA, Harrison JJ, Turner RJ (2013) Antimicrobial activity of metals: mechanisms, moleculartargets and applications. Nat Rev Microbiol 11:371–384

Leuba KD, Durmus NG, Taylor EN, Webster TJ (2013) Short communication: carboxylate functionalized superparamagnetic iron oxide nanoparticles (SPION) for the reduction of S. aureus growth post biofilm formation. Int J Nanomedicine 8:731–736

Leung YH, Ng AMC, Xu X, Shen Z, Gethings LA, Wong MT, Chan CM, Guo MY, Ng YH, Djurišić AB, Lee PK (2014) Mechanisms of antibacterial activity of MgO: non-ROS mediated toxicity of MgO nanoparticles towards *Escherichia coli*. Small 10:1171–1183

Li XF, Feng XQ, Yang S, Wang TP (2008) Effects of molecular weight and concentration of chitosan on antifungal activity against *Aspergillus niger*. Iran Polym J 17:843–852

Li Y, Zhang W, Niu J, Chen Y (2012a) Mechanism of photogenerated reactive oxygen species and correlation *with* the antibacterial properties of engineered metal-oxide nanoparticles. ACS Nano 6:5164–5173

Li Y, Yu S, Wu Q, Tang M, Pu Y, Wang D (2012b) Chronic Al_2O_3-nanoparticle exposure causes neurotoxic effects on locomotion behaviors by inducing severe ROS production and disruption of ROS defense mechanisms in nematode *Caenorhabditis elegans*. Hazard Mater J 219–220:221–230

Li B, Liu BP, Shan CL, Ibrahim M, Lou YH, Wang YL, Xie GL, Li HY, Sun GC (2013a) Antibacterial activity of two chitosan solutions and their effect on rice bacterial leaf blight and leaf streak. Pest Manag Sci 69:312–320

Li C, Wang X, Chen F, Zhang C, Zhi X, Wang K, Cui D (2013b) The antifungal activity of graphene oxide –silver nanocomposites. Biomaterials 34:3882–3890

Li JG, Zhu WH, Zhang M, Zheng X, Di Z et al (2014) Antibacterial activity of large area monolayer graphene film manipulated by charge transfer. Sci Rep 4:4359–4367

Lin X, Li J, Ma S, et al (2014) Toxicity of TiO2 nanoparticles to *Escherichia coli*: effects of particle size, crystal phase and water chemistry. Rozhkova EA, ed. PLoS One. 2014;9(10):e110247. doi:https://doi.org/10.1371/journal.pone.0110247

Liu XF, Guan YL, Yang DZ, Li Z, Yao KD (2001) Antibacterial action of chitosan and carboxymethylated chitosan. J Appl Polym Sci 79:1324–1335

Liu H, Du YM, Wang XH, Sun LP (2004) Chitosan kills bacteria through cell membrane damage. Int J Food Microbiol 95:147–155

Liu J, Qiao SZ, Hu QZ, Lu GQ (2011a) Magnetic nanocomposites with mesoporous structures: synthesis and applications. Small 7:425–443

Liu S, Zeng TH, Hofmann M, Burcombe E, Wei J, Jiang R, Kong J, Chen Y (2011b) Antibacterial activity of graphite, graphite oxide, graphene oxide, and reduced graphene oxide: membrane and oxidative stress. ACS Nano 5:6971–6980

Liu J, Zhao Z, Feng H, Cui FJ (2012) One-pot synthesis of $Ag–Fe_3O_4$ nanocomposites in the absence of additional reductant and its potent antibacterial properties. Mater Chem 22:13891–13894

Lok CN, Ho CM, Chen R, He QY, Yu WY, Sun H et al (2006) Proteomic analysis of the mode of antibacterial action of silver nanoparticles. J Proteome Res 5:916–924

Ma L, Li J, Y Y, Yu CM, Wang Y, Li XM, Li N (2014) Germination and physiological response of wheat (*Triticum aestivum*) to pre-soaking with oligochitosan. Int J Agric Biol 16:766–770
Makhluf S, Dror R, Nitzan Y et al (2005) Microwave-assisted synthesis of nanocrystalline MgO and its use as bacteriocide. Adv Funct Mater 15:1708–1715
Malarkodi C, Rajeshkumar S, Paulkumar K, Gnanajobitha G, Vanaja M, Annadurai G (2013) Biosynthesis of semiconductor nanoparticles by using sulfur reducing bacteria Serratia nematodiphila. Adv Nano Res 1:83–91
Mansilla AY, Albertengo L, Rodríguez MS, Debbaudt A, Zúñiga A, Casalongué CA (2013) Evidence on antimicrobial properties and mode of action of a chitosan obtained from crustacean exoskeletons on *Pseudomonas syringae* pv. tomato DC3000. Appl Microbiol Biotechnol 97:6957–6966
Marambio-Jones C, Hoek EMV (2010) A review of the antibacterial effects of silver nanomaterials and potential implications for human health and the environment. J Nanopart Res 12:1531–1551
Markowska-Szczupak A, Ulfig K, Morawski AW (2011) The application of titanium dioxide for deactivation of bioparticulates: an overview. Catal Today 169:249–257
Marquez IG, Akuaku J, Cruz I, Cheetham J, Golshani A, Smith ML (2013) Disruption of protein synthesis as antifungal mode of action by chitosan. Int J Food Micobiol 164:108–112
Matˇeejka V, Tokarsk´y J (2014) Photocatalytical nanocomposites: a review. J Nanosci Nanotechnol 14:1597–1616
Mathur A, Raghavan A, Chaudhury P, Johnson JB, Roy R, Kumari J, Chaudhuri G, Chandrasekaran N, Suraishkumar GK (2015) Cytotoxicity of titania nanoparticles towards waste water isolate *Exiguobacterium acetylicum* under UVA, visible light and dark conditions. Environ J Chem Eng 3:1837–1846
Matsumura Y, Yoshikata K, Kunisaki S, Tsuchido T (2003) Mode of bactericidal action of silver zeolite and its comparison with that of silver nitrate. Appl Environ Microbiol 69:4278–4281
Meng XH, Tian S (2009) Effects of preharvest application of antagonistic yeast combined with chitosan on decay and quality of harvested table grape fruit. J Sci Food Agric 89:1838–1842
Meng XH, Qin GZ, Tian SP (2010) Influences of preharvest spraying *Cryptococcus laurentii* combined with chitosan coating on postharvest disease and quality of table grapes in storage. LWT – Food Sci Technol 43:596–601
Moghimi R, Ghaderi L, Rafati H, Aliahmadi A, McClements DJ (2016) Superior antibacterial activity of nanoemulsion of Thymus daenensis essential oil against *E. coli*. Food Chem 194:410–415
Morones JR, Elechiguerra JL, Camacho A, Holt K, Kouri JB, Ramirez JT et al (2005) The bactericidal effect of silver nanoparticles. Nanotechnology 16:2346–2353
Mueller NC, Nowack B (2008) Exposure modeling of engineered nanoparticles in the environment. Environ Sci Technol 42:4447–4453
Nair R, Kumar D (2013) Plant diseases—control and remedy through nanotechnology. In: Tuteja N, Gill S (eds) Crop improvement under adverse conditions. Springer, New York, pp 231–243
Nambiar PR, Gupta RR, Misra V (2010) An “omics” based survey of human colon cancer. Mutat Res 693:3–18
Navrotsky A (2000) Technology and applications nanomaterials in the environment agriculture and technology (NEAT). J Nanopart Res 2:321–323
Niazi JH, Gu MB (2009) Toxicity of metallic nanoparticles in microorganisms- a review. In. 452. Atmospheric and biological environmental monitoring, Kim YJ, Platt U, Gu MB, Iwahashi H, 453. Eds. Springer Netherlands: Dordrecht, 2009, pp 193–206
Ocsoy I, Paret LM, Ocsoy AM, Kunwar S, Chen T, You M, Tan W (2013) Nanotechnology in plant disease management: DNA-directed silver nanoparticles on graphene oxide as an antibacterial against *Xanthomonas perforans*. ASC Nano 7:8972–8980
Pandian CJR, Palanivel S, Dhanasekaran (2016) Screening antimicrobial activity of nickel nanoparticles synthesized using *Ocimum sanctum* leaf extract. J Nanoparticle Article ID 4694367, 13. https://doi.org/10.1155/2016/4694367
Parry JM, Parry EM (2012) Genetic toxicology: principles and methods. Humana Press, New York

Paspaltsis I, Kotta K, Lagoudaki R, Grigoriadis N, Poulios I, Sklaviadis T (2006) Titanium dioxide photocatalytic inactivation of prions. J Gen Virol 87:3125–3130
Pelgrift RY, Friedman AJ (2013) Nanotechnology as a therapeutic tool to combat microbial resistance. Adv Drug Deliv Rev 65:1803–1815
Perez Espitia PJP, Soares NFF, Coimbra JSR, Andrade NJ, Cruz RS, Medeiros EAA (2012) Zinc oxide nanoparticles: synthesis antimicrobial activity and food packaging applications. Food Bioprocess Technol 5:1447–1464
Peter KS, Rosalyn L, George LM, Klabunde JK (2002) Metal oxide nanoparticles as bactericidal agents. Langmuir 18:6679–6686
Petrus E, Petrus E, Tinakumari S, Chai L, Ubong A, Tunung R, Elexson N, Chai LF, Son R (2011) A study on the minimum inhibitory concentration and minimum bactericidal concentration *of* nano colloidal silver on food borne pathogens. Int Food Res J 18:55–66
Pichyangkura R, Chadchawan S (2015) Biostimulant activity of chitosan in horticulture. Sci Hortic 196:49–65
Pontón J (2008) La pared celular de los hongos y el mecanismo de accio'n de la anidulafungina. Rev Iberoam Micol 25:78–82
Prasad R (2014) Synthesis of silver nanoparticles in photosynthetic plants. J Nanopart:963961 https://doi.org/10.1155/2014/963961
Prasad R, Swamy VS, Varma A (2012) Biogenic synthesis of silver nanoparticles from the leaf extract of *Syzygium cumini* (L.) and its antibacterial activity. Int J Pharma Bio Sci 3(4):745–752
Prasad R, Kumar V, Prasad KS (2014) Nanotechnology in sustainable agriculture: present concerns and future aspects. Afri J Biotechnol 13:705–713
Prasad R, Pandey R, Barman I (2016) Engineering tailored nanoparticles with microbes: quo vadis. WIREs Nanomed Nanobiotechnol 8:316–330. https://doi.org/10.1002/wnan.1363
Prasad R, Bhattacharyya A, Nguyen QD (2017a) Nanotechnology in sustainable agriculture: recent developments, challenges, and perspectives. Front Microbiol 8:1014. https://doi.org/10.3389/fmicb.2017.01014
Prasad R, Gupta N, Kumar M, Kumar V, Wang S, Abd-Elsalam KA (2017b) Nanomaterials act as plant defense mechanism. In: Prasad R, Kumar V, Kumar M (eds) Nanotechnology. Springer, Singapore, pp 253–269
Priyanka KP, Harikumar VS, Balakrishna KM, Varghese T (2017) Inhibitory effect of TiO2 NPs on symbiotic arbuscular mycorrhizal fungi in plant roots. IET Nanobiotechnol 111:66–70
Rabea EI, Badawy MEI, Steurbaut W, Stevens CV (2009) In vitro assessment of N-(benzyl) chitosan derivatives against some plant pathogenic bacteria and fungi. Eur Polym J 45:237–245
Raghupathi KR, Koodali RT, Manna AC (2011) Size dependent bacterial growth inhibition and mechanism of antibacterial activity of zinc oxide nanoparticles. Langmuir 27:4020–4028
Rai M, Ingle A (2012) Role of nanotechnology in agriculture with special reference to management of insect pests. Appl Microbiol Biotechnol 94:287–293
Rai M, Yadav A, Gade A (2009) Silver nanoparticles as a new generation of antimicrobials. Biotech Adv 27:76–83
Rai A, Prabhune A, Perry CC (2010) Antibiotic mediated synthesis of gold nanoparticles with potent antimicrobial activity and their application in antimicrobial coatings. J Mater Chem 20:6789–6798
Raimondi S, Zanni E, Talora C, Rossi M, Palleschi C, Uccelletti D (2008) SOD1, a new *Kluyveromyces lactis* helper gene for heterologous protein secretion. Appl Environ Microbiol 74:7130–7137
Roda F, Al-Thani PNK, Al-Maadeed MA (2014) Garphene oxide as antimicrobial against two gram-positive and two gram-negative bacteria in addition to one fungus. Online J Biol Sci 14:230–239
Roopan SM, Surendra TV, Elango G et al (2014) Biosynthetic trends and future aspects of bimetallic nanoparticles and its medicinal applications. Appl Microbiol Biotechnol 98:5289–5300
Ruparelia JP, Chatterjee A, Duttagupta SP, Mukherji S (2008) Strain specificity in antimicrobial activity of silver and copper nanoparticles. Acta Biomater 4:707–716

Sadiq IM, Chowdhury N, Chandrasekaran A, Mukherjee (2009) Antimicrobial sensitivity of *E. coli* to alumina NPs. Nanomed Nanotechnol 5:282–286

Sadiq IM, Pakrashi S, Chandrasekaran N, Mukherjee A (2011) Studies on toxicity of aluminium oxide (Al2O3) nanoparticles to microalgae species: *Scenedesmus* sp and *Chlorella* sp. J Nanopart Res 13:3287–3299

Sadjad S, Ali K, Hossein M (2017) Nano-bio control of bacteria: a novel mechanism for antibacterial activities of magnetic nanoparticles as a temporary nanomagnets. J Mol Liquids 251. https://doi.org/10.1016/j.molliq.2017.12.036

Saharan V, Mehrotra A, Khatik R, Rawal P, Sharma SS, Pal A (2013) Synthesis of chitosan based nanoparticles and their *in vitro* evaluation against phytopathogenic fungi. Int J Biol Macromol 62:677–683

Saharan V, Sharma G, Yadav M, Choudhary MK, Sharma SS, Pal A, Raliya R, Biswas P (2015) Synthesis and in vitro antifungal efficacy of Cu–chitosan nanoparticles against pathogenic fungi of tomato. Int J Biol Macromolecules 75:346–353

Salata OV (2004) Applications of nanoparticles in biology and medicine. J Nanobiotechnol 2:3–3

Samberg ME, Orndorff PE, Monteiro-Riviere NA (2011) Antibacterial efficacy of silver nanoparticles of different sizes, surface conditions and synthesis methods. Nanotoxicology 5(2):244–253

Sang WK, Jin HJ, Kabir L et al (2012) Antifungal effects of silver nanoparticles (Ag NPs) against various plant pathogenic fungi. Mycobiology 40:415–427

Santos CMJ, Mangadlao F, Ahmed A, Advincula LRC et al (2012) Graphene nanocomposite for biomedical applications: fabrication, antimicrobial and cytotoxic investigations. Nanotechnology 23:395101–395101

Saraf M, Pandya U, Thakkar A (2014) Role of allelochemicals in plant growth promoting rhizobacteria for biocontrol of phytopathogens. Microbiol Res 169:18–29

Sawai J, Yoshikawa T (2004) Quantitative evaluation of antifungal activity of metallic oxide powders (MgO, CaO and ZnO) by an indirect conductimetric assay. J Appl Microbiol 96:803–809

Sawai JE, Kawada F, Kanou H, Igarashi A, Hashimoto T, Kokugan M, Shimizu (1996) Detection of active oxygen generated from ceramic powders having antibacterial activity. J Chem Eng Jpn 29:627–633

Sawai J, Kojima H, Igarashi H, Hasimoto A, Shoji S, Takehara A, Sawaki T, Kokugan T, Shimizu M (1997) *Escherichia coli* damage by ceramic powder slurries. J Chem Eng Jpn 30:1034–1039

Sawai J, Kojima H, Igarashi H, Hasimoto A, Shoji S, Sawaki T, Hakoda A, Kawada E, Kokugan T, Shimizu M (2000) Antibacterial characteristics of magnesium oxide powder. World J Microbiol Biotechnol 16:187–194

Sawai J, Shuga S, Kojima H (2001) Kinetic analysis of death of bacteria in CaO powder slurry. Int Biodeterior Biodegradation 47:23–26

Sawangphruk M, Srimuk P, Chiochan P, Sangsri T, Siwayaprahm P (2012) Synthesis antifungal activity of reduced graphene oxide nanosheets. Carbon 50:5156–5161

Sayed HH, Shamroukh AH, Rashad AE (2006) Synthesis and biological evaluation of some pyrimidine, pyrimido [2,1-b] [1,3] thiazine and thiazolo [3,2-a] pyrimidine derivatives. Acta Pharma 56:23–144

Schrand AM, Rahman MF, Hussain SM, Schlager JJ, Smith DA, Syed AF (2010) Metal-based nanoparticles and their toxicity assessment. WIREs Nanomed Nanobiotechnol 2:554–568

Shah MA, Towkeer A (2010) Principles of nanosciences and nanotechnology. Narosa Publishing House, New Delhi

Shankramma K, Yallappa S, Shivanna MB, Manjanna J (2016) Fe2O3 magnetic nanoparticles to enhance *S. lycopersicum* (tomato) plant growth and their biomineralization. Appl Nanosci 6:983–990

Shao X, Wang H, Xu F, Cheng S (2013) Effects and possible mechanisms of tea tree oil vapor treatment on the main disease in postharvest strawberry fruit. Postharvest Biol Technol 7:94–101

Sharma P, Sharma A, Sharma M, Bhalla N, Estrela P, Jain A, Thakur P, Thakur A (2017) Nanomaterial fungicides: in vitro and in vivo antimycotic activity of cobalt and nickel nanoferrites on phytopathogenic fungi. Glob Chall 1(1700041):1–7

Sharon M, Choudhary AK, Kumar R (2010) Nanotechnology in agricultural diseases and food safety. J Phytology 2:83–92

Shrivastava S, Prasad R, Varma A (2014) Anatomy of root from eyes of a microbiologist. In: Morte A, Varma A (eds) Root engineering, vol 40. Springer, Berlin/Heidelberg, pp 3–22

Sierra-Fernandez A, De la Rosa-García SC, Gomez-Villalba LS, Gómez-Cornelio S, Rabanal ME, Fort R, Quintana P (2017) Synthesis, photocatalytic, and antifungal properties of MgO, ZnO and Zn/Mg oxide nanoparticles for the protection of calcareous stone heritage. ACS Appl Mater Interfaces 929:24873–24886

Singh N, Manshian B, Jenkins GJ, Griffiths SM, Williams PM, Maffeis TG, Wright CJ, Doak SH (2009) Nanogenotoxicology: the DNA damaging potential of engineered nanomaterials. Biomaterials 30:3891–3914

Sirelkhatim A, Mahmud S, Seeni A, Kaus NHM, Ann LC, Bakhori SKM, Hasan H, Mohamad D (2015) Review on zinc oxide nanoparticles: antibacterial activity and toxicity mechanism. Nanomicro Lett 7:219–242

Sonawane RS, Hegde SG, Dongare MK (2003) Preparation of titanium (VI) oxide thin film photocatalyst by sol–gel dip coating. Mater Chem Phys 77(3):744–746

Sondi I, Salopek-Sondi B (2004) Silver nanoparticles as antimicrobial agent: a case study on *E. coli* as a model for gram-negative bacteria. J Colloid Interface Sci 275:177–182

Stohs SJ, Bagchi D (1995) Oxidative mechanisms in the toxicity of metal ions. Free Radic Biol Med 18:321–336

Stoimenov PK, Klinger RL, Marchin GL, Klabunde KJ (2002) Metal oxide nanoparticles as bactericidal agents. Langmuir 18:6679–6686

Sudarshan NR, Hoover DG, Knorr D (1992) Antibacterial action of chitosan. Food Biotechnol 6:257–272

Tang Y (2015) Non-genomic omic techniques. In: Tang Y, Sussman M, Liu D, Poxton I, Schwartzman J (eds) Molecular Medical Microbiology. Academic Press, London, pp 399–406

Tang Z-X, Lv B-F (2014) MgO nanoparticles as antibacterial agent: preparation and activity. Braz J Chem Eng 31(3):591–601

Tarafdar JC, Sharma S, Raliya R (2013) Nanotechnology: interdisciplinary science of applications. Afr J Biotechnol 12(3):219–226

Tavaria FK, Costa EM, Gens EJ, Malcata FX, Pintado ME (2013) Influence of abiotic factors on the antimicrobial activity of chitosan. J Dermatol 40:1014–1019

Theron J, Walker JA, Cloete TE (2008) Nanotechnology and water treatment: applications and emerging opportunities. Crit Rev Microbiol 34:43–69

Torres R, Valentines MC, Usall J, Vinas I, Larrigaudiere C (2003) Possible involvement of hydrogen peroxide in the development of resistance mechanisms in "golden delicious" apple fruit. Postharvest Biol Technol 27:235–242

Tu Y, Lv M, Xiu P, Huynh T, Zhang M, Castelli M, Liu Z, Huang Q, Fan C, Fang H, Zhou R (2013) Destructive extraction of phospholipids from *Escherichia coli* membranes by graphene nanosheets. Nat Nanotechnol 8:594–601

Usha R, Prabu E, Palaniswamy M, Venil CK, Rajendran R (2010) Synthesis of metal oxide nanoparticles by Streptomyces sp for development of antimicrobial textiles. Global J Biotechnol Biochem 5:153–160

Usman MS, Ibrahim NA, Shameli K, Zainuddin N, Junus WMZW (2012) Copper nanoparticles mediated by chitosan: synthesis and characterization via chemical methods. Molecules 17:14928–14936

Vahabi K, Mansoori GA, Karimi S (2011) Biosynthesis of silver nanoparticles by fungus *Trichoderma reesei*. Inscience J 1:65–79

Van SN, Minh HD, Anh DN (2013) Study on chitosan nanoparticles on biophysical characteristics and growth of *Robusta coffee* in green house. Biocatal Agric Biotechnol 2(4):289–294

Van-Loon LC, Van-Strien EA (1999) The families of pathogenesis-related proteins, their activities, and comparative analysis of PR-1 type proteins. Physiol Mol Plant Pathol 55:85–97

Vidic J, Stankic S, Haque F, Ciric D, Goffic R, Vidy A, Jupille J, Delmas BJ (2013) Selective antibacterial effects of mixed ZnMgO *nanoparticles*. J Nanopart Res 15:1–10

Wang SY, Gao H (2012) Effect of chitosan-based edible coating on antioxidants, antioxidant enzyme system, and postharvest fruit quality of strawberries (*Fragaria x ananassa* Duch.). LWT – Food Sci Technol 52:71–79

Wang SR, Lawson PC, Ray H, Yu (2010) Toxic effects of gold nanoparticles on *Salmonella typhimurium* bacteria. Toxicol Ind Health 27:547–554

Wang X, Liu X, Chen J, Han H, Yuan Z (2014) Evaluation and mechanism of antifungal effects of carbon nanomaterials in controlling plant fungal pathogen. Carbon 68:798–806

Wang L, Hu C, Shao L (2017a) The antimicrobial activity of nanoparticles: present situation and prospects for the future. Int J Nanomedicine 12:1227–1249

Wang X, Zhou Z, Chen F (2017b) Surface modification of carbon nanotubes with an enhanced antifungal activity for the control of plant fungal pathogen. Materials 10:1375. https://doi.org/10.3390/ma10121375

Wani AH, Shah MA (2012) A unique and profound effect of MgO and ZnO nanoparticles on some plant pathogenic fungi. J App Pharm Sci 2:40–44

Waters KM, Masiello LM, Zangar RC et al (2009) Macrophage responses to silica nanoparticles are highly conserved across particle sizes. Toxicol Sci 107:553–569

Xie Y, He Y, Irwin PL, Jin T, Shi X (2011) Antibacterial activity and mode of action of ZnO. Appl Environ Microbiol 77:2325–2331

Xing K, Chen XG, Liu CS, Cha DS, Park HJ (2009) Oleoyl-chitosan nanoparticles inhibits Escherichia coli and *Staphylococcus aureus* by damaging the cell membrane and putative binding to extracellular or intracellular targets. Int J Food Microbiol 132:127–133

Xing K, Zhu X, Peng X, Qin S (2015) Chitosan antimicrobial and eliciting properties for pest control in agriculture a review. Agron Sustain Dev 35:569–588

Xu FF, Imlay JA (2012) Silver (I), mercury (II), cadmium (II), and zinc (II) target exposed enzymic iron-sulfur clusters when they toxify Escherichia coli. Appl Environ Microbiol 78:3614–3621

Yamamoto O, Sawai J, Sasamoto T (2000) Change in antibacterial characteristics with doping amount of ZnO in MgO-ZnO solid solution. Int J Inorg Mater 2:451–454

Yamanaka M, Hara K, Kudo J (2005) Bactericidal actions of a silver ion solution on *Escherichia coli* studied by energy filtering transmission electron microscopy and proteomic analysis. Appl Environ Microbiol 71:7589–7593

Yates JR, Ruse CI, Nakorchevsky A (2009) Proteomics by mass spectrometry: approaches, advances, and applications. Annu Rev Biomed Eng 11:49–79

Yen MT, Yang JH, Mau JL (2008) Antioxidant properties of chitosan from crab shells. Carbohydr Polym 74:840–844

Yin JJ, Liu J, Ehrenshaft M, Roberts JE, Fu PP, Mason RP, Zhao B (2012) Phototoxicity of nano titanium dioxides in HaCaT keratinocytes generation of reactive oxygen species and cell damage. Toxicol Appl Pharmacol 263:81–88

Yu JC, Tang HY, Yu JG (2002) Bactericidal and photocatalytic activities of TiO2 thin films prepared by sol–gel and reverse micelle methods. J Photochem Photobiol A Chem 3:211–219

Yu Q, Li J, Zhang Y, Wang Y, Liu L, Li M (2016) Inhibition of gold nanoparticles (AuNPs) *on* pathogenic biofilm formation and invasion to host cells. Sci Rep 26667(36):6

Zakrzewska A, Boorsma A, Brul S, Hellingwerf KJ, Klis FM (2005) Transcriptional response of *Saccharomyces cerevisiae* to the plasma membrane-perturbing compound chitosan. Eukaryot Cell 4:703–715

Zhang L, Jiang Y, Ding Y, Povey M, York D (2007) Investigation into the antibacterial behaviour of suspensions of ZnO nanoparticles (ZnO nanofluids). J Nanopart Res 9:479–489

Zhang H, Li R, Liu (2011) Effects of chitin and its derivative chitosan on postharvest decay of fruits: a review. Int J Mol Sci 12:917–934

Zhang A, Sun H, Wang P, Han Y, Wang X (2012) Modern analytical techniques in metabolomics analysis. Analyst 137:293–300

Zhao D, Wang J, Sun BH, Sun BH, Gao JQ, Xu R (2000) Development and application of TiO2 photocatalysis as antimicrobial agent. J Liaoning Univ (Nat Sci Ed) 2:173–174

Zhao Y, Tian Y, Cui Y, Liu W, Ma W, Jiang X (2010) Small molecule-capped gold nanoparticles as potent antibacterial agents that target gram-negative bacteria. J Amer Chem Soc 132(35):12349–11256

Zhou R, Gao H (2014) Cytotoxicity of graphene: recent advances and future perspective. Wiley Interdiscip Rev Nanomed Nanobiotechnol 6:452–474

Chapter 12
Sustainable Nanotechnology: Mycotoxin Detection and Protection

Velaphi C. Thipe, Marshall Keyster, and Kattesh V. Katti

12.1 Introduction

Fungi have the ability to manifest in a wide range of temperatures and predominantly appear to be a serious problem in agriculture. Fungi contamination poses a greater threat to the economy due to reduced input and exports of commodities (such as crops, cereals and peanuts) affected by fungal growth. Mycotoxins are toxins produced as secondary metabolites by certain fungi and released into the crops, cereals, peanuts and animal feed. There are over 300 known mycotoxins that have been identified, and these mycotoxins can be produced at any point during harvesting, processing, distribution and storage of crops (Milicevic et al. 2010; Zain 2011; Mejri-Omrani et al. 2016). The United Nations of Food and Agriculture Organization reported that 25% of food production worldwide is affected by mycotoxin contamination. To date, we are faced with the challenge of safe food security due to population growth (Godfray et al. 2010). Over the past half-century, world hunger continues to rise from 777 million to 815 million in 2016 undernourished people and increased food insecurity because of climate change due to hot and humid conditions which favours fungal manifestation. This leads to increased mycotoxin contamination with detrimental effects, where 1/4 of children under the

V. C. Thipe
Department of Chemistry, University of Missouri, Columbia, MO, USA

Institute of Green Nanotechnology, University of Missouri, Columbia, MO, USA

M. Keyster
Department of Biotechnology, University of the Western Cape, Bellville, South Africa

K. V. Katti (✉)
Department of Physics and Astronomy, University of Missouri, Columbia, MO, USA

Institute of Green Nanotechnology, University of Missouri, Columbia, MO, USA
e-mail: KattiK@health.missouri.edu

K. A. Abd-Elsalam, R. Prasad (eds.), *Nanobiotechnology Applications in Plant Protection*, Nanotechnology in the Life Sciences,
https://doi.org/10.1007/978-3-319-91161-8_12

age of 5 years suffer from stunting because of the intake of contaminated foods (Godfray et al. 2010; FAO 2017).

The concept of safe food security is threatened by mycotoxin contamination, the inability to provide food that is safe for human and animal consumption. Mycotoxins are toxigenically and chemically heterogeneous and some displace overlapping toxicities that affect human and animal health. All mycotoxins are produced by fungi; however not all mycotoxins are toxic, and some are useful such as penicillin which is an antibiotic used against bacteria (Roberts 1995). Mycotoxins are small molecular weight compounds (MW ~ 700 g/mol) that are structurally diverse (Actis et al. 2010). Group 1 carcinogen because human and animal health, and this significantly affects agro-economy. The mycotoxins that are of concern include aflatoxins (AF), ochratoxins (OT), fumonisins (FUM), trichothecenes and zearalenone (ZEN). These mycotoxins are responsible for annual economic losses worldwide due to their manifestation in the food chain (condemned agricultural products) associated with human and animal health issues.

The principal classes of mycotoxins include aflatoxin B_1 (AFB_1), the most potent hepatocarcinogenic substance amongst all mycotoxins and is classified as a Group 1 human carcinogen, followed by OT and FUM as Group 2B carcinogen, trichothecenes and ZEN, as human carcinogens (Group 3). Mycotoxins determination and quantification has challenges that include biosynthetic origins, biological effects, chemical structures and their production that can be facilitated by a wide range of fungal species and new strains. Therefore, classification of mycotoxins is different based on professions, for example, clinicians would classify mycotoxins based on the organ they target (e.g. hepatotoxins, nephrotoxins, neurotoxins, immunotoxins, etc.). Cell biologists on the other hand would classify mycotoxins into generic groups (e.g. allergens, carcinogens, mutagens and teratogens); organic chemists classify mycotoxins based on their chemical structures (e.g. coumarins and lactones); biochemists use their biosynthetic origins for classification (amino acid-derived, polyketides, etc.) and mycologists by producing fungi (e.g. *Aspergillus* toxins, *Fusarium* toxins, *Penicillium* toxins). In this regard, mycotoxins are unavoidable, and there is a consist requirement for their detection in foods and feedstuff. The control and monitoring of mycotoxins are critical at different steps within the food sector from raw materials, food processing, final product and during storage. Advanced technologies are required to assess the safety and quality of the foods to achieve a multi-sensing system that integrates analytical requirements of sensitivity and selectivity in mycotoxin detection.

The field of nanotechnology has shown to have the potential for improving food quality and safety for proving reliable qualitative and quantitative data. Nanotechnology is the holistic combination of biology, chemistry, engineering, physics, material science and toxicology (mycotoxicology) all integrated to form nanostructured materials to produce novel devices for the detection of mycotoxins and treatment modalities against toxigenic fungi. The increase in mycotoxin occurrence and their derivatives has attracted interest in developing multiplexed sensors based on nano-platforms. The application of nanotechnology will provide a system that can facilitate rapid on-site analysis with high sensitivity and reproducibility; this gets rid of the necessity to send samples for laboratory analysis that requires expensive instrumentation that is time consuming. New novel systems (e.g. nanosensors, microarrays, lab-on-a-chip devices, etc.) are being developed that harness the power of nanomaterials applied for multiplex mycotoxin analysis and nanotracking towards precision agricultural practices for logistic food surveillance (Prasad 2016; Prasad et al. 2017a).

Lab-on-a-chip microfluidics for the analysis of mycotoxins is an example of nanotechnological systems that can be an invaluable tool in the food industry for ensuring high safety and quality of food. The high surface-to-volume ratio allows for increased ligand(s) loading that allows for multiple functionality for greater sensitivity by using nanowire transducers (Putzbach and Ronkainen 2013). A number of nanomaterials (dendrimers, carbon nanotubes, conducting and semiconducting polymer nanowires, gold nanoshells, polymeric nanoparticles, noble metals nanoparticles, quantum dots, silica nanoparticles and superparamagnetic nanoparticles) have been investigated in biosensor fabrication. These biosensors exhibit signal amplification in miniaturized devices with high precision and accuracy, increased sensitivity and selectivity that allows multiplexing.

There are many benefits in exploiting nanotechnology for mycotoxin detection, where fluorophores have been replaced by quantum dots (luminescent nanocrystals) for labelling biomolecules for the detection of different analytes such as mycotoxins. Quantum dots exhibit a narrow emission peak and a high quantum yield resulting in improved signal-to-noise ratio, but their stability is still an issue that needs to be resolved. On the other hand, metallic nanoparticles have been extensively used for analyte detection and gold nanoparticles conjugated probes for detecting mycotoxins because of ease in functionalization and immobilization in nanosensors. The ultra-small size of nanoparticles allows for multiple biomolecule conjugation due to the increased surface-to-volume ratio, and the reduced band gap facilitates ease electron transfer to achieve high sensitivity for low-level detection (Logrieco et al. 2005).

The present chapter describes the role of nanotechnology in the fabrication of biosensors for the detection of mycotoxins. The use of different types of nanomaterials is explored for detecting mycotoxins. We also discuss the advancements in nanotechnology for applications in detection and surveillance of mycotoxins in precision agriculture. The application of green nanotechnology is also discussed by using phytochemical from plants to synthesize nanoparticles that can be used against toxigenic fungi and subsequently reduce the manifestation of mycotoxins (Prasad 2014). The chapter reports on the current nano-platforms that are commercially available in the market for the detection and monitoring of mycotoxins.

12.2 Types of Mycotoxins

Mycotoxins that are of great significance and pose threat to public health and agroeconomics due to their carcinogenicity and toxicities include:

12.2.1 Aflatoxins

Aflatoxins (AF) are composed of aflatoxin B (AFB_1 and AFB_2), aflatoxin G (AFG_1 and AFG_2) and aflatoxin M (AFM_1 and AFM_2). The AFB_1 is the most potent hepatocarcinogenic substance. Aflatoxins are mostly found in maize and peanuts. The transformation of AFB_1 and AFB_2 results in the production of their hydroxylated metabolites, AFM_1 and AFM_2, due to ingestion of contaminated feed consumed by livestock. Milk and milk products (e.g. cheese) are often contaminated with AFM_1 and AFM_2; aflatoxins are produced by members of the *Aspergillus* spp. (*A. flavus* and *A. parasiticus*) (Zain 2011). The International Agency for Research on Cancer (IARC) classified AFB_1 and AFM_1 as a Group 1 carcinogen because they have been linked to cause human primary liver cancer synergistically with hepatitis B virus (HBV) (Mosiello and Lamberti 2009). The high consumption of aflatoxins by children shows evidence of growth impairment and stunting, which make these children more susceptible to other illnesses. High exposure levels of aflatoxins can lead to acute poisoning and even deaths.

12.2.2 Fumonisins

Fumonisins (FUM) manifestation is ubiquitous in corn, produced by *Fusarium* spp. (*F. verticilloides* and *F. proliferatum* which is responsible for high production of FUM). Fumonisins B_1 (FB_1), B_2 (FB_2) and B_3 (FB_3) are predominantly found in contaminated corn. FB_1 is abundant, representing 70% concentration in contaminated foods and feeds compared to FB_2 and FB_3 (Direito et al. 2008). FUMs are hepatotoxic and nephrotoxic and affect the immune system. FUMs are structurally similar to sphingosine, a component of sphingolipids that are composed in myelin and in certain nerve tissues. FB_1 is classified as a Group 2B carcinogen with associated cases of oesophageal cancer reported in Egypt, China, South Africa and the United States (Milicevic et al. 2010).

12.2.3 Ochratoxins

Ochratoxins are another class of mycotoxins produced by several species of *Aspergillus* and *Penicillium* (Muñoz et al. 2010). Ochratoxins consist of ochratoxin A (OTA), ochratoxin C (OTC), ochratoxin B (OTB) and ochratoxin α (OTα) (Ringot et al. 2006). OTA is the most toxic in this group; it is a nephrotoxic and nephrocarcinogenic (Milicevic et al. 2010) and is prevalently found in cereals, coffee, dried fruits, spices, grape juice and animal feeds. Heussner and Binge (2015) reported once ingested OTA has a high affinity to bind serum albumin and is associated with Balkan endemic nephropathy which is a renal disease that can cause kidney and liver failure (Roberts 1995; Ringot et al. 2006). The IARC classified OTA as a Group 2B carcinogen based on its possibility to induce oxidative DNA damage that causes immunosuppression and toxicity (Ringot et al. 2006; Kaushik et al. 2009).

12.2.4 Trichothecenes

Trichothecenes (TCs) is a group of immunosuppressive mycotoxins that causes gastrointestinal adverse effects such as diarrhoea, vomiting and to some extent abortion. They are mostly produced by *Fusarium* genus, and the TCs are divided into four groups based on their chemical properties and producing fungi (Yazar and Omurtag 2008). Amongst the TCs, mycotoxins T-2 toxin and deoxynivalenol are frequently occurring toxins. T-2 toxin is the most toxic and predominantly contaminates barley, corn, wheat and rice crops both in the field and during storage. T-2 toxin has some detrimental effects to human and animal health; this includes irreversible damage to the bone marrow, inhibition of protein synthesis and reduction in white blood cells (Zain 2011).

12.2.5 Zearalenone

Milicevic et al. (2010) reported that zearalenone (ZEN) has the ability to cause osteogenesis and reproductive system toxicity in animals; ZEN is produced by *Fusarium* spp. (*F. graminearum and F. culmorum*) as a naturally occurring endocrine-disrupting chemical. These species evade maize, wheat, oats and sorghum. Most mycotoxins have thermal stability ranging between 80 and 121 °C cooking conditions, and even pasteurization to some extent causes less destruction (Bosco and Mollea 2012).

12.3 Conventional Techniques for Mycotoxin Detection

The conventional techniques for the detection and monitoring of mycotoxins rely on analytical methods which require the use of expensive instruments such as chromatographic methods which are time consuming and require trained personnel to carry out the analysis. The Association of Official Agricultural Chemists (AOAC) has established methodologies that make use of inter-laboratory-validated methods of analysis towards improved reliability of mycotoxin data analysis. There are approximately 45 analytical methods used for the determination of mycotoxins, which are constantly being updated and reviewed by the AOAC (2005). However, analytical quality assurance and quality control protocols are prerequisite for adequate food safety and security. Emerging mycotoxins pose a challenge of being masked; therefore the determination of these mycotoxins by conventional analytical procedure is limited due to modified compounds becoming un-extractable by the extraction solvents used for their parent compounds.

12.3.1 Chromatographic-Based Techniques

This technique is frequently used for routine analysis; chromatographic-based methods include thin-layer chromatography (TLC) primarily used for the screening and detection of mycotoxins; OTA concentrations ranging from 0.2 to 136.7 μg/kg have been detected in coffee using TLC. Liquid chromatography (LC) which comprises of the high performance liquid chromatography (HPLC) and gas chromatography (GC) often coupled to an ultraviolet (UV), fluorescence detector (FLD) or mass spectrometric (MS) detectors for the quantification of mycotoxin (Anfossi et al. 2016) as shown in Table 12.1. These sophisticated and expensive instruments allow for the determination of mycotoxins with high sensitivity. Instruments coupled with tandem mass spectrometry (MS/MS) have an advantage of accuracy, selectivity and high throughput (Capriotti et al. 2014; Sarkar et al. 2009). The mass spectrometry provides that benefit of multiple analysis of different mycotoxins and their derivatives according to their molecular weight and can provide structural data for the identification of unknown compounds. Mycotoxins can be measured with liquid chromatography with mass spectrometry LC-MS/MS analysis; however this approach is time-consuming and requires additional steps for extraction, pretreatment and clean-up that can affect the result (Mosiello and Lamberti 2009; Muñoz et al. 2010).

12.3.2 Immunochemical-Based Techniques

The growing demand for rapid, portable, affordable and easy-to-operate systems for mycotoxin detection was introduced by immunochemical-based tests (Tang et al. 2014; Zhang et al. 2015). Commercially available enzyme-linked immunosorbent assay (ELISA) and electrochemical immunoassay for the detection of mycotoxins have been widely used due to their increased stability, on-site detection and ability to immobilize specific antibodies for mycotoxin detection (Anfossi et al. 2016). ELISA has been a gold standard for toxin detection for decades; mycotoxins are low molecular weight molecules that fail to induce significant refractive index and chemiluminescent/bioluminescent signal upon binding to the surface of the sensor. Biosensors are a form of immunochemical-based technique that can be used with synthetic ligands for binding mycotoxins (Tothill 2011). They consist of the following (Fig. 12.1):

1. A ligand (e.g. antibodies, enzymes, nucleic acids, etc.) integrated on a surface platform of the sensor that have a high binding affinity to a mycotoxin(s)
2. A transducer element (e.g. calorimetric, electrochemical, magnetic, mass sensitive) that converts mycotoxin binding to an optical and electrical signal
3. An electronic system for easy display of the results

Table 12.1 Analytical methods used for the detection of mycotoxins

Methods	Advantages	Disadvantages
Chromatography		
TLC	Inexpensive, fast and simple Simultaneous detection of multiple mycotoxins	Poor sensitivity (only suitable for OTs and AFs) and precision Mainly used for screening, quantification is possible with a densimeter
GC	Good sensitivity Mass detection provides confirmation Automated simultaneous detection of mycotoxins	Requires expertise for operation and derivatization Expensive instrumentation Variation in reproducibility Drifting response that can cause carry-over effect Non-linear calibration curve
HPLC	Good sensitivity and selectivity Automation allows for short analysis time	Expensive instrumentation that requires expertise
LC-MS	Good sensitivity Simultaneous detection of mycotoxins No derivatization required	Requires expertise Expensive instrumentation Matrix-assisted calibration curve for quantification
Enzyme-based assays		
Enzyme-linked immunosorbent assay (ELISA)	Inexpensive assay with simple sample preparation Visual assessment for screening Less organic solvents are used	Matrix interface problem that can lead to cross-reactivity of related mycotoxins Reliability issue with possible positive/negative results
Real-time polymerase chain reaction (RT-PCR)	Good sensitivity and accuracy for high-throughput analysis	Time consuming and errors can occur during polymerization
Biosensors		
Electrochemical biosensor	Cost-effective and rapid Portable for on-site analysis with good selectivity and sensitivity No expertise required	Choice of immobilizing matrix is crucial to obtain enhanced signal amplification

12.3.3 Microarrays for Mycotoxin Detection

Microarrays are a multiplex lab-on-a-chip device that provides the ability to simultaneously detect multiple mycotoxins in a single experiment. This is achieved by immobilizing two or more different types of ligands that can specifically and selectively bind to their respective mycotoxins. Lamberti et al. (2010) demonstrated the detection of AFB_1 and FB_1 using a microarray. Immunochemical-based methods are limited to the detection of a single mycotoxin and detection/identification of unknown toxins. One of the most important parameters in mycotoxin detection is the use of certified reference standards that warrants reliable mycotoxin determination.

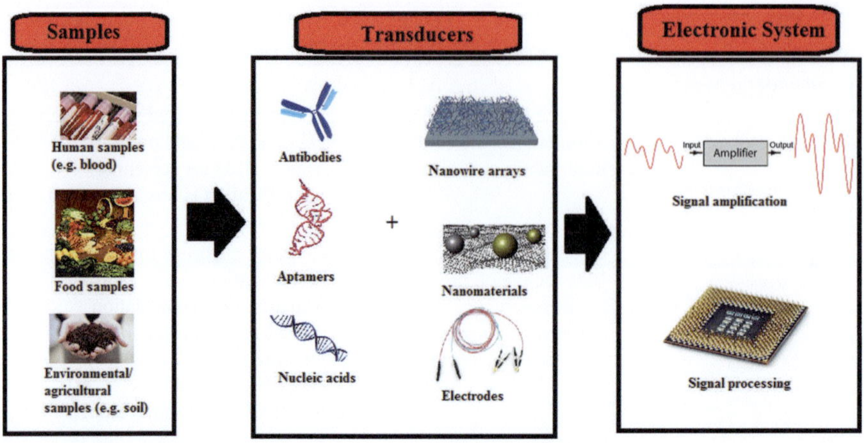

Fig. 12.1 Components of a biosensor include a ligand, transducer and electronic system

12.4 Nanotechnologies for Mycotoxin Detection

Nanotechnology has presented exceptional advancements in a variety of applications due to the fabrication of nanomaterials exhibiting unprecedented properties which include high surface-to-volume ratio that provides improved matrix for the immobilization of desired biomolecules (e.g. antibodies, peptides, DNA, RNA and aptamers) to increase loading capacity per nanoparticle (Kaushik et al. 2009; Sertova 2015). Antibodies are wildly used as molecular recognition receptors for toxin detection due to their specificity and sensitivity (Tothill 2011). Recently, synthetic receptors such as aptamers, peptides, proteins and imprinted polymers have been coupled with nanomaterials for the development of devices for mycotoxin detection (Peters et al. 2014).

Aptamers are synthetic oligonucleotides made via synthetic evolution of ligands (SELEX); they exhibit superior specific affinity towards target molecules than antibodies. Aptamers have been used for OTA detection which was achieved by a fluorescence polarization immunoassay (Cruz-Aguado and Penner 2008). Work by Tothill (2010), at Cranfield University, used computational approaches to design specific peptides for mycotoxin detection (AFM_1 and OTA) through molecular modelling software to optimize binding affinities. Imprinted polymers have been used as clean-up media (Turner et al. (2004); Boulanouar et al. (2018)) because they can be engineered as selective sorbents for the solid phase extraction of mycotoxins such as OTA, FB and T-2 (De Smet et al. 2009).

This has the potential to improve food quality/safety and surveillance through the tracking nanosystems. Moreover, multiplex ability for the conjugation of multiple enzyme molecules to the surface of nanomaterial increases stability and allows for the detection of more than one mycotoxin in situ on one platform. High surface reactivity for improved sensitivity; high catalytic efficiency for increased selectivity and strong absorption capability for signal amplification. The fabrication of existing methods with nanomaterials has shown to improve sensitivity and performance of the sensors by enhanced mycotoxin detection that allows for signal amplification.

Yin et al. (2018) engineered a label-free method for the detection of AFM_1 and OTA by utilizing DNA strand displacement and G-quadruplex-specific fluorescence probe amplification; briefly, the assay uses double-stranded aptamers, where one of the strands is specific to target mycotoxin (either AFM_1 or OTA). Detection of the target mycotoxin triggers strand displacement DNA amplification that assembles to G-quadruplex to amplify the fluorescence of N-methyl mesoporphyrin IX (NMM). Increase in DNA amplification causes an increase in fluorescence, which is directly proportional to mycotoxin detection. The limit of detection (LOD) for the system was 17.79 ng/kg and 18.98 ng/kg for AFM_1 and OTA, respectively.

12.4.1 Nano-Based Immunosensors

The application of nanotechnology with the incorporation of nanomaterials in biosensor and immunosensor systems for mycotoxin detection has improved signal amplification. Nanoparticles can be functionalized with a variety of biomolecules (e.g. antibodies, peptides, proteins and nucleic acids). Immunosensors fabricated with nanomaterials have been used for mycotoxin detection due to biocompatibility, robustness, low production cost with no toxicity, easily modified to aid immobilization, integrating labels as markers for detection and better thermal, chemical and physical properties. Multiplex analysis is achieved towards nanofluidics through advanced fabrication (Tothill 2011); the nanofluidics make use of nanoelectrode array chips with multi-array working electrodes that can immobilize different antibodies to detect specific mycotoxins. Nanofluidics are composed of small capacitive charging current that allows faster diffusion rate through capillary motions for improved sensitivity and response time.

12.4.1.1 Metallic Nanoparticles

Metallic nanoparticles have been widely used for immobilization of biomolecules in the construction of sensors. Gold nanoparticles (AuNPs) are extensively incorporated in bioelectric devices due to their high biocompatibility, ease in surface functionalization/modification and surface plasmon resonance (SPR) for increased signal. Owino et al. (2008) exploited the use of polythionine-modified electrode AuNP conjugate in electrochemical immunosensors for AFB_1 absorption. This approach requires no secondary labelled antibodies because the AuNP signal is directly proportional to the amount of bound AFB_1.

Metal oxide nanomaterials have attracted interest because they are chemically inert and have high mechanical and thermal strength and high adsorption capability

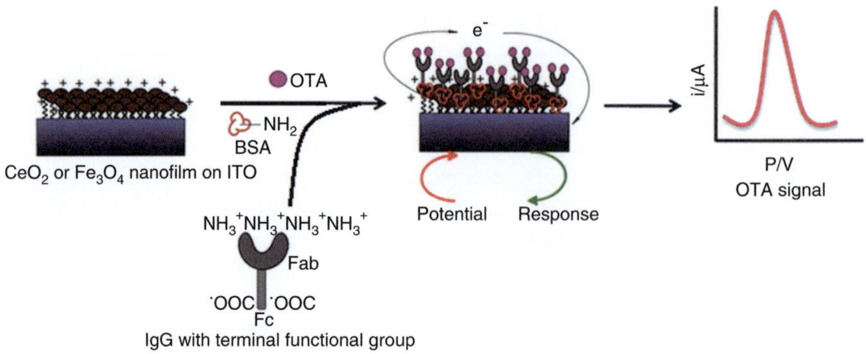

Fig. 12.2 Biochemical reaction CeO_2 or Fe_3O_4 – BSA/r-IgGs/ITO immunoelectrode for detecting OTA molecules

with increased isoelectric point (IEP) for increased immobilized, tunable optical, chemical and physical properties. These nanostructures include cerium oxide (CeO_2), tin oxide (SnO_2), titanium oxide (TiO_2) (Khan and Dhayal 2008), zinc oxide (ZnO) and zirconium oxide (ZrO_2). Electrochemical immunosensor has been developed by Kaushik et al. (2009) which uses nanomaterials of chitosan (CH)–iron oxide (Fe_3O_4) nanoparticles and nanostructured CeO_2, film immobilized with rabbit-immunoglobulin antibodies (r-IgGs) and bovine serum albumin (BSA) that was fabricated onto an indium–tin-oxide (ITO)-coated glass plate for the detection of OTA as shown in Fig. 12.2 (Kaushik et al. 2013). Their work demonstrated that the immunoelectrode with CeO_2 nanoparticles performed better than Fe_3O_4 nanoparticles with improved LOD (from 0.5 (Fe_3O_4) to 0.2 (CeO_2) ng/dl), high sensitivity (36 to 1.27 μA/ng/dl cm^{-2}, respectively) and a high association constant (K_a, 0.9×10^{11} l/mol) and a 30s response time.

Sharma et al. (2018) developed a platform for the rapid electrochemical detection of AFB_1, the platform is composed of poly(3, 4-ethylenedioxythiophene) and graphene oxide doped with AuNPs by electrochemical deposition onto a glass carbon electrode, and this was covalently linked with anti-aflatoxin B_1 antibody (anti-AFB_1) via EDC/NHS coupling. The immunosensors exhibited high sensitivity between 0.397 μA ng/mL and 0.989 μA ng/mL. A nanoporous cobalt (NPCo)/cobalt (II, III) oxide alloy functionalized with AuNPs was used to detect deoxynivalenol (DON) using an electrochemiluminescence (ECL)-modified electrode (Lv et al. 2015). Microfluidic device has been extensively studied for their application in mycotoxin detection. Biosensors immobilized with antibodies suffer from specificity and sensitivity. The integration of nanomaterials in the fabrication of biosensors has vastly improved the performance of biosensors. In the construction of aptasensors, nanomaterials provide an immobilization support, to increase signal amplification and a platform for artificial enzyme labelling (Rhouati et al. 2017).

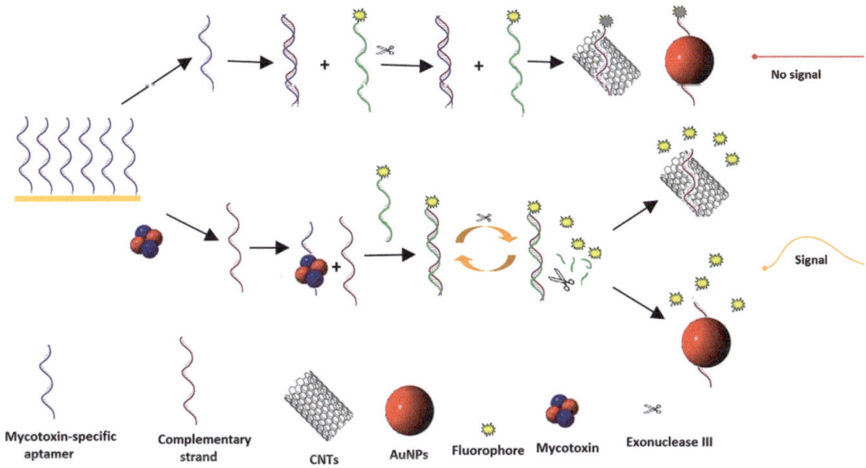

Scheme 12.1 Nano-platform for the detection of mycotoxins. In the absences of mycotoxins, the nanomaterials (e.g. CNTs or AuNPs) quenches the fluorophore, and in the presence of mycotoxins, the fluorophore is released to give a signal signifying mycotoxin detection

Aptasensors provided the advantage of high specificity for the detection of AFB_1 and OTA. Aptamer-A and aptamer-O have affinity to bind AFB_1 and OTA, respectively (Yuan Zhao et al. 2015), with the detection limits at pictogram and femtogram levels. However, their drawback is poor signal amplification that can lead to false-positive results. The incorporation of nanoparticles can increase surface-enhanced Raman scattering that can improve signal amplification for mycotoxin detection. Zhao et al. (2015) reported the use of bimetallic Ag@Au core shell nanoparticles synthesized through galvanic replacement-free deposition for multiplex detection of OTA and AFB_1. The method exploits the properties of both metals, Ag functions as a plasmonic enhancer and the Au shell that also functions to improve the stability prevents the oxidation of Ag. A strong plasmonic coupling, which greatly amplifies the SERS signal, is achieved due to the Ag@Au nanoparticles demonstrated 0.03 ng/mL and 0.006 ng/mL LOD for AFB_1 and OTA in maize meal. Scheme 12.1 demonstrated an exonuclease III assisted system for the detection of mycotoxins and the signal amplification is enhanced by the presence of nanoparticles.

Carbon nanotubes in the form of single-walled carbon nanohorns (SWCNHs) have been used as quenchers of fluorescence in a fluorometric aptamer assay for the detection of OTA (Wu et al. 2018). The system is composed of two probes; the signal probe (SP) is labelled at its 5′-end with carboxyfluorescein (FAM) with an excitation/emission maxima at 495/518 nm quenched by SWCNHs. The hairpin probe (HP) contains OTA-specific aptamer sequence, complementary strand sequence to the SP, and in the presence of OTA, the aptamer binds OTA and the single-stranded sequence hybridizes with SP. Exonuclease III digests the SP to fluorescent since it is no longer quenched by SWCNHs and the LOD was determined to be 4.2 nM. Similar work by Zhao et al. (2018) was carried by using AuNPs instead of SWCNHs for quenching and they reported a LOD of 4.82 nM.

12.4.1.2 Magnetic Nanoparticles

Radoi et al. (2008) demonstrated that super paramagnetic nanoparticles could be used to detect AFM_1. Magnetic nanoparticles coated with antibody (anti-AFM_1 or anti-OTA) were used to bind antigen (AFM_1 or OTA) in AFM_1 and OTA contaminated milk and cereal with a detection limit of 0.01 μg/L and 40 nM with a good reproducibility, respectively (Selvaraj et al. 2015; Man et al. 2017). Sol-gel aflatoxin-oxidase conjugated multi-walled carbon nanotubes (MWCNTs)-modified Pt electrode biosensor was developed by Tsai and Hsieh (2007) to respond to the oxidation of AFB_1 with a LOD (1.6 nmol/L), high sensitivity (0.33×10^2 A/ng/dl cm^{-2}, respectively) and a 44 s response time.

The collection of mycotoxins plays a role in their determination; a study by Sang-ho et al. (2010) used magnetic nanoparticles (MNPs) for the collection and determination of AFB_1 and ZEN in feed. The surface of the nanoparticles was coated with silica (SiO_2) and modified with aminoorganosilanes for the conjugation of monoclonal antibodies (kj-AFB and kk-ZEN) against AFB_1 and ZEN, respectively. The C-AFB_1/MNP@SiO_2 system showed a 90.28% and 83.7% recovery in 10 ppb AFB_1 spiked corn and feed while the C-ZEN/MNP@SiO_2 system recovered 99.9% and 86.0% 50 ppb ZEN in spiked corn and feed, respectively. These results demonstrated that this approach was more efficient in collection of mycotoxin in feed compared to conventional immunoaffinity chromatography (IAC) methods. This novel collection system is useful in monitoring and regulating mycotoxin-contaminated feed.

Another study by Mak et al. (2010) investigated the use of a magnetic nanotag-based immunoassay for simultaneous detection of mycotoxins (AFB_1, ZEN and HT-2) at sub-picomolar concentration levels (Actis et al. 2010). Briefly, ~ 10 nm MNPs are coated with streptavidin-biotinylated antibodies (anti-AFB_1, anti-ZEN and anti-HT-2) that are fabricated on a chip in a close proximity with giant magnetoresistive to measure the change in resistance upon mycotoxin binding with LOD at 50 pg/mL for AFB_1 and ZEN and 333 pg/mL for HT-2. This immune-MNP assay can be adapted as a point of use for mycotoxin testing.

Actis et al. (2010) developed a new signal transduction by ion nano-gating (STING) technology to detect HT-2 toxin. This system uses functionalized quartz nanopipette as an electrochemical biosensor; the nanopipette is immobilized with poly-L-lysine (PLL) physisorbed on a negatively charged quartz surface, which is cross-linked with sulfosuccinimidyl 4-(N-malemidomethyl)cyclohexane-1-carboxylate (Sulfo-SMCC) and the maleimide group of the PLL bind through a thio-ether bond with anti-HT-2 monoclonal antibody. The STING platform's LOD was 100 fg/ml for HT-2 detection. The advantage of incorporating nanoparticles in biosensors is enhanced binding specificity that enables portable, cheap and fast analysis.

12.5 Nanotechnology in Precision Agriculture

The era of nanotechnology in agriculture is still at its early stages with the aim to maximize crop production while minimizing the use of pesticides and herbicides (Peters et al. 2014; Sertova 2015). Application of nanotechnology in medicine has been used for precision medicine; this rationale also applies to the pursuit of precision agriculture. Precision agriculture is the continuous measurement of inter- and intra-field variability using biosensors for detecting pesticides, plant diseases and mycotoxin contamination towards smart farming (Duhan et al. 2017). The permits of environmental changes through climate change have undoubtedly affected agriculture in a significant way. The use of nanomaterials in equipment for increased sensitivity allows early response to these environmental changes. There is a growing need for the formulation of nanopesticides through the encapsulation of pesticides that are triggered in the presences of toxigenic fungi to minimize mycotoxin production. The steady increase in the application of nanomaterials in the food sector relates to concerns about their safety since there is limited knowledge of nanotoxicology towards human, animal and plant health.

Nanosensors based applications to monitor crops at a cellular level to alert farmers when the crops require nutrients, water, fertilizer and surveillance of pathogens (e.g. bacteria, viruses, fungi, pests, etc.) and to some extent monitor soil conditions in order to optimize growth and production of individual crops are required. The quality of agricultural products is currently being monitored by biosensors which make use of nanotechnology (Sertova 2015). There is a growing interest of using integrating nanoparticles in innovative packaging materials to monitor the freshness of the food and optimize the shelf life of product by a colour change response (Hosseini et al. 2011; Soleimanpour et al. 2012), and this will be beneficial during food storage. The potential application of smart packaging is to respond to moisture and temperature changes and alert the customer if the food is contaminated with mycotoxins and/or other pathogens.

12.5.1 Nanopackaging

The increased application of nanoscience and nanotechnology has found prospects in the food industry towards novel packaging materials by using nanocomposite systems with antimicrobial properties to prevent microbial contamination for extending shelf life and securing food safety (Prasad et al. 2014; Deng et al. 2017; Fortunati et al. 2017) as shown in Fig. 12.3. Nanomaterials that have received attention in the development of new-generation packaging include metal nanoparticles (AgNPs and CuNPs), metal oxide nanomaterials (ZnONPs and TiO_2NPs), CNTs and organic and inorganic polymers (Llorens et al. 2012). Fungi are susceptible to AgNPs and CuNPs that reduce mycotoxin contamination in food and feedstuff; this is due to the antimicrobial actions of these nanoparticles by affecting the

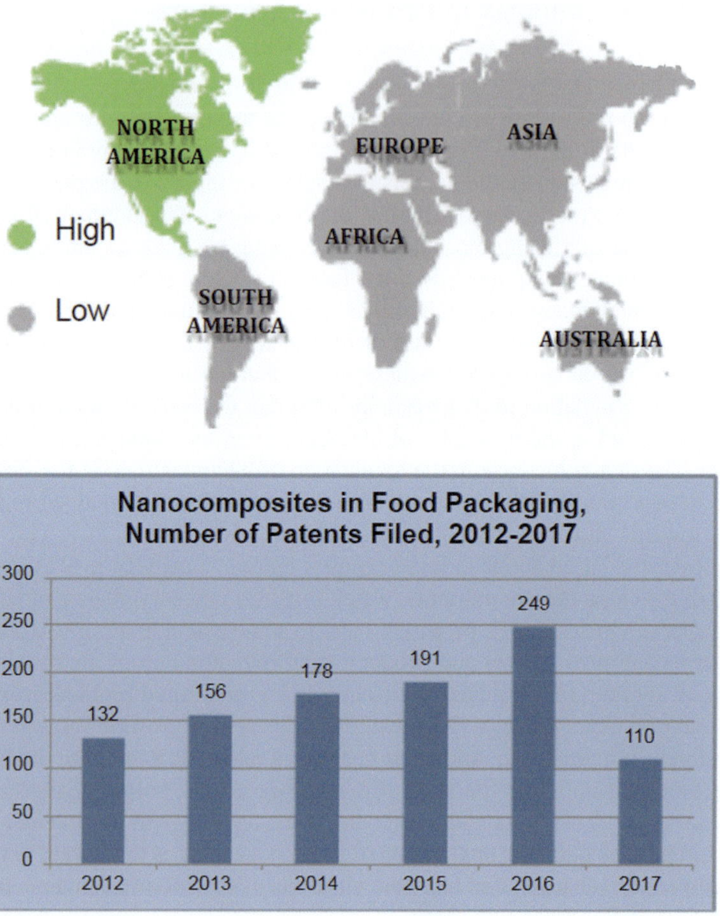

Fig. 12.3 Nanopackaging by using nanocomposite materials in the food packaging industry. North America dominates the development of these nanocomposite solutions, and novel metallic nanocomposites are explored in China. The number of patents filed in the area of nanopackaging include clay, silica and ZnO nanocomposites dispersion in various matrixes (Adapted from Frost and Sullivan 2017a)

permeability and functionality of the outer cell membrane (Chatterjee et al. 2012; Llorens et al. 2012). The FDA has approved ZnO and TiO_2 in cosmetics, drugs and food packaging (Othman et al. 2014); however, the antimicrobial activity of TiO_2NPs requires photocatalysis in the presence of UV irradiation (Kim et al. 2003; Llorens et al. 2012).

The use of CNTs in packaging is not approved because they are cytotoxic to human cells. The use of nanocomposites has been the choice in the fabrication of antimicrobial packaging polymer-based nanoparticles such as polylactic acid (PLA), polyvinylpyrrolidone (PVP), low-density polyethylene (LDPE), gelatin and biopolymer chitosan doped with AgNPs and ZnONPs because of ease in availability, biocompatibility, biodegradability and low cost (Brandelli et al. 2017). Organic nanoparticles include lipid-based nanoparticles composed of fatty acids, phospholipids, steroids and triacylglycerides. The most common are liposomes for the encapsulation of antimicrobial agents; Silva et al. (2014) demonstrated the encapsulation of nisin into nanovesicles containing chitosan that showed efficient inhibition of *L. monocytogenes*. Composite polymer nanofibers can be functionalized with AgNPs to inhibit microbial proliferation on solid surfaces used for the storage of food and feedstuff.

There are various methods used for incorporating, grafting antimicrobial nanoparticles within and/or on packaging materials/surfaces which include:

1. Solvent casting method: uses a thin polymer film which is dissolved in an organic solvent, and nanoparticles are added into the solution to produce a nanocomposite
2. Thermal method: incorporating nanoparticles in LDPE by melt bending, hot pressing at 140 °C and extrusion to produce a nanocomposite film
3. Pure wet chemical pretreatment: film functionalized with certain functional groups or charges to bind nanoparticles by covalent bonding, electrostatic force and hydrogen bonding
4. Plasma pretreatment: surface treatment to create a dielectric barrier discharge for coating with nanomaterials
5. Deposition: simultaneous sputtering nanoparticles on a support matrix (e.g. organosilicon matrix) and plasma polymerization to produce a nanocomposite

Appell and Jackson (2012) reported the use of nanosponge materials, which are a class of novel sorbents, composed of β-cyclodextrin-polyurethane. Their studies suggested that the β-cyclodextrin nanosponge absorbed approximately 1–10 μg/L of OTA from red wine samples. This revealed that nanosponges could be potentially used in the remediation of mycotoxins from foodstuff.

12.5.2 Green Nanofungicides to Control Toxigenic Fungi and Mycotoxins

12.5.2.1 Phytochemical Properties for Mycotoxin Protection

Plants have a defensive mechanism to protect themselves agents pests, insects and to some extent fungi. The plant releases secondary metabolites called phytochemicals that are active chemical compounds with protective properties (Prasad et al. 2017b). OMAF (2004) eluded that these phytochemicals can be used against toxigenic fungi and mycotoxins. Phytofungicides are formulations developed from phytochemicals extracted from plant materials with antibacterial and antifungal activity. A number of botanical extracts have been reported to exhibit fungicidal properties attributed to phytochemicals present in the extracts. This presents an alternative to synthetic fungicides that improves protection but also has health hazards to humans and animals due to toxic chemicals used in their formulations (Reddy et al. 2007).

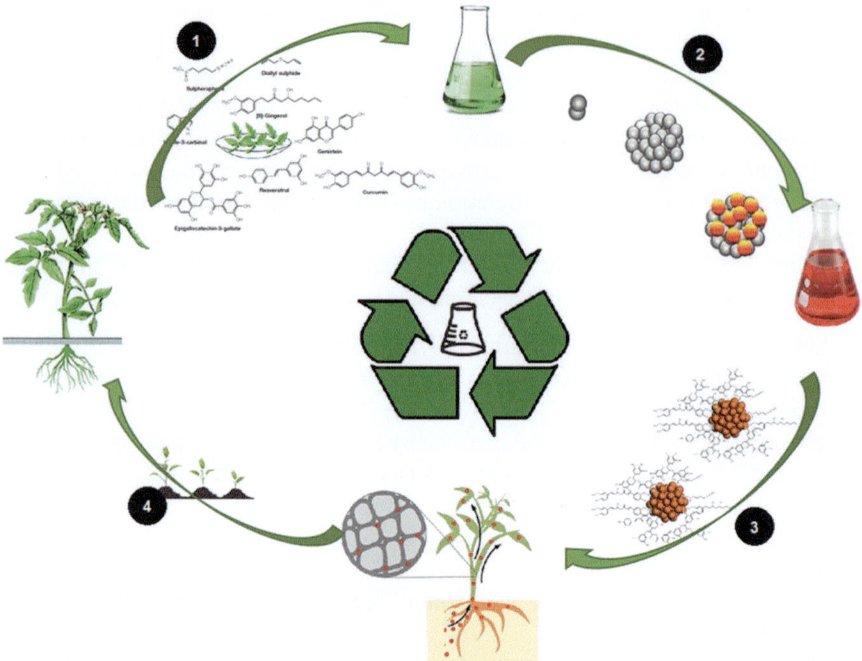

Fig. 12.4 Green nanofungicides engineering for sustainable agriculture, (1) the active phytochemicals in plant materials that can be used (2) to synthesize nanoparticles for the formation of (3) nanofungicides to protect plants from pathogen and (4) ensure food production and public health

The approach of using phytochemicals has great potential because they are easy and cheap to prepare; and most importantly, they are natural constituents from plants that present no toxicity to human and animal health. A variety of phytochemicals (catechols, eugenol, essential oils, phloretin, hexanal, d-limonene, menthol, caffeic acid, thymol, tannins, etc.) have been reported to have fungicidal properties (Juglal et al. 2002; Hasan et al. 2005; Bugno et al. 2006; Reddy et al. 2007; Prasad 2014; Llana-Ruiz-Cabello et al. 2015; Perricone et al. 2015 and Fortunati et al. 2017). Recent studies have demonstrated the power of phytochemicals against toxigenic fungi that subsequently resulted in reduced mycotoxin production. Mondali et al. (2009) reported extracts of garlic, ginger, neem leaf and onion bulb significantly inhibited of *A. flavus*. A number of studies have used phytochemicals for the synthesis of nanomaterials, and these nanomaterials are loaded with phytochemicals with antimicrobial properties that can be used to protect plants from toxigenic fungi and reduce mycotoxin production in the food chain (Fig. 12.4) (Shukla et al. 2008; Nune et al. 2009; Mallikarjuna et al. 2014; Kuppusamy et al. 2016).

Challenges in the Use of Phytofungicidal Agents

The practical application of fungicidal agents from plant materials is hampered by lack of infrastructure and instruments that allows the high extraction of these active phytochemicals from plant materials (Akunyili and Ivbijaro 2006). The effects of climate change tremendously cause scarcity of certain plant materials that have high concentrations of phytochemicals and the low shelf life of the phytochemicals due to thermal and photodegradation (Singh et al. 2017). The large-scale production of the phytofungicidal agents is also a problem that hinders their application in agriculture.

12.5.3 Nanofungicidal and Nanofertilizers Agents

Professor Kattesh Katti has pioneered the field of green nanotechnology, by using phytochemicals from plant materials to produce nanomaterials that are eco-friendly and present no toxic effects towards human and animal health (Katti et al. 2009). This approach is feasible because no toxic waste is generated in the production and the synergistic effect of loading more phytochemical molecules onto the surface of nanoparticles due to exposed surface atoms increases their antifungal activity. Nanofungicidal agents include nanofertilizers and nanopesticides; their formulation contains nanomaterials (such as gold, liposomes, iron, titanium dioxide, silica, zinc nanoparticles, etc.) that exhibit antifungal activities (Prasad et al. 2014, 2017a; Bhattacharyya et al. 2016).

The field of nano-phytopathology has emerged which combines the knowledge of understanding the plant disease state and applying nanotechnology for controlling and/or prevents fungal growth that reduces mycotoxin contamination (Hussain 2017). ZnONP application in agricultural sciences is vastly investigated because they are biocompatible and present no health hazards in human and their antimicrobial activity is independent on photoactivation (Abd-Elsalam et al. 2017). A number of studies have demonstrated that ZnONPs can inhibit the proliferation of toxigenic strains of *Aspergillus* spp. (*A. flavus, A. fumigatus* and *A. ochraceus*), *Fusarium* spp. (*F. graminearum* and *F. oxysporum*) and *Penicillium* spp. (*P. citrinum* and *P. expansum*) that subsequently reduced the production of AFB_1, OTA and FB_1 (Hassan et al. 2013; Nabawy et al. 2014).

Islam et al. (2015) synthesized AuNPs functionalized with *Salix alba L.* leaves extract (Au-WAs). The active phenolic compounds (salicin, quercetin and tannins) in this plant that are highly electron-rich were responsible for reducing the gold precursor to AuNPs. The Au-WAs exhibited antifungal activity against *A. solani*, *A. flavus* and *A. niger*. Certain metabolites from *Salix alba* have been reported to induce apoptosis in various tumour cells (El-Shemy et al. 2007). Kumar and Sudha (2013) used aqueous extract of brown seaweed *Dictyola bartayresiana* as reducing agents for the synthesis of AgNPs. This resulted in the production of AgNPs ranging from 15 to 40 nm in size, and the AgNPs showed considerable antifungal activity against *Fusarium dimerum* and *Humiclo insulans*. Carnivorous plants (*Drosera indica*, *Drosera binata*, *Drosera spatulata* and *Dionaea muscipula*) were used to produce AgNPs that exhibited antifungal activity against *Candida albicans* (Banasiuk et al. 2017). *Citrullus colocynthis* is a medicinal plant that grows in some African countries that has been used in folk medicine because of its antidiabetic, antioxidant and anti-inflammatory activities. Shawkey et al. (2014) used aqueous extract of the plant to generate AgNPs that showed antifungal activity against *A. fumigatus* and *Candida albicans* with inhibition zones from 15.1 ± 0.44 to 25.2 ± 0.37 mm.

One of the major constituents that influence the high yield of agricultural produce is the availability of essential growth-simulating micronutrients (Co, Cu, Fe, Mn and Zn). These elements are responsible for increasing yield and plant reproduction. Nanofertilizers can be utilized for the delivery of essential micronutrient in deficient soil composition (Gholami-Shabani et al. 2017). Iron oxide and zinc oxide nanoparticles improved seed germination by supplying requisite amount of Fe and Zn and the increased absorption of phosphorus and potassium to plants (Adhikari et al. 2016; Askary et al. 2016). In addition, Deshpande et al. (2017) synthesized zinc complex chitosan nanoparticles (Zn-CNP) for the improvement of Zn content in durum wheat cultivars. Through foliar application, the authors observed an increase in seed Zn content. Furthermore, Jeyasubramanian et al. (2016) applied iron oxide (Fe_2O_3) nanoparticles onto spinach plants and observed improved plant growth as well as an increase in Fe uptake. Peralta-Videa et al. (2014) applied CeO_2 and ZnO nanoparticles to soybean plants and observed that the nanoparticles improved the nutritional quality of soybean.

Climate change has resulted in increased temperatures and reduced rainfall which compromises the irrigation of crops in agricultural farming. A study by Vundavalli et al. (2015) investigated the use of biodegradable nano-hydrogels as super absorbent polymers for water absorption and retention under high temperatures. The nano-hydrogels were composed of cross-linked polyacrylamide polymers coated with Ag nano-clay composite, and their data demonstrated revealed that soil with this nano-hydrogels retained 7.5% water compared to soils without the nano-hydrogels. That is, approximately 190 times more irrigation or rain water storage, the AgNPs and hydrogels presented a synergistic effect of water absorption and plant growth enhancement (Montesano et al. 2015), and this can be useful in areas experiencing drought to increase crop production. Furthermore, Dimkpa et al. (2017) tested whether ZnO, B_2O_3 and CuO nanoparticles can alleviate drought stress (at 50% field moisture capacity) in soybean plants. The authors observed an increase in soybean shoot growth in the nanoparticle and drought combination treatments. Dimkpa et al. (2017) then concluded that the nanoparticles improve soybean drought stress tolerance by increasing shoot N, K, Zn, B and Cu content. Green nanotechnology can thus

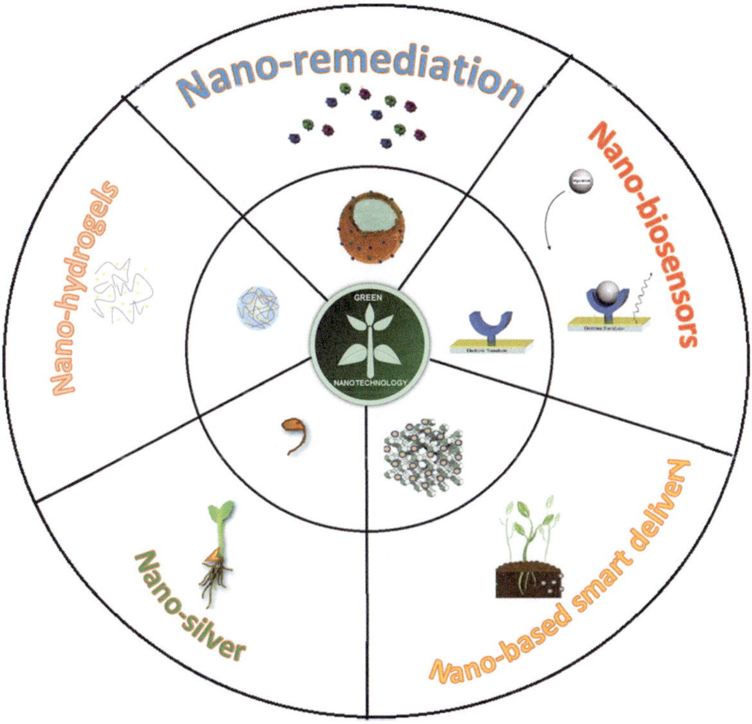

Fig. 12.5 Green nanotechnology in agriculture, nano-biosensors used for the detection of toxigenic fungi and mycotoxin detection; nano-based smart delivery of nutrients to improve availability for high yields; nano-silver for climate resilience to increased seed longevity for improved germination; nano-hydrogels for enhanced water absorption and release and nano-remediation for the reduction of mycotoxins

increase the quality of agricultural products and improve quantity yields by protecting crops from toxigenic fungi, add nutrients to the crops, detect mycotoxins and remediate them that contributes to the food safety and security (Fig. 12.5).

The use of green nanofungicides exploits the high surface-to-volume ratio of the nanoparticles to increase loading and encapsulation of phytochemicals to amplify the magnitude of phytochemical efficacy against toxigenic fungi and other plant pathogens. A research team from the Korea Advanced Institute of Science and Technology (KAIST) has demonstrated the use of plant-derived nanocoating spray with colloidal copper to extend the shelf life of agricultural produce (Frost and Sullivan 2017b). Park et al. (2017) used supramolecular Fe (III)-tannic acid nanocoating spray to prolong the shelf life of fresh produce. Kumar et al. (2016) converted two commercial fungicides [trifloxystrobin 25% + tebuconazole 50% (75 WG)] into a nanoform by using the ball milling method. The authors observed an enhanced fungicidal activity against the plant pathogen, namely, *Macrophomina phaseolina*. Furthermore, Mondal et al. (2017) converted azomethine fungicides into nanoform and observed a ± twofold activity increase against pathogenic fungi, namely, *Rhizoctonia solani*, *Rhizoctonia bataticola* and *Sclerotium rolfsii*. Plants do not pose circulating immune cells that destroy pathogens but have a rudimentary immune system; therefore, the application of green nanofungicides will assist the plant to fight back evading pathogens like fungi to prevent their manifestation.

12.6 Nanotoxicology

The regulatory standards of nanotechnology in the food industry are still at initial stages, and the leaching of nanoparticles from the nanocomposite food packaging into food during storage is a major concern (Faroodi 2015; Brandelli et al. 2017). Therefore, it is of paramount importance to evaluate the health and environmental safety of nanocomposite materials used in food packaging. Purohit et al. (2017) highlighted some of the ethical, environmental and social implications of nanotechnology that need to be taken into consideration. Nanoparticles have been reported to cause immunosuppression and promote inflammatory response due to the surface area which results in higher specific interaction with blood proteins, cells and tissues which increase the production of reactive oxygen species. Size of the nanoparticles plays an important role, where nanoparticles larger than 200 nm accumulate in the liver and spleen and are later processed by mononuclear phagocytic system cells (Cauerhiff et al. 2014).

The shape of nanoparticles is also under investigation since rod-shaped nanoparticles have been reported to have the highest cellular uptake and toxicity in in vitro studies (Tang et al. 2012) and the surface charge of nanoparticles influence their stability; cationic nanoparticles can cause the aggregation of hemolyte and platelets (Dokka et al. 2000). The use of green nanotechnology (phytochemicals and stabilizing agents from plant materials) in the formulation of nanoparticles eliminates these concerns because nanoparticles less than 6 nm will be eliminated through the kidneys (Cauerhiff et al. 2014).

12.7 Current Nanotechnologies for Mycotoxin in the Market

There are a number of commercially available assays, nanoformulations that are used in agriculture for a range of mycotoxins that use antibodies for detection as shown in Table 12.2.

Table 12.2 Commercial products with nanoformulations used in agriculture

Commercial brand	Constituents	Manufacturer
Nano-pro™	Nanofertilizer for agriculture and better farming	Aqua-Yield, Utah, United States
Biozar Nano-fertilizer	Combination of organic materials, micronutrients and macromolecules	Fanavar Nano-Pazhoohesh Markazi Company, Iran
CelluForce NCC™	Cellulose matrix for heat stability to prevent microbial growth	CelluForce Inc. Montreal, Canada
MycoFlex BAT	Multiplex mycotoxin detection device	Randox Food Diagnostics, Crumlin, United Kingdom
Nano-ag answer®	Microorganism, sea kelp and mineral electrolyte	Urth Agriculture, CA, United States
Nano-Gro™	Plant growth regulator and immunity enhancer	Agro Nanotechnology Corp., FL, United States
Nano green	Extracts of corn, grain, soybeans, potatoes, coconut and palms	Nano Green Sciences, Inc., India
Nanolok™	Nanodispersed silicates to prolong food shelf life	InMat, Inc, Raritan, United States
Nano max NPK fertilizer	Multiple organic acids chelated with major nutrients, amino acids, organic carbon, organic micronutrients/trace nutrients, vitamins and probiotics	JU Agri Sciences Pvt. Ltd, Janakpuri, New Delhi, India
Master Nano chitosan organic fertilizer	Water soluble liquid chitosan, organic acid and salicylic acid, phenolic compounds	Pannaraj Intertrade, Thailand
TAG NANO (NPK, PhoS, zinc, Cal, etc.) fertilizers	Proteino-lacto-gluconate chelated with micronutrients, vitamins, probiotics, seaweed extracts, humic acid	Tropical Agrosystem India (P) Ltd, India

Adapted from Prasad et al. (2017a)

12.8 Mycotoxins as Therapeutic Compounds

The recent advances in the fields of biochemistry, cellular biology, mycotoxicology, immunology and nanotechnology have shed light on how to better understand the biosynthesis of mycotoxins and their mechanism of action. This acquired knowledge has provided us with strategies of detecting mycotoxins and how these mycotoxins can be exploited to benefit society through the development of novel technologies for converting mycotoxins into therapeutic agents. This similar approach has been used for drug development from toxins such as bee, snake and frog venom. As mentioned above, some mycotoxins and their derivatives are used as antibiotics. Ergotamine, ergocryptine, ergometrine and the semi-synthetic methylergometrine derivatives of ergot alkaloid have been used for relieving migraine attacks, affecting dopaminergic activity and preventing and treating excessive bleeding of the uterus to initiating delivery and following childbirth. The bromation derivative is used against Parkinsonism (De Costa 2002; Eadie 2004).

12.9 Conclusion

The application of nanotechnology in agriculture shows tremendous potential in reducing mycotoxin production. The incorporation of nanomaterials such as AgNPs, CuNPs, TiO_2NPs and ZnONPs in nanocomposites for the fabrication of food packaging has inhibited the growth of fungi and other pathogens, thereby improving food safety and food quality for extended shelf life. The minimal toxicology data of nanomaterials still limits their application in agricultural practices. However, the use of green approaches through green nanotechnology warrants the production of nanoparticles that are eco-friendly and can find application in the agricultural sector.

Reference

Abd-Elsalam KA, Hashim AF, Alghuthaymi MA, Said-Galiev E (2017) Nanobiotechnological strategies for toxigenic fungi and mycotoxin control. In: Grumezescu AM (ed) Nanotechnology in the Agri-food industry, pp 337–364. https://doi.org/10.1016/B978-0-12-804303-5.00010-9

Adhikari T, Kundu S, Rao AS (2016) Zinc delivery to plants through seed coating with nano - zinc oxide particles. J Plant Nutr 39:136–146

Actis P, Jejelowo O, Pourmand N (2010) Ultrasensitive mycotoxin detection by STING sensors. Biosensors and Bioelectronics 26(2):333–337.

Akunyili D, Ivbijaro MFA (2006) Pesticide regulations and their implementation in Nigeria. In: MFA I, Akintola F, Okechukwu RU (eds) Sustainable environmental Management in Nigeria. Mattivi Production, Ibadan, pp 187–210

Anfossi L, Giovannoli C, Baggiani C (2016) Mycotoxin detection. Curr Opin Biotechnol 37:120–126

AOAC International (ed) (2005) Official methods of analysis of AOAC international, 18th edn. AOAC International, Gaithersburg

Appell M, Jackson MA (2012) Sorption of ochratoxin a from aqueous solutions using β-Cyclodextrin-polyurethane polymer. Toxins (Basel) 4(2):98–109

Askary M, Amirjani MR, Saberi T (2016) Comparison of the effects of nano-iron fertilizer with iron chelate on growth parameters and some biochemical properties of *Catharanthus roseus*. J Plant Nutr 40(7):974–982

Banasiuk R, Krychowiak M, Swigon D, Tomaszewicz W, Michalak A, Chylewska A, Ziabka M, Lapinski M, Koscielska B, Narajczyk M, Krolicka A (2017) Carnivorous plants used for green synthesis of silver nanoparticles with broad-spectrum antimicrobial activity. Arab J Chem. https://doi.org/10.1016/j.arabjc.2017.11.013

Bhattacharyya A, Duraisamy P, Govindarajan M, Buhroo AA, Prasad R (2016) Nano-biofungicides: emerging trend in insect pest control. In: Prasad R (ed) Advances and applications through fungal Nanobiotechnology. Springer International Publishing, Switzerland, pp 307–319

Bosco F, Mollea C (2012) In: Valdez B (ed) Mycotoxins in food, food industrial processes - methods and equipment ISBN: 978-953-307-905-9, InTech, Available from: https://www.intechopen.com/books/food-industrial-processes-methods-and-equipment/mycotoxins-in-food

Boulanouar SS, Mezzache A, Combes A, Pichon V (2018) Molecularly imprinted polymers for the determination of organophosphorus pesticides in complex samples. Talanta 176:465–478

Brandelli A, Lopes NA, Boelter JF (2017) Food applications of nanostructured antimicrobials. In: Grumezescu AM (ed) Nanotechnology in the Agri-food industry, pp 35–74. https://doi.org/10.1016/B978-0-12-804303-5.00002-X

Bugno A, Almodovar AAB, Pereira TC (2006) Occurrence of toxigenic fungi in herbal drugs. Braz J Microbiol 37:1–7

Capriotti AL, Cavaliere C, Foglia P, Samperi R, Stampachiacchiere S, Ventura S, Lagana A (2014) Multiclass analysis of mycotoxins in biscuits by high performance liquid chromatography-tandem mass spectrometry. Comparison of different extraction procedures. J Chromatogr A 1343:69–78

Cauerhff A, Martinez Y, Islan GA, Castro GR (2014) Nanostability. In: Durán N, Guterres SS, Alves OL (eds) Nanotoxycology: materials, methodologies, and assessments. Springer Science + Business Media, New York, pp 57–95

Chatterjee AK, Sarkar RK, Chattopadhyay AP, Aich P, Chakraborty R, Basu T (2012) A simple robust method for synthesis of metallic copper nanoparticles of high antibacterial potency against *E. coli*. Nanotechnology 23:085103

Cruz-Aguado JA, Penner G (2008) Fluorescence polarization based displacement assay for the determination of small molecules with aptamers. Anal Chem 80(22):8853–8855

De Costa C (2002) St Anthony's fire and living ligatures: a short history of ergometrine. Lancet 359(9319):1768–1770

De Smet D, Dubruel P, Van Peteghem C, Schacht E, De Saeger S (2009) Molecularly imprinted solid-phase extraction of fumonisin B analogues in bell pepper, rice and corn flakes. Foof Addit Contam Part A 26(6):874–884

Deng X, Nikiforov AY, Leys C (2017) Antimicrobial nanocomposites for food packaging. In: Grumezescu AM (ed) Nanotechnology in the Agri-food industry. Elsevier, San Diego, pp 1–30

Deshpande P, Dapkekar A, Oak MD, Paknikar KM, Rajwade JM (2017) Zinc complexed chitosan/tpp nanoparticles: a promising micronutrient nanocarrier suited for foliar application. Carbohydr Polym 165:394–401

Dimkpa CO, Bindraban PS, Fugice J, Agyin-Birikorang S, Singh U, Hellums D (2017) Composite micronutrient nanoparticles and salts decrease drought stress in soybean. Agron Sustain Dev 37(1):5

Direito GM, Almeida AP, Aquino S, Alves dos Reis T, Pozzi CR, Correa B (2008) Evaluation of sphingolipids in wistar rats treated to prolonged and single oral doses of fumonisin B1. Int J Mol Sci 10:50–61

Dokka S, Toledo D, Shi X, Castranova V, Rojanasakul Y (2000) Oxygen radical- mediated pulmonary toxicity induced by some cationic liposomes. Pharm Res 17:521–525

Duhan JS Kumar R, Kumar N, Kaur P, Nehra K, Duhan S (2017) Nanotechnology: the new perspective in precision agriculture. Biotechnol Rep 15:11–23

Eadie MJ (2004) Ergot of rye - the first specific for migraine. J Clin Neurosci 11(1):4–7

El-Shemy HA, Aboul-Enein AM, Aboul-Enein KM, Fujita K (2007) Willow leaves' extracts contain anti-tumor agents effective against three cell types. PLoS One 2:e178

FAO IFAD UNICEF, WFP & WHO (2017) Fao The State of Food Security and Nutrition in the World. http://www.fao.org/state-of-food-security-nutrition/en/. Accessed February 2017

Farhoodi M (2015) Nanocomposite materials for food packaging applications: charactization and safety evaluation. Food Eng Rev 8(1):35–51

Fortunati E, Puglia D, Armentano I, Valdés A, Ramos M, Juárez N, Garrigós M, Kenny JM (2017) Multifunctional antimicrobial nanocomposites for food packaging applications. In: Grumezescu AM (ed) Nanotechnology in the Agri-food industry, pp 265–303. https://doi.org/10.1016/B978-0-12-804303-5.00008-0

Frost and Sullivan (2017a) Advances of Nanotechnology in F&B Packaging. TechVision Opportunity Engines. https://cds.frost.com/p/50959/#!/ppt/c?id=D787-00-5C-00-00&hq=Nanolok. Accessed February 2017

Frost and Sullivan (2017b) Advances in nanotechnology for coatings, additives, healthcare, and materials. TechVision Opportunity Engines. https://cds.fro st.com/p/50959/#!/ppt/c?id=D787–00–54-00-00&hq=nanotechnology%20in%20agriculture. Accessed February 2017

Gholami-Shabani M, Gholami-Shabani Z, Shams-Ghahfarokhi M, Jamzivar F, Razzaghi-Abyaneh M (2017) Green nanotechnology: biomimetic synthesis of metal nanoparticles using plants and their application in agriculture and forestry. In: Prasad R, Kumar M, Kumar V (eds) Nanotechnology. Springer, Singapore

Godfray H, Beddington JR, Crute IR, Haddad L, Lawrence D, Muir JF, Pretty J, Robinson S, Thomas SM, Toulmin C (2010) Food security: the challenge of feeding 9 billion people. Science 327(5967):812–818

Hasan MM, Chowdhury SP, Alam S, Hossain B, Alam MS (2005) Antifungal effects of plant extracts on seed-borne fungi of wheat seed regarding seed germination, seedling health and vigour index. Pak J Biol Sci 8:1284–1289

Hassan AA, Howayda M, El-Shafei A, Mahmoud HH (2013) Effect of zinc oxide nanoparticles on the growth of some mycotoxigenic moulds. J Stud Chem Process Technol (SCPT) ASSE 1:16–25

Heussner AH, Bingle LEH (2015) Comparative ochratoxin toxicity: a review of the available data. Toxins (Basel) 7(10):4253–4282

Hosseini SJ, Jamal S, Esmaeeli S, Ansari B (2011) Challenges in commercialization of nano and biotechnologies in agricultural sector of Iran. Afr J Biotechnol 10(34):6516–6521

Hussain T (2017) Nanotechnology: diagnosis of plant diseases. ARTOAJ 10(1):1–2

Islam NU, Jalil K, Shahid M, Rauf A, Muhammad N, Khan A, Shah MR, Khan MA (2015) Green synthesis and biological activities of gold nanoparticles functionalized with *Salix alba*. Arab J Chem. https://doi.org/10.1016/j.arabjc.2015.06.025

Jeyasubramanian K, Thoppey UUG, Hikku GS, Selvakumar N, Subramania A, Krishnamoorthy K (2016) Enhancement in growth rate and productivity of spinach grown in hydroponics with iron oxide nanoparticles. RSC Adv 6(19):15451–15459

Juglal S, Govinden R, Odhav B (2002) Spice oils for the control of co-occurring mycotoxin producing fungi. J Food Prot 65:683–687

Katti K, Chanda N, Shukla R, Zambre A, Suibramanian T, Kulkarni RR, Kannan R, Katti KV (2009) Green nanotechnology from cumin phytochemicals: generation of biocompatible gold nanoparticles. Int J Green Nanotechnol Biomed 1(1):B39–B52

Kaushik A, Solanki PR, Ansari AA, Ahmad S, Malhotra BD (2009) A nanostructured cerium oxide film-based immunosensor for mycotoxin detection. Nanotechnology 20(5)

Kaushik A, Arya SK, Vasudev A, Bhansali S (2013) Recent advances in detection of ochratoxin-A. OJAB 2:1): 1–1):11

Khan R, Dhayal M (2008) Nanocrystalline bioactive TiO_2-chitosan impedimetric immunosensor for ochratoxin-A. Electrochem Commun 10(3):492–495

Kim B, Kim D, Cho D, Cho S (2003) Bactericidal effect of TiO_2 photocatalyst on selected food-borne pathogenic bacteria. Chemosphere 52:277–281

Kumar PS, Sudha S (2013) Biosynthesis of silver nanoparticles from *Dictyota Bartayresiana* extract and their antifungal activity. Nano Biomed Eng 5(2):72–75

Kumar GD, Natarajan N, Nakkeeran S (2016) Antifungal activity of nanofungicide Trifloxystrobin 25%+ Tebuconazole 50% against Macrophomina phaseolina. Afr J Microbiol Res 10(4):100–105

Kuppusamy P, Yusoff MM, Maniam GP, Govindan N (2016) Biosynthesis of metallic nanoparticles using plant derivatives and their new avenues in pharmacological applications – an updated report. Saudi Pharm J 24(4):473–484

Lamberti I, Mosiello L, Cenciarelli C, Antoccia A, Tanzarella C (2010) A novel based protein microarray for the simultaneous analysis of activated caspases. In: Malcovati P, Baschirotto A, d'Amico A, Natale C (eds) Sensors and microsystems. Lecture notes in electrical engineering, vol 54. Springer, Dordrecht

Llana-Ruiz-Cabello M, Pichardo S, Maisanaba S, Puerto M, Prieto AI, Gutiérrez-Praena D, Jos A, Cameán AM (2015) In vitro toxicological evaluation of essential oils and their main compounds used in active food packaging: a review. Food Chem Toxicol 81:9–27

Llorens A, Lloret E, Picouet PA, Trbojevich R, Fernandez A (2012) Metallic- based micro and nanocomposites in food contact materials and active food packaging. Trends Food Sci Technol 24:19–29

Logrieco A, Arrigan DW, Brengel-Pesce K, Siciliano P, Tothill I (2005) DNA arrays, electronic noses and tongues, biosensors and receptors for rapid detection of toxigenic fungi and mycotoxins: a review. Food Addit Contam 22(4):335–344

Lv X, Li Y, Yan T, Pang X, Cao W, Du B, Wu D, Wei Q (2015) Electrochemiluminescence modified electrodes based on RuSi@Ru(bpy)32+ loaded with gold functioned nanoporous CO/Co3O4 for detection of mycotoxin deoxynivalenol. Biosens Bioelectron 70:28–33

Mak AC, Osterfeld SJ, Yu H, Wang SX, Davis RW, Jejelowo OA, Pourmand N (2010) Magnetic nanotag-based immunoassay for multiplex mycotoxin and protein detection. Department of Materials Science and Engineering. Stanford University 3:27–30

Mallikarjuna K, Sushma NJ, Narasimha G, Manoj L, Rajue DP (2014) Phytochemical fabrication and characterization of silver nanoparticles by using pepper leaf broth. Arab J Chem 7(6):1099–1103

Man Y, Liang G, Li A, Pan L (2017) Recent advances in mycotoxin determination for food monitoring via microchip. Toxins (Basel) 9(10):324

Mejri-Omrani N, Miodek A, Zribi B, Marrakchi M, Hamdi M, Marty JL, Korri-Youssoufi H (2016) Direct detection of OTA by impedimetric aptasensor based on modified polypyrrole-dendrimers. Analytica Chimica Acta 920:37–46

Milicevic DR, Skrinjar M, Baltic T (2010) Real and perceived risks for mycotoxin contamination in foods and feeds: challenges for food safety control. Toxins (Basel) 2(4):572–592

Mondal P, Kumar R, Gogoi R (2017) Azomethine based nano-chemicals: development, in vitro and in vivo fungicidal evaluation against Sclerotium rolfsii, Rhizoctonia bataticola and Rhizoctonia solani. Bioorg Chem 70:153–162

Mondali NK, Mojumdar A, Chatterje SK, Banerjee A, Datta JK, Gupta S (2009) Antifungal activities and chemical characterization of Neem leaf extracts on the growth of some selected fungal species in vitro culture medium. J Appl Sci Environ Manage 13(1):49–53

Montesano FF, Parente A, Santamaria P, Sannino A, Serio F (2015) Biodegradable superabsorbent hydrogel increases water retention properties of growing media and plant growth. Agric Agric Sci Procedia 4:451–458

Mosiello L, Lamberti I (2009, 60) Biosensors for aflatoxins detection. IntechopenCom:147 http://www.intechopen.com/source/pdfs/22037/InTech-Biosensors_for_aflatoxins_detection.pdf. Accessed February 2017

Muñoz K, Blaszkewicz M, Degen GH (2010) Simultaneous analysis of ochratoxin a and its major metabolite ochratoxin alpha in plasma and urine for an advanced biomonitoring of the mycotoxin. J Chromatogr B Anal Technol Biomed Life Sci 878(27):2623–2629

Nabawy GA, Hassan AA, El-Ahl HS, Refai MK (2014) Effect of metal nanoparticles in comparison with commercial antifungal feed additives on the growth of *Aspergillus flavus* and aflatoxin b1 production. J Global Biosci 3:954–971

Nune SK, Chanda N, Shukla R, Katti K, Kulkarni RR, Thilakavathi S, Mekapothula S, Kannan R, Katti KV (2009) Green nanotechnology from tea: phytochemicals in tea as building blocks for production of biocompatible gold nanoparticles. J Mater Chem 19(19):2912–2920

OMAF (2004) Pesticide storage, handling and application. Ontario Ministry of Agriculture, Food and Rural Affairs, California

Othman SH, Abd Salam NR, Zainal N, Kadir Basha R, Talib RA (2014) Antimicrobial activity TiO2 nanoparticle-coated film for potential food packaging applications. Int J Photoenergy:1–6

Owino JHO, Arotiba OA, Hendricks N, Songa EA, Jahed N, Waryo TT, Ngece RF, Baker P, Iwuoha EI (2008) Electrochemical immunosensor based on polythionine/gold nanoparticles for the determination of aflatoxin B1. Sensors (Basel):8262–8274

Park J, Choi S, Moon H, Seo H, Kim J, Hong S, Lee B, Kang E, Lee J, Ryu D, Cho IS (2017) Antimicrobial spray nanocoating of supramolecular Fe(III)-tannic acid metal-organic coordination complex: applications to shoe insoles and fruits. Sci Rep 7:6980

Peralta-Videa JR, Hernandez-Viezcas JA, Zhao L, Diaz BC, Ge Y, Priester JH, Holden PA, Gardea-Torresdey JL (2014) Cerium dioxide and zinc oxide nanoparticles alter the nutritional value of soil cultivated soybean plants. Plant Physiol Biochem 80:128–135

Perricone M, Arace E, Corbo MR, Sinigaglia M, Bevilacqua A (2015) Bioactivity of essential oils: a review on their interaction with food components. Front Microbiol 6

Peters R, Brandhoff P, Weigel S, Marvin H, Bouwmeester H, Aschberger K, Rauscher H, Amenta V, Arena M, Moniz FB, Gottardo S, Mech A (2014) Inventory of nanotechnology applications in the agricultural, feed and food sector. EFSA Supporting Publications 11(7)

Prasad R (2014) Synthesis of silver nanoparticles in photosynthetic plants. J Nanoparticles:963961. https://doi.org/10.1155/2014/963961

Prasad R (2016) Advances and applications through fungal nanobiotechnology. Springer. In: International publishing Switzerland (ISBN: 978–3–319-42989-2)

Prasad R (2017) Fungal nanotechnology: applications in agriculture, industry, and medicine. Springer Nature Singapore Pte Ltd. (ISBN 978-3-319-68423-9

Prasad R, Kumar V, Prasad KS (2014) Nanotechnology in sustainable agriculture: present concerns and future aspects. Afr J Biotechnol 13(6):705–713

Prasad R, Bhattacharyya A, Nguyen QD (2017a) Nanotechnology in sustainable agriculture: recent developments, challenges, and perspectives. Front Microbiol 8(JUN):1–13

Prasad R, Gupta N, Kumar M, Kumar V, Wang S, Abd-Elsalam KA (2017b) Nanomaterials act as plant defense mechanism. In: Prasad R, Kumar V, Kumar M (eds) Nanotechnology. Springer, Singapore, pp 253–269

Purohit R, Mittal A, Dalela S, Warudkar V, Purohit K, Purohit S (2017) Social, environmental and ethical impacts of nanotechnology. Mater Today (4, 4):5461–5467

Putzbach W, Ronkainen NJ (2013) Immobilization techniques in the fabrication of nanomaterial-based electrochemical biosensors: a review. Sensors (Basel) 13(4):4811–4840

Radoi A, Targa M, Marty J (2008) Enzyme-linked immunosorbent assay (ELISA) based on superparamagnetic nanoparticles for aflatoxin M1 detection. Talanta 77:138–143

Reddy CS, Reddy KRN, Prameela M, Mangala UN, Muralidharan K (2007) Identification of antifungal component in clove that inhibits *Aspergillus spp.* colonizing rice grains. J Mycol Plant Pathol 37(1):87–94

Rhouati A, Bulbul G, Latif U, Hayat A, Li ZH, Marty JL (2017) Nano-aptasensing in mycotoxin analysis: recent updates and progress. Toxins (Basel) 9(11):1–23

Ringot D, Chango A, Schneider YJ, Larondelle Y (2006) Toxicokinetics and toxicodynamics of ochratoxin A, an update. Chem Biol Interact 159(1):18–46

Roberts SL (1995) Mycotoxin list. Boletim Científico (46):1–13

Sang-Ho C, Jin-Kyu L, Hak-Sung J, Sung-Hee K, Hyun-Jung K, Cheong-Up C, Hwan-Goo K (2010) Novel mycotoxin collection system using magnetic nanoparticles for determination of aflatoxin B1 and zearalenone in feed. J Vet Sci 3:569–571

Sarkar PK, Prajapati PK, Shukla VJ, Ravishankar B, Choudhary AK (2009) Toxicity and recovery studies of two ayurvedic preparations of iron. Indian J Ecp Biol 47(12):987–992

Selvaraj JN, Lu Z, Yan W, Yue-ju Z, Fu-guo X, Xiao-feng D, Yang L (2015) Mycotoxin detection-recent trends at global level. J Integr Agric 14(11):2265–2281

Sertova NM (2015) Application of nanotechnology in detection of mycotoxins and in agricultural sector. JCEA 16(2):117–130

Sharma A, Kumar A, Khan R (2018) A highly sensitive amperometric immunosensor probe based on gold nanoparticle functionalized poly (3, 4-ethylenedioxythiophene) doped with graphene oxide for efficient detection of aflatoxin B1. Synth Met 235:136–144

Shawkey AM, Abdulall AK, Rabeh MA, Abdellatif AO (2014) Enhanced biocidal activities of *Citrullus colocynthis* aqueous extracts by green nanotechnology. Int J Appl Res Nat Prod 7(2):1–10

Shukla R, Nune SK, Chanda N, Katti K, Mekapothula S, Kulkarni RR, Welshons WV, Kannan R, Katti KV (2008) Soybeans as a phytochemical reservoir for the production and stabilization of biocompatible gold nanoparticles. Small (9):1425–1436

Silva IM, Boelter JF, Silveira NP, Brandelli A (2014) Phosphatidylcholine nanovesicles coated with chitosan or chondroitin sulfate as novel devices for bacteriocin delivery. J Nanopart Res 16:2479

Singh T, Shukla S, Kumar P, Wahla V, Bajpai VK (2017) Application of nanotechnology in food science: perception and overview. Front Microbiol 8(AUG):1–7

Soleimanpour MR, Hosseini SJF, Mirdamadi SM (2012) Exploring the mocel of nanotechnology dewvelopment in agriculture sector of Iran. ARPN J Agric Biol Sci 7(12):1002–1006

Tang F, Li L, Chen D (2012) Mesoporus silica nanoparticles: synthesis, biocompatibility, and drug delivery. Adv Mater 24:1504–1534

Tang X, Li X, Li P, Zhang Q, Li R, Zhang W, Ding X, Lei J, Zhang Z (2014) Development and application of an immunoaffinity column enzyme immunoassay for mycotoxin zearalenone in complicated samples. PLoS One 9(1):1–7

Tothill IE (2010) Peptides as molecular receptors. In: Zourob M (ed) Recognition receptors in biosensors. Springer New York, pp 249–274 ISBN: 978-1-4419- 0918-3

Tothill IE (2011) Biosensors and nanomaterials and their application for mycotoxin determination. World Mycotoxin J 4(4):361–374

Tsai WC, Hsieh CK (2007) QCM-based immunosensor for the determination of ochratoxin A. Analytical Letters 40(10):1979–1991

Turner NW, Piletska EV, Karim K, Whitcombe M, Malecha M, Magan N, Baggiani C, Piletsky SA (2004) Effect of the solvent on recognition properties of molecularly imprinted polymer specific for ochratoxin A. Biosens Bioelectron 20(6):1060–1067

Vundavalli R, Vundavalli S, Nakka M, Rao DS (2015) Biodegradable nano-hydrogels in agricultural farming – alternative source for water resources. Procedia Mater Sci 10:548–554

Wu H, Liu R, Kang X, Liang C, Lv L, Guo Z (2018) Fluorometric aptamer assay for ochratoxin a based on the use of single walled carbon nanohorns and exonuclease III-aided amplification. Microchim Acta 185(1)

Yazar S, Omurtag GZ (2008) Fumonisins, trichothecenes and zearalenone in cereals. Int J Mol Sci 9(11):2062–2090

Yin J, Liu Y, Wang S, Deng J, Lin X, Gao J (2018) Engineering a universal and label-free evaluation method for mycotoxins detection based on strand displacement amplification and G-quadruplex signal amplification. Sens Actuator B-Chem 256:573–579

Zain ME (2011) Impact of mycotoxins on humans and animals. J Saudi Chem Soc 15(2):129–144

Zhang Z, Tang X, Wang D, Zhang Q, Li P, Ding X (2015) Rapid on-site sensing aflatoxin B1 in food and feed via a chromatographic time-resolved fluoroimmunoassay. PLoS One 10(4):1–14

Zhao Y, Yang Y, Luo Y, Yang X, Li M, Song Q (2015) Double detection of mycotoxins based on SERS labels embedded Ag@Au core-shell nanoparticles. ACS Appl Mater Interfaces 7(39):21780–21786

Zhao Y, Liu R, Sun W, Lv L, Guo Z (2018) Ochratoxin a detection platform based on signal amplification by exonuclease III and fluorescence quenching by gold nanoparticles. Sens Actuator B-Chem 255:1640–1645

Chapter 13
Chitosan-Based Nanostructures in Plant Protection Applications

Fahad A. Al-Dhabaan, Manal Mostafa, Hassan Almoammar, and Kamel A. Abd-Elsalam

13.1 Introduction

Chitosan is a nontoxic and biodegradable natural biopolymer resulting from the deacetylation of chitin. Chitosan is a linear semicrystalline polysaccharide obtained by deacetylation of chitin and composed by N-acetyl D-glucosamine and D-glucosamine units, linked through β (1→4) glycosidic bonds, Chitin or poly(β-(1→4)-N-acetyl-D-glucosamine) is prepared by some living organisms, being the structural component of the shells of crustaceous, cell walls of fungi, and exoskeletons of insects (Alves and Mano 2008; Rinaudo 2006; Mano 2008). Most of the naturally occurring polysaccharides are acidic in nature, but chitosan is a one of the naturally occurring basic polymers (Prasad et al. 2017a). Chitosan can be dissolved in water under acidic conditions after the amino protonation to discuss positive charges, gelations, and membrane-forming properties (Berscht et al. 1994).

F. A. Al-Dhabaan
Department of Biology, Science and Humanities College, Shaqra University, Alquwayiyah, Saudi Arabia

M. Mostafa
CIHEAM IAMB – Mediterranean Agronomic Institute of Bari, Valenzano, BA, Italy

Microbiology Department, Faculty of Agriculture, Cairo University, Giza, Egypt

Plant Pathology Research Institute, Agricultural Research Center (ARC), Giza, Egypt

H. Almoammar
National Centre for Biotechnology, King Abdulaziz City for Science and Technology (KACST), Riyadh, Saudi Arabia

K. A. Abd-Elsalam (✉)
Plant Pathology Research Institute, Agricultural Research Center (ARC), Giza, Egypt

K. A. Abd-Elsalam, R. Prasad (eds.), *Nanobiotechnology Applications in Plant Protection*, Nanotechnology in the Life Sciences,
https://doi.org/10.1007/978-3-319-91161-8_13

The physical and chemical properties of chitosan depend mainly on its molecular weight and degree of deacetylation (Shukla et al. 2013). Based on molecular weight chitosan can be classified into three types including; low molecular weight (LMWC), medium molecular weight (MMWC), and high molecular weight (HMWC). Some studies indicated that chitosan with shorter oligomers, such as LMWC, has greater antifungal activity (Rhoades and Roller 2000; Park et al. 2008). Chitosan-based nanoparticles (CHT NPs) can be easily prepared through self-aggregation. The appliction of nanotechnology is less discovered in agriculture in general, but some substantial research has been done in crop protection (Park et al. 2006; Jo et al. 2009; Nair et al. 2010; Sharon et al. 2010; Ghormade et al. 2011; He et al. 2011; Lamsal et al. 2011; Kim et al. 2012; Perez-de-Luque et al. 2012; Jayaseelan et al. 2013; Wani and Ahmad 2013; Saharan et al. 2013; Prasad 2014; Saharan et al. 2015; Wang et al. 2016; Saharan et al. 2016; Abd-Elsalam et al. 2017; Rubina et al. 2017; Prasad et al. 2014, 2017a; Sangeetha et al. 2017). Chitosan has garnered extensive interest because of its antimicrobial and antifungal properties (Kendra and Hadwiger 1984; Aziz et al. 2015, 2016). In future, there is a urgently need to discover chitosan biopolymer usage not only for its innovative antimicrobial agent but also for its plant defense supporter property to combat future challenges of mutating plant pathogenic population under climate change (Saharan et al. 2014; Prasad et al. 2017b). This is because nanochitosan as compared to bulk chitosan has superior physicochemical characteristics that provide enhanced biological activities (Saharan et al. 2013; Van et al. 2013). This fact has promoted the study of CHT NPs at using them as new antimicrobial agents (Qi et al. 2004; Wazed et al. 2011). Copper, zinc, and other metal hybrids with chitosan are active components of many agrochemicals like fertilizers and pesticides and serve as cofactors of several enzymes in plant. Thus, Cu and Zn ultimately lead to higher plant growth and contribute in plant disease control (Gornik et al. 2008; Cabrera et al. 2013; Saharan et al. 2016; Rubina et al. 2017). Chitosan-based nanomaterials hold enormous promise with regard to their application in plant protection and growth (Shukla et al. 2013). Although, specific reports has been introduced to evaluate chitosan biopolymer/nanoparticle for use in plant growth and protection by using nanotechnology tools (Shukla et al. 2013; Saharan et al. 2014; Abd-Elsalam et al. 2017: Choudhary 2017; Prasad et al. 2017b). Some pioneering uses of CHT polymer include synthesis of CHT NPs as a valuable delivery system for fertilizers, herbicides, pesticides, and micronutrients for crop growth promotion by a balanced and sustained nutrition (Siddiqui et al. 2015). In addition, CHT NPs can safely deliver genetic material for plant transformation.

13.2 Synthesis Chitosan Nanomaterials

13.2.1 Synthesis Chitosan Nanoparticles

However, several methods have been used to produce chitosan nanoparticulate. Choice of method depends on the shape and particle size requirements (Agnihotri et al. 2004). Chitosan is a hydrophilic polymer with positive charge that comes from

Table 13.1 Synthesis of chitosan-based nanoparticles: advantages and disadvantages

Methods	Advantages	Disadvantages
Emulsion cross-linking	1. Easy control of particle size	1. Need to remove oil and surfactant
	2. Good loading efficiency	2. Cross-linker can react with active ingredient
	3. Good stability of nanomaterials	3. Purification of nanoparticle is tedious process
Emulsion-droplet coalescence method	1. Higher encapsulation efficiency	1. Excess alkali induces more precipitation and leads to increased particle size
	2. No cross-linker leads to higher zeta potential	2. Due to absence of cross-linker nanoparticles stability decreases
	3. No reactivity of cross-linker to active ingredient	
Ionic gelation method	1. Reduced the chemical side effect	1. Partial size, distribution, and stability strongly affected by degree of deacetylation, MW of chitosan and molar ratio of chitosan and TPP
	2. Better control on physicochemical feature of nanoparticles	
	3. Easy and fast	
Reverse Micellar Method	1. Stable and smaller and monodispersed nanoparticles with suitable polydispersity index	1. Cumbersome procedure
		2. Chance of side effect of reaction components (solvent, surfactant, etc.)
Sieving method	1. Very simple and rapid procedure	1. Need specialized sieve with particular size for desirable size of nanoparticles
	2. Easy mass synthesis	
Spray drying method	1. Mechanized method for mass production	1. Size of nanoparticles depends on nozzle size, flow rate, and temperature of air
		2. Temperature-sensitive component could not be encapsulated

Agnihotri et al. (2004), Kashyap et al. (2015)

weak basic groups, which provides it specific characteristics from the technological point of view (Lopez-Leon et al. 2005). Recently, chitosan nanoparticles have received great attention in numerous fields because of their physicochemical and biological properties (Racovita et al. 2008; Shi et al. 2011). Some advantages and disadvantages of the methods used for synthesis of nanochitosan are summarized in Table 13.1.

13.2.1.1 Emulsion Cross-Linking Method

This method utilizes the reactive functional amine group of chitosan cross links with aldehyde groups of the cross-linking agent. Actually, chitosan solution is emulsified in oil phase (water-in-oil emulsion) and the aqueous droplets are stabilized using an appropriate surfactant. The stable emulsion was reacted with an appropriate cross-linking agent such as glutaraldehyde to stabilize the chitosan droplets. The nanoparticles are then washed and dried (Agnihotri et al. 2004).

13.2.1.2 Ionotropic Gelation Method

Ionotropic gelation method is the modified method linking multiphase for the production of biopolymeric oil-core microcapsules due to its nontoxic and slight processing conditions. This technique is based on the electrostatic interactions between the chitosan amine group and a polyanion such as tripolyphosphate. In this method, chitosan is dissolved in water or in weak acidic medium. This solution is then added dropwise under continuous stirring to the solutions containing other counterions. Due to the complexation between oppositely charged species, chitosan goes through ionic gelation and precipitate to produce spherical CHT NPs (Racovita et al. 2008).

13.2.1.3 In Coacervation/Precipitation Method

In the coacervation/precipitation method, chitosan solution is sprayed into sodium hydroxide, NaOH methanol, or ethanediamine alkaline solutions using compressed air, which in turn creates coacervated chitosan droplets in the form of nanoparticles (Shi et al. 2011). Emulsion–bead combination strategy incorporates both emulsion cross-connecting and precipitation. A stable emulsion containing the aqueous chitosan solution in oil and a second emulsion containing an NaOH solution are formed. By mixing the two emulsions under high magnetic stir, droplets of each emulsion collide at random, coalesce, and lastly precipitate as small size particles (Shikata et al. 2002).

13.2.1.4 In the Reverse Micelles Technique

In the reverse micelles technique, a surfactant is dissolved in organic solvent to formulate reverse micelles. The aqueous phase containing the chitosan is mixed to this emulsion with constant vortexing and the chitosan nanoparticles forms in the core of the reverse micelles (Agnihotri et al. 2004).

13.2.1.5 Self-Assembly

Self-assembly strategy depends on cationic and hydrophobic properties of chitosan, cationic chitosan derivatives can be simply adsorbed onto the colloid surface of anionic inorganic (bentonite) suspensions because of electrostatic attraction. This strategy is described by dispersion took after by particular relationship of atoms through noncovalent interactions, including electrostatic as well as hydrophobic connections (Ichikawa et al. 2005).

13.2.2 Synthesis Chitosan Nanocomposites

Chitosan nanocomposite generally refers to chitosan polymer containing dispersed nanofillers having an average particles size less than 100 nm. Thus this composite retains exceptionally enhanced properties pertaining to both of polymer and nanoparticles. By and large, regular polymers are considered as great facilitating materials for nanoparticles especially for natural applications, because of their maintainability, eco-accommodating property, nontoxicity, biodegradability, and biocompatibility. The natural inorganic half-breed nanocomposites have been seen to be valuable in numerous fields like medication conveyance, tissue building, bundling material, covering, and sensors (Rhim et al. 2006; Di Carlo et al. 2012).The primary methods that have been utilized to deliver chitosan nanocomposites, in particular dissolvable throwing are solidify drying, layer-by-layer, and electrospinning (Moura et al. 2016).

13.2.2.1 Solvent Casting

Dissolvable throwing is a standout amongst the most widely recognized strategies for planning of CHI nanocomposites films and membranes (Caridade et al. 2013; Zheng et al. 2015). The polymer is broken down into a dissolvable and then cast onto a surface, for example, glass Petri dish. The dissolvable is accordingly permitted to dissipate at room temperature or in air stove, and from that point onward, the films/membranes are separated (Ambrosio 2009).

13.2.2.2 Freeze Drying

Stop drying method was utilized for the arrangement of exceedingly porous scaffolds by inducing thermal phase separation. Usually, the solution temperature is brought down until solid– fluid demixing happens, forming two unique stages: solidified dissolvable and polymer stage. At this point, the solidified dissolvable, through sublimation, leaves the polymeric structure framing a pore. The resultant structure can be controlled by changing the kind of polymer and its concentration (Sarmento and Neves 2012).

13.2.2.3 Layer-by-Layer Assembly

Layer-by-layer (LbL) meeting, proposed by way of Iler in 1966, is a technique able to enhance surfaces and fabricate especially ordered polymeric films and nanocomposites over extraordinary varieties of substrates (Decher 1997; Borges et al. 2014). The current technique is based on the sequential adsorption of different macromolecular components, which are attracted to each other due to electrostatic interactions, hydrogen bonding, van der Waals forces, and electron exchange, among others (Borges et al.

2014; Borges and Mano 2014). Various LbL tactics can be used to build up a multilayer film, including dip coating, spin coating, and spraying coating (Richardson et al. 2015). Due to its versatility and notable availability of building blocks (e.g., CNTs, clays, NPs, polymers), this technology permits fabrication of multilayered gadgets of any nature, length, shape, and chemical composition, assuring the improvement of nanostructures with favored geometries and functionalities (Costa and Mano 2014). Furthermore, the properties of multilayered devices may be tuned through solution pH, temperature, or ionic electricity (Borges and Mano 2014).

13.2.2.4 Electrospinning

Electrospinning represents an appropriate technique to produce fibers with diameters in the nm–μm period scale, since it permits morphology, porosity, and composition to be managed by the use of relatively unsophisticated equipment. Typically, the electrospinning process uses an electric field created between the polymer solution and the collector, which generates internal repulsive forces in the polymer solution and, at a critical point, causes the expulsion of the polymer solution in shape of fibers towards the collector (Martins et al. 2008; Teo and Ramakrishna 2006). There are three special kinds of electrospinning: wet–dry electrospinning, moist–wet electrospinning, and coaxial electrospinning, The main distinction between the first two techniques is that the moist–dry approach makes use of a risky solvent that evaporates because the fibers are spun via the collector, while the moist–wet approach spins a nonunstable solvent to a collector with a 2D solvent. Regarding the final technique, it is possible to achieve fibers with a center-sheath shape, as two exclusive components may be spun at the same time (Zheng et al. 2015).

13.2.2.5 Template Polymerization

In the present method, chitosan is firstly dissolved in an acrylic monomer solution below magnetic stirring. Due to the electrostatic interplay, the negatively charged acrylic monomers align along the chitosan molecules. After whole dissolution of chitosan, the polymerization is commenced through adding potassium persulfate initiator (K2S2O8) underneath stirring at 70 °C. The entire polymerization ends in the appearance of an opalescent answer, indicating the nanoparticles formation (Fang et al. 2009; Shi et al. 2011).

13.3 Nanochitosan Characterization

13.3.1 X-Ray Powder Diffraction

X-ray Powder Diffraction (XRPD) analysis was performed to confirm the crystalline structures of magnetite present in chitosan/Fe3O4 nanocomposites, The samples were analyzed by X-ray powder diffractometer Xpert Pro MPD (Panalytical) using

Bragg–Brentano geometry in the range 15–70° with a rate of 1° min^{-1}. CuK _ radiation (k = 1.54059 Å) was used and the tube operated at 40 kV and 30 mA (Freire et al. 2016).

13.3.2 Fourier Transform Infrared Spectroscopy

Fourier Transform Infrared Spectroscopy (FTIR) evaluation was carried out in a PerkinElmer 2000 spectrophotometer used to report spectra within the range between 4000 and 400 cm^{-1}. In previous measurements, the samples were dried and level-headed to be powdered and pressed (~10 mg of sample to 100 mg of KBr) in disk format (Freire et al. 2016).

13.3.3 Transmission Electron Microscopy (TEM)

In Transmission Electron Microscopy (TEM), the snap shots were recorded through a JEOL JEM-1400 electron microscope operating at an accelerating voltage of 120 kV. The samples had been prepared by way of diluting nanoparticles dispersion in distillated water. Then, one droplet of the sample was placed on 300-mesh carbon-covered copper grids and dried overnight under ambient conditions. The size distribution was decided by means of 50 randomly selected debris in exclusive regions of the increasing TEM micrograph (Freire et al. 2016).

13.3.4 Thermogravimetric Analysis

Thermogravimetric analysis (TGA) is a useful method for examining thermal stability of materials as well as for compositional information. Five miligram of nanoparticles were executed in nitrogen atmosphere by employing a Thermogravimetric Analyzer Q50 V20. The lack of mass was monitored heating up samples from 25 to 900 °C in a fee of 10 °C min^{-1}. The zero time for the thermal degradation observe became taken after temperature stabilization (Freire et al. 2016).

13.3.5 Dynamic Light Scattering

Dynamic light scattering (DLS) was used for the size of particle length, polydispersity index (PDI), and zeta capacity of NPs via Zetasizer model PSS0012–22 (Malvern, UK). The evaluation was completed at a scattering angle of 90° at RT. Sample was accurately diluted with distilled water prior to size (Kheiri et al. 2016).

13.4 Application of Nanochitosan in Plant Protection

Natural defense reaction of plants against pathogenesis depends upon early detection of pathogens. In fact, during the process of evolution, plant researchers have evolved numerous techniques to combat some plant pathogens. This induction of herbal protection response consists of over expression of various defense-associated genes and enzymes, increased accumulation of phenolic compounds, cellular wall synthesis, etc. (Sánchez et al. 2009; Chandra et al. 2015). Recently, chitosan nanoparticles have shown significant antimicrobial activity against fungal plant pathogens (Saharan et al. 2013). Chitosan nanoparticles serve as the precise choice for the control of pests which might eliminate the toxic pesticide use and additionally enhances the yield of soybean (Kendra and Hadwiger 1984). Application of chitosan-based nanostructures in plant protection is shown in Fig. 13.1 and Table 13.2.

13.4.1 Antimicrobial

Chitosan nanoparticles specific extra affinity toward pathogen's outer membrane and therefore without difficulty enter into the pathogens' cell (Van et al. 2013). Several research confirmed that chitosan isn't best an antimicrobial agent however additionally an effective elicitor of plant systemic received resistance to pathogens (Sharp 2013; Katiyar et al. 2015; Xing et al. 2014), enhancer and regulator of plant growth, development and increase crop yield (Gornik et al. 2008; Cabrera et al.

Fig. 13.1 The potential application of nanochitosan in plant protection

Table 13.2 Chitosan-based nanomaterials used in plant growth and crop protection

Chitosan NMs	Applications	Findings	References
Chitosan NPs	Antifungal activity	Significantly delayed mycelia growth of *Rhizopus* spp., *Colletotrichum* spp. (*C.capsici*, *C. gloesporoides*) and *Aspergillus niger*	Chookhongkha et al. (2012)
	Antifungal activity	Effective against *A. alternata*, *M. phaseolina* and *R. solani*	Saharan et al. (2013)
	Plant promotion	Study on chitosan nanoparticles on biophysical characteristic and growth of Robusta coffee in green house	Van et al. (2013)
	Controlled delivery system/bioactivity in plants	Effective delivery system for sustainable agriculture	Kashyap et al. (2015)
	Defense activity	Positive modulator of innate immune responses in plants	Chandra et al. (2015)
	Defense activity	biological pesticide for controlling Fusarium head blight	Kheiri et al. (2016)
	Plant promotions Antifungal activity	Chitosan nanoparticle induced defense responses in Finger millet plants against blast disease caused by *Pyricularia grisea*	Manikandan and Sathiyabama (2016)
	Plant promotions Antifungal activity	promote seed germination and seedling vigor, induced systemic and durable resistance against *Sclerospora graminicola*	Siddaiah et al. (2018)
Ag-Chitosan NCs	Antifungal activity	Seed-borne fungi	Kaur et al. (2012)
Cu-Chitosan NCs	Antifungal activity	*Fusarium graminearum*	Brunel et al. (2013)
	Seedling growth promotions, Antifungal activity	Efficacy against *Alternaria solani* and *Fusarium oxysporum*	Saharan et al. (2015)
	Seedling growth promotions, Antifungal activity	*Rhizoctonia solani*	Abd-Elsalam et al. (2017)
	Growth promontory effect on maize seedling growth Antifungal activity	*Fusarium verticillioids*	Choudhary (2017)
	Antifungal activity	Sclerotia-forming plant pathogenic fungi	Rubina et al. (2017)
Chitosan and methyljasmonate	Defense activity	Efficacy against *Alternaria alternata* and enhancing activity of cherry tomato fruit defense mechanisms	Chen et al. (2014)

(continued)

Table 13.2 (continued)

Chitosan NMs	Applications	Findings	References
Chitosan/boehmite	Antifungal activity	*Monilinia laxa*	Cindi et al. (2015)
Cu-chitosan Hydrogel	Plant growth promoting	Positive effects on tomato growth	Juárez-Maldonado et al. (2016)
Oleoyl-chitosan nanoparticle	Antifungal activity	Synthesis and in vitro antifungal efficacy of oleoyl-chitosan nanoparticles against plantpathogenic fungi	Xing et al. (2016)

2013; Wang et al. 2016). Chitosan NPs treatment of leaves and seeds produced significant improvement in the plant growth and innate immune response through induction of defense enzyme activity, upregulation of defence related genes including that of several antioxidant enzymes as well as elevation of the levels of total phenolics (Chen et al. 2014; Chandra et al. 2015).

However, very few reports have been conducted on the biological activity of chitosan nanoparticles to control of plant pathogenic bacteria. An in vitro observe in which chitosan nanoparticles and chitosan nanocomposites with lime essential oil and thyme critical oil was accomplished. The effects indicated an inhibition of the *Pectobacterium carotovorum* via the treatment the usage of chitosan nanocomposites with thyme essential oil (Sotelo-Boyás et al. 2016). Concentration 1000 ppm of Cu-chitosan NPs have antibacterial effect against *Pseudomonas syringae* pv. *glycinea* that cause bacterial blight of soybean (Swati et al. 2017).

Chitosan was proven to inhibit the systemic propagation of viruses and viroids throughout the plant and to enhance the host's hypersensitive response to infection (Pospieszny et al. 1991; Faoro et al. 2001; Chirkov 2002). The degree of suppression of viral infections diverse based on the type of chitosan molecular weight (Kulikov et al. 2006). However, not one of the studies that investigated this effect has in reality proven the capacity of chitosan in absolutely inactivating viruses or viroids. Most literature i.e, (Kulikov et al. 2006) Reported on the inactivation of replication, which cause the stoppage of multiplication and spread. This will be linked to the reality that upon penetration into plant tissues, chitosan nanoparticles tightly bind nucleic acids and purpose a ramification of damages and selective inhibitions. Diverse time requirements by means of chitosan remedies for improvement of most appropriate defense stimulation ability was recorded in advance research in specific plant-pathogen systems wherein tobacco-TMV interactions (Zhao et al. 2007).

13.4.2 Antifungal

Chitosan NP were stated to have antifungal activity against specific plant pathogens. Nanoparticles on their own can negotiate mobile partitions and membranes at some distance more effectively than compared to the middle molecules they are organized from. This partially explains why chitosan NPs were discovered to demonstrate higher immune stimulation in comparison to chitosan itself (Chandra et al. 2015). Chitosan can effectively inhibit the development of phytopathogenic fungi at different life cycle ranges. For instance, chitosan completely inhibited spore germination, germ tube elongation, and mycelial growth of *Alternaria kikuchiana* Tanaka and *Physalospora piricola* Nose at 5 g/ L in vitro assay (Meng et al. 2010). In the pear fruit, treatments with chitosan reduced the ailment prevalence and inhibited the lesion growth as a result of the two aforementioned phytopathogenic fungi (Meng et al. 2010). The antifungal effect of rhodamine-labeled chitosan on *Fusarium oxysporum* was greater than on *Pochonia chlamydosporia*. Chitosan penetrated cell membranes of *F. oxysporum* by plasma membrane permeabilization, resulting in cell lysis (Palma-Guerrero et al. 2010). In commercial wine grapes, chitosan efficaciously inhibited growth of *Botrytis cinerea* in liquid and suppressed gray mold on detached grapevine leaves and bunch rot (Reglinski et al. 2010). The chitosan TPP NPs with sizes between 80 nm and 20 nm were evaluated for their antifungal efficacy against *Aspergillus parasiticus*, and an improved antifungal potential was found in comparison to the chitosan solution (Cota-Arriola et al. 2013). The antifungal activity of chitosan nanoparticles in controlling both *A. niger* and *F. solani* was recorded (Ing et al. 2012). Chitosan–silver nanoparticles were investigated for control of seed-borne pathogens in chickpea. Significant antifungal activity against *Aspergillus flvus*, *R. solani*, and *A. alternate* was evaluated (Kaur et al. 2012). Chitosan has strong antifungal activity against *Rhizoctonia solani* in rice. Two kinds of acid-soluble chitosan (with different degrees of deacetylation) caused a 60–91% inhibition in mycelial growth, 31–84% inhibition of disease incidence, and 66–91% inhibition in lesion length (Liu et al. 2012). The antifungal impact of chitosan on fungal growth of *Rhizopus* sp., *C. capsici*, *C. gloeosporioides*, and *A. niger* in seeds of chilli pepper (*Capsicum* sp.) was confirmed, showing at the least mycelial increase of 2.8, 2.2, 2.4, and 5.5 mm, respectively (Chookhongkha et al. 2013).

The antifungal properties of chitosan nanoparticles with silver functionalized with 4(E)-2-(3-hydroxynaphthalene-2-yl) diazen-1-yl) and benzoic acid for controlling *A. flavus* and *Aspergillus terreus* was evaluated. The fungal increase inhibition was proven between 20.2 and 27.0 mm, respectively (Mathew and Kuriakose 2013). Chitosan nanoparticles displayed robust inhibition of mycelial increase of *M. phaseolina*. A maximum 87.6% inhibition degree was recorded at 0.1% of chitosan nanoparticles, so specifically chitosan and Cu-chitosan nanoparticles proved their uniform size and stability which may additionally make a contribution to their higher antifungal efficacy against both of *A. Alternata*, *M. phaseolina* and *R. solani* in vitro assay. Cu-chitosan nanoparticles also confirmed maximum inhibition rate of spore germination of *A. alternate* (Saharan et al. 2013). Chitosan nanoemulsion was used for controlling *C. gloeosporioides* in vitro; the effects showed that the low molecular weight chitosan at a concentration of 1.0% had the best results in phrases of inhibition of conidial germination of the fungus (Zahid et al. 2013). The effect of peppermint vital oil (*Mentha piperita*) encapsulated in chitosan nanogels with cinnamic acid on *A. flavus* was studied. Inhibitory effect at the mycelial growth of the fungus at a concentration of 800 mg/mL was observed in vitro (Beyki et al. 2014). The antifungal impact of chitosan combined with silver for controlling *C. gloeospo-*

Fig. 13.2 Effect of different treatments on disease severity of FHB, 4 weeks after fungus inoculation. (**a**) Control (distilled water), (**b**) CS/NPs, (**c**) Control (0.5% v/v acetic acid aqueous solution), (**d**) CS, (**e**) Positive control (Tilt fungicide). (Reprinted from Kheiri et al. 2016)

rioides was evaluated using mango fruit, the varied concentrations was decreased the anthracnose by means of 45.7 and 71.3%, respectively (Chowdappa et al. 2014).

The polyethylene terephthalate punnets containing thyme oil and wrapped with chitosan/boehmite nanocomposite lidding films drastically reduced the incidence and severity of brown rot because of *Monilinia laxa* in artificially inoculated peach fruit (cv. Kakawa) held at 25 °C for 5 days and considerably reduced brown rot occurrence additionally at decrease temperatures (Cindi et al. 2015). Encapsulation of thyme EO in chitosan–benzoic acid nanogel has greater antimicrobial property in *A. flavus* for the protection of tomatoes (Khalili et al. 2015). As a broad-spectrum fungicide, primarily chitosan-based nanoparticles have been showed to be fungicidal against many pathogenic fungi. Cu–chitosan nanoparticles at 0.12% concentration caused 70.5 and 73.5% inhibition of mycelia growth in *Alternaria solani* and *Fusarium oxysporum*, respectively (Saharan et al. 2015). The management of Fusarium head blight (FHB) by using chitosan (CS) and chitosan nanoparticles (CS/NPs) has been studied. CS/NPs prevented fungal growth and CS/NPs may be a beneficial organic pesticide for controlling FHB. Spikelets treated with CS and CS/NPs showed gradual infection (Kheiri et al. 2016). The results confirmed that the percentage of disease severity reduced when CS and CS/NPs were employed before the fungus inoculation on the host (Fig. 13.2). Thus, CS and specifically CS/NPs, synthetized via an appropriate technique, can be used as biological pesticide in controlling fungal plant pathogens. Mycelial growth showed that there were considerable differences in tolerance to primarily chitosan-based nanoparticles among the various fungi examined. Plant pathogenic fungi including; *N. sphaerica*, *B. dothidea*, *N. oryzae*, and *A. tenuissima* was sensetive to O-chitosan nanoparticles, whilst *G. zeae* and *F. culmorum* was resistant. The antifungal activity of nanoparticles on four chitosan-sensitive fungi was concentration-dependent. The maximum value of antifungal index was discovered in medium containing nanoparticles at 2.0 mg/mL for each fungus (Xing et al. 2016). Synthesize and describe Cu-chitosan NCPs and evaluates its antifungal activities against *Fusarium verticillioids* causing publish flowering stalk rot (PFSR) disorder of maize. All collectively Cu-chitosan NCs has exquisite potential as antifungal agent against PFSR of maize in pot condition as well as in field condition (Choudhary 2017). The antifungal activity of α-chitin nanoformulations with sizes ranging from 80–100 nm diameters became evaluated on *A. niger* fungal increase, and they showed 87% of fungal growth inhibition (Salaberria et al. 2017).

Chitosan NPs have been synthesized from low-molecular-weight chitosan having better degree of acetylation was evaluated for thier efficacy aganist downy mildew disease of pearl millet resulting from *Sclerospora graminicola* (Siddaiah et al. 2018). This report confirmed that pearl millet seed remedy with nanochitosan particles was appreciably promoted seed germination and seedling vigor and, in addition, effectively induced systemic and durable resistance against downy mildew disease under greenhouse conditions. Nanochitosan has a broad-spectrum activity and an extensive antifungal activity on plant pathogenic fungi (Table 13.2).

13.4.3 Postharvest

Plant pathogenic bacteria negatively affect a wide variety of vital fruit and vegetables during the growing season and the duration of postharvest storage (Li et al. 2009a, b, 2010).

Traditional antimicrobials have been extensively used for decades. Recently, chitosan coating has emerged as a super alternative to chemically synthesized pesticides. In postharvest production, to prevent unwanted losses, chitosan biopolymer has emerged as a promising coating agent for end-product and vegetables due to its antimicrobial and defense eliciting properties. Some studies confirm that the in vitro growth of *Pseudomonas fluorescens* causing bacterial head rot of Broccoli (Li et al. 2009a, b), and *Burkholderia seminalis* inflicting bacterial fruit rot of apricot (Lou et al. 2011), have been markedly inhibited by chitosan under various environmental conditions. In addition, the antibacterial agent comprising 1–2%, also chitosan has accurate results in inhibiting the bacteria in the fruit and vegetable in reserving process and has inhibiting effect to the rot of fruit and vegetable. Besides, it has simple processing technology and low fee and is in accordance with the developmental path of modern antibacterial agent (Feng et al. 2009). The most efficient treatment to reduce disease occurrence has been the ones based on *Cinnamomum zeylanicum* essential oil (CEO)-loaded chitosan NPs. This treatment become not simplest powerful in controlling cucumber decay, however additionally in delaying disease symptoms and slowing down *P. drechsleri* increase during the storage stage. Uncoated cucumbers began to decay from the fourth day of storage (Mohammadi et al. 2015). Chitosan can also help to preserve the safety of fit to be eaten products.

The protection of fresh cut broccoli aganist *E. coli* and *Listeria monocytogenes* was confirmed with the aid of chitosan alongside bioactive components which includes bee pollen and extracts from propolis and pomegranate. Coating of chitosan on fruits and greens induces various defense enzymes, together with phenylalanine ammonia lyase (PAL), chitinase, and glucanase. Introduction of chitosan nanofilm composites made from chitosan and other antimicrobial agents could be more powerful in controlling postharvest losses of end-product and vegetables.

13.4.4 Food Preservation

Food packaging nanoparticles with antimicrobial interest which include chitosan (Qi et al. 2004; Tan et al. 2013), chitosan nanoparticles, and different inorganics. Combination of silver nanoparticles to the polymer matrix made the nanocomposites more appealing to be used in packaging (Duncan 2011; Fernández-Saiz and Lagaron 2011; Molina and Mejía 2016). Incorporation of those substances into the polymer matrix renders it lighter, stronger, fire resistance, better thermal properties, much less permeable to gases. Development of nanocomposites (up to 5% w/w nanoparticles) has been reported (Llorens et al. 2012). Chitosan nanoparticles (89 nm) and pectin (used as plasticizer) have been added into banana puree films (Martelli et al. 2013). The chitosan derivative coating was effective in reducing the decay of green asparagus caused by *F. concentricum*. In vitro assay, L-chitosan and H-chitosan inhibited the radial growth of *F. concentricum*, with an extremely good impact at concentration of 4 mg/ml, and totally inhibited spore germination at a concentration of 0.05 mg/ml, indicating that chitosan derivatives have been either fungistatic or fungicidal. The found inhibition of degradation and augmented lignification became attributed to the fungistatic activity of chitosan and its ability to induce a protection response (Qiu et al. 2014). The use of nanoparticles might assist in manufacturing of recent meals packaging materials with advanced mechanical, barrier and antimicrobial properties to growth shelf life (Chaudhry et al. 2008; Mihindukulasuriya and Lim 2014). Incorporated ZnO nanomaterial at diverse concentrations in a chitosan polymer with neem critical oil to enhance the properties of the bionanocomposite films (Sanuja and Agalya 2015).

13.4.5 Antimicrobial Mechanism

Chitosan has been validated to be a natural molecule that induces numerous organic responses in plant life, dependent on its structure, concentration, and on species and developmental stage of the plant (Malerba and Cerana 2016). Chitosan NPs remedy of leaves and seeds produced giant development inside the plant increase and innate immune reaction via induction of defense enzyme interest, up-regulation of defense-associated genes together with that of several antioxidant enzymes as well as elevation of total phenolics (Chen et al. 2014; Chandra et al. 2015). Natural protection response of plant life depends upon early detection of pathogens during the pathogenesis. Chitosan, whilst realised to plant tissues, often agglutinate across the penetration sites and has two main effects. The first one is the isolation of the penetration plant host via the formation of a physical barrier stopping the pathogen from spreading and invading other wholesome tissues. This phenomenon resembles the abscission zones regularly found on leaves stopping several necrotrophic pathogens from spreading further. It is widely found on potato tubers for instance (El Hadrami et al. 2009).

13.4.6 Induce Resistance

Furthermore, El Hassni et al. (2004) observed the impact of chitosan in date palm in reaction to *Fusarium oxysporum* f. sp. *albedinis*, the causal agent of a fusarium wilt on this crop. Beside an instantaneous toxicity of the molecule on the fungus, the authors confirmed an enhancement of critical additives of the host resistance. When injected into the roots at different concentrations, chitosan elicited date palm peroxidase and polyphenoloxidase activities, and increased the extent of phenolic compounds. Among the accrued phenolics, there was growth in content of unique nonconstitutive hydroxycinnamic acid derivatives, regarded to be of great importance in the resistance of this plant to this vascular fusariosis. Similarly, remedy of wheat seeds with chitosan found a growth in hydroxycinnamic (i.e., p-coumaric, caffeic, and ferulic) and benzoic (i.e., benzoic, protocatechuic, and gallic) acid derivatives, key to an increase in lignin synthesis and accumulation (Reddy et al. 1999). Chitosan is thought to behave as powerful inducer, enhancing a battery of plant responses each regionally around the infection sites and systemically to alert healthy elements of the plant. These encompass early signaling events in addition to the buildup of defense-related metabolites and proteins including phytoalexins and PR-proteins (Reddy et al. 1999; El Hadrami et al. 2009; Hammerschmidt 1999; Vander et al. 1998; Wang et al. 2008). Hence, as an exogenous elicitor, chitosan can stimulate resistance in plant host by way of increasing a few defense-related enzymes activities, together with phenylalanine ammonia-lyase (PAL), peroxidase(POD), catalase (CAT), superoxide dismutase (SOD), and polyphenol oxidase (PPO) activity (Xing et al. 2015). Finally, Chandra et al. (2015) have mentioned that accumulation of chitosan NP will increase plant protection through increasing the stages of SOD and CAT. Chitosan NP binds extracellularly across the cellular wall of leaves. One of the most crucial signaling molecules is nitric oxide (NO), which is also related to a variety of physiological procedures including induction of defense mechanism in plants. Plants treated with chitosan NP showed increased degrees of NO, in comparison to nontreated plants with chitosan NP (Raho et al. 2011; Malerba et al. 2012).

Chitosan NP treated sets resulted in up-regulation of PAL leading to a higher level of phenolic compound accumulation. These accumulated compounds help in alteration to distinct environmental conditions and offer resistance against pathogen. In the presence of NADPH, ANR uses anthocyanidins as substrates to synthesize EC. EC, sooner or later, transforms into proanthocyanidins, which are broadly dispensed as plant defense compounds having great toxicity toward pathogens. High levels of flavonoid accumulation are probably an illustration of superior resistance to flowers. Chitosan NP-treated flora, higher expression of SOD and CAT was observed, resulting in increase of these enzymes. SOD and CAT are the important antioxidant enzymes involved with ROS scavenging (Chandra et al. 2015). Polyphenol oxidase, catalyzing the phenolic materials to provide lignin, is everywhere among angiosperms and said to be involved in plant protection by means of helping the formation of lignin that contributes to the reinforcement of the cell wall structure heading off the penetration of pathogen (Li and Zhu 2013). ROS, Ca2+, nitric oxide (NO), ethylene (ET), jasmonic acid (JA), salicylic acid (SA), and abscisic acid (ABA) are all engaged in chitosan-mediated sign pathway (Xing et al. 2015). More details related to antimicrobial mechanism for nanochitosin was reviewed in Chap. 11.

13.4.7 Anti-insects

Chitosan has been found to expose robust insecticidal activity in a few plant pests (Zheng et al. 2005; Rabea et al. 2005). Chitosan (i.e., N-alkyl-, N-benzylchitosans) are made to be had through chemical synthesis, their insecticidal activities are being reported using an oral larvae feeding bioassay (Rabea et al. 2005; Badawy et al. 2005). Encapsulated microcrystals of the insecticide imidacloprid (IMI) with the aid of LbL assembly the use of chitosan and sodium alginate accompanied by way of addition of photocatalytic NPs (Guan et al. 2008). The insecticide etofenprox was encapsulated the use of a nanosized chitosan carrier in three types according to a difference in release patterns by adjusting the molecular weight and concentration of chitosan. Release properties of etofenprox and its organic activity against *Spodoptera litura* suggested that such managed-release method is used as a technique for preventing loss of etofenprox, increasing its activity against the target pest (Hwang et al. 2011).

The entomopathogenic fungi *Nomuraea rileyi* were investigated against *S. Litura*, and chitosan nanoparticle coated fungal metabolite (CNPCFM) showed better pesticidal activity, as compared with Uncoated Fungal Metabolite (UFM) and Fungal Spores (FS) (Chandra et al. 2013). Chitosan nanoparticles integrated insecticidal protein beauvericin (CSNp-BV) became organized by using ionic gelation technique to enhance insecticidal activity against *S. litura*. Pesticidal interest found out that all lifestyles stages have been susceptible to the CSNp-BV formulation and the maximum mortality was recorded in early larval instars. CSNp-BV treatment reduced pupal and adult emergence (Bharani et al. 2014).

Chitosan (CS)-g-poly (acrylic acid) PAA nanoparticles reduced egg laying of *Aphis gossypii* (20.9 ± 9.1 and 28.9 ± 9.2 eggs/woman for laboratory and below semi-discipline situations, respectively) than manage (97.3 ± 4.9 and 90.34. Nine eggs/female for laboratory and below semi-subject situations, respectively (Sahab et al. 2015). Chitosan nanoparticles decreased egg laying of *Callosobruchus maculatus* (10.9 ± 9.9 and 19.9 ± 9. Nine eggs/female laboratory and under semi-storage conditions, respectively (Sahab et al. 2015). Under semi subject situations, the wide variety of *Schistocerca gregaria* had been notably reduced after the chitosan and nanochitosan remedy, the quantity of infestations with *S. gregaria* reduced to 29 ± 3.6 and 8 ± 1.1 individuals after 120 days of remedies (Sabbour 2016). Chapter 1 was tested the application information of nanotechnology for insect's management.

13.4.8 Growth Promotions

During the past 100 years, chemical fertilizers and pesticides were used to face these problems and increase yield. The huge utilization of these products raised productivity significantly, but it also led to reduced biological diversity and degraded natural and agricultural systems. In addition, residue accumulation led to environmental pollution and public health problems, with development of resistant pests (Sun et al. 2012). Therefore, alternative techniques that include nanomaterials are important to address the issues of lowering the environmental impact without affecting agricultural productivity and economic viability for farmers (Ghormade et al. 2011; Kah and Hofmann 2014).

Micro-chitosan remedies have plant growth promoting activities, resulting in increased yields and plant health in numerous plants and fruits. The activation of defensive mechanisms in plant tissues with chitosan inhibited the growth of taxonomically unique pathogens (Vasyukova et al. 2001). Therefore, radicles, after penetrating the seed coatings may want to contact the NPs directly. The presence of chitosan NPs in huge quantities on the root floor should adjust the surface chemistry of the root and block the root openings and both hydraulic and nutrient uptake in roots is inhibited. Therefore, plant growth is negatively affected because of NPs (Behboudi et al. 2017). In the case of tomato seeds, Saharan et al. (2016) observed that Cu-chitosan NPs at high concentration showed more inhibitory effect on seedling growth. The seed quality of minitubers derived from chitosan treatments in vitro was improved, giving rise to field plants with increased tuber numbers and yields (Kowalski et al. 2006). Chitosan and chitooligosaccharides are as molecular signals to induce plant promoter and develop disease resistance system in plants (Bueter et al. 2013). Chitosan oligomer has large influences on biophysical feature, growth, and improvement of coffee such as increasing nutrient uptake, content material of chlorophyll and information of growth and yield (Dzung et al. 2011). Chitosan has remarkable film-forming property making it a clean agent to form a semipermeable coat at the seed surface that may keep the seed moisture and take in soil moisture which hence can promote seed germination. Moreover, it restrains the seed respiration through preventing oxygen entry, proscribing loss of CO_2 and keeping high concentration of CO_2 inside the film (Furbank et al. 2004). Chitosan also can increase soluble sugar content and improves the activity of protease to increase loose amino acid content (Zeng et al. 2012).

The effect of chitosan NPs on seed germination and seedling vigor has been evaluated in diverse vegetation for promotion of plant growth through increasing the uptake of vitamins and water through adjusting cellular osmotic pressure (Guan et al. 2009; Katiyar et al. 2015). Insufficient studies were carried out on using Cu/Zn chitosan NPs in seed germination and seedling growth. Wheat growth was promoted in terms of germination ability, root period, and seedling top by way of the motion of oligochitosan (Ma et al. 2014). Positive response of micro-chitosan on seed germination and seedling increase results in improvement of nanoformulation of chitosan. Chitosan nanoparticles have nanometer size and they can inter into the cells of the plants simply and improve their better bioactivities (Van et al. 2013). Chitosan nanoparticles effected strongly on nutrient uptake of the coffee seedlings in the green house condition. In the same cultivation condition, application of chitosan nanoparticles enhanced significantly the uptake of nitrogen, phosphorus, potassium, calcium and magnesium compared to the control (Van et al. 2013). In Robusta coffee nanochitosan substantially expanded chlorophyll content material, photosynthetic depth, nutrient uptake, and seedling growth (Van et al. 2013). Zn–chitosan formulation has been evaluated for growth and development in dry bean (Ibrahima and Ramadan 2015). The effect of foliar application of nanochitosan NPK fertilizer at the chemical composition of wheat grains. Foliar application of nanofertilizers showed a sizable increase in total saccharide, while induced significant decrease in protein content and nitrogen content of the wheat grains (Abdel-Aziz et al. 2018).

The growth promoting impact of micro-chitosan has been verified to be significantly lower as compared to chitosan NPs. The growth promotory effect of bulk chitosan has been recorded signifiantly lower as compared to chitosan NPs. Similarly, as compared to control and $CuSO_4$, bulk chitosan has been reported to have higher value for all parameters except for percent germination (Saharan et al. 2016). Similar efforts are needed for nanomaterials intended for use as fertilizers, so that nanofertilizers are advanced from the current mostly pristine products easily manipulated by the test environment to more functional products. To this quit, upgrades to this point made to nanoscale nutrients to generate improved nanofertilizers consist of the ones already stated in following sections concerning surface modifications along with alginate and chitosan (Saharan et al. 2016; Abdel-Aziz et al. 2016). There is an pressing neediness to make the most chitosan-based nanoparticles alone or in conjugation with various organic and inorganic compounds. Chitosan biopolymer based totally nanoparticles have large plant growth stimulatory activity (Table 13.2).

Very few studies have implemented chitosan particles in the agricultural context including incorporation of NPK fertilizer into the chitosan nanoparticles to make fertilizer consumption more efficient (Corradini et al. 2010; Wu et al. 2008; Hasaneen et al. 2014). Nanoparticles composed with an internal coating of CHT, an outer coating of cross-linked poly(acrylic acid)/diatomite-containing urea, and a middle of water-soluble granular nitrogen (N), phosphorus (P), and potassium (K) (NPK) fertilizer showed controlled release of the nutrients with no adverse impact on the soil (Wu et al. 2008). Recently, biodegradable polymeric chitosan NPs (~78 nm) were used for controlled release of the NPK fertilizer sources, which includes urea, calcium phosphate, and potassium chloride (Corradini et al. 2010), while Hussain et al. (2012) encapsulated urea in CHT microspheres acquiring a controlled release of the nutrient.

Chitosan NPs of the mean diameter 20 nm for loading NPK fertilizers were prepared and showed that the stability of the colloidal chitosan nanoemulsion was better with the addition of nitrogen and potassium than with the addition of phosphorus because of the higher anion rate from the calcium phosphate than the anion charges from the potassium chloride and urea (Hasaneen et al. 2014). Biodegradable and biocompatible nanoparticles in a green manner originated from locally available raw materials and natural excipients addressing the said risks which will ultimately lead to development of eco-friendly nanofertilizers to release nutrients gradually in a controlled manner was produced. A herbal pass linker, Genipin was extracted from tender fruit of *Gardenia jasminoides* (Shantha Siri et al. 2017).

13.4.9 Pesticides Delivery and Remediation

Nano-pesticide transport systems have a series of critical advantages over their classical bulk counterparts. Chitosan has emerged as one of the most promising polymers for the green transport of agrochemicals along with micronutrients, insecticides, and herbicides in nanoparticles. Chitosan simply absorbs to plant surfaces (e.g., leaves and stems), which helps to prolong the touch time between agrochemicals and the goal absorptive surface. Chitosan nanoparticles are seen to facilitate active molecule or compound uptake through the cellular membrane. The absorption improving impact of chitosan nanoparticles improves the molecular bioavailability of the active elements contained inside the nanoparticles (Tiyaboonchai 2003).

Polyvinylpyridine and polyvinylpyridine-co-styrene nanoparticles were investigated to control release of tebuconazole and chlorothalonil fungicides for better wood maintenance (Liu et al. 2001). Boehm et al. (2003) developed strong polymeric nanospheres (135 nm) with 3.5% encapsulation rate, and apart from the low active ingredient content, this formulation yielded substantial upgrades in the bioavailability of the insecticide (RPA 107382) to crops. It has additionally been reported that aluminosilicate-crammed nanotubes stick to plant surfaces whilst the nanoscale aluminosilicate particles leach from the nanotubes and in the end stick to the surface hair of insect pests. These particles ultimately enter the body and influence certain physiological functions (Bhattacharyya et al. 2010; Kashyap et al. 2013). A new type from amphiphilic derivative of chitosan, N-(octadecanol-1-glycidyl ether)-O-sulfate chitosan was produced by Lao et al. (2010). Nanoparticulate structures based totally on herbal polysaccharides (chitosan and cyclodextrin) have been prepared for use in the ionic gelation technique and have been used by companies for botanical pesticides, including carvacrol and linalool (Campos et al. 2018).

Some authors synthesized and tested changed CHT nanoparticles to hold paraquat, the most widely used herbicide. Cea et al. (2010) included atrazine, a herbicide used for broadleaf weed control, into ethyl cellulose managed release formulations (CRFs) with the aid of solvent evaporation. Allophanic clays and nano-clay changed the matrix. All CRFs multiplied the atrazine activity and decreased leaching loss. Silva et al. (2011) verified that alginate/CHT nanoparticles regulate the discharge of the herbicide and its interaction with the soil. In addition, CHT/tripolyphosphate nanoparticles decreased paraquat toxicity (Grillo et al. 2014), improved herbicidal activity against Eichhornia crassipes was observed when paraquat was encapsulated in silver/CHT nanoparticles (Namasivayam et al. 2014). Decomposable chitosan-lactide copolymer was implemented as a hydrophobic provider for pyraclostrobin, a wide-spectrum foliar fungicide. In comparison with 25% pyraclostrobin emulsifiable listen, the NPs confirmed higher fungicidal interest against *Colletotrichum gossypii* (Xu et al. 2014). More recently, nanoparticles of CHT and sodium tripolyphosphate have been prepared and evaluated for suppression of *Pyricularia grisea* (Manikandan and Sathiyabama 2016) and oleoyl-chitosan nanoparticles have been synthesized and used to disperse antifungal products (Xing et al. 2016).

The authors concluded that these novel absorbent materials may be implemented as an alternative biocompatible and green method for pesticide removal, and also applied in water remedy systems. Bin Hussein et al. (2009) had already evolved a unique nanohybrid pesticide controlled release machine from 4-(2, four-dichlorophenoxy) butyrate and a Zn–Al-layered double hydroxide inorganic interlayer with the aid of specific methods. The bioavailability of the chiral herbicide dichlorprop to the green alga *Chlorella pyrenoidosa*, in the absence and presence of chitosan nanoparticles was reported by Wen et al. (2010). The capacity of debris primarily based on polymers and cyclodextrin to adsorb insecticides is indicative of their capability for use in water remediation (Liu et al. 2011). Chitosan complexes and nanoparticles of chitosan containing the chiral herbicide dichlorprop were synthesized and characterized with the aim of decreasing potential for leaching and contamination of subterranean waters (Wen et al. 2011). Celis et al. (2012) used specific methods to prepare a bionanocomposite cloth based totally on chitosan and clay (montmorillonite), used as an adsorbent for the herbicide clopyralid found in an aqueous solution or in a mixture of water and soil. A nanocomposite material of CHT and montmorillonite was used to adsorb and dispose of the herbicide clopyralid existing in water and soil (Grillo et al. 2014). Chitosan–zinc oxide (CS–ZnO) nanoparticles as an absorbent for the elimination of the pesticide permethrin from water. ZnO nanoparticles possessed an nearly spherical morphology with a size of 58 nm. Researchers investigated the influence of the quantity of absorbent, agitation time, pesticide initial concentration, and pH on the sorption of the pesticide by using CS–ZnO absorbents (Dehaghi et al. 2014). In addition, CHT/tripolyphosphate nanoparticles decreased paraquat toxicity (Grillo et al. 2014), (Namasivayam et al. 2014). Alginate/chitosan and chitosan/tripolyphosphate nanoparticles were developed for encapsulation of the herbicides imazapic and imazapyr. Comparison was made between the nanoparticle systems and the free herbicides in phases of their

Table 13.3 Applications of nanochitosan for removal of pesticides

Carrier/polysaccharide	Active principle	Class	Reference
Alginate-chitosan nanoparticles	Paraquat	Herbicide	Silva et al. (2011)
chitosan/tripolyphosphate	imazapic and imazapyr	Herbicide	
Chitosan nanoparticles and chitosan	Dichlorprop	Herbicide	Wen et al. (2011)
Chitosan	Dichlorprop	Herbicide	Wen et al. (2010)
Montmorillonite-chitosan bionanocomposites	Clopyralid	Herbicide	Celis et al. (2012)
Copper chitosan nanocomposites	organophosphorous	Pesticide	Jaiswal et al. (2012)
Chitosan nanoparticles	Hexavalent chromium	Metal	Geng et al. (2009)
Chitosan/ β-cyclodextrins Films	Carvacrol	Insecticide	Higueras et al. (2013)
Chitosan	–	–	Alves and Mano (2008)
Chitosan	NPK	Fertilizer	Wu and Liu (2008)
Chitosan	–	–	El-Sawy et al. (2010)
Chitosan	Rotenone	Insecticide	Lao et al. (2010)
Chitosan	1-Naphthylacetic acid	Hormone	Tao et al. (2012)
Chitosan-silver nanoparticles	–	pesticide	Saifuddin et al. (2011)
Chitosan NPs/ β-cyclodextrins	Carvacrol and linalool	pesticide	Campos et al. (2018)

cytotoxicity and genotoxicity that allows to verify that the encapsulation procedure led to a reduction in toxicity (Maruyama et al. 2016). These observations delivered a clear suggestion that chitosan was able to modify the enantioselective bioavailability of the pesticide and herbicide, which could be safe for agricultural ecology and environment (Table 13.3).

13.4.10 Plant Genetic Transformation

Nanotechnology will possibly play a critical role in the improvement of genetically modified plants. The improvement of biotic and abiotic kinds of crop plant involves the transport of genetic cloth of both DNA and RNA resulting in the alteration of gene expression (Palerice and Gatehouse 2008). There are many obstacles for gene transfer to transform plant (Ghormade et al. 2011). Gene transfer is a technique for plant transformation using nanoparticle through nonviral-mediated transport vehicles including chitosan nanoparticle. And it's better than traditional methods in plants such as Agrobacterium-mediated gene transfer, electroporation, PEG-mediated gene transfer, particle gun bombardment, etc., are costly, labor intensive and cause significant perturbation to the growth of cells. In addition, these methods have very low efficiency (0.01–20% performance). However, it has been particularly a success for genetic transformation of dicots (Sivamani et al. 2009).

Biodegradable nanoparticles with common diameters of approximately 0.2 μm can be a product of chitosan and polyglutamic acid and used as microprojectiles (Lee et al. 2008). These particles had been effective for encapsulating and protecting DNA for transdermal gene transport by way of acceleration with a low-strain gene gun. Yu-qin et al. (2012) have screened the benefits of using nanoparticles over conventional companies. Firstly, nanoparticles are applicable to both monocotyledons and dicotyledonous plants and any types of organs. Secondly, this type of gene carriers can effectively overcome transgenic silencing via controlling the copies of DNA combined to nanoparticles. Thirdly, nanoparticles can be easily functionalized so as to further enhance transformation efficiency. Finally, nanoparticles-mediated multigene transformation can be achieved without involving traditional building method of complex carrier.

Biodegradable polymers like chitosan can report an advantageous charge on DNA conjugate nanomaterial surface (Li et al. 2011). Chitosan is a polymer that has been used extensively both in nucleic acid delivery and tissue engineering packages (Raftery et al. 2013). CHT nanoparticles can appropriately deliver genetic material for plant transformation (Malerba and Cerana 2016). Chitosan/DNA nanoparticles can be effectively formed by using coacervation among the definitely charged amine corporations on chitosan and negatively charged phosphate groups on DNA. However, the transfection performance of chitosan is low. The transfection performance has been shown to rely on the chitosan molecular weight, degree of deacetylation, pH of the transfecting medium, and cell kind (Mao et al. 2010). A pH of 6.8–7.0 is vital for transfection (Sailaja et al. 2013), and proof shows that DNA complexes shaped through shorter and near monodisperse chitosan oligomers (24-mer) have extra applicable properties than ultrapure chitosan and are therefore extra attractive as gene shipping systems than the traditional high molecular weight chitosans (Koping-Hoggard et al. 2003). Besides this, the degree of chitosan deacetylation additionally acts as an important aspect in chitosan–DNA nanoparticle method because it impacts DNA binding, launch, and gene transfer performance in vitro and in vivo assays (Kiang et al. 2004).

In particular, the ability of use of CHT NPs for genetic transformation is proposed by using its capability to shape, via electrostatic interactions, a complicated where DNA is included from nuclease degradation (Mao et al. 2010; Pichyalriti and Varoni 2015). Interesting effects were obtained by Wang et al. (2013), who devolped QD-labeled CHT-DNA complexes to display nanoparticle-mediated genetic transformation of cultured cells of *Jatropha curcas*. This approach gave an upward push to solid transformants with higher performance than other conventional strategies of gene transfer. Recent data confirmed that primarily CHT-based nanoparticles with highly positive surface coatings can passively penetrate throughout the chloroplast membrane.

Once in the chloroplast, these nanoparticles showcase both restricted diffusion and convection before attaining an irreversibly trapped state (Wong et al. 2016). These results propose that CHT nanoparticles may be used as viable molecular transporters into plastids, just like the chloroplast. Fungal resistance of transgenic potato flowers expressing thionin genes isolated from Brassicaceae species

(*Arabidopsis thaliana*) was evaluated against the phytopathogenic fungi. Thionins inhibit the increase in vitro of approximately 20 one-of-a-kind fungal plant pathogens that include *Botrytis cinerea*, *Fusarium* spp., *Phytophthora infestans*, and *Rhizoctonia solani* (Cammue et al. 1992). It is obvious that protein extract of five replicates of transgenic potato that incorporate thionin proteins showed antifungal activity against some pathogenic fungi with reduction in mycelial growth, illustrate the inhibitory effect of protein extract of the transgenic potato cultivars on radial growth of the PDA plates of four different pathogenic fungi after incubation for 7 days at 28 °C compared with control treatment (Abdel-Razik et al. 2017).

Two thionin resistance genes have been transferred into potato cultivars using chitosan nanoparticle, and then examined the resistance of transgenic species with the fungal pathogens *Alternaria alternata* and *Rhizoctonia solani* (Abdel-Razik et al. 2017). Recently, chitosan has attracted sizeable attention for use in formulations with small interfering RNA (siRNA). Because of the cationic nature, chitosan could make complex with siRNA in nanoparticles profile. Some studies suggest the utility of chitosan nanoparticle-entrapped siRNA as a service for siRNA delivery (Ragelle et al. 2013). Zhang et al. (2010) have proven that chitosan nanoparticles successfully added dsRNA (against chitin synthase genes) in stabilized shape to mosquito larvae through feeding. Chitosan nanoparticles must verify to be effective in dsRNA delivery due to their efficient binding with RNA, protection, and the ability to penetrate through the cellular membrane. These effects really suggest that chitosan nanoparticles primarily based siRNA formulations may additionally make contributions to plant pathogen and pest control while warding off the lengthy method of conventional plant transformations. There are a few mixed consequences with reference to gene transport through chitosan nanoparticles, but given the potential benefits of chitosan nanoparticles to assist within the transfer of genetic material to design new and enhanced plant genotypes (Kashyap et al. 2015).

13.5 Future Prospects

Therefore the incorporation of natural antimicrobial compounds in based-chitosan release matrices as micro- and nanoparticles is a topic of interest for future research. This may represent a promising alternative for use in the control of bacteria, fungi and insects in pre- and put up-harvest foods, contributing to the eradication of the troubles related to the usage of chemicals on food crops (Cota-Arriola et al. 2013). However, nanotechnology can both enhance crop productivity and reduce nutrient losses. As demonstrated in the course of this chapter, the nano-chitosan have been widely used in research in order to develop new formulations with active compounds of interest in plant protection. These formulations not only release the active compound in a slow manner but also after degradation increase the crop output besides proving the water-holding capacity of the soil. Cu and Zn have traditionally been used in agrochemicals. Cu and Zn, blended with chitosan may additionally reduce the threat of hazardous agrochemicals for crop improvement and protection.

Chitosan nanofertilizer assessments should be done using mixtures of nanoscale nutrients to mimic conventional fertilizer application regimens typically involving multiple nutrients applied simultaneously (e.g., NPK).

13.6 Conclusions

The role of chitosan NPs in the agricultural sector is highlighted as insecticides and growth promoters for plants, and additionally as preservatives in the course of post-harvest and packaging of agriproducts. Studies of those nanomaterials are specifically focused on components of in vitro control, so it is important to perform in situ control assessments to offer answers and alternatives to the problems that the agricultural field faces. Additionally, the numerous makes use of nanochitosan inside the plant protection turned into mentioned, beginning with the utility of it as a pesticide remediation and gene delivery for the plants and controlling post-harvest diseases. The shape of chitosan NPs and their impact is determined by their type, concentration, mixture (monomer or combined with other compounds), and response (temperature and time). However, this biopolymer requires investigation and further exploration to better understand its multifunctionality, properties, and mechanisms. Use of nanochitosan for transport of agrochemicals (insecticides, micronutrients, fertilizers, and plant growth hormones) would be the most promising discipline in the coming years for nanotechnology utility in agriculture. In conclusion, chitosan formulations may be used for more secure use of nano-agrochemicals.

Acknowledgment This research was supported by the Science and Technology Development Fund (STDF), Joint Egypt (STDF)–South Africa (NRF) Scientific Cooperation, Grant ID. 27837 to Kamel Abd-Elsalam.

References

Abdel-Aziz HMM, Hasaneen MNA, Omar AM (2018) Effect of foliar application of nano chitosan NPK fertilizer on the chemical composition of wheat grains. Egypt J Bot. https://doi.org/10.21608/EJBO.2018.1907.1137

Abdel-Razik AB, Hammad IA, Tawfik E (2017) Transformation of Thionin genes using chitosan nanoparticle into potato plant to be resistant to fungal infection. IOSR J Biotechnol Biochem 3(3):1–13

Abd–Elsalam KA, Vasil'kov AY, Said–Galiev EE, Rubina MS, Khokhlov AR, Naumkin AV, Shtykova EV, Alghuthaymi MA (2017) Bimetallic and chitosan nanocomposites hybrid with trichoderma: novel antifungal agent against cotton soil–borne fungi. Eur J Plant Pathol. https://doi.org/10.1007/s10658–017–1349–8

Agnihotri AA, Mallikarjuna NN, Aminabhavi TM (2004) Recent advances on chitosan-based micro- and nanoparticles in drug delivery. J Control Release 100(1):5–28

Alves NM, Mano JF (2008) Chitosan derivatives obtained by chemical modifications for biomedical and environmental applications. Int J Biol Macromol 43:401–414

Ambrosio L (2009) Biomedical composites. Woodhead Publishing, Cambridge

Aziz N, Faraz M, Pandey R, Sakir M, Fatma T, Varma A, Barman I, Prasad R (2015) Facile algae-derived route to biogenic silver nanoparticles: synthesis, antibacterial and photocatalytic properties. Langmuir 31:11605–11612. https://doi.org/10.1021/acs.langmuir.5b03081

Aziz N, Pandey R, Barman I, Prasad R (2016) Leveraging the attributes of *Mucor hiemalis*-derived silver nanoparticles for a synergistic broad-spectrum antimicrobial platform. Front Microbiol 7:1984. https://doi.org/10.3389/fmicb.2016.01984

Badawy MEI, Rabea EI, Rogge TM, Stevens CV, Steurbaut W, Höfte M, Smagghe G (2005) Fungicidal and insecticidal activity of O-acyl chitosan derivatives. Polym Bull 54:279–289

Behboudi F, Tahmasebi SZ, Kassaee MZ, Modares Sanavi SAM, Sorooshzadeh A (2017) Phytotoxicity of chitosan and SiO2 nanoparticles to seed germination of wheat (Triticum aestivum L.) and barley (*Hordeum vulgare* L.) plants. Not Sci Biol 9(2):242–249

Berscht PC, Nies B, Liebendorfer A, Kreuter J (1994) Incorporation of basic ibroblast growth factor into methylpyrrolidinone chitosan leeces and determination of the in vitro release characteristics. Biomaterials 15:593–600

Beyki M, Zhaveh S, Tahere S, Rahmani-Cherati T, Abollahi A, Mansour B, Bayat Tabatabaei M, Mohsenifarc A (2014) Encapsulation of *Mentha piperita* essential oils in chitosan–cinnamic acid nanogel with enhanced antimicrobial activity against *Aspergillus flavus*. Ind Crop Prod 54:310–319

Bharani RA, Namasivayam SKR, Shankar SS (2014) Biocompatible chitosan nanoparticles incorporated pesticidal protein beauvericin (CSNp-BV) preparation for the improved pesticidal activity against major groundnut defoliator *Spodoptera litura* (Fab.) (Lepidoptera; Noctuidae). Int J Chem Tech Res 6:5007–5012

Bhattacharyya A, Bhaumik A, Usha Rani P, Suvra Mandal S, Epidi TT (2010) Nano- particles - a recent approach to insect pest control. Afr J Biotechnol 9(24):3489–3493

Bin Hussein MZ, Hashim N, Yahaya AH, Zainal Z (2009) Controlled release formulation of agrochemical pesticide based on 4-(2,4-dichlorophenoxy)butyrate nanohybrid. J Nanosci Nanotechnol 9:2140–2147

Boehm AL, Martinon I, Zerrouk R, Rump E, Fessi H (2003) Nanoprecipitation technique for the encapsulation of agrochemical active ingredients. J Microencapsul 20:433–441

Borges J, Mano JF (2014) Molecular interactions driving the layer-by-layer assembly of multilayers. Chem Rev 114:8883–8942

Borges J, Rodrigues LC, Reis RL et al (2014) Layer-by-layer assembly of light-responsive polymeric multilayer systems. Adv Funct Mater 24:5624–5648

Brunel F, El Gueddari NE, Moerschbacher BM (2013) Complexation of copper(II) with chitosan nanogels: Toward control of microbial growth. Carbohydr Polym 92(2):1348–1356

Bueter CL, Specht CA, Levitz SM (2013) Innate sensing of chitin and chitosan. PLoS Pathog 9(1):e1003080

Cabrera JC, Wégria G, Onderwater RCA, González G, Nápoles MC, Falcón-Rodríguez AB, Costales D, Rogers HJ, Diosdado E, González S, Cabrera G, González L, Wattiez R (2013) In: Saa Silva S et al (eds) Proc. 1st world Congresson the use of biostimulants in agriculture, Acta horticultural 1009 ISHS

Cammue BPA, De Bolle MFC, Terras FRG, Proost P, Van Damme J, Rees SB, Vanderleydenand J, Broekaert WF (1992) Isolation and characterization of a novel class of plant antimicrobial peptides from *Mirabilis jalapa* L. seeds. J Biol Chem 267:2228–2233

Campos EVR, Proença PLF, Oliveira JL, Melville CC, Della Vechia JF, de Andrade DJ, Fraceto LF (2018) Chitosan nanoparticles functionalized with β-cyclodextrin: a promising carrier for botanical pesticides. Sci Rep 8:2067. https://doi.org/10.1038/s41598-018-20602-y

Caridade SG, Merino EG, Alves NM, Bermudez VZ, Boccaccini AR, Manoa JF (2013) Chitosan membranes containing micro or nano-size bioactive glass particles: evolution of biomineralization followed by in situ dynamic mechanical analysis. J Mech Behav Biomed Mater 20:173–183

Cea M, Cartes P, Palma G, Mora ML (2010) Atrazine efficiency in an andisol as affected by clays and nanoclays in ethylcellulose controlled release formulations. R C Suelo Nutr Veg 10:62–77
Celis R, Adelino MA, Hermosín MC, Cornejo J (2012) Montmorillonite–chitosan bionanocomposites as adsorbents of the herbicide clopyralid in aqueous solution and soil/water suspensions. J Hazard Mater 209:21067–21076
Chandra JH, Raj LFAA, Namasivayam SKR, Bharani RSA (2013) Improved pesticidal activity of fungal metabolite from nomureae rileyi with chitosan nanoparticles. Proceedings of the international conference on advanced nanomaterials and emerging engineering technologies. Chennai. pp 387–390
Chandra S, Chakarborty N, Dasgupt A, Sarkar J, Panda K, Acharya K (2015) Chitosan nanoparticle: a positive modulator of innate immune responses in plants. Sci Rep 5:1–13
Chaudhry Q, Scotter M, Blackburn J, Ross B, Boxall A, Castle L, Aitken R, Watkins R (2008) Applications and implications of nanotechnologies for the food sector. Food Addit Contam Part A 25:241–258
Chen J, Zou X, Liu Q, Wang F, Feng W, Wan W (2014) Combination effect of chitosan and methyl jasmonate on controlling *Alternaria alternata* and enhancing activity of cherry tomato fruit defense mechanisms. Crop Prot 56:31–36
Chirkov SN (2002) The antiviral activity of chitosan (review). Appl Biochem Microbiol 38:1–8
Chookhongkha N, Sopondilok T, Photchanachai S (2012) Effect of chitosan and chitosan nanoparticles on fungal growth and chilliseed quality. International conference on postharvest pest and diseas e management in exporting horticultural crops-PPDM2012 973:231–237
Chookhongkha N, Sopondilok T, Photchanachai S (2013) Effect of chitosan and chitosan nanoparticles on fungal growth and chilli seed quality. Acta Hortic 973:231–237
Choudhary MK (2017) Development and evaluation of Cu chitosan nanocomposite for its antifungal activity against post flowering stalk rot (PFSR) disease of maize caused by *Fusarium verticillioids* (Sheldon). Ph.D. Thesis, Maharana Pratap University of Agriculture and Technology, Udaipur, India. 79 pages
Chowdappa P, Shivakumar C, Chethana S, Madhura S (2014) Antifungal activity of chitosan-silver nanoparticles composite against *Colletotrichum gloeosporioides* associated with mango anthracnose. Afr J Microbiol Res 81:1803–1812
Cindi MD, Shittu T, Sivakumar D, Bautista-Baños S (2015) Chitosan boehmite alumina nanocomposite films and thyme oil vapour control brown rot in peaches (*Prunus persica* L.) during postharvest storage. Crop Prot 72:127–131
Corradini E, De Moura M, Mattoso L (2010) A preliminary study of the incorporation of NPK fertilizer into chitosan nanoparticles. Express Polym Lett 4(8):509–515
Costa RR, Mano JF (2014) Polyelectrolyte multilayered assemblies in biomedical technologies. Chem Soc Rev 43:3453–3479
Cota-Arriola O, Cortez-Rocha MO, Ezquerra-Brauer JM, Lizardi-Mendoza J, Burgos-Hernández A, Robles-Sánchez RM (2013) Ultrastructural, morphological, and antifungal properties of micro and nanoparticles of chitosan crosslinked with sodium tripolyphosphate. J Polym Environ 21:971–980
Decher G (1997) Fuzzy nanoassemblies: toward layered polymeric multicomposites. Science 277:1232–1237
Dehaghi SM, Rahmanifar B, Moradi AM, Azar PA (2014) Removal of permethrin pesticide from water by chitosan–zinc oxide nanoparticles composite as an adsorbent. J Saudi Chem Soc 18:348–355
Di Carlo G, Curulli A, Toro RG, Bianchini C, De Caro T, Padeletti G, Zane D, Ingo GM (2012) Green synthesis of gold? Chitosan nanocomposites for caffeic acid sensing. Langmuir 28:5471–5479
Duncan TV (2011) Applications of nanotechnology in food packaging and food safety: barrier materials, antimicrobials and sensors. J Colloid Interface Sci 363(1):1–24
Dzung NA, Khanh VTP, Dung TT (2011) Research on impact of chitosan oligomer on biophysical characteristics, growth, development and drought resistance of coffee. Carbohydr Polym 84:751–755

El Hadrami A, El Hadrami I, Daayf F (2009) Suppression of induced plant defense responses by fungal pathogens. In: Bouarab K, Brisson N, Daayf F (eds) Molecular-plant microbe interactions. CABI, Wallingford Chapter 10, pp 231–268

El Hassni M, El Hadrami A, Daayf F, Chérif M, Ait Barka E, El Hadrami I (2004) Chitosan, antifungal product against fusarium oxysporum f. Sp. albedinis and elicitor of defence reactions in date palm roots. Phytopathol Mediterr 43:195–204

El-Sawy NM, Abd El-Rehim HA, Elbarbary AM, Hegazy E-SA (2010) Radiation-induced degradation of chitosan for possible use as a growth promoter in agricultural purposes. Carbohydr Polym 79:555–562

Fang H, Huang J, Ding L, Li M, Chen Z (2009) Preparation of magnetic chitosan nanoparticles and immobilization of laccase. J Wuhan Univ Technol Mater Sci Ed 24:42–47. https://doi.org/10.1007/s11595-009-1042-7

Faoro F, Sant S, Iriti M, Appiano A (2001) Chitosan-elicited resitance to plant viruses: a histochemical and cytochemical study. In: Muzzarelli RAA (ed) Chitin enzymology. Atec, Grottammare, pp 57–62

Feng J, He J, Ma Z, Wang Z, Zhang X (2009) Plant source fruit and vegetable fresh-keeping agent and its preparation method. Patent number: CN101305747–A

Fernández-Saiz P, Lagaron JM (2011) Chitosan for film and coating applications. In: Plackett D (ed) Biopolymers: new materials for sustainable films and coatings. Wiley, West Sussex, pp 87–105

Freire TM, Dutra LMU, Queiroz DC, Ricardo NMPS, Barreto K, Denardin JC, Wurm FR, Sousa CP, Correia AN, de Lima-Neto P, Fechine PBA (2016) Fast ultrasound assisted synthesis of chitosan-based magnetite nanocomposites as a modified electrode sensor. Carbohydr Polym 151:760–769

Furbank RT, White RJ, Palta A, Turner NC (2004) Internal recycling of respiratory CO2 in pods of chickpea (*Cicerarietinum* L.): the role of pod wall, seed coat, and embryo. J Exp Bot 55:1687–1696

Geng B, Jin Z, Li T, Qi X (2009) Preparation of chitosan-stabilized Fe(0) nanoparticles for removal of hexavalent chromium in water. Sci Total Environ 407:4994–5000

Ghormade V, Deshpande MV, Paknikar KM (2011) Perspectives for nano-biotechnology enabled protection and nutrition of plants. Biotechnol Adv 29:792–803

Gornik K, Grzesik M, Duda BR (2008) The effect of chitosan on rooting of gravevine cuttings and on subsequent plant growth under drought and temperature stress. J Fruit Ornamental Plant Resour 16:333–343

Grillo R, Pereira AES, Nishisaka CS, de Lima R, Oehlke K, Greiner R, Fraceto LF (2014) Chitosan/tripolyphosphate nanoparticles loaded with paraquat herbicide: an environmentally safer alternative for weed control. J Hazard Mater 278:163–171

Guan H, Chi D, Yu J, Li X (2008) A novel photodegradable insecticide: preparation, characterization and properties evaluation of nano-Imidacloprid. Pestic Biochem Physiol 92:83–91

Guan YJ, Hu J, Wang XJ, Shao CX (2009) Seed priming with chitosan improves maize germination and seedling growth in relation to physiological changes under low temperature stress. J Zhejiang Univ Sci B 10:427–433

Hammerschmidt R (1999) Phytoalexins: what have we learned after 60 years? Annu Rev Phytopathol 37:285–306

Hasaneen MNA, Abdel-Aziz HMM, El-Bialy DMA, Omer AM (2014) Preparation of chitosan nanoparticles for loading with NPK. Afr J Biotech 13:3158–3164

He L, Liu Y, Mustapha A, Lin M (2011) Antifungal activity of zinc oxide nanoparticles against *Botrytis cinerea* and *Penicillium expansum*. Microbiol Res 166:207–215

Higueras L, López-Carballo G, Cerisuelo JP et al (2013) Preparation and characterization of chitosan/HP-β-cyclodextrins composites with high sorption capacity for carvacrol. Carbohydr Polym 97:262–268

Hussain MR, Devi RR, Maji TK (2012) Controlled release of urea from chitosan microspheres prepared by emulsification and cross-linking method. Iran Polym J 21:473–479

Hwang IC, Kim TH, Bang SH, Kim KS, Kwon HR, Seo MJ, Youn YN, Park HJ, Yasunaga-Aoki C, Yu YM (2011) Insecticidal effect of controlled release formulations of etofenprox based on nano-bio technique. J Fac Agric Kyushu Univ 56:33–40
Ibrahima EA, Ramadan WA (2015) Effect of zinc foliar spray alone and combined with humic acid or/and chitosan on growth, nutrient elements content and yield of dry bean (*Phaseolus vulgaris* L.) plants sown at different dates. Sci Hortic 184:101–105
Ichikawa S, Iwamoto S, Watanabe J (2005) Formation of biocopmpatible nanoparticles by selfassembly of enzymatic hydrolysates of chitosan and carboxymethyl cellulose. Biosci Biotechnol Biochem 69:1637–1642 PMID: 16195579
Iler RK (1966) Multilayers of colloidal particles. Colloid Interf Sci J 21:569–594
Ing LY, Zin NM, Sarwar A, Katas H (2012) Antifungal activity of chitosan nanoparticles and correlation with their physical properties. Int J Biomater 2012:632698. https://doi.org/10.1155/2012/632698
Jaiswal M, Chauhan D, Sankararamakrishnan N (2012) Copper chitosan nanocomposites: synthesis, characterization, and application in removal of organophosphorous pesticide from agricultural runoff. Environ Sci Pollut Res 19:2005–2062
Jayaseelan C, Ramkumar R, Rahuman AA, Perumal P (2013) Green synthesis of gold nanoparticles using seed aqueous extract of *Abelmo schusesculentus* and its antifungal activity. Ind Crop Prod 45:423–429
Jo YK, Kim BH, Jung G (2009) Antifungal activity of silver ions and nanoparticles on phytopathogenic fungi. Plant Dis 93:1037–1043
Juárez-Maldonado A, Ortega-Ortiz, H, Pérez-Labrada F, Cadenas-Pliego G, Benavidez-Mendoza A (2016) Cu nanoparticle absorbed on chitosan hydrogels positively alter morphological production and quality characteristics of tomato. J Appl Bot Food Qual 89:183–189
Kah M, Hofmann T (2014) Nanopesticide research: current trends and future priorities. Environ Int 63:224–235
Kashyap PL, Kumar S, Srivastava AK, Sharma AK (2013) Myconanotechnology in agriculture: a perspective. World J Microbiol Biotechnol 29(2):191–207
Kashyap PL, Xiang X, Heiden P (2015) Chitosan nanoparticle based delivery systems for sustainable agriculture. Int J Biol Macromol 77:36–51
Katiyar D, Hemantarajan A, Sing B (2015) Chitosan as a promising natural compound to enhance potential physiological responses in plant: a review. Indian J Plant Physiol 20(1):1–9
Kaur P, Thakur R, Choudhary A (2012) An in vitro study of the antifungal activity of silver/chitosan nanoformulations against important seed borne pathogens. Int J Sci Technol Res 1:83–86
Kendra DF, Hadwiger LA (1984) Characterization of the smallest chitosan oligomer that is maximally antifungal to *Fusarium solani* and elicits pisatin formation by *Pisum sativum*. Exp Mycol 8:276–281
Khalili ST, Mohsenifar A, Beyki M, Zhaveh S, Rahmani-Cherati T, Bayat M et al (2015) Encapsulation of thyme essential oils in chitosan–benzoic acid nanogel with enhanced antimicrobial activity against *Aspergillus flavus*. LWT Food Sci Technol 60:502–508
Kheiri A, Moosawi Jorf SA, Malihipour A, Saremi H, Nikkhah M (2016) Application of chitosan and chitosan nanoparticles for the control of fusarium head blight of wheat (*Fusarium graminearum*) in vitro and greenhouse. Int J Biol Macromol 93:1261–1272
Kiang T, Wen J, Lim HW, Leong KW (2004) The effect of the degree of chitosan deacetylation on the efficiency of gene transfection. Biomaterials 25(22):5293–5301
Kim SW, Jung JH, Lamsal K, Kim YS, Min JS, Lee YS (2012) Antifungal effects of silver nanoparticles (AgNPs) against various plant pathogenic fungi. Mycobiology 40:53–58
Koping-Hoggard M, Mel'nikova YS, Varum KM, Lindman B, Artursson B (2003) Relationship between the physical shape and the efficiency of oligomeric chitosan as a gene delivery system in vitro and in vivo. J Gene Med 5:30–141
Kowalski B, Terry FJ, Herrera L, Peñalver DA (2006) Application of soluble chitosan in vitro and in the greenhouse to increase yield and seed quality of potato minitubers. Potato Res 49:167–176

Kulikov SN, Chirkov SN, Il'ina AV, Lopatin SA, Varlamov VP (2006) Effect of the molecular weight of chitosan on its antiviral activity in plants. Prikl Biokhim Mikrobiol 42(2):224–228

Lamsal K, Kim SW, Jung JH, Kim YS, Kim KS, Lee YS (2011) Inhibition effects of silver nanoparticles against powdery mildews on cucumber and pumpkin. Mycobiology 39:26–32

Lao S-B, Zhang Z-X, Xu H-H, Jiang G-B (2010) Novel amphiphilic chitosan derivatives: synthesis, characterization and micellar solubilization of rotenone. Carbohydr Polym 82:1136–1142

Lee P-W, Peng S-F, Su C-J, Mi F-L, Chen H-L, Wei M-C, Lin H-J, Sung H-W (2008) The use of biodegradable polymeric nanoparticles in combination with a low-pressure gene gun for transdermal DNA delivery. Biomaterials 29:742–751

Li SJ, Zhu TH (2013) Biochemical response and induced resistance against anthracnose (*Colletotrichum camelliae*) of camellia (*Camellia pitardii*) by chitosan oligosaccharide application. For Pathol 43:67–76. https://doi.org/10.1111/j.1439-0329.2012.00797

Li B, Wang GL, Wu ZY, Qiu W, Tang QM, Xie GL (2009a) First report of bacterial head rot of broccoli caused by *Pseudomonas fluorescens* in China. Plant Dis 93:12–19

Li B, Yu RR, Yu SH, Qiu W, Fang Y, Xie GL (2009b) First report on bacterial heart rot of garlic caused by *Pseudomonas fluorescens* in China. Plant Pathol J 25:91–94

Li B, Fang Y, Zhang GQ, Yu RR, Lou MM, Xie GL, Wang YL, Sun GC (2010) Molecular characterization of Burkholderia cepacia complex isolates causing bacterial fruit rot of apricot. Plant Pathol J 26:223–230

Li C, Guo T, Zhou D, Hu Y, Zhou H, Wang S, Chen J, Zhang Z (2011) A novel glutathione modified chitosan conjugate for efficient gene delivery. J Control Release 154:177–188

Liu Y, Yan L, Heiden P, Laks P (2001) Use of nanoparticles for controlled release of biocides in solid wood. J Appl Polym Sci 79:458–465

Liu H, Cai X, Wang Y, Chen J (2011) Adsorption mechanism-based screening of cyclodextrin polymers for adsorption and separation of pesticides from water. Water Res 45:3499–3511

Liu H, Tian WX, Li B, Wu GX, Ibrahim M, Tao ZY, Wang YL, Xie GL, Li HY, Sun GC (2012) Antifungal effect and mechanism of chitosan against the rice sheath blight pathogen, *Rhizoctonia solani*. Biotechnol Lett 34:2291–2298

Llorens A, Lloret E, Picouet PA, Trbojevich R, Fernandez A (2012) Metallic-based micro and nanocomposites in food contact materials and active food packaging. Trends Food Sci Technol 24:19–29

Lopez-Leon T, Carvalho ELS, Seijo B, OrtegaVinuesa JL, Bastos-Gonzalez D (2005) Physicochemical characterization of chitosan nanoparticles: Electrokinetic and stability behavior. Colloid Interf Sci J 283:344–351. https://doi.org/10.1016/j.jcis.2004.08.186

Lou MM, Zhu B, Muhammad I, Li B, Xie GL, Wang YL, Li HY, Sun GC (2011) Antibacterial activity and mechanism of action of chitosan solutions against apricot fruit rot pathogen *Burkholderia seminalis*. Carbohydr Res 346:1294–1301. https://doi.org/10.1016/j.carres.2011.04.042

Ma L, Li J, Yy YCM, Wang Y, Xm L, Li N (2014) Germination and physiological response of wheat (*Triticum aestivum*) to pre-soaking with oligochitosan. Int J Agric Biol 16:766–770

Malerba M, Cerana R (2016) Chitosan effects on plant systems. Int J Mol Sci 17:996. https://doi.org/10.3390/ijms17070996

Malerba M, Crosti P, Cerana R (2012) Defense/stress responses activated bychitosan in sycamore cultured cells. Protoplasma 249:89–98. https://doi.org/10.1007/s00709–011–0264–7

Manikandan A, Sathiyabama M (2016) Preparation of chitosan nanoparticles and its effect on detached rice leaves infected with *Pyricularia grisea*. Int J Biol Macromol 84:58–61

Mano JF (2008) Stimuli-responsive polymeric systems for biomedical applications. Adv Eng Mater 10:515–527

Mao S, Sun W, Kissel T (2010) Chitosan-based formulations for delivery of DNA and siRNA. Adv Drug Deliv Rev 62:12–27

Martelli MR, Barros TT, de Moura MR, Mattoso LH, Assis OB (2013) Effect of chitosan nanoparticles and pectin content on mechanical properties and water vapor permeability of banana puree films. J Food Sci 78:N98–N104

Martins A, Reis RL, Neves NM (2008) Electrospinning: processing technique for tissue engineering scaffolding. Int Mater Rev 53:257–274

Maruyama CR, Guilger M, Pascoli M et al (2016) Nanoparticles based on chitosan as carriers for the combined herbicides Imazapic and Imazapyr. Sci Rep 6:19768. https://doi.org/10.1038/srep19768

Mathew T, Kuriakose S (2013) Photochemical and antimicrobial properties of silver nanoparticle-encapsulated chitosan functionalized with photoactive groups. Mater Sci Eng C 33:4409–4415

Meng XH, Yang LY, Kennedy JF, Tian SP (2010) Effects of chitosan and oligochitosan on growth of two fungal pathogens and physiological properties in pear fruit. Carbohydr Polym 81:70–75

Mihindukulasuriya SDF, Lim LT (2014) Nanotechnology development in food packaging: a review. Trends Food Sci Technol 40:149–167

Mohammadi A, Hashemi M, Hosseini SM (2015) Chitosan nanoparticles loaded with *Cinnamomum zeylanicum* essential oil enhance the shelf life of cucumber during cold storage. Postharvest Biol Technol 110:203–213

Molina EB, Mejía LZ (2016) New bioactive biomaterials based on chitosan A2 – Baños B, Silvia. In: Chitosan in the preservation of agricultural commodities. Chapter 2, pp 33–64, Academic Press, Elsevier, USA

Moura D, Mano JF, Paiva MC, Alves NM (2016) Chitosan nanocomposites based on distinct inorganic fillers for biomedical applications. Sci Technol Adv Mater 17(1):626–643

Nair R, Varghese SH, Nair BG, Maekawa T, Yoshida Y, Sakthi Kumar D (2010) Nanoparticulate material delivery to plants. Plant Sci 179:154–163

Namasivayam SKR, Aruna A, Gokila G (2014) Evaluation of silver nanoparticles-chitosan encapsulated synthetic herbicide paraquat (AgNp-CS-PQ) preparation for the controlled release and improved herbicidal activity against *Eichhornia crassipes*. Res J Biotechnol 9:19–27

Nguyen VS, Dinh MH, Nguyen AD (2013) Study on chitosan nanoparticles on biophysical characteristics and growth of Robusta coffee in green house. Biocatal Agric Biotechnol 2:289–294

Palerice DRG, Gatehouse JA (2008) RNAi-mediated crop protection against insects. Trends Biotechnol 26:393–309

Palma-Guerrero J, López-Jiménez JA, Pérez-Berná AJ, Huang IC, Jansson HB, Salinas J, Villalaín J, Read ND, Lopez-Llorca LV (2010) Membrane fluidity determines sensitivity of fiamentous fungi to chitosan. Mol Microbiol 75:1021–1032

Park HJ, Kim SH, Kim HJ, Choi SH (2006) A new composition of nanosized silica–silver for control of various plant diseases. Plant Pathol J 22:295–302

Park Y, Kim MH, Park SC, Cheong H, Jang MK, Nah JW, Hahm KS (2008) Investigation of the antifungal activity and mechanism of action of LMWS-chitosan. J Microbiol Biotechnol 18:1729–1734

Perez-de-Luque A, Cifuentes Z, Beckstead JA, Sillero JC, Anila C, Rubio J, Ryan RO (2012) Effect of amphotericin B nanodisks on plant fungal disease. Pest Manag Sci 68:67–74

Pichyalriti M, Varoni EM (2015) Chitosan-induced antiviral activity and innate immunity in plants. Environ Sci Pollut Res 22:2935–2944

Pospieszny H, Chirkov S, Atabekov J (1991) Induction of antiviral resistance in plants by chitosan. Plant Sci 79:63–68

Prasad R (2014) Synthesis of silver nanoparticles in photosynthetic plants. J Nanopart:963961. https://doi.org/10.1155/2014/963961

Prasad R, Kumar V, Prasad KS (2014) Nanotechnology in sustainable agriculture: present concerns and future aspects. Afr J Biotechnol 13(6):705–713

Prasad R, Bhattacharyya A, Nguyen QD (2017a) Nanotechnology in sustainable agriculture: recent developments, challenges, and perspectives. Front Microbiol 8:1014. https://doi.org/10.3389/fmicb.2017.01014

Prasad R, Gupta N, Kumar M, Kumar V, Abd-Elsalam KA (2017b) Nanomaterials acts as plant defense mechanism. In: Prasad R, Kumar V, Kumar M (eds) Nanotechnology: food and environmental paradigm. Springer Pvt Ltd, Cham, pp 253–269

Qi L, Xu Z, Jiang X, Hu C, Zou X (2004) Preparation and antibacterial activity of chitosan nanoparticles. Carbohydr Res 339:2693–2700

Qiu M, Wu C, Ren G, Liang X, Wang X, Huang X (2014) Effect of chitosan and its derivatives as antifungal and preservative agents on postharvest green asparagus. Food Chem155:105–111

Rabea EI, Badawy MEI, Rogge TM, Stevens CV, Hofte M, Steurbaut W, Smagghe G (2005) Insecticidal and fungicidal activity of new synthesized chitosan derivatives. Pest Manag Sci 61:951–960

Racovita S, Vasiliu S, Popa M, Luca C (2008) Polysaccharides based on micro-and nanoparticles obtained by ionic gelation and their applications as drug delivery systems. Rev Roum Chim 54:709–718

Raftery R, O'Brien FJ, Cryan SA (2013) Chitosan for gene delivery and orthopedic tissue engineering applications. Molecules 18:5611–5647

Ragelle H, Vandermeulen G, Préa V (2013) Chitosan-based siRNA delivery systems. J Control Release 172:207–218

Raho N, Ramirez L, Lanteri ML, Gonorazky G, Lamattina L, ten Have A, Laxalt AM (2011) Phosphatidic acid production in chitosan-elicited tomato cells, via both phospholipase D and phospholipase C/diacylglycerol kinase, requires nitric oxide. J Plant Physiol 168:534–539. https://doi.org/10.1016/j.jplph.2010.09.004

Reddy MV, Arul J, Angers P, Couture L (1999) Chitosan treatment of wheat seeds induces resistance to *Fusarium graminearun* and improves seed quality. J Agric Food Chem 47:1208–1216

Reglinski T, Elmer PAG, Taylor JT, Wood PN, Hoyte SM (2010) Inhibition of *Botrytis cinerea* growth and suppression of botrytis bunch rot in grapes using chitosan. Plant Pathol 59:882–890

Rhim JW, Hong SI, Park HM, Ng PKW (2006) Preparation and characterization of chitosan-based nanocomposite films with antimicrobialactivity. J Agric Food Chem 54:5814–5822

Rhoades J, Roller S (2000) Antimicrobial actions of degraded and native chitosan against spoilage organisms in laboratory media and foods. Appl Environ Microbiol 66:80–86

Richardson JJ, Bjornmalm M, Caruso F (2015) Multilayer assembly. Technology-driven layer-by-layer assembly of nanofilms. Science (New York, NY) 348:2348–2491

Rinaudo M (2006) Chitin and chitosan: properties and applications. Prog Polym Sci 31:603–632

Rubina RS, Vasil'kov AY, Naumkin AV, Shtykova EV, Abramchuk SS, Alghuthaymi MA, Abd–Elsalam KA (2017) Synthesis and characterization of chitosan–copper nanocomposites and their fungicidal activity against two sclerotia–forming plant pathogenic fungi. J Nanostruct Chem. https://doi.org/10.1007/s40097–017–0235–4

Sabbour MM (2016) Observations of the effect of Nano chitosan against the locust *Schistocerca gregaria* (Orthoptera: Acrididae). J Nanosci Nanoengin 2:28–33

Sahab AF, Waly AI, Sabbour MM, Lubna SN (2015) Synthesis, antifungal and insecticidal potential of chitosan (CS)-g-poly (acrylic acid) (PAA) nanoparticles against some seed borne fungi and insects of soybean. Int J ChemTech Res 8(2):589–598

Saharan V, Mehrotra A, Khatik R, Rawal P, Sharma SS, Pal A (2013) Synthesis of chitosan based nanoparticles and their in vitro evaluation against phytopathogenic fungi. Int J Biol Macromol 62:677–683

Saharan V, Khatik R, Choudhary MK, Mehrotra A, Jakhar S, Raliya R, Nallamuthu I, Pal A (2014) Nano-materials for plant protection with special reference to Nano chitosan. In: Proceedings of 4th annual international conference on advances in biotechnology. GSTF, Dubai, pp 23–25

Saharan V, Sharma G, Yadav M, Choudhary MK, Sharma SS, Pal A, Biswas P (2015) Synthesis and in vitro antifungal efficacy of cu–chitosan nanoparticles against pathogenic fungi of tomato. Int J Biol Macromol 75:346–353

Saharan V, Kumaraswamy RV, Choudhary RC, Kumari S, Pal A, Raliya R, Biswas P (2016) Cu-chitosan nanoparticle mediated sustainable approach to enhance seedling growth in maize by mobilizing reserved food. J Agric Food Chem 64(31):6148–6155

Saifuddin N, Nian CY, Zhan LW, Ning KX (2011) Chitosan-silver nanoparticles composite as point of-use drinking water filtration system for household to remove pesticides in water. Asian J Biochem 6:142–159

Sailaja AK, Amareshwar P, Chakravarty P (2013) Chitosan nanoparticles as a drug delivery system. Res J Pharm Biol Chem Sci 1(3):474–484

Salaberria AM, Diaz RH, Andrés MA, Fernandes SCM, Labidi J (2017) The antifungal activity of functionalized chitin nanocrystals in poly (Lactid acid) films. Anzai J, ed. Materials 10(5):546

Sánchez EA, Tiznado HME, Ojeda CAJ, Valenzuela-Quintanar AI, Troncoso-Rojas R (2009) Induction of enzymes and phenolic compounds related to the natural defence response of netted melon fruit by a bio-elicitor. J Phytopathol 157:24–32

Sangeetha J, Thangadurai D, Hospet R, Harish ER, Purushotham P, Mujeeb MA, Shrinivas J, David M, Mundaragi AC, Thimmappa AC, Arakera SB, Prasad R (2017) Nanoagrotechnology for soil quality, crop performance and environmental management. In: Prasad R, Kumar M, Kumar V (eds) Nanotechnology. pp 73–97, Springer Nature, Singapore Pte Ltd.

Sanuja S, Agalya A, Umapathy MJ (2015) Synthesis and characterization of zinc oxide–neem oil–chitosan bionanocomposite for food packaging application. Int J Biol Macromol 74:76–84

Saharan V, Mehrotra A, Khatik R, Rawal P, Sharma SS, Pal A (2013). Synthesis of chitosan based nanoparticles and their in vitro evaluation against phytopathogenic fungi. Int J Biol Macromol 62:677–683

Sarmento B, Neves J (2012) Chitosan-based systems for biopharmaceuticals: deliver, targeting and polymer therapeutics. John Wiley & Sons, Ltd.

Shantha Siri JG, Fernando CAN, De Silva N (2017) Eco-friendly chitosan nanoparticles cross linked with genipin: basis to develop control release nanofertilizer. J SciTech Res 7:26–31

Sharon M, Choudhary A, Kumar R (2010) Nanotechnology in agricultural diseases and food safety. J Phytology 2:83–92

Sharp RG (2013) A review of the applications of chitin and its derivatives in agriculture to modify plant-microbial interactions and improve crop yields. Agronomy 3:757–793

Shi LE, Fang XJ, Xing LY, Chen M, Zhu DS et al (2011) Chitosan nanoparticles as drug delivery carriers for biomedical engineering. J Chem So Pak 33:929–934

Shikata F, Tokumitsu H, Ichikawa H, Fukumori Y (2002) In vitro cellular accumulation of gadolinium incorporated into chitosan nanoparticles designed for neutron-capture therapy of cancer. Eur J Pharm Biopharm 53:57–63 PMID: 11777753

Shukla SK, Mishra AK, Arotiba OA, Mamba BB (2013) Chitosan-based nanomaterials: a state-of-the-art review. Int J Biol Macromol 59:46–58

Siddaiah CN, Prasanth KVH, Satyanarayana NR, Mudili V, Gupta VK, Kalagatur NK, Satyavati T, Dai XF, Chen JY, Mocan A, Singh BP, Srivastava RK (2018) Chitosan nanoparticles having higher degree of acetylation induce resistance against pearl millet downy mildew through nitric oxide generation. Sci Rep 8(1):2485. https://doi.org/10.1038/s41598-017-19016-z

Siddiqui MH, Al-Whaibi MH, Firoz M, Al-Khaishany MY (2015) Role of nanoparticles in plants. Nanotech Plant Sci:19–35. https://doi.org/10.1007/978-3-319-14502-0-2

Silva MS, Cocenza DS, Grillo R, de Melo NFS, Tonello POS, deOliveira LC, Cassimiro DL, Rosa AH, Fraceto LF (2011) Paraquat-loaded alginate/chitosan nanoparticles: preparation, characterization and soil sorption studies. J Hazard Mater 190:366–374

Sivamani E, DeLong RK, Qu R (2009) Protamine-mediated DNA coating remarkably improves bombardment transformation efficiency in plant cells. Plant Cell Rep 28:213–221

Sotelo-Boyás ME, Bautista-Baños S, Correa-Pacheco ZN, Jiménez-Aparicio A, Sivakumar D (2016) Biological activity of chitosan nanoparticles against pathogenic fungi and bacteria. Chapter 13. In: Bautista-Banos S, Romanazzi G, Jiménez-Aparicio A (eds) Chitosan in the preservation of agricultural commodities, pp 339–349, Academic Press, Elsevier, USA

Sun B, Zhang L, Yang L, Zhang F, Norse D, Zhu Z (2012) Agricultural non-point source pollution in China: causes and mitigation measures. Ambio 41:370–337

Swati, Choudhary MK, Joshi A, Saharan V (2017) Assessment of cu-chitosan nanoparticles for its antibacterial activity against *Pseudomonas syringae* pv. *glycinea*. Int J Curr Microbiol App Sci 6(11):1335–1350

Tan H, Ma R, Lin C, Liu Z, Tang T (2013) Quaternized chitosan as an antimicrobial agent: antimicrobial activity, mechanism of action and biomedical applications in orthopedics. Int J Mol Sci 14:1854–1869

Tao S, Pang R, Chen C et al (2012) Synthesis, characterization and slow release properties of O-naphthylacetyl chitosan. Carbohydr Polym 88:1189–1194

Teo WE, Ramakrishna S (2006) A review on electrospinning design and nanofibre assemblies. Nanotechnolgy 17:R89–R106

Tiyaboonchai W (2003) Chitosan nanoparticles: a promising system for drug delivery. Naresuan Univ J 11:51–66

Vander P, Vaêrum KM, Domard A, El Gueddari NE, Moerschbacher BM (1998) Comparison of the ability of partially N-acetylated chitosans and chitooligosaccharides to elicit resistance reactions in wheat leaves. Plant Physiol 118:1353–1359

Vasyukova NI, Zinovèva SV, Ilìnskaya LI, Perekhod EA, Chalenko GI, Gerasimova NG, Il'ina AV, Varlamov VP, Ozeretskovskaya OL (2001) Modulation of plant resistance to diseases by water-soluble chitosan. App Biochem Microbiol 37:103–109

Wang X, El Hadrami A, Adam LR, Daayf F (2008) Differential activation and suppression of potato defence responses by *Phytophthora infestans* isolates representing US-1 and US-8 genotypes. Plant Pathol 57:1026–1037

Wang Q, Chen JN, Zhan P, Zhang L, Kong QQ (2013) Establishment of a suspension cell system for transformation of *Jatropha curcas* using nanoparticles. Adv Mater Res 608–609:314–319

Wang P, Lombi E, Zhao FJ, Kopittke PM (2016) Nanotechnology: a new opportunity in plant sciences. Trends Plant Sci 21:699–712

Wani IA, Ahmad T (2013) Size and shape dependant antifungal activity of gold nanoparticles: a case study of Candida. Colloids Surf B101:162–170

Wazed AS, Joshi M, Rajendran S (2011) Novel, selfassembled antimicrobial textile coating containing chitosan nanoparticles. AATCC Rev 11:49–55

Wen Y, Yuan Y, Chen H et al (2010) Effect of chitosan on the enantioselective bioavailability of the herbicide dichlorprop to Chlorella pyrenoidosa. Environ Sci Technol 44:4981–4949

Wen Y, Chen H, Yuan Y et al (2011) Enantioselective ecotoxicity of the herbicide dichlorprop and complexes formed with chitosan in two fresh water green algae. J Environ Monitor JEM 13:879–88587

Wong MH, Misra RP, Giraldo JP, Kwak SY, Son Y, Landry MP, Swan JW, Blankschtein D, Strano MS (2016) Lipid exchange envelope penetration (LEEP) of nanoparticles for plant engineering: a universal localization mechanism. Nano Lett 16:1161–1172

Wu L, Liu M (2008) Preparation and properties of chitosan-coated NPK compound fertilizer with controlled-release and water retention. Carbohydr Polym 72:240–247

Wu L, Liu M, Liang R (2008) Preparation and properties of a double-coated slow-release NPK compound fertilizer with superabsorbent and water-retention. Bioresour Technol 99:547–554

Xing K, Zhu X, Peng X, Qin S (2014) Chitosan antimicrobial and eliciting properties for pest control in agricultural: a review. Agron Sustain Dev 35(2):569–588

Xing K, Zhu X, Peng X, Qin S (2015) Chitosan antimicrobial and eliciting prop erties for pest control in agriculture: a review. Agronomy for sustainable development. Springer Verlag/EDP Sci/INRA 35(2):569–588

Xing K, Shen X, Zhu X, Ju X, Miao X, Tian J, Feng Z, Peng X, Jiang J, Qin S (2016) Synthesis and in vitro antifungal efficacy of oleoyl-chitosan nanoparticles against plant pathogenic fungi. Int J Biol Macromol 82:830–836

Xu L, Cao LD, Li FM, Wang XJ, Huang QLJ (2014) Utilization of chitosan-lactide copolymer nanoparticles as controlled release pesticide carrier for pyraclostrobin against *Colletotrichum gossypii* Southw. Dispers Sci Technol 35:544–550

Yu-qin F, Lu-hua L, Pi-wu W, Jing Q, Yong-ping F, Hui W, Jing-ran S, Chang-li L (2012) Delivering DNA into plant cell by gene carriers of ZnS nanoparticles. Chem Res Chin Univ 28(4):672–676

Zahid N, Alderson P, Ali A, Maqbool M, Manickam S (2013) *In vitro* control of *Colletotrichum gloeosporioides* by using chitosan loaded nanoemulsions. Acta Hortic 1012:769–774

Zeng D, Luo X, Tu R (2012) Application of bioactive coatings based on chitosan for soybean seed protection. Int J Carbohydr Chem 1:1–5

Zhang X, Zhang J, Zhu KY (2010) Chitosan/double-stranded RNA nanoparticle-mediated RNA interference to silence chitin synthase genes through larval feeding in the African malaria mosquito (*Anopheles gambiae*). Insect Mol Biol 19(5):683–693

Zhao X, She X, Du Y, Liang X (2007) Induction of antiviral resistance and stimulary effect by oligochitosan in tobacco. Pestic Biochem Phys 87:78–84

Zheng L, Hong F, Lu S, Liu C (2005) Effect of nano-TiO2 on strength of naturally aged seeds and growth of spinach. Biol Trace Elem Res 104:83–91

Zheng YY, Monty J, Linhardt RJ (2015) Polysaccharide-based nanocomposites and their applications. Carbohydr Res 405:23–32

Index

K. A. Abd-Elsalam, R. Prasad (eds.), *Nanobiotechnology Applications in Plant Protection*, Nanotechnology in the Life Sciences,
https://doi.org/10.1007/978-3-319-91161-8

O

P

MIX
Papier aus verantwortungsvollen Quellen
Paper from responsible sources
FSC® C105338
FSC
www.fsc.org

If you have any concerns about our products,
you can contact us on
ProductSafety@springernature.com

In case Publisher is established outside the EU,
the EU authorized representative is:
Springer Nature Customer Service Center GmbH
Europaplatz 3, 69115 Heidelberg, Germany

Printed by Libri Plureos GmbH
in Hamburg, Germany